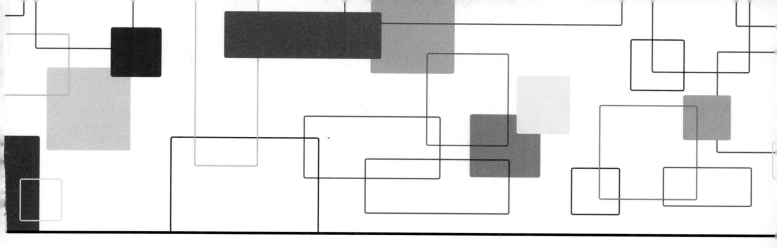

MANAGEMENT INFORMATION SYSTEMS

SYSTEMS

Sixth Edition

EFFY OZ

The Pennsylvania State University, Great Valley

COURSE TECHNOLOGY
CENGAGE Learning™

Australia • Brazil • Japan • Korea • Mexico • Singapore • Spain • United Kingdom • United States

COURSE TECHNOLOGY
CENGAGE Learning

Management Information Systems, Sixth Edition
Effy Oz

Acquisitions Editor: Maureen P. Martin

Product Manager: Kate Hennessy

Development Editor: Deb Kaufmann

Editorial Assistant: Patrick Frank

Marketing Manager: Bryant Chrzan

Marketing Coordinator: Victoria Ortiz

Content Project Manager: Aimee Poirier

Compositor: GEX Publishing Services

Print Buyer: Justin Palmeiro

Cover Photo: Walter Pietsch/Alamy Images

For product information and technology assistance, contact us at
Cengage Learning Customer & Sales Support, 1-800-354-9706

For permission to use material from this text or product, submit all requests online at **www.cengage.com/permissions**
Further permissions questions can be emailed to
permissionrequest@cengage.com

ISBN-13: 978-1-4239-0178-5

ISBN-10: 1-4239-0178-9

Course Technology
25 Thomson Place
Boston, Massachusetts 02210
USA

Cengage Learning is a leading provider of customized learning solutions with office locations around the globe, including Singapore, the United Kingdom, Australia, Mexico, Brazil, and Japan. Locate your local office at:
international.cengage.com/region

Cengage Learning products are represented in Canada by Nelson Education, Ltd.

To learn more about Course Technology, visit **www.cengage.com/coursetechnology**

To learn more about Cengage Learning, visit **www.cengage.com**

Purchase any of our products at your local bookstore or at our preferred online store **www.ichapters.com**

Microsoft, Windows 95, Windows 98, Windows 2000, and Windows XP are registered trademarks of Microsoft® Corporation. Some of the product names and company names used in this book have been used for identification purposes only and may be trademarks or registered trademarks of their manufacturers and sellers. SAP, R/3, and other SAP product/services referenced herein are trademarks of SAP Aktiengesellschaft, Systems, Applications and Products in Data Processing, Neurottstasse 16, 69190 Walldorf, Germany. The publisher gratefully acknowledges SAP's kind permission to use these trademarks in this publication. SAP AG is not the publisher of this book and is not responsible for it under any aspect of press law.

Printed in the United States of America
2 3 4 5 6 7 12 11 10 09 08

To Narda, Sahar, Adi, Noam, Ron, Jess, and Lily
and in memory of my sister, Miry Herzog

BRIEF CONTENTS

TABLE OF CONTENTS

TABLE OF CONTENTS

TABLE OF CONTENTS

TABLE OF CONTENTS

TABLE OF CONTENTS

PREFACE

The goal of *Management Information Systems, Sixth Edition* is to provide a real-world understanding of information systems (ISs) for business and computer science students. Like its predecessor, this Sixth Edition provides students with a firm foundation in business-related information technology (IT) on which they can build successful careers regardless of the particular fields they choose. They may find themselves formulating strategic plans in executive suites, optimizing operations in businesses or on factory floors, fine-tuning plans for their own entrepreneurial ventures, designing ISs to optimize their organization's operations, working as consultants, augmenting business activities on the Web, or creating valuable new information products in any number of industries.

This Sixth Edition is organized in fourteen chapters that contain the most important topics for business students. The fundamental principle guiding this book is that ISs are everywhere in business. Information systems are pervasive because information is the single most powerful resource in every business function in every industry. Knowledge of IT is not always explicitly stated as a job requirement, but it is an essential element of success in virtually any position. Not everyone in business needs to have all the technical skills of an IT professional, but everyone needs a deep-enough understanding of the subject to know how to use IT in his or her profession. This is especially so in the increasingly digital and networked business world.

Management Information Systems provides students with the proper balance of technical information and real-world applications. No matter what field they undertake, students will enter the business world knowing how to get information to work for them. They will know enough about IT to work productively with IT specialists, and they will know enough about business applications to get information systems to support their work in the best way possible.

APPROACH

Part Cases Show IS Principles in Action

In this edition Part Cases were carefully updated to integrate all the IT principles that arise in business, to give students an opportunity to view IS issues in action, and to solve business problems related to IT just as they arise in the real world. The cases are built around companies that range in size from the entrepreneurial start-up to the multimillion-dollar corporate giant, reflecting a wide variety of industries. These cases were created to show students how the full range of business functions operate within virtually every business setting. The Part Cases are integrated into the text in four ways:

- ***The Case:*** Each part of the text (made up of between two and four chapters) opens with the Part Case: the story of a business, including the business's IS challenges, the characters involved, and the issues. Everyone in business knows that almost every business problem has a human element; this aspect of managing IT-related challenges is realistically represented in each case.

- ***The Business Challenge:*** The presentation of each case is immediately followed by a succinct statement of the business challenge of the case and the ways the information in each chapter in the case will help the reader meet that challenge.

- ***Case Installments:*** Each chapter opens with an installment of the Part Case that focuses and expands on an aspect of the original story that relates most closely to the chapter content.

- ***Case Revisited Sections:*** Each chapter ends with a Case Revisited section, which includes a concise summary of the challenge in the case installment; a section called **What Would You Do?**, a series of questions that asks the readers to play a role in the case and decide how they would handle a variety of challenges inherent in the case; and **New Perspectives**, a series of questions that introduces a wide variety of "what ifs" reaching beyond the original scope of the case and again asking the students to play different roles to meet business challenges.

Emphasis on the Real World

Management Information Systems is not afraid to warn about the limitations of ISs. The text also explains the great potential of many information technologies, which many organizations have not yet unleashed. Of course, this book includes chapters and features that provide a thorough, concise—and refreshingly clear—grounding in the technology of information systems, because all professionals in successful organizations are involved in making decisions about hardware, software, and telecommunications. But, through current, detail-rich, real-world case studies throughout the book, and a dedication to qualifying each presentation with the real-world factors that may affect business, this book stays close to the workplace in its presentation.

Attention to New Business Practices and Trends

Large parts of the text are devoted to discussing innovative uses of information technology and its benefits and risks. Contemporary concepts such as supply chain management systems, data warehousing, business intelligence systems, knowledge management, Web-based electronic data interchange, and software as a service are explained in plain, easy-to-understand language.

Illustration of the Importance of Each Subject to One's Career

Business students often do not understand why they have to learn about information technology. The reason many students are frustrated with introductory MIS courses is that they do not fully understand how information technology works or why it is important for them to understand it. One of the primary goals of this book is for its entire presentation to make the answers to these questions apparent. First, all subjects are explained so clearly that even the least technically oriented student can understand them. Technology is never explained for technology's sake, but to immediately demonstrate how it supports businesses. For instance, networking, database management, and Web technologies (Chapters 6 through 8), which are often confusing topics, are presented with clear, concise, and vivid descriptions to paint a picture of technology at work. In addition, each chapter includes a feature titled **Why You Should**, which explains to students how being well-versed in that chapter's aspect of IT is important to their careers.

Emphasis on Ethical Thinking

The book puts a great emphasis on some of the questionable and controversial uses of information technology, with special treatment provided in the **Ethical & Societal Issues** boxes. The students are required to weigh the positive and negative impacts of technology and to convincingly argue their own positions on important issues such as privacy, free speech, and professional conduct.

Emphasis on Critical Thinking

Critical thinking is used throughout the text as well as in the book's many features. For instance, the students are put in the midst of a business dilemma relating to the running case of each chapter and required to answer **What Would You Do?** questions. The questions motivate students to evaluate many aspects of each situation and to repeatedly consider how quickly IT evolves. Similarly, many of the **Discussion Questions** at the end of chapters call for their evaluation and judgment.

ADDITIONAL EMPHASES IN THE SIXTH EDITION

Building on the success of the Fifth Edition, *Management Information Systems, Sixth Edition* includes a uniquely effective combination of features.

Updated and New Part and Chapter Case Studies

This Sixth Edition highlights again the well-received, powerful pedagogical tool: five **Part Cases** that clearly incorporate a wide array of real-world events and challenges that dramatize how information technology is integrated into everyday business.

Strong Foundation in Strategic ISs in Business Functions

In addition to a complete chapter on strategic uses of ISs (Chapter 2), strategic thinking is an underlying theme throughout the book. Current examples are used to illustrate how information systems can give businesses a strategic advantage.

Up-to-date Coverage of Web Technologies and Web-Enabled Commerce

Reflecting the use of Web technologies in so many business activities, the book integrates the topic seamlessly throughout the text, just as it has become integrated into business in general. But the text goes beyond the well-worn discussions of the topic (and the handful of sites everyone knows about) to tell the students what works about e-commerce and what doesn't work.

Thorough Discussion of Supply Chain Management Systems

As SCM systems are becoming pervasive in the business world, supply chains and their management are discussed both in a dedicated chapter (Chapter 3) and throughout the text. Related technologies, such as RFID, are clearly explained. In text and diagrams, the importance of these systems is underscored.

Current Real-world Examples Reflect a Wide Variety of Businesses

The text incorporates more applications, cases, and projects in the full range of business functions and industries throughout the book. The cases at the end of the chapter, in the **From Ideas to Application: Real Cases** sections, have been carefully selected to include critical thinking questions to guide students to apply what they have learned. Most of these cases are new to this edition and others have been updated and reflect current technology and trends. In addition, for strong pedagogical reinforcement, examples are embedded throughout the book.

Coverage of Global Issues

Globalization has become an important issue both economically and technologically. An entire chapter, Chapter 9, is devoted to discussing challenges to global information systems, from legal discrepancies through cultural issues to time zone issues. The chapter also discusses how the challenges can be met successfully. This topic receives little coverage in similar textbooks. The breadth and depth of coverage of challenges to global uses of IT in this book has been enthusiastically received by adopters.

New Aspects of Ethical and Societal Issues

The coverage of **Ethical & Societal Issues** in *Management Information Systems* builds on the strong foundation started in the first five editions. However, new issues have emerged, such as phishing and offshoring, which are discussed in this edition. This is a powerful feature provided by an author who is internationally recognized as a researcher in the field of IT Ethics.

New Student Assignments for Reinforcement of Material

This Sixth Edition continues to provide a large selection of assignments at the ends of chapters, mainly assignments that require the use of relevant software and the Web. Many of these assignments, including **Applying Concepts**, **Hands-On Activities**, and **Team Activities**, have been updated for the this Edition. Responding to instructors' recommendations, more assignments require research involving the Web. In addition to the hands-on exercises in each chapter, students and instructors will find a host of additional new hands-on work available at the Student Companion Web site, which is discussed later in this Preface.

More Points of Interest

Responding to instructors' enthusiastic reception of **Points of Interest**, we added a wealth of new sidebar statistics, anecdotes, and short stories that add an interesting and entertaining aspect to the main chapter text. Except for a few entries, all are new in this edition.

ASSESSMENT OPTIONS FOR INSTRUCTORS

To further enhance student learning, Course Technology offers SAM (Skills Assessment Manager), the worldwide leader in online assessment and proven to be the most effective tool to assess and train students in Microsoft Office tasks, Computer Concepts, Windows, the Internet, and more. SAM is a hands-on, simulated computer assessment and training tool that gives students the feeling of working live in the computer application.

Want More? SAM 2007

Inject a wider breadth of applications, as well as additional Excel, Access, and Computer Concepts coverage into your MIS course with SAM 2007! Visit *http://samcentral.course.com* to learn more.

Please contact your Course Technology Sales Representative for more information regarding these assessment options.

PREFACE

STUDENT COMPANION WEB SITE

We have created an exciting online companion for students to utilize as they work through the Sixth Edition of *Management Information Systems*. In the back of this text you will find a key code that provides full access to a robust Web site, located at *www.course.com/mis/mis6*. This Web resource includes the following features:

PowerPoint Slides

Direct access is offered to the book's PowerPoint presentations, which cover the key points from each chapter. These presentations are a useful study tool.

Videos

Twelve topical video clips, linked to chapters throughout the book, can be found on this Web site. Questions to accompany the respective video clips are featured on the Student Companion Web site. These exercises reinforce the concepts taught and provide the students with more critical thinking opportunities.

Glossary of Key Terms

Students can view a PDF file of the glossary from the book.

Part Case Resources from the Sixth Edition

Gain access to a multitude of online resources tied to the five Part Opening Cases which have been updated from the previous edition.

Sixth Edition Part Case Projects

Unique hands-on projects associated with the five Part Cases have been created to allow for first-hand participation in the businesses introduced in each Part. For each Part Case, there is a selection of hands-on projects that asks the user to become a "character" in the cases and perform small tasks to help meet business needs. The solution files for these activities are available to instructors at *www.course.com*, via the password-protected Instructor Downloads page for this textbook.

"Bike Guys" Business Cases

For more examples of MIS concepts in action, we have supplied the popular "Bike Guys" cases from the Third Edition of the text.

Further Case Offerings

Course Technology now offers cases from Harvard Business School Publishing and other leading case-writing institutions. Create the ideal casebook for your course by selecting cases, adding your own materials, and combining it with our best-selling Course Technology titles. For further information, please contact your instructor.

Additional business articles and cases are offered through InfoTrac, the popular Journal Database, made up of more than 15 million full-text articles from over 5000 scholarly and popular periodicals. Please speak with your instructor about accessing this database.

Additional Content

Here you will find the following additional material:

- Organizing Information Technology Resources
- Measurement Units

Test Yourself on MIS

Brand new quizzes, created specifically for this site, allow users to test themselves on the content of each chapter and immediately see what answers were answered right and wrong. For each question answered incorrectly, users are provided with the correct answer and the page in the text where that information is covered. Special testing software randomly compiles a selection of questions from a large database, so students can take quizzes multiple times on a given chapter, with some new questions each time.

Additional Exercises

Also created just for this Student Companion Web site, a selection of exercises asks users to apply what they have learned in each chapter and further explore various software tools. The solution files for these activities are also available to instructors at *www.course.com*.

Useful Web Links

Access a repository of links to the home pages of the primary Web sites relative to each chapter for further research.

INSTRUCTOR'S PACKAGE

Management Information Systems, Sixth Edition, includes teaching tools to support instructors in the classroom. The ancillaries that accompany the textbook include an Instructor's Manual, Solutions, Test Banks and Test Engine, Distance Learning content, PowerPoint presentations, and Figure Files. This textbook is one of the few accompanied by an Instructor's Manual written by the text author, ensuring compatibility with the textbook in content, pedagogy, and philosophy. All teaching tools available with this book are provided to the instructor on a single CD-ROM and also available on the Web at *www.course.com*.

The Instructor's Manual

The text author has created this manual to provide materials to help instructors make their classes informative and interesting. The manual offers several approaches to teaching the material, with sample syllabi and comments on different components. It also suggests alternative course outlines and ideas for term projects. For each chapter, the manual includes teaching tips, useful Web sites, and answers to the Review Questions, Discussion Questions, and Thinking about the Case questions. Having an Instructor's Manual created by the text author is particularly valuable, as the author is most familiar with the topical and pedagogical approach of the text.

Solutions

We provide instructors with solutions to Review Questions and Discussion Questions as well as for quantitative hands-on work in each chapter. If appropriate, we will also provide solution files for various activities. Solutions may also be found on the Course Technology Web site at *www.course.com*. The solutions are password protected.

ExamView®

This objective-based test generator lets the instructor create paper, LAN, or Web-based tests from test banks designed specifically for this Course Technology text. Instructors can use the QuickTest Wizard to create tests in fewer than five minutes by taking advantage of Course Technology's question banks—or create customized exams.

PowerPoint Presentations

Microsoft PowerPoint slides are included for each chapter. Instructors might use the slides in a variety of ways, including as teaching aids during classroom presentations or as printed handouts for classroom distribution. Instructors can add their own slides for additional topics introduced to the class.

Figure Files

Figure files allow instructors to create their own presentations using figures taken directly from the text.

Distance Learning Content

Course Technology, the premiere innovator in management information systems publishing, is proud to present online courses in WebCT and Blackboard.

- *Blackboard and WebCT Level 1 Online Content.* If you use Blackboard or WebCT, the test bank for this textbook is available at no cost in a simple, ready-to-use format. Go to *www.course.com* and search for this textbook to download the test bank.

- *Blackboard and WebCT Level 2 Online Content.* Blackboard Level 2 and WebCT Level 2 are also available for *Management Information Systems.* Level 2 offers course management and access to a Web site that is fully populated with content for this book.

For more information on how to bring distance learning to your course, instructors should contact their Course Technology sales representative.

ORGANIZATION

Management Information Systems, Sixth Edition is organized into five parts, followed by a glossary and an index. It includes the following major elements.

Part One: The Information Age

Part One of the book includes three chapters. Chapter 1, "Business Information Systems: An Overview," provides an overview of information technology (IT) and information systems (ISs) and a framework for discussions in subsequent chapters. Chapter 2, "Strategic Uses of Information Systems," discusses organizational strategy and ways in which ISs can be used to meet strategic goals. Chapter 3, "Business Functions and Supply Chains," provides a detailed discussion of business functions, supply chains, and the systems that support management of supply chains in various industries. Together, these three chapters address the essence of all overarching ideas that are discussed at greater depth in subsequent chapters.

Part Two: Information Technology

To understand how ISs enhance managerial practices, one must be well versed in the technical principles of information technology, which are covered in Part Two. Chapters 4, "Business Hardware," 5, "Business Software," and 6, "Business Networks and Telecommunications," provide a concise treatment of state-of-the-art hardware, software, and networking technologies in business. Chapter 7, "Databases and Data Warehouses," covers database management systems and data warehousing, which provide the technical foundation for a discussion of business intelligence and knowledge management in Chapter 11.

Part Three: Web-Enabled Commerce

Part Three is devoted to networked businesses and their use of the Internet. Chapter 8, "The Web-enabled Enterprise," is fully devoted to a thorough discussion of relevant Web technologies for business operations. Chapter 9, "Challenges of Global Information Systems," highlights cultural and other challenges organizations face in planning and using the Web and international information systems.

Part Four: Decision Support and Business Intelligence

Part Four provides a view of state-of-the-art decision support and expert systems in Chapter 10 and business intelligence in Chapter 11. Electronic decision aids have been integrated into other systems in recent years, but understanding of their fundamentals is important. Business intelligence applications, such as data mining and online analytical processing, are essential tools in a growing number of businesses. Plenty of examples are provided to demonstrate their power.

Part Five: Planning, Acquisition, and Controls

Part Five is devoted to planning, acquisition, and controls of information systems to ensure their successful and timely development and implementation, as well as their security. Chapter 12, "Systems Planning and Development," discusses how professionals plan information systems. It details traditional and agile methods of software development. Chapter 13, "Choices in Systems Acquisition," presents alternative acquisition methods to in-house development: outsourcing, purchased applications, end-user systems development, and software as a service. Chapter 14, "Risks, Security, and Disaster Recovery," discusses the risks that information systems face and ways to minimize them, as well as approaches to recovering from disasters.

NEW FEATURES OF THIS EDITION

We listened carefully to our adopters, potential adopters, and reviewers in planning and writing this Sixth Edition of *Management Information Systems*. We kept the number and organization of chapters the same as in the previous edition to suit optimal coverage, pedagogy, and allow for flexibile term management. The major changes and improvements in this edition are:

- More brief, real-life examples within the text of chapters

- Updated and extended coverage of the latest technologies and trends in MIS, including information security

- New Point of Interest boxes throughout

- All-new end-of-chapter case studies

- New or revised end-of-chapter exercises

- A wealth of online, video, and lab resources to accompany the text

Some instructors would like students to consider careers in IT. Therefore, the discussion of IT careers was moved to Chapter 1, "Business Information Systems: An Overview." This allows the students to learn what IT professionals do early on.

Supply chain management (SCM) systems and customer relationship management (CRM) systems have become important staples in businesses. Therefore, they are now introduced early in Chapter 1, thoroughly explained in Chapter 3, "Business Functions and Supply Chains," and discussed widely throughout the text in various contexts. While we still discuss information systems by business function in Chapter 3, a large part of the chapter is devoted to enterprise applications such as SCM, CRM, and ERP systems.

Chapter 4, "Business Hardware," now includes shorter discussions of the innards of computers and extensive discussions on external memory devices and networked storage technologies such as SAN and NAS.

In Chapter 5, "Business Software," the discussion of programming language generations was significantly cut to make room for more important discussions of software that all students will encounter in most organizations. The growing trend of using open source software is extensively discussed and no longer focuses only on Linux. The students are exposed to a plethora of open source applications.

Chapter 6, "Business Networks and Telecommunications," no longer includes discussions of modulation and demodulation, and the technical aspect has been toned down. Most of the chapter now focuses on the use of various networking technologies in business. A new section covers the latest wireless technologies, as this is the future of networking in communities, businesses, and homes. A detailed discussion of RFID technologies is included to provide the technical foundation for further discussion of current and future application of this technology in business.

The major Web technologies are discussed and demonstrated in Chapter 8, "The Web-Enabled Enterprise." The entire chapter was rewritten to reflect new technologies. The section on alternatives in establishing commercial Web sites reflects the latest array of hosting options. Chapter 9, "Challenges of Global Information Systems," is devoted to illuminating the challenges and efficiencies of managing business information systems on a global scale.

Many current examples of decision support systems and artificial intelligence are provided in Chapter 10, "Decision Support and Expert Systems." Chapter 11, "Business Intelligence and Knowledge Management," combines discussions that were included in different chapters in earlier editions. The concept of employee knowledge networks is explained and demonstrated in examples.

Chapter 12, "Systems Planning and Development," discusses the traditional "waterfall" approaches such as the systems development life cycle, but also devotes a thorough discussion to agile methods, which have become so popular among software developers.

Chapter 13, "Choices in Systems Acquisition," discusses alternatives to in-house software development, such as Software as a Service.

Security and disaster recovery are discussed in Chapter 14, "Risks, Security, and Disaster Recovery," with more attention to increasingly severe risks, such as phishing. Discussion of threats to privacy were updated to address new technologies such as RFID tags.

Except for very few entries, all the *Point of Interest* box features are new. All *Ethical & Societal Issues* discussions have been updated.

Nearly all of the end-of-chapter Real Cases are new. As in previous editions, all are real-world examples reported in a wide range of major business and technology journals. About 90 percent of all the examples given in chapter discussions are new and recent. The only examples that are older than 2 years are those that are classic stories of strategic use of IT. Thus, the pedagogy of this edition is significantly enhanced.

ACKNOWLEDGMENTS

This book is the fruit of a great concerted effort. A project such as this could not be successful without the contribution of many people. I would first like to thank my colleagues in the business and IT fields whose ideas and opinions over all these years have helped me understand the educational needs of our students. I also recognize the indirect contribution of the many students I have taught. Their comments helped me understand the points that need extra emphasis or a different presentation to make subjects that are potentially overwhelming clearer and more interesting.

Many thanks go to Kate Hennessy for being so enthusiastic about this project. She was always there for me with advice and encouragement. Kate exerted much energy when heading this project. Her active guidance and constant involvement made an immense contribution to this edition. Kate also handled the smooth coordination of the instructor's package, Web materials, and more. Aimee Poirier, the production editor, shepherded the book through production, managing the process in a very orderly and timely manner. The design and art managers at GEX Publishing Services made sure the text and photos were visually appealing, and the team of artists there skillfully rendered our ideas. Abby Reip ensured that the text concepts were supported with photos. She was knowledgeable and agile. I applaud all of them.

Deb Kaufmann, the developmental editor, has demonstrated again her excellent skills and high integrity. It was wonderful to work with an editor who excels not only in improving style and organization but who is also so knowledgeable in the subject matter. Her broad perspective while still attending to the details were essential ingredients supporting my work.

My thanks also to Dr. Carlos Ferran and Dr. Ricardo Salim for their help in updating the opening cases for this edition.

Reviewers are the most important aides to any writer, let alone one who prepares a text for college students. I would like to thank the reviewers who carefully read every chapter of this edition and/or reviewed the revision proposal for this edition:

Mary Astone, *Troy State University*

Efrem Mallach, *University of Massachusetts, Dartmouth*

John Moreno, *Golden Gate University*

G. Shankaranarayanan, *Boston University*

Elizabeth Sigman, *Georgetown University*

Howard Sundwall, *West Chester University*

I also thank the following reviewers for their candid and constructive feedback on the previous editions:

Gary Armstrong, *Shippensburg University*

Karin Bast, *University of Wisconsin/La Crosse*

Siddhartha Bhattacharya, *Southern Illinois University/Carbondale*

Douglas Bock, *Southern Illinois University/Edwardsville*

George Bohlen, *University of Dayton*

Sonny Butler, *Eastern Kentucky University*

Jane Carey, *Arizona State University*

Judith Carlisle, *Georgia Institute of Technology*

Jason Chen, *Gonzaga University*

Paul Cheney, *University of South Florida*

Jim Danowski, *University of Illinois/Chicago*

Sergio Davalos, *University of Portland*

Robert Davis, *Southwest Texas State University*

Glenn Dietrich, *University of Texas/San Antonio*

James Divoky, *University of Akron*

Charles Downing, *Boston College*

Richard Evans, *Rhode Island College*

Karen Forcht, *James Madison University*

Jeff Guan, *University of Louisville*

Constanza Hagmann, *Kansas State University*

Bassam Hassan, *Univeristy of Toledo*

Sunil Hazari, *University of West Georgia*

Jeff Hedrington, *University of Phoenix*

Charlotte Hiatt, *California State University/Fresno*

Ellen Hoadley, *Loyola College*

Joan Hoopes, *Marist College*

Andrew Hurd, *Hudson Valley Community College*

Anthony Keys, *Wichita State University*

Al Lederer, *University of Kentucky*

Jo Mae Maris, *Arizona State University*

Kenneth Marr, *Hofstra University*
Patricia McQuaid, *California Polytechnic State University*
John Melrose, *University of Wisconsin/Eau Claire*
Lisa Miller, *University of Central Oklahoma*
Jennifer Nightingale, *Duquesne University*
Pat Ormond, *Utah Valley State College*
Denise Padavano, *Peirce College*
Leah Pietron, *University of Nebraska/Omaha*
Floyd Ploeger, *Texas State Univeristy – San Marcos*
Jack Powell, *University of South Dakota*
Leonard Presby, *William Paterson University*
Colleen Ramos, *Bellhaven College*
Raghav Rao, *State University of New York/Buffalo*
Lora Robinson, *St. Cloud State University*
Subhashish Samaddar, *Western Illinois University*
William Schiano, *Bentley College*
Shannon Taylor, *Montana State University*
Barbara Warner, *University of South Florida*
Wallace Wood, *Bryant College*
Zachary Wong, *Sonoma State University*
Amy Woszczynski, *Kennesaw State University*

Lastly, I would like to thank the members of my family for their encouragement and support. Narda, my wife of 33 years, as well as our children—Sahar, Adi, Noam, and Ron, and our daughter-in-law, Jess. Adi was instrumental in finding rich business cases and materials for our *Points of Interest.*

As always, I welcome suggestions and comments from our adopters and their students.

Effy Oz
effyoz@psu.edu

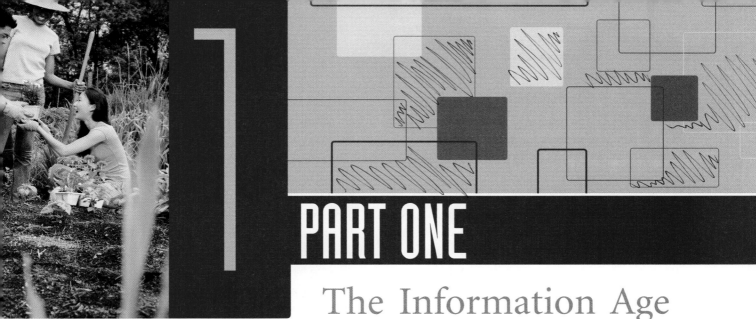

PART ONE

The Information Age

CASE I: GARDENERS+

Mary Jones and Amanda Moreno had a problem. Like many others in their neighborhood, they had a need for simple gardening services that went beyond mere lawn mowing but fell short of the full landscaping projects developed by professional (and usually expensive) landscapers. Mary and Amanda, and other homeowners they knew, had questions about which plants would thrive in specific parts of their gardens, the potential viability of new plantings alongside the ones already in place, transplanting bushes and shrubs, clearing the garden of weeds, selecting the appropriate fertilizer or insecticide, and the proper amount of mulch to place in the flower beds. These services were beyond the capabilities of the neighborhood teenagers, but were of little interest to landscaping firms since they would not bring enough revenue to pay their professionally certified staff.

Mary and Amanda had long joked about converting that problem into a business, but this time Mary was not joking: "I think that the problems of people like us can be solved by companies run by people like us." Both were college graduates who had been

in the workforce before they became stay-at-home mothers. Now they were feeling ready and able to return to the workforce.

To begin, Amanda and Mary decided to have a brainstorming session. They invited Julian, a professional gardener who had helped them with some of their projects, a few of their neighbors, and some friends with entrepreneurial business experience. The object of the brainstorming session was to (a) appropriately define the market niche; (b) get a few ideas on how to approach such a market; (c) establish a price range that homeowners would be willing to pay for such services; and (d) determine how much startup capital they would need. The plan was to establish the business with existing resources, to operate from their houses, and to hire gardeners but no clerical help.

The market segment was clearly defined by Julian. "My customers keep requesting additional services that cannot be done by an inexperienced gardener. I don't always have the knowledge and experience to do them, and even if they offer to pay me more I don't always have the time. And if I refer the job to someone else I worry that I'll lose the client. I could go to work for a landscaping company, but what I earn there in a day I can make in just two hours working for myself!"

Ed Goldstein, a young CPA who lived a few houses down the street, suggested an "open gardeners association." An "open" organizational structure would allow each gardener to remain independent. Each would still have his or her own clients and charge their standard fees without paying any commission to the association. The gardeners would only transact with the association when referring a client to the association or getting a client from the association. They would pay a small member fee, which would be credited toward any fees owed the association for referrals.

Under this arrangement, the cost to the customer would be the same with or without the association. This would minimize the potential competition between the gardeners and the association, but would offer the customer a reliable entity (the association) that could provide replacement or supplemental gardening services. At the same time, participating gardeners could assume additional work when available, and could also benefit from offers of work for which they don't have the time or the skill—all without the risk of losing any steady clients.

Ed suggested that the association could explore renting gardening and transportation equipment to participating members. Amanda noted that the structure proposed by Ed could also incorporate designers, architects, horticulturists, and landscapers. However, since these professionals tend to offer more occasional services, they might require a different commission model. Mary added that the model could also include snow removal.

Under the model suggested by Ed, the association's revenue stream would consist of the members' monthly fees and commissions, while the expenses would be the salaries paid to Mary and Amanda, phone costs, office rental, utilities, equipment amortization, and marketing. Since the plan was to operate from their homes, use their current phone and equipment, and not hire any other personnel, the only "real" expense was marketing. Ed offered his services as CPA once the association started operating.

Mary observed that this open association reflected her initial idea, but Amanda said that she had been thinking of a more "closed" enterprise, one in which gardeners worked for her and she would pay them for their services. Julian said he felt that an open association was more likely to succeed, since gardeners who had clients would not likely give them up for an hourly wage. Furthermore, since Mary and Amanda could not be with the gardeners at all times, they would always try to persuade the client to call them directly in the future, instead of operating through the association. That way they would both benefit—the customer would pay less and the gardener would keep it all. Nonetheless, Julian was willing to consider joining Amanda's closed association if he was to be the general supervisor.

Their market research suggested that the open association was more appealing to gardeners and to potential clients. They all agreed and decided to model the business with 200 customers. They developed two spreadsheets: one that assumed that the 200 customers had already been acquired, and one in which all customer acquisition costs were included. The idea was to evaluate the business from an ongoing point of view and to calculate the startup costs. In a third spreadsheet they modeled the business assuming that they would purchase and own all the necessary equipment (trucks and machinery). The results showed a viable business with 200 houses, but an acceptable profit margin began with 600 houses.

Assessing Business Needs

At the first meeting of all the partners, Mary, Amanda, and Ed (who opted to become a partner instead of an outside consultant) made a list of all the startup requirements. They assigned responsibilities to each partner based on their business experiences. Mary had experience in marketing and sales, Amanda knew how to implement information systems in business settings, and Ed had expertise in finance, accounting, and legal issues.

They assigned tasks due in two weeks. Mary was in charge of the launch plan, while Amanda prepared a basic information system that could track customers, gardeners and service suppliers, and requests for services from customers, as well as match service requests to service providers. Meanwhile, Ed would do all the legal paperwork needed to create the association.

Writing a Business Plan

They based their business plan on the spreadsheet they'd developed to determine whether the idea was viable. They assumed that the 200 customers would come from the housing developments located in their own township. A survey of the region had shown that the township had over 20 developments, each with more than 50 houses.

Mary, Amanda, and Ed turned to the task of writing a business plan. They knew that a good business plan was the key to obtaining the necessary seed capital. They opted for a bank loan instead of trying to get venture capital from friends and family. However, they knew that bank loan officers would scrutinize every detail of the business plan to ensure that the three partners were worth the risk and were reliable.

A good business plan needs to catch the interest of the lender. It must generate excitement so that it stands out from other loan applicants. They began work, fleshing out the plan to provide an overview of their business. The Executive Summary identified the three partners (who they were and why they were qualified to own and run a business) and the business (why there was a need for it, where they planned to offer services, and when they would start). It also explained the concept of an "open association." The Introduction described the business in more detail, explaining its purpose and general objectives, the services offered, and its initial geographical coverage. The Marketing section described the target market, their main competitors, and their plans for advertising and pricing their services. The partners also included information and statistics on the growing need for gardening services as well as a survey of existing service providers. The financial section detailed the projected revenues and expenses as well as the expected cash flow based on the spreadsheet projections. The partners explained that they would perform their own clerical work to avoid additional fixed costs. They included a budget forecast, the estimated total gardening needs in the area, the market share that the association could capture, the amount of required startup capital, and a plan for spending the funds. Finally, the Résumé section listed all three partners' backgrounds, experience, and references.

JoAnn Petrini, the local bank manager, reviewed their plan and moved it forward to the loan analysis department. She ordered viability and risk analyses. In a later meeting between the business partners and a bank risk analyst, they learned that their plan was lacking several important elements: (1) the marketing and promotion plan; (2) a list of all necessary permits and a plan for how they would obtain them; (3) a more detailed forecast of the first year's cash flow and of the profit and loss; and (4) the pro forma contract for both customers and suppliers (gardeners and other independent service providers).

Mary, Amanda, and Ed added all the requested material to the business plan. They hired the services of a small but well respected law firm that inspected the pro forma contracts, and they also included a list of gardeners that had already agreed to participate. Their hard work paid off. The local bank approved their loan. They obtained a line of credit that would support the cash flow described in the business plan and an additional 10 percent for unexpected expenses. *Gardeners+* was ready to become a reality. The three partners realized that they needed to work hard to ensure that in six months, at the start of the spring season, they would be ready for business.

BUSINESS CHALLENGES

In the next three chapters, you will learn what Mary, Amanda, and Ed will need to know to get started: how to harness information technology to help build and grow their gardening business.

- In Chapter 1, "Business Information Systems: An Overview," you learn what types of information systems businesses use and why familiarity with information technology is important for your career. You also are introduced to some of the major ethical and societal concerns about acquiring, storing, and reporting potentially sensitive information.

- In Chapter 2, "Strategic Uses of Information Systems," you learn how to use information strategically, and how to harness information technology for competitive advantage.

- In Chapter 3, "Business Functions and Supply Chains," you learn how you might best use information technology to help manage a business, whether you need to order inventory and track sales, generate financial statements, or automate payroll systems. You also learn how supply chain management systems serve whole enterprises.

© Paul Burns/Getty Images

1

ONE

Business Information Systems:

AN OVERVIEW

LEARNING OBJECTIVES

It is likely that you are carrying or using an information system. This is so if you have an advanced mobile phone, a handheld electronic device, or a laptop computer. Information systems pervade almost every aspect of our lives. Whether you are withdrawing money from a bank's automatic teller machine or surfing the Web on your cell phone, hardly a day goes by without our feeding data into, or using information generated by, an information system. In business especially, digital information systems generate most of the information we use. These systems have become essential to successful business operations.

When you finish this chapter, you will be able to:

■ Explain why information technology matters.

■ Define digital information and explain why digital systems are so powerful and useful.

■ Explain why information systems are essential to business.

■ Describe how computers process data into useful information for problem solving and decision making.

■ Identify the functions of different types of information systems in business.

■ Describe careers in information technology.

■ Identify major ethical and societal concerns created by widespread use of information technology.

GARDENERS+:
Business Systems and Information

Mary, Amanda, and Ed could not believe what they accomplished in the three months since they obtained their small business loan for their gardening business, *Gardeners+*. They had made many decisions and solved many problems.

Solving Problems and Making Decisions

Mary and Ed set up a small office in Ed's garage, with a telephone and a personal computer equipped with a software suite for office use. Marketing to residential clients would primarily consist of flyers left in the doors of houses in their targeted area, but they also planned to run ads in the local newspapers. Gardeners would be approached by means of relationship marketing: Julian would distribute business cards to the gardeners that he knew, and then as new gardeners joined the association, they would in turn distribute cards to their own acquaintances.

Amanda purchased a relatively small software package to handle their information processing needs: it would record information about gardeners, clients, and service requests; match service requests and gardener's skills and availability; and generate and track contracts. The system was very simple but scalable, in case business boomed.

The first clients were a small group of near neighbors, and the first gardeners were close friends of Julian. Mary handled all the first transactions personally and took very detailed notes of all client and gardener feedback: what they liked and disliked, what was missing, and their ideas on how to manage service arrangements.

After a week of pilot testing, the partners met to evaluate the results. They decided to add a new type of service: a single-job contract for a service that would be performed once rather than on a rolling basis. They also decided to add a free confirmation call the day before scheduled work to remind the client but also to ask if there was anything else the client wanted.

Generating Business Information

Mary distributed the flyers to several hundred houses in the surrounding area. She also placed ads in three local newspapers and magazines. Julian passed out a few dozen business cards to friends and acquaintances.

Amanda made some additional adjustments to the software configuration, and Mary continued to use the business suite's word processing program to create ads, basic forms, and the business stationery. Ed prepared a few spreadsheets to help him keep track of sales, revenues, expenses, taxes, and profit. One critical piece of software was Amanda's system, which processed the business transactions and tracked clients' subscriptions and gardeners' contracts.

Amanda tested the system with mock data. She then tweaked some of it, and retested the system. All worked well. The system was now ready.

Managing Data

After a month of operations, the cash flow was as expected. The contract and subscription systems operated by Mary and Amanda were functioning well, and Ed's spreadsheet was sufficient for their needs. However, data transcription was starting to take a toll. Client, subscription, and contract data were first entered into Amanda's system by Mary or Amanda. Then, Ed had to manually transcribe a large part of the data sets from the printed contracts and receipts into his spreadsheet program.

Soon they realized that they were falling behind on their paperwork. The business was running fine, but the back office could not keep up. It was inefficient to input the transactions into Amanda's system and later transcribe them into Ed's accounting and financial spreadsheets. As the daily transactions and client backlog grew, Ed had to spend more and more time every night entering all data so that he could keep the cash flow under control, generate sales tax reports, and make timely loan and rental payments.

Gathering Useful Information from Customers

Mary noticed that the one-time service sold well, but the rolling monthly contract did not sell as well as expected. She also noticed that they had a much higher than expected number of commissions for referrals from gardeners. And with summer nearing they wanted to consider adding or modifying seasonal services. Mary, Amanda, and Ed had to consider the costs and potential benefits of adding, modifying, dropping, and repricing services. To do this, they went back to their initial business models and fed them with real historical rather than projected data. They revised the models to include the services they'd already added as well as the ones they were planning to add to respond to the upcoming summer demand.

The models with this new data would provide a detailed forecast of the demand for each service and improve the "matching" between clients and gardeners. They would also use the models to determine if the occasional failures to properly match clients and gardeners were the result of startup problems, system problems, or structural business problems. They could not allow the current percentage of matching failures to extend over the summer season. Dissatisfied customers not only meant lost sales and fewer profits but, more importantly, bad word-of-mouth. Therefore they needed to generate reports that analyzed "matching"; reports that would show which types of services, areas, and gardeners had larger or smaller failure rates.

DOES INFORMATION TECHNOLOGY MATTER?

The Hackett Group, a strategic advisory firm, used data from 2,100 companies and published a report titled "Does IT Matter? Hackett Concludes the Answer is Yes." The firm found that the world's best performing companies spent 7 percent more per employee on information technology (IT) than typical companies, but recouped the investment fivefold in lower operational costs.

This report, as well as many other observations, show that IT is no longer the sole domain of IT professionals. Business professionals can no longer count solely on IT specialists to make decisions on development, purchasing, and deployment of information systems. Today's business professionals are expected to know how to develop and use IT significantly more than just a few years ago. Regardless of their major field of expertise, those who have the proper IT knowledge and skills stand a better chance of receiving more lucrative job offers and faster promotions.

THE POWER OF DIGITAL SYSTEMS

We are accustomed to using 10 digits to represent quantities. We call it the decimal counting system. However, we could also use a system consisting of only two digits, zero and one, to represent quantities. This is the binary counting system. Because computers and related devices use the binary system—a system that uses two *digits*—they are referred to as **digital systems**. However, digital systems are not used only to represent information that contains numbers, or quantities. They can also represent any information as combinations of zeroes and ones, or, more accurately, the two states that represent zeroes and ones.

Digital information consists of zeroes and ones representing two states. When you have a mechanism that can represent two states, such as electrically charged and uncharged elements, magnetized and nonmagnetized areas, light and no light, you have a way to represent the zeroes and ones. Based on such signals, information can be represented, stored, communicated, and processed *digitally*.

Unlike analog systems (systems based on a continuous signal that varies in strength or quantity), digital systems are capable of delivering data and information—quantities, text, sound, pictures, video, and any other type of information—so that the original information can be re-created with complete accuracy. That is, a digital copy is an exact copy of the original. For example, an analog copy machine reproduces images by reflection or a similar technique. The copy may be good, but it is never as good as the original. And as you make a copy from the copy, the quality deteriorates. When you make a copy of a digital file, such as an image file or a musical file, the system you use first captures the combinations of signals (the digits, zeroes and ones) that make up the file. When processed by the proper hardware and software, the digits are transformed back into the image, or music, or whatever other information you copied. As long as your computer or other digital device can capture all the digits that make up the information, the original information can be re-created fully.

Digital information is stored and communicated by way of electromagnetic signals—electricity, magnetism, and light. These processes involve little or no moving parts. Therefore, storage, retrieval, processing, and communication of digital information are extremely fast. These capabilities—accuracy and speed—make digital systems powerful and therefore useful and important in so many fields: business, education, entertainment, and many others.

POINT OF INTEREST

Information at the Tip of Your... Umbrella

Shall I or shall I not take the umbrella? You don't want to carry an umbrella for nothing, but you also don't want to get wet, right? Perhaps you should buy a smart umbrella, such as the Ambient Forecasting Umbrella. Through a radio receiver, the umbrella receives weather information from AccuWeather.com. A small display in the handle pulses light according to the probability of rain. If the probability is 60 percent, the handle pulses once per second. If the probability is 100 percent, it pulses 100 times per minute.

Source: Bermudez, A., "The Smart Umbrella," *PC Magazine*, February 20, 2007, p. 23.

THE PURPOSE OF INFORMATION SYSTEMS

People require information for many reasons and in varied ways. For instance, you probably seek information for entertainment and enlightenment by viewing television, watching movies, browsing the Internet, listening to the radio, and reading newspapers, magazines, and books. In business, however, people and organizations seek and use information mainly to make sound decisions and to solve problems—two closely related practices that form the foundation of every successful company.

What is a problem? A *problem* is any undesirable situation. When you are stuck in the middle of nowhere with a flat tire, you have a problem. If you know that some customers do not pay their debts on time, but you don't know who or how much they owe, you have a problem. You can solve both problems with the aid of information. In the first case, you can call a towing company, which might use a computerized tracking system to send the tow truck closest to your location; in the second case, simple accounting software can help.

An organization or individual that identifies more than one way to solve a problem or a dilemma must make a *decision*. The problem "2 + 2 = ?" does not require decision making because it has only one solution. However, as a manager, you might face a dilemma such as "Which is the best way to promote the company's new car?" There are many potential ways to promote the new car—television advertising, radio advertising, newspaper advertising, Web advertising, auto shows, direct mail, or any combination of these methods. This dilemma calls for decision making.

Both problem solving and decision making require information. Gathering the right information efficiently, storing it so that it can be used and manipulated as necessary, and using it to help an organization achieve its business goals—all topics covered in this book—are the keys to success in business today. The purpose of information systems is to support these activities. In addition to solving problems and making decisions, businesses use information systems to support daily operations, such as electronic commerce, making airline reservations, and many other activities. As a professional, you need to understand and apply information fundamentals to succeed.

Why You Should Be Well-Versed in Information Systems

You might be surprised at how much information technology (IT) knowledge your prospective employer will expect of you when you interview for your next job, even if the position you seek is not in the IT area. Today's corporations look for IT-savvy professionals, and with good reason. Information is the lifeblood of any organization, commercial or nonprofit; it is essential to sound problem solving and decision making, upon which business success is built. In fact, the main factor limiting the services and information that computers can provide within an organization is the budget.

Because of rapid changes in technology, information systems, unlike many other business components, are quickly changing in form and content. A computer considered fast and powerful today will be an outdated machine in 18–24 months. In 12–24 months, a better program will surpass one that is considered innovative right now. The dynamic nature of information technology is like a moving target. A professional who does not stay informed is of diminishing value to an organization. All knowledge workers—professionals, scientists, managers, and others who create new information and knowledge in their work—must be familiar with IT. Moreover, they must know which IT is relevant for their work and what information they can obtain with a certain technology or networked resource.

Professionals must at all times maintain a clear picture of their organizations and the outside business environment. They must know what resources are available to them and to their competitors. Information technology provides excellent tools for collecting, storing, and presenting facts. But to be truly effective, those facts must be manipulated into useful information that indicates the best allocation of various resources, including personnel, time, money, equipment, and other assets. Regardless of the operations being managed, information systems (ISs) are important tools. Successful professionals must know which ISs are available to their organizations and what systems might be developed in the future.

DATA, INFORMATION, AND INFORMATION SYSTEMS

We use the words "data," "information," and "system" almost daily. Understanding what these terms mean, both generally and in the business context, is necessary if you are to use information effectively in your career.

Data vs. Information

The terms "data" and "information" do not mean the same thing. The word **data** is derived from the Latin *datum*, literally a given or fact, which might take the form of a number, a statement, or a picture. Data is the raw material in the production of information. **Information**, on the other hand, is facts or conclusions that have meaning within a context. Raw data is rarely meaningful or useful as information. To become information, data is manipulated through tabulation, statistical analysis, or any other operation that leads to greater understanding of a situation.

Data Manipulation

Here's a simple example that demonstrates the difference between data and information. Assume that you work for a car manufacturer. Last year, the company introduced a new vehicle to the market. Because management realizes that keeping a loyal customer base requires continuously improving products and services, it periodically surveys large samples of buyers. It sends out questionnaires that include 30 questions in several categories, including demographic data (such as gender, age, and annual income); complaints about different performance areas (such as ease of handling, braking, and the quality of the sound system); features that satisfy buyers most; and courtesy of the dealer's personnel.

Reading through all this data would be extremely time consuming and not very helpful. However, if the data is manipulated, it might provide highly useful information. For example, by categorizing complaints by topic and totaling the number of complaints for each type of dissatisfaction and each car model, the company might be able to pinpoint a car's weaknesses. The marketing analysts then can pass the resulting information along to the appropriate engineering or manufacturing unit.

Also, the company might already have sufficient data on dealers who sold cars to the customers surveyed, the car models they sold, and the financing method for each purchase. But with the survey results, the company can generate new information to improve its marketing. For instance, by calculating the average age and income of current buyers and categorizing them by the car they purchased, marketing executives can better target advertising to groups most likely to purchase each car. If the majority of buyers of a particular type of car do not ask for financing, the company might wish to drop this service option for that car and divert more loan money to finance purchases of other cars. In this way, the company generates useful information from data.

Generating Information

In the examples just cited, calculating totals and averages of different complaints or purchasers' ages may reveal trends in buying habits. These calculations are processes. A **process** is any manipulation of data, usually with the goal of producing information. Hence, while data is essentially raw materials, information is output. Just as raw materials are processed in manufacturing to create useful end products, so raw data is processed in information systems to create useful information (see Figure 1.1). Some processes, however, produce yet another set of data.

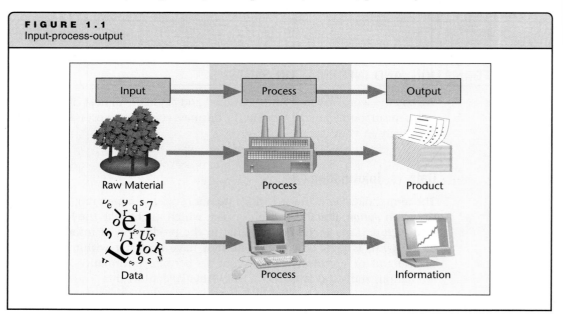

FIGURE 1.1
Input-process-output

Input → Process → Output

Raw Material → Process → Product

Data → Process → Information

Sometimes, data in one context is considered information in another context. For example, if an organization needs to know the age of every person attending a basketball game, then a list of that data is actually information. But if that same organization wants to know the average price of tickets each age group purchases, the list of ages is only data, which the organization must process to generate information.

Information in Context

Information is an extremely important resource for both individuals and organizations, but not all information is useful. Consider the following story. Two people touring in a hot-air balloon encountered unexpected wind that soon blew them off course. When they managed to lower their balloon, they shouted to a farmer on the ground, "Where are we?" The farmer answered, "You are right above a cornfield!" The balloonists looked at each other, and one groaned, "Some information! Highly accurate and totally useless!" To be useful, information must be relevant, complete, accurate, and current. And in business, information must also be obtained economically, that is, cost effectively. Figure 1.2 lists characteristics of useful information.

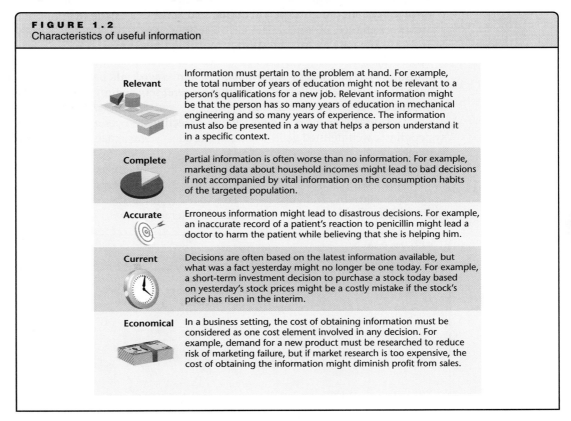

FIGURE 1.2
Characteristics of useful information

Relevant — Information must pertain to the problem at hand. For example, the total number of years of education might not be relevant to a person's qualifications for a new job. Relevant information might be that the person has so many years of education in mechanical engineering and so many years of experience. The information must also be presented in a way that helps a person understand it in a specific context.

Complete — Partial information is often worse than no information. For example, marketing data about household incomes might lead to bad decisions if not accompanied by vital information on the consumption habits of the targeted population.

Accurate — Erroneous information might lead to disastrous decisions. For example, an inaccurate record of a patient's reaction to penicillin might lead a doctor to harm the patient while believing that she is helping him.

Current — Decisions are often based on the latest information available, but what was a fact yesterday might no longer be one today. For example, a short-term investment decision to purchase a stock today based on yesterday's stock prices might be a costly mistake if the stock's price has risen in the interim.

Economical — In a business setting, the cost of obtaining information must be considered as one cost element involved in any decision. For example, demand for a new product must be researched to reduce risk of marketing failure, but if market research is too expensive, the cost of obtaining the information might diminish profit from sales.

What Is a System?

Simply put, a **system** is an array of components that work together to achieve a common goal, or multiple goals, by accepting input, processing it, and producing output in an organized manner. Consider the following examples:

- A sound system consists of many electronic and mechanical parts, such as a laser head, an amplifier, an equalizer, and so on. This system uses input in the form of electrical power and sound recorded on a medium such as a CD or DVD, and processes the input to reproduce music and other sounds. The components work together to achieve this goal.

- Consider the times you have heard the phrase "to beat the system." Here, the term "system" refers to an organization of human beings—a government agency, a commercial company, or any other bureaucracy. Organizations, too, are systems; they consist of components—people organized into departments and divisions—that work together to achieve common goals.

Systems and Subsystems
Not every system has a single goal. Often, a system consists of several **subsystems**—components of a larger system—with subgoals, all contributing to meeting the main goal. Subsystems can receive input from, and transfer output to, other systems or subsystems.

Consider the different departments of a manufacturing business. The marketing department promotes sales of the organization's products; the engineering department designs new products and improves existing ones; the finance department plans a budget and arranges for every unused penny to earn interest by the end of the day. Each department is a subsystem with its own goal, which is a subgoal of a larger system (the company), whose goal is to maximize profit.

Now consider the goals of a manufacturing organization's information system, which stores and processes operational data and produces information about all aspects of company operations. The purpose of its inventory control subsystem is to let managers know what quantities of which items are on hand and which may soon have to be reordered. The purpose of the production control subsystem is to track the status of manufactured parts. The assembly control subsystem presents the bill of material (a list of all parts that make up a product) and the status of assembled products. The entire system's goal is to help deliver finished goods at the lowest possible cost within the shortest possible time.

Figure 1.3 shows an example of a system found in every business: an accounting system. An accounting system consists of several subsystems: accounts payable, records information about money that the organization owes to suppliers and service providers; accounts receivable, records sums owed to the organization and by whom; a general ledger, records current transactions; and a reporting mechanism, generates reports reflecting the company's financial status. Each subsystem has a well-defined goal. Together, the subsystems make up the organization's accounting system.

All professionals must understand systems, both organizational and physical. They need to understand their position in an organization so they can interact well with coworkers, employees of business partners, and customers. They need to understand information systems so that they can utilize them to support their work and interactions with other people.

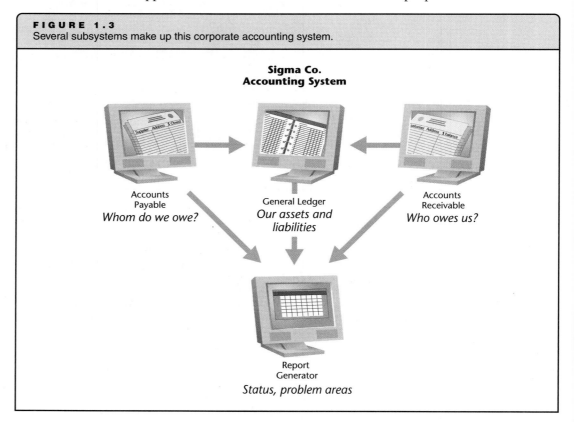

FIGURE 1.3
Several subsystems make up this corporate accounting system.

**Sigma Co.
Accounting System**

Accounts Payable
Whom do we owe?

General Ledger
Our assets and liabilities

Accounts Receivable
Who owes us?

Report Generator
Status, problem areas

Closed vs. Open Systems

Systems are closed or open, depending on the nature of the information flow in the system. A **closed system** stands alone, with no connection to another system: nothing flows in from another system, nothing flows out to another system. For example, a small check-producing system that prints and cuts checks when an employee enters data through a keyboard is a closed system. The system might be isolated for security purposes. An **open system** interfaces and interacts with other systems. For example, an accounting system that records accounts receivable, accounts payable, and cash flow is open if it receives its payroll figures from the payroll system. Subsystems, by definition, are always open, because as components of a bigger system, they must receive information from, and give information to, other subsystems. Increasingly, companies are implementing open—interfaced—information systems. Each system may then be referred to as a module of a larger system, and the modules are interconnected and exchange data and information. For better cooperation, many organizations have interconnected their information systems to those of their business partners, mainly suppliers and clients.

Information Systems

With an understanding of the terms "information" and "system," the definition of an information system is almost intuitive: an **information system (IS)** consists of all the components that work together to process data and produce information. Almost all business information systems consist of many subsystems with subgoals, all contributing to the organization's main goal.

Information and Managers

Thinking of an organization in terms of its suborganizations or subsystems—called systems thinking—is a powerful management approach because it creates a framework for excellent problem solving and decision making. To solve problems, managers need to identify them, which they do by recognizing the subsystems in which the problems occur and solving the problems within those subsystems' constraints and strengths.

Systems thinking can also help keep managers focused on the overall goals and operations of a business. It encourages them to consider the entire system, not only their specific subsystem, when solving problems and making decisions. A satisfactory solution for one subsystem might be inadequate for the business as a whole. For example, when the sales department creates a Web site to take online customer orders, it automates a formerly labor-intensive activity of the sales subsystem. This saves cost. However, increased orders may cause understocking of finished goods. With systems thinking, improving the sales process could also improve other company processes. Without systems thinking, managers from other departments aren't involved in the decision, so they don't benefit. In the case of the sales department, if other managers are involved in planning for automated online ordering, they could suggest that sales data recorded on a shared **database**—a large collection of electronic records—connected to the Web also be accessible to other departments such as shipping and manufacturing. The shipping department could use the records to expedite packaging and shipping, thanks to the information that appears on a computer monitor rather than a sheet of paper. The manufacturing units could use the order records for planning resources such as laborers and inventory. Figuratively, by applying systems thinking, effective managers view their areas of responsibility as puzzle pieces. Each piece is important and should fit well with adjacent pieces, but the entire picture should always be kept in view.

Consider the different approaches Wal-Mart and Kmart took in the 1980s and 1990s. Kmart spent millions of dollars on information systems that helped it advertise and market products. Wal-Mart, on the other hand, spent money on developing information systems that support the entire supply chain—the processes from purchasing through stocking and selling. Kmart succeeded in creating more demand, but often could not satisfy it. Wal-Mart's systems thinking helped it adjust inventories based on demand, saving the costs involved in overstocking and avoiding lost sales due to understocking. Kmart later filed for bankruptcy while Wal-Mart became the world's largest company.

One of an information system's most important contributions to the sound workings of an organization is the automation of information exchange among subsystems (such as departments and divisions). Consider the earlier example: customer orders taken via a Web site by the sales department could be automatically routed to the manufacturing and shipping units and processed by their own information systems for their specific purposes. In fact, such information exchanges make up a major portion of all interactions among business subsystems.

The **information map** of a modern business—that is, the description of data and information flow within an organization—shows a network of information subsystems that exchange information with each other and with the world outside the system. In an ideal organization, no human would need to retrieve information from one IS and transfer it to another. The organization would capture only new raw data, usually from its operations or from outside the organization. Then, data captured at any point in the system would automatically become available to any other subsystem that needs it. Thus, systems thinking is served well by **information technology (IT)**, a term that refers to all technologies that collectively facilitate construction and maintenance of information systems. Systems thinking is the basic reasoning behind equipping organizations with enterprise software applications. Enterprise software applications are systems that serve many parts of the organization by minimizing the need for human data entry and ensuring timely, useful information for the organization's entire supply chain, including taking customer orders, receiving raw materials, manufacturing and shipping, and billing and collection. In the service sector, companies often use document management systems, enabling workers from many departments to add information and signatures to a document from request to approval, or from draft to a final document. You will learn about these systems throughout this book.

The Benefits of Human-Computer Synergy

It is important to remember that computers can only carry out instructions that humans give them. Computers can process data accurately at far greater speeds than people can, yet they are limited in many respects—most importantly, they lack common sense. However, combining the strengths of these machines with human strengths creates synergy.

Some people call synergy the "2 + 2 = 5" rule. **Synergy** (from the Greek "work together") occurs when combined resources produce output that exceeds the sum of the outputs of the same resources employed separately. A computer works quickly and accurately; humans work relatively slowly and make mistakes. A computer cannot make independent decisions, however, or formulate steps for solving problems, unless programmed to do so by humans. Even with sophisticated artificial intelligence, which enables the computer to learn and then implement what it learns, the initial programming must be done by humans. Thus, a human-computer combination allows the results of human thought to be translated into efficient processing of large amounts of data. For example, when you use a Web search engine to find articles about a topic, you, the human, enter a keyword or a series of keywords. By clicking the Search button you shift control to a computer program that quickly finds the articles for you. A human programmed a computer to perform an extremely fast search in a huge database of Web links; another human entered keywords and triggered the program; and the computer performed the matching of keywords with the links at a speed that is way beyond the capability of any human. The result is an efficient search that takes only seconds, which no human would be able to complete in a lifetime. Humans aided by computers increases productivity, producing more while spending less on labor. Figure 1.4 presents qualities of humans and computers that result in synergy. It is important to notice not only the potential benefits of synergy but also what computers should not be expected to do independently.

Information Systems in Organizations

In an organization, an information system consists of data, hardware, software, telecommunications, people, and procedures, as summarized in Figure 1.5. An information system has become synonymous with a computer-based information system, a system with one or more computers at its center, and which is how the term is used in this book. In a computer-based information system, computers collect, store, and process data into information according to instructions people provide via computer programs.

FIGURE 1.4
Qualities of humans and computers that contribute to synergy

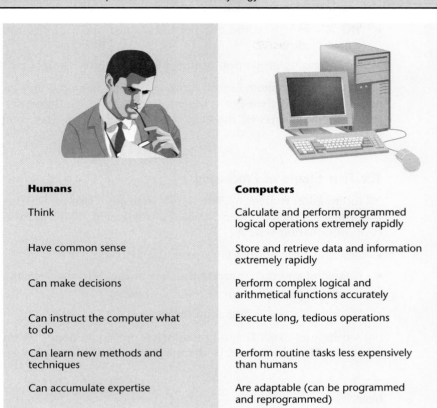

Humans	Computers
Think	Calculate and perform programmed logical operations extremely rapidly
Have common sense	Store and retrieve data and information extremely rapidly
Can make decisions	Perform complex logical and arithmetical functions accurately
Can instruct the computer what to do	Execute long, tedious operations
Can learn new methods and techniques	Perform routine tasks less expensively than humans
Can accumulate expertise	Are adaptable (can be programmed and reprogrammed)

FIGURE 1.5
Components of an information system

Data	Input that the system takes to produce information
Hardware	A computer and its peripheral equipment: input, output, and storage devices; hardware also includes data communication equipment
Software	Sets of instructions that tell the computer how to take data in, how to process it, how to display information, and how to store data and information
Telecommunications	Hardware and software that facilitate fast transmission and reception of text, pictures, sound, and animation in the form of electronic data
People	Information systems professionals and users who analyze organizational information needs, design and construct information systems, write computer programs, operate the hardware, and maintain software
Procedures	Rules for achieving optimal and secure operations in data processing; procedures include priorities in dispensing software applications and security measures

Several trends have made the use of information systems (ISs) very important in business:

- The power of computers has grown tremendously while their prices have dropped.
- The capacity of data storage devices has grown while their prices have decreased.

- The variety and ingenuity of computer programs have increased.

- Quick and reliable communication lines and access to the Internet and the Web have become widely available and affordable.

- The fast growth of the Internet has opened opportunities and encouraged competition in global markets.

- An increasing proportion of the global workforce is computer literate.

In this environment, organizations quickly lag behind if they do not use information systems and skills to meet their goals. Moreover, they must continuously upgrade the features of their information systems and the skills of their employees to stay competitive.

The Four Stages of Processing

All information systems operate in the same basic fashion whether they include a computer or not. However, the computer provides a convenient means to execute the four main operations of an information system:

- Entering data into the IS **(input)**.

- Changing and manipulating the data in the IS **(data processing)**.

- Getting information out of the IS **(output)**.

- Storing data and information **(storage)**.

A computer-based IS also uses a logical process to decide which data to capture and how to process it. This process will be discussed later.

Input

The first step in producing information is collecting and introducing data, known as input, into the IS. Most data an organization uses as input to its ISs are generated and collected within the organization. These data elements result from transactions undertaken in the course of doing business. A **transaction** is a business event: a sale, a purchase, a payment, the hiring of a new employee, and the like. These transactions can be recorded on paper and later entered into a computer system; directly recorded through terminals of a **transaction processing system (TPS)**, such as a point-of-sale (POS) machine; or captured online when someone transacts through the Web. A TPS is any system that records transactions. Often, the same system also processes the transactions, summarizing and routing information to other systems; therefore, these systems are transaction *processing* systems, not just transaction *recording* systems.

Input devices (devices used to enter data into an IS) include the keyboard (currently the most widely used), infrared devices that sense bar codes, voice recognition systems, and touch screens. Chapter 4, "Business Hardware," describes these and other means to input data. The trend has been to decrease the time and effort of input by using devices that allow scanning or auditory data entry.

Processing

The computer's greatest contribution to ISs is efficient data processing. The computer's speed and accuracy enable organizations to process millions of pieces of data in several seconds. For example, managers of a national retail chain can receive up-to-date information on inventory levels of every item the chain carries and then order accordingly; in the past, obtaining such information would take days. The huge gains in the speed and affordability of computing have made information the essential ingredient for an organization's success.

Output

Output is the information an IS produces and displays in the format most useful to an organization. The most widely used output device is the video display, or video monitor, which displays output visually. Another common output device is the printer, used to print hard copies of information on paper. However, computers can communicate output through speakers in the form of music or speech and also can transmit it to another computer or electronic device in computer-coded form, for later interpretation.

Storage

One of the greatest benefits of using IT is the ability to store vast amounts of data and information. Technically, storing a library of millions of volumes on magnetic or optical storage media is feasible. Publishers, libraries, and governments have done that. For example, close to 8 million patents registered in the United States are stored on storage devices accessible through the Web.

Computer Equipment for Information Systems

To support the four data processing functions, different types of technologies are used. Figure 1.6 illustrates the five basic components of the computer system within an IS:

- Input devices introduce data into the IS.
- The computer processes data through the IS.
- Output devices display information.
- Storage devices store data and information.
- Networking devices and communications lines transfer data and information over various distances.

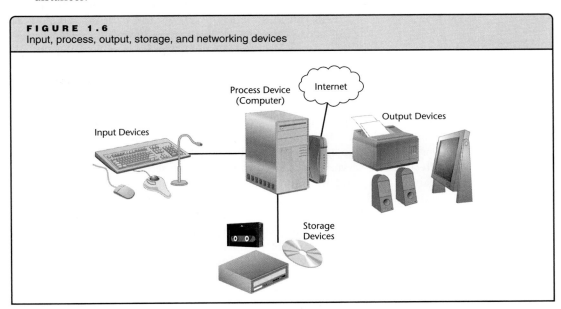

FIGURE 1.6
Input, process, output, storage, and networking devices

In addition to communication that takes place between computer components, communication occurs between computers over great distances (called **telecommunications**). Communications technology lets users access data and other electronic resources of many computers, all connected in a network. This way, the capabilities of a single computer might be augmented with the power of an entire network.

Different types of information systems serve different functions—for particular types of organizations, functions within organizations, business needs, and management levels of an organization. Business enterprises differ in their objectives, structure, interests, and approaches. However, ISs can be generally categorized based on the level of a system's complexity and the type of functions it serves. ISs in business range from the basic transaction processing system that records events such as sales to sophisticated expert systems, which provide advice and reduce the need for the expensive services of a human expert. In recent years the capabilities of many applications have been combined and merged. It is less likely that you will find any of the following applications as stand-alone systems with a single capability. Managers and other professionals plan, control, and make decisions. As long as a system supports one or more of these activities, it may be referred to as a **management information system (MIS)**.

Transaction Processing Systems

Point-of-sale (POS) machines are a ubiquitous type of transaction processing system.

© Anderson Ross/Getty Images

Transaction processing systems (TPSs) are the most widely used information systems. The predominant function of TPSs is to record data collected at the boundaries of organizations, in other words, at the point where the organization transacts business with other parties. They also record many of the transactions that take place inside an organization. For example, they record the movement of parts from one phase of manufacturing to another, from raw materials to finished products. TPSs include POS machines, which record sales; automatic teller machines, which record cash withdrawals, deposits, and transfers; and purchase order systems, which record purchases. A typical example would be the purchase of gasoline at a pump, using a credit card. The purchase is recorded by the gasoline company and later at the credit card-processing bank. After these data elements are collected, the IS can automatically process the data immediately and store it for later access on demand. Transaction processing systems provide most of the data in organizations for further processing by other ISs.

Supply Chain Management Systems

The term "supply chain" refers to the sequence of activities involved in producing and selling a product or service. In industries that produce goods, the activities include marketing, purchasing raw materials, manufacturing and assembly, packing and shipping, billing, collection, and after-the-sale services. In service industries, the sequence might include marketing, document management, and monitoring customer portfolios. Information systems that support these activities and are linked to become one large IS providing information on any stage of a business process are called **supply chain management (SCM) systems**.

Often, such systems are called **enterprise resource planning (ERP) systems**, because the information they provide supports the planning of shipping resources such as personnel, funds, raw materials, and vehicles. However, ERP is a misnomer for the systems, because they mainly serve managers in monitoring and modifying business processes as they occur, and not only for planning. The term "supply chain," too, is somewhat misleading. Business processes do not always take the form of a sequence; some processes take place in parallel. This is true in manufacturing, where two or three teams work on different parts of a product, and in services, where two or three different people peruse a document online and add their input to it within a certain period of time rather than sequentially. In the production of goods and services, some modules of SCM systems provide support to the major processes. These components include human resources (HR) information systems and cost accounting systems.

SCM systems are the result of systems thinking and support systems thinking. They eliminate the need to reenter data that has already been captured somewhere else in the organization. An

SCM is an **enterprise application** because the systems that support each business process are connected to each other to form one large IS. Technically, anyone with access to the system can know the status of every part of an order received by the business: whether the raw materials have been purchased, which subassemblies are ready, how many units of the finished product have been shipped, and how much money has been billed or collected for the order. HR managers can tell which workers are involved in any of the processes of the order. Accountants can use their module of the system to know how much money has been spent on the order and what the breakdown of the cost is in labor, materials, and overhead expenditures.

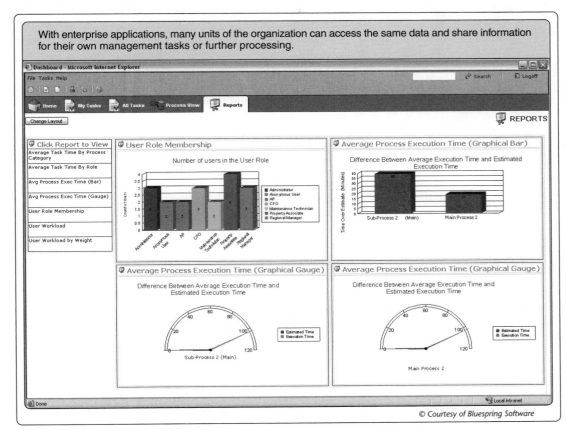

With enterprise applications, many units of the organization can access the same data and share information for their own management tasks or further processing.

© Courtesy of Bluespring Software

Customer Relationship Management Systems

Customer relationship management (CRM) systems help manage an organization's relationships with its customers. The term refers to a large variety of information systems, from simple ones that help maintain customer records to sophisticated systems that dynamically analyze and detect buying patterns and predict when a specific customer is about to switch to a competitor. Many CRM systems are used by service representatives in combination with a telephone. When a customer telephones, the representative can view the entire history of the customer's relationship with the company: anything that the customer has purchased, deliveries made, unfulfilled orders, and other information that can help resolve a problem or help the customer find the desired product or service. The main goals of CRM systems are to increase the quality of customer service, reduce the amount of labor involved in serving customers, and learn as much as possible about the buying habits and service preferences of individual customers.

CRM systems are often linked to Web applications that track online shopping and process online transactions. Using sophisticated applications, a company can learn what makes a customer balk just before submitting an online order, or what a customer prefers to see displayed on Web pages. Online retailers such as Amazon.com, Buy.com, and Target.com use applications that construct different Web pages for different customers, even when they search on the same

Customer relationship management systems help support customers and glean business intelligence.

© Tom Grill/Getty Images

keywords. The pages are constructed to optimally suit the individual customer's interests as inferred from previous visits and purchases. CRM systems provide important data that can be accumulated in large databases and processed into business intelligence.

Effective CRM systems are accessible to both sales and service people. They enable continuous and smooth interaction with everyone from prospective customers to buyers who need after-the-sale service. Both sales people and service crews can view the entire record of a customer and the product purchased and fit the service according to the product service schedule. Because retaining loyal customers is significantly less expensive than acquiring new ones, CRM systems may increase an organization's profitability.

Business Intelligence Systems

ISs whose purpose is to glean from raw data relationships and trends that might help organizations compete better are called **business intelligence (BI)** systems. Usually, these applications consist of sophisticated statistical models, sometimes general and sometimes tailored for an industry or an organization. The applications access large pools of data, usually transactional records stored in large databases called **data warehouses**. With proper analysis models, BI systems might discover particular buying patterns of consumers, such as combinations of products purchased by a certain demographic group or on certain days; products that are sold at faster cycles than others; reasons for customer's churns, that is, customers leaving a service provider for a competitor; and other valuable business intelligence that helps managers quickly decide on changing a strategy.

Decision Support and Expert Systems

Professionals often need to select one course of action from many alternatives. Because they have neither the time nor the resources to study and absorb long, detailed reports of data and information, organizations often build information systems specifically designed to help make decisions. These systems are called **decision support systems (DSSs)**. While DSSs rely on models and formulas to produce concise tables or a single number that determines a decision, **expert systems (ESs)** rely on artificial intelligence techniques to support knowledge-intensive decision-making processes.

Decision support systems help find the optimal course of action and answer "What if?" questions. "What if we purchase raw materials overseas? What if we merge our warehouses? What if we double our shifts and cut our staff?" These questions seek answers like, "This is how this action will impact our revenue, or our market share, or our costs." DSSs are programmed to process raw data, make comparisons, and generate information to help professionals glean the best alternatives for financial investment, marketing strategy, credit approval, and the like. However, it is important to understand that a DSS is only a decision aid, not an absolute alternative to human decision making.

Many environments are not sufficiently structured to let an IS use data to provide the one best answer. For instance, stock portfolio management takes place in a highly uncertain environment. No single method exists to determine which securities portfolio is best, that is, which one will yield the highest return. Medical care is another unstructured environment. There might be many methods of diagnosing a patient on the basis of his or her symptoms. Indeed, a patient with a particular set of symptoms might receive as many different diagnoses as the number of doctors he or she visits.

Using ESs preserves the knowledge of retiring experts and saves a company the high cost of employing human experts. After gathering expertise from experts and building a program, the program can be distributed and used repeatedly. The expertise resides in the program in the form of a knowledge base consisting of facts and relationships among the facts. You will learn about DSS and ES in detail in Chapter 10, "Decision Support and Expert Systems."

Geographic Information Systems

In some cases, the information decision makers need is related to a map or floor plan. In such cases, special ISs called **geographic information systems (GISs)** can be used to tie data to physical locations. A GIS application accesses a database that contains data about a building, neighborhood, city, county, state, country, or even the entire world. By representing data on a map in different graphical forms, a user is able to understand promptly a situation taking place in that part of the world and act upon it. Examples of such information include flood-prone regions, population levels, the number of police officers deployed, probabilities of finding minerals, transportation routes, and vehicle allocation for transportation or distribution systems. Thus, when a supermarket chain considers locations for expansion, executives look at a map that reflects not only geographic attributes but also demographic information such as population growth by age and income groups. GISs are often used to manage daily operations as well as for planning and decision making. They also have been used to provide service via the Web, such as helping residents find locations of different services on a city map or plan travel routes. Some GISs that support operations use information from global positioning system (GPS) satellites, especially to show the current location of a vehicle or person on a map or to provide directions or information on traffic congestion, alternate routes, or various services along a route. This nonstationary type of GIS has become popular, preinstalled in vehicles or sold as a portable device.

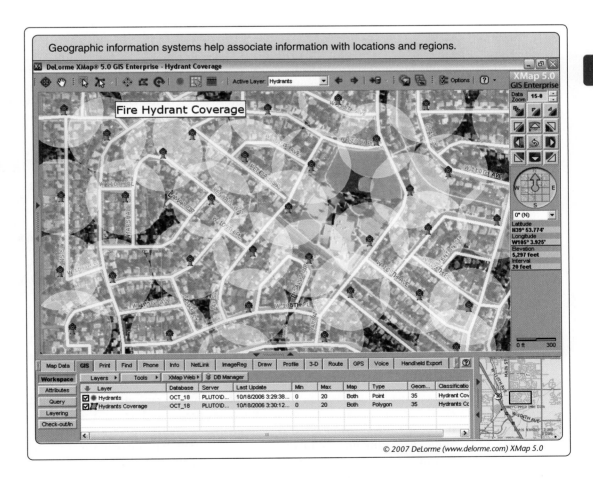

Geographic information systems help associate information with locations and regions.

© 2007 DeLorme (www.delorme.com) XMap 5.0

Commonly used GISs on the Web are Google Earth and Mapquest. They combine maps with street addresses, directions, distances, and travel time calculations. Other Web-based GISs provide real estate information. One such popular system is Zillow (www.zillow.com), which provides maps and information about homes for sale, recent sales, and price estimates.

INFORMATION SYSTEMS IN BUSINESS FUNCTIONS

ISs serve various purposes throughout an organization in what are known as functional business areas—in-house services that support an organization's main business. Functional business areas include, but are not limited to, accounting, finance, marketing, and human resources. As previously mentioned, in a growing number of organizations these systems are modules of a larger enterprise system, an SCM, or ERP system. Chapter 3, "Business Functions and Supply Chains," discusses business functions and their systems in detail.

Accounting

In accounting, information systems help record business transactions, produce periodic financial statements, and create reports required by law, such as balance sheets and profit-and-loss statements. In the United States, the Sarbanes-Oxley Act of 2002 has forced companies to modify their ISs or install new systems to comply with more demanding accounting rules. ISs also help create reports that might not be required by law, but that help managers understand changes in an organization's finances. Accounting ISs contain controls to ascertain adherence to standards, such as double entry.

Finance

While accounting systems focus on recording and reporting financial changes and states, the purpose of financial systems is to facilitate financial planning and business transactions. In finance, information systems help organize budgets, manage cash flow, analyze investments, and make decisions that could reduce interest payments and increase revenues from financial transactions.

POINT OF INTEREST

Protecting Women in Malawi

Information technology can help solve social problems. In Malawi, a southeastern African country, women often face difficulties opening bank accounts because they are illiterate and cannot sign their names. According to local culture, widows often lose their property to the family of their deceased husbands. To prevent financial ruin for these widows, the Bill and Melinda Gates Foundation dispensed debit cards that work with a simple but useful information technology: fingerprint readers. This way, widows in this AIDS-plagued country can retain control of their money.

Source: Gates, B., "The Way We Give," *Fortune*, January 22, 2007, pp 41-46.

Marketing

Marketing's purpose is to pinpoint the people and organizations most likely to purchase what the organization sells and to promote the appropriate products and services to them. For instance, marketing information systems help analyze demand for various products in different regions and population groups in order to more accurately market the right product to the right consumers. Marketing ISs provide information that helps management decide how many sales representatives to assign to specific products in specific geographical areas. The systems identify trends in the demand for the company's products and services. They also help answer such questions as, "How can an advertising campaign affect our profit?" The Web has created excellent opportunities both to collect marketing data and to promote products and services by displaying information about them. That is why organizations conduct so much of their marketing efforts through ISs linked to the Web.

Human Resources

Human resource (HR) management systems help mainly in record-keeping, employee evaluation, and employee benefits. Every organization must maintain accurate employee records. Human resource management systems maintain such records, including employees' pictures, marital status, tax information, and other data that other systems, such as payroll, might use.

Performance evaluation systems provide essential checklists that managers can use to assess their subordinates. These systems also offer a scoring utility to quantify workers' strengths and weaknesses.

Human resource management systems help users track and promote employees and allow employees to select benefits plans.

Courtesy of IRIS Software Ltd

HR management systems have evolved to serve many purposes: recruiting, selection, placement, benefits analysis, requirement projections (how many employees with certain skills will be required in so many months?), and other services. Many companies enable employees to use online systems to compare and select benefit packages such as health insurance and pension plans.

WEB-EMPOWERED ENTERPRISES

The most exciting intersection of IT and business in recent years has been networked commerce—buying and selling goods and services via a telecommunications network—or as it is popularly called, **e-commerce**. The development of the Web and the opening of the Internet to commercial activities spawned a huge surge in business-to-business and business-to-consumer electronic trade. Now, every individual and small business can afford to use a network for business: the Internet.

The Internet is a vast network of computers connected across the globe that can share both information and processing. The Web is capable of displaying text, graphics, sounds, and moving images. It has enticed thousands of businesses to become involved in commercial, social, and educational initiatives. Social networking through sites such as MySpace and Facebook provide a virtual meeting place for people and an opportunity to advertise before many eyes. Thus, the Web is not only a place to conduct e-commerce, but also an emerging advertising medium, gradually replacing other media such as television and newspapers. Almost every brick-and-mortar business has extended its operations to the Web. Chapter 8, "The Web-Enabled Enterprise," discusses Web technologies and how they are used in business activities. Because of its great influence on the use of information technology, the Web's impact on the use of information systems is discussed throughout the book.

POINT OF INTEREST

Left the Laptop on the Bus

The common misperception is that most personal records that fall into the wrong hands are stolen by some electronic criminal hacking into corporate databases from a foreign country. The reality is quite different. Of the personal records compromised in the United States, 35 percent fall into the wrong hands when an employee loses a laptop or another device; 21 percent are lost by a third party with whom the firm works, 19 percent are lost backup records, 9 percent are misplaced paper records, another 9 percent are the result of an inside job or malicious code, and only 7 percent of records fall into hackers' hands.

Source: Di Justo, P., "Your Secret Is Out: Data breaches cost companies billions each year," *Wired*, February 2007, p. 50.

CAREERS IN INFORMATION SYSTEMS

Regardless of the career you choose, you are almost certain to interact with IT professionals. The IT trade is made up of people engaged in a wide variety of activities. According to a forecast by the U.S. Bureau of Labor Statistics, demand for IT professionals in the United States will continue to grow. The Bureau estimates an increase of 23 percent in demand for computer support specialists for the decade 2004–2014. The estimated growth in demand for computer systems analysts is 31.4 percent, for database administrators it is 38.2 percent, for network and computer systems administrators it is 38.4 percent, and for network systems and data communications analysts it is 54.6 percent. All of these occupations will continue to be among the top 25 percent of the best-paying jobs. The following sections review the responsibilities of IT professionals in typical areas of specialization and show parts of posted online help wanted ads from Monster. com, the largest online source for employers seeking IT professionals.

Help Desk Technician

Help desk technicians support end users in their daily use of IT, especially applications. They may be part of an organizational help desk group or employees of an organization that provides help desk to other organizations. In both cases, but especially in the latter, they often provide help via the telephone. They may also communicate directly with a user's PC via a network and special software that gives them control of the user's PC. Help desk technicians are often required to have knowledge of a wide variety of PC applications.

New technology almost always improves lives. But it often also has undesirable effects. This was true of the labor-saving machines that prompted the industrial revolution (introducing 16-hour workdays and child labor under harsh conditions), and it is also true about information technology. Think of the bliss of IT: it makes our work more productive because a few keystrokes on a computer keyboard prompt the computer to calculate and print what would otherwise take many human hours. It educates us via technologies such as multimedia classes delivered online. It opens new economic opportunities such as trading with overseas consumers via the Internet. It makes the world smaller by letting people work and socialize together over great distances via networks such as the Web. It democratizes the business community by making important business tools affordable to both established and start-up companies. And it puts at our fingertips information on practically every imaginable subject. So, what's the dark side? There are quite a few dark sides, which we will discuss in the following chapters. Here is a sample of the main issues and the questions they raise.

- **Consumer Privacy.** The ability to inexpensively and quickly collect, maintain, manipulate, and transfer data enables every individual and organization to collect millions of personal records. When visiting a commercial Web site, chances are the site installs a little file, a "cookie," on your computer's hard disk. This file helps track every click you make on that site, so companies specializing in consumer profiling can learn your shopping and buying habits. When you purchase drugs, the druggist collects details about you. Every time you pay with a credit card, the purchase is recorded to a personally identifiable record. All these data are channeled into large databases for commercial exploitation. Your control of such data is minimal. While consumers, patients, and employees might consent to the collection of information on one aspect of their lives by one party and on another aspect by another party, the combination of such information might reveal more than they would like. For example, a firm can easily and inexpensively purchase your data from a druggist and several consumer goods companies, combine the data into larger records, and practically prepare a dossier about you: your name, age, and gender; your

shopping habits; the drugs you take (and through this information, the diseases you might have); the political party to which you contributed; and so on.

Civil rights advocates argue that IT has created a Big Brother society where anyone can be observed. U.S. business leaders oppose European-style legislation to curb collection and dissemination of private data because this limits target marketing and other economic activities. Are you willing to give up some of your privacy to help companies better market to you products and services you might be interested in? Do you accept the manipulation and selling of your personal data?

- **Employee Privacy.** IT helps employers monitor their employees, not only via the ubiquitous video camera, but also through the personal computers they use. Employers feel it is their right to monitor keystrokes, e-mail traffic, the Web sites employees visit, and the whereabouts of people whose wages they pay while on the job. So, while IT increases productivity, it might violate privacy and create stress. Which is more important: your employer's right to electronically monitor you, or your privacy and mental well-being?

- **Freedom of Speech.** On the Web anyone can become a publisher without censorship. Blogging and other technologies encourage netizens (Internet users) to opine about anything, from products to their employers' misdeeds. Much of the material published is of violent and pornographic nature. If someone posts slurs about your ethnic group at a Web site, do you want the government to step in and ban such postings? To what extent should Web server operators be responsible for what others publish through their sites? Is unsolicited commercial e-mail (spam) a form of free speech?

- **Online Annoyances.** Over 80 percent of all e-mail is spam. Do you accept this? And if you own a new small business and want to advertise via e-mail (because it is the least expensive advertising method), wouldn't you want the freedom to do so? While surfing the Web you encounter pop-up windows and pop-under windows. Your computer contracts spyware. Sometimes special software hijacks your browser and automatically takes you to a commercial site that you do not care for. Are these annoyances legitimate, or should they be stopped by legislation?

- **Phishing and Identity Theft.** Millions of people have fallen prey to phishing, the practice of enticing netizens to provide personal information to imposters. E-mail recipients are directed to copycat sites that purport to be legitimate sites of banks and other businesses where they are requested to "update" or "correct" their social security numbers, credit card account numbers, passwords, and other information. This information is used by the phishers to make fraudulent purchases and obtain loans. Victims do not only lose money. In many cases when phishers steal an identity, the victims experience a long nightmare with authorities and businesses.

- **IT Professionalism.** IT specialists play an increasing role in the lives of individuals and the operations of organizations. The information systems they develop and maintain affect our physical and financial well-being tremendously. If IT specialists are considered professionals, why don't they comply with a mandatory code of ethics as other professionals, such as physicians and lawyers, do?

We will discuss these and other ethical and social issues throughout this book. As you will see, these issues are not easy to resolve. The purpose of these discussions is to make you aware of issues and provoke your thoughts. Remember that the purpose of education is not only to develop skilled professionals but also to remind professionals of the impact of their work on the welfare of other people, and to encourage professionals to be socially responsible.

Systems Analyst

Many IT professionals start their careers as programmers, or **programmer/analysts**, and then are promoted to **systems analysts**, positions that require a broad range of skills. A programmer/analyst is partly involved in the analysis of business needs and ISs, but the greater part of the job involves setting up business applications. A systems analyst is responsible for researching, planning, and recommending software and systems choices to meet an organization's business requirements. Systems analysts are normally responsible for developing cost analyses, design considerations, implementation timelines, and feasibility studies of a computer system before making recommendations to senior management. A big part of this job includes developing alternative system plans based on (1) analyzing system requirements provided by user input, (2) documenting development efforts and system features, and (3) providing adequate specifications for programmers.

To succeed, systems analysts must possess excellent communication skills to translate users' descriptions of business processes into system concepts. They must understand a wide range of business processes and ways in which IT can be applied to support them.

Most importantly, systems analysts must always keep in mind that they are agents of change, and that most people resist change. Unlike many other occupations, theirs often involves the creation of new systems or the modification of existing ones. Because new or modified systems often affect human activities and organizational cultures, systems analysts must be able to convince both line workers and managers that change will benefit them. Thus, these IS professionals must possess good persuasive and presentation skills.

Senior systems analysts often advance to become project leaders. In this capacity, they are put in charge of several analysts and programmers. They seek and allocate resources, such as funds, personnel, hardware, and software, that are used in the development process, and they use project management methods to plan activities, determine milestones, and control use of resources.

Database Administrator

The **database administrator (DBA)** is responsible for the databases and data warehouses of an organization—a very sensitive and powerful position. Since access to information often connotes power, this person must be astute not only technologically but politically as well. He or she must evaluate requests for access to data from managers to determine who has a real "need to know." The DBA is responsible for developing or acquiring database applications and must carefully consider how data will be used. In addition, the DBA must adhere to federal, state, and corporate regulations to protect the privacy of customers and employees.

A growing number of organizations link their databases to the Web for use by employees, business partners, and consumers. Attacks on corporate databases by hackers and computer viruses have made the DBA's job more difficult. In addition to optimizing databases and developing data management applications, this person must oversee the planning and implementation of sophisticated security measures to block unauthorized access but at the same time to allow easy and timely access to authorized users. The DBA is also highly involved in the implementation of SCM systems, because they access corporate databases.

Network Administrator

Among the many IT areas, the one that has seen the most exciting developments in recent years is networks and telecommunications. Not surprisingly, this area has also seen the greatest increase in corporate allocation of IT resources in many organizations. The emergence of new technologies, such as Voice over Internet Protocol and Wi-Fi, which are discussed in Chapter 6, "Business Networks and Communications," is expected to sustain this trend for some years, allowing network professionals to be in great demand and to command high salaries.

An excerpt from a help wanted ad for a network administrator

MAJOR RESPONSIBILITIES:

- Install, configure, and maintain the company's network.
- Maintain user account information including e-mail, rights, security, and system groups.
- Perform system backups and data recovery.
- Monitor system configuration to ensure data integrity.
- Implement, maintain, and troubleshoot network and server security, including file/folder permissions and enterprise anti-virus system.
- Establish and maintain server-based storage for user data and application files.
- Evaluation and installation of new hardware and software.
- Consult with and advise management on operational system problems.
- Perform application support functions for various accounting software (ProFx, CCH, time & billing).
- Provide accounting software training support to accounting professionals.
- Resolve network connectivity issues.

MANDATORY REQUIREMENTS:

- Bachelor's degree in a computer-related discipline or accounting.
- Minimum 3 years current and relevant experience.
- Current working knowledge of computer networks.
- Current working knowledge of accounting software applications (i.e., ProFx, CCH, time & billing systems).
- Current extensive knowledge of routers, switches, servers, TCP/IP & VPN.
- Windows 2000 Active Directory administration skills (users, groups, printers, NTFS, shares, etc.).
- Excellent written and verbal communication skills.
- Ability to manage multiple tasks and work in a fast-paced, highly professional team environment.

Network administrators plan and supervise the organization's local area networks and their connections to the Internet and other external networks.

© Erik Von Weber /Getty Images

The **network administrator** is responsible for acquiring, implementing, managing, maintaining, and troubleshooting local area networks throughout the organization and their interfaces with the wide area networks such as the Internet. He or she is also often involved in selecting and implementing network security measures such as firewalls and access codes.

System Administrator

A **system administrator**—often referred to as "sys admin"—is responsible for managing an organization's computer operating systems. System administrators often manage and maintain several operating systems, such as UNIX and Microsoft Windows Vista, and ensure that the operating systems work together, support end-users' business requirements, and function properly. System administrators are also responsible for the day-to-day maintenance of an organization's operating systems, including backup and recovery, adding and deleting user accounts, and performing software upgrades.

Webmaster

The rapid spread of the Web, intranets, and extranets has increased the responsibility and stature of the organizational Webmaster. A **Webmaster** is responsible for creating and maintaining the organization's Web site as well as its intranet and extranet. Webmasters are increasingly involved in creatively deciding how to represent the organization on the Web. These decisions involve elements of marketing and graphic design. Since many organizations use the Web for commerce, Webmasters must also be well-versed in Web transaction software, payment-processing software, and security software. In small organizations, the Web site may be the responsibility of a single person. In large organizations, the Webmaster often manages a crew of programmers who specialize in developing and updating code specifically for Web pages and their links with other organizational ISs.

Chief Security Officer

Because of the growing threat to information security, many organizations have created the position of **chief security officer (CSO)**, or chief information security officer (CISO). In most organizations, the person in this position reports to the chief information officer (CIO) (see next section), but in some cases the two executives report to the same person, usually the chief executive officer (CEO). The rationale is that security should be a business issue, not an IT issue. A major challenge for CSOs is the misperception of other executives that IT security is an inhibitor rather than an enabler to operations.

Chief Information Officer and Chief Technology Officer

The fact that a corporation has a position titled **chief information officer (CIO)** reflects the importance that the company places on ISs as a strategic resource. The CIO, who is responsible for all aspects of an organization's ISs, is often, but not always, a corporate vice president. Some companies prefer to call this position **chief technology officer (CTO)**. However, you might find organizations where there are both a CIO and a CTO and one reports to the other. There is no universal agreement on what the responsibility of each should be. Yet, in most cases when you encounter both positions in one organization, the CTO reports to the CIO.

An excerpt from a help wanted ad for a chief technical officer

SKILLS/QUALIFICATIONS:

- BS or BA in fields of Information Technology or Operations Management, MS or MBA preferred.
- 10+ years to include experience in direct IT management; experience to include technology, product, and vendor assessment and evaluation; technology vision and strategy.
- Experience with infrastructure support services to include data center management and application support and development.
- Industry knowledge – IT Managed services, customers, marketplace, solution selling.
- Strong professional network.
- Ability to develop creative solutions for all areas of IT operations.
- Good personality – ability to work with all people.
- Strong communication and presentation skills – clear/concise; ability to summarize into clear message.
- Good negotiating skills.
- Decisive, fact-based decision maker.

A person who holds the position of CIO must have both technical understanding of current and developing information technologies and business knowledge. As Figure 1.7 shows, the CIO plays an important role in integrating the IS strategic plan into the organization's overall strategic plan. He or she must not only keep abreast of technical developments but also have a keen understanding of how different technologies can improve business processes or aid in the creation of new products and services.

FIGURE 1.7
Traits of a successful CIO

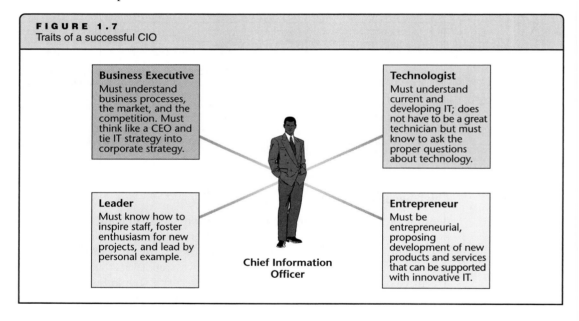

Business Executive
Must understand business processes, the market, and the competition. Must think like a CEO and tie IT strategy into corporate strategy.

Technologist
Must understand current and developing IT; does not have to be a great technician but must know to ask the proper questions about technology.

Leader
Must know how to inspire staff, foster enthusiasm for new projects, and lead by personal example.

Entrepreneur
Must be entrepreneurial, proposing development of new products and services that can be supported with innovative IT.

Chief Information Officer

- Today's business professionals are expected to know how to develop and use IT significantly more than just a few years ago, regardless of their major field of expertise.

- Digital systems quickly and accurately store, process, and communicate information of any type.

- Computer-based information systems pervade almost every aspect of our lives. Their ability to help solve problems and guide decisions makes them indispensable in business and management. Computer-based information systems take data as raw material, process the data, and produce information as output. While data sometimes can be useful as is, it usually must be manipulated to produce information that is useful for reporting and decision making.

- A system is a set of components that work together to achieve a common goal. An information system (IS) consists of several components: hardware, software, data, people, and procedures. The components' common goal is to produce the best information from available data.

- Often, a system performs a limited task that produces an end result, which must be combined with other products from other systems to reach an ultimate goal. Such a system is called a subsystem. Several subsystems might make up a system. Sometimes, systems are also classified as closed or open. A stand-alone system that is not interfaced with other systems is called a closed system. A system that interfaces with other systems is an open system.

- Data processing has four basic stages. In the input stage, data elements are collected and entered into the computer. The computer then performs the next stage, data processing, which is the manipulation of data into information using mathematical, statistical, and other tools. The subsequent stage, output, displays or presents the information. We often also want to maintain data and information for later use. This activity is called storage.

- Any information system that helps in management may be referred to as a management information system (MIS). MISs use recorded transactions and other data to produce information for problem solving and decision making.

- There are several types of information systems. They include transaction processing systems (TPSs), supply chain management (SCM) systems, customer relationship management (CRM) systems, business intelligence (BI) systems, decision support systems (DSSs) and expert systems (ESs), and geographic information systems (GISs). Often, some or all of these systems are linked to each other or to other information systems.

- Enterprise application systems, such as SCM or ERP systems, are information systems that tie together the different functional areas of a business, such as order entry, inventory management, accounting and finance, and manufacturing. Such systems allow businesses to operate more efficiently by avoiding reentry and duplication of information. The systems can provide an up-to-the-minute picture of inventory, work-in-progress, and the status of an order to be fulfilled.

- ISs are used in many business functions, most commonly accounting, finance, marketing, and human resources. These systems aid in the daily operations of organizations by maintaining proper accounting information and producing reports, assisting in managing cash and investments, helping marketing professionals find the most likely buyers for their products and services, and keeping accurate employee records and assisting with their performance evaluations.

- The job prospects for IT professionals are bright. Among the typical careers in this field are systems analyst, database administrator, network administrator, system administrator, Webmaster, chief security officer, chief information officer, and chief technology officer.

- IT has many advantages, but it also has created societal concerns. Issues such as privacy, phishing and identity theft, free speech on the Web, spam, and Web annoyances are viewed by many people as serious ethical issues. And while IT professionals increasingly affect our lives through the systems they develop and maintain, they are not required to adhere to any code of ethics as other professionals are. These and related issues are discussed throughout the book.

GARDENERS+ REVISITED

Now that Chapter 1 has helped you understand how businesses use data, information, and information systems, let's revisit *Gardeners+*. Mary, Amanda, and Ed are trying to improve their gardening business. How would you cope with their challenges?

What Would You Do?

1. Ed is bogged down in entering all the financial data. He is swamped at the end of the month with loan and rental payments, sales tax reports, and paying expenses. He needs a better system. What would you do to improve his efficiency? Examine the business's inputs, processing, and outputs. Formulate a method to streamline the business transactions. What type of reports does Ed need? How would you alter the back-office work to better suit his needs?

2. Mary noticed that some services sold better than expected while others did poorly. What sales information does she need to optimize revenues, costs, and profits when adding, modifying, dropping, or repricing? What is the best method for getting that information?

3. Currently, *Gardeners+* does not collect information on the services that are ordered by an individual customer. Do you think that they would benefit from such information? How might they gather and use such information?

New Perspectives

1. The landscaping industry is not static; new challenges and opportunities always arise. The township recently adopted the following new standards for tree and garden maintenance in both public and private areas:

   - *Tree compatibility.* Due to new pests, a large public investment in pine trees was lost. Some homeowners planted nonindigenous trees that commonly host insects that can be fatal to local pines. Therefore, the township is releasing a list of permitted and prohibited tree species; the prohibitions apply to new as well as existing trees.

   - *Minimum distance from the sidewalk and house foundations by tree type.* The township has mandated that a tree's root system should not damage nearby construction, and that trees must not block drivers' views of the street as they exit their driveways.

   - *Prohibition of certain chemical fertilizers.* A few pets died due to chemical fertilizer poisoning, and tests to several local ponds have revealed high content of extraneous chemicals.

   - *Mandatory gardening licenses.* Homeowners can take care of their own gardens; however, hired gardeners must be licensed, just as electrical and plumbing professionals. The township has recently begun running certification programs.

2. Explain how personal computers and the Internet can help *Gardeners+* to comply with these rules.

   - Several vendors of gardening equipment and supplies are setting up Web sites to allow customers to order supplies online. Explain how these Web sites could affect the current operations of *Gardeners+*. List both benefits and challenges.

business intelligence (BI), 20
chief information officer
    (CIO), 29
chief security officer (CSO), 29
chief technology officer
    (CTO), 29
closed system, 12
customer relationship
    management (CRM), 19
data, 9
data processing, 16
data warehouse, 20
database, 13
database administrator
    (DBA), 26
decision support system
    (DSS), 20

digital systems, 7
e-commerce, 23
enterprise application, 19
enterprise resource planning
    (ERP) system, 18
expert system (ES), 20
geographic information system
    (GIS), 21
information, 9
information map, 14
information system (IS), 13
information
    technology (IT), 14
input, 16
management information
    system (MIS), 18
network administrator, 28

open system, 13
output, 16
process, 10
programmer/analyst, 26
storage, 16
subsystem, 11
supply chain management
    (SCM) system, 18
synergy, 14
system, 11
system administrator, 28
systems analyst, 26
telecommunications, 17
transaction, 16
transaction processing system
    (TPS), 16
Webmaster, 29

1. What does the word "processing" in data processing mean?

2. Give three examples in which raw data also serves as useful information.

3. Give three business examples (not mentioned in the text) of data that must be processed to provide useful information.

4. Give three examples of subsystems not operating in the context of IT. Why are these considered subsystems and not systems?

5. How do TPSs and DSSs differ?

6. What is a problem? Give an example of a business problem and discuss how a computer-based information system could solve it.

7. What is synergy? How is synergy accomplished when a person uses a computer? Explain the connection between synergy and increased productivity.

8. "An information system consists of hardware and software." Why is this statement inadequate?

9. In which situations does one need to make a decision? Give three examples not mentioned in the chapter.

10. How can a DSS help make decisions?

11. Note the word "support" in decision support systems. Why are these applications not called decision-*making* systems?

12. Who is considered a knowledge worker? Will you have a career as a knowledge worker? Explain.

13. What is the most prevalent type of information system? Why is this type of IS so ubiquitous?

14. TPSs are usually used at the boundaries of the organization. What are boundaries in this context? Give three examples of boundaries.

15. Among IT professionals, the greatest demand is for network administrators and analysts. Why?

16. No longer the domain of technical personnel, information systems are the business of every professional. Why?

17. Assume that computers can recognize voices easily and detect their users' exact meaning when talking. Will the necessity for written language be reduced to zero? Why or why not?

18. Information systems cannot solve some business problems. Give three examples and explain why technology cannot help.

19. Practically all knowledge workers must know how to use information systems. Why?

20. Often, computer illiteracy is likened to reading illiteracy. Is this realistic? Is computer illiteracy as severe a handicap as reading illiteracy? (Note that "computer literacy" refers not only to the ability to use a computer, but also to the ability to use software applications, find information on the Web, and share information and files through the Internet.)

21. Think of two examples of fully Web-based businesses. What made the Web so attractive for these entrepreneurs?

22. We will soon stop talking of e-commerce and simply speak of commerce. Why?

23. Help wanted advertisements do not use the term "computer specialists"; rather, they use the term "information system professionals" or "information technology professionals." Why?

24. How do traditional commerce and Web-based commerce differ? What aspects of traditional shopping do you prefer over online shopping?

25. What changed the average citizen's life more, the industrial revolution or the information revolution? How and why?

26. Information technology might bring people together, but it also isolates them. Explain the latter claim and give an example.

27. Give two examples of phenomena that are a social concern because of information technology. Explain.

28. What irritates you about the Web? What would you do to minimize this irritation?

29. Do you foresee an IT-related societal or ethical concern that is not a current concern? Explain.

30. If you chose a career in IT apart from CIO or CTO, which position would you choose, and why?

31. Identity theft existed before the advent of the Internet. However, increased identity theft is one of the unintended, undesirable results of using the Internet. What is the role of educating the public in containing this crime?

APPLYING CONCEPTS

32. Recall what you did yesterday from the moment you got up until the moment you went to bed. How many times did you use a computer or receive data or information from someone who used a computer? (Do not forget ATMs, POS machines, automated kiosks, personal devices, etc.) Write a two-page essay on your daily experience with IT and on society's dependency on computers.

33. Contact a business organization and ask permission to observe a business process. Pinpoint the segments in the process that a computer-based information system could aid. Write a report detailing your observations and suggestions.

34. Observe activities in a supermarket: shoppers looking down aisles for specific products; lines forming at the POS machines; workers putting new prices on items. Prepare a list of shoppers' and workers' activities that could be carried out with less use of human time and more accuracy if they were aided by IT. Explain how you would change those activities.

35. Scientists are researching a contagious disease. They found that, on average, each person who is infected transmits the disease to three other people within one year. Currently, 3,000 people in the country are infected. Use Excel or another spreadsheet application to find out (1) how many people will contract the disease each year over the next decade, and (2) how many infected people will there be each year if no medication is administered. (Do not worry: there is a medication for this disease.) "Currently" means in the first year of your calculation. Calculate for the next nine years. Explain why this is a modeling problem. What is your model in the spreadsheet?

36. Use a résumé template in your word-processing program to type your résumé. If you don't have a lot of direct work experience, remember to include all types of work, whether it's babysitting, camp counseling, mowing the lawn, or volunteer work. Now turn your résumé into one that can be displayed well as a Web page.

37. Prepare a list: what information that you currently receive through other means could you receive through your computer? The list should include text, images, audio, and animated information. Would you prefer to receive this information on the computer or as you do now?

38. Form a team with two other students. Each team member should play the role of a vice president in charge of a business function: human resources, accounting, marketing, finance, and so on. Each vice president should enumerate information he or she needs to perform his or her function. Now list information that two or more of the functions must share and data produced by one function that another function uses.

39. Team up with another two students. Brainstorm and try to think of a new business opportunity that you would like to pursue in which you will not need IT. You should be able to convince your professor that IT cannot improve the operations of this business.

FROM IDEAS TO APPLICATION: REAL CASES

The Personal Touch

FedEx is an organization that never sleeps and for which every minute counts. Each business day the company's more than 275,000 employees and independent contractors handle an average of 6 million packages, using over 669 aircraft and 71,000 trucks. On the peak days between Christmas and New Years' Eve, it typically ships more than 8 million packages. The company, which generates $33 billion of revenues per year, serves more than 220 countries and territories. Inevitably, some packages miss their delivery time, some miss their destination, and some are damaged. When that happens, FedEx's 4,000 customer service reps in 56 call centers are the people customers call.

Prompt, efficient customer service is extremely important for staying in this highly competitive global shipping industry, let alone doing so with a satisfactory profit. Incoming telephone calls at the FedEx customer service center in Fullerton, California, never stop, and FedEx reps never have an idle moment on shift. Sitting in front of computer monitors in a cluster of cubicles with headsets on, these agents barely have time to stretch their limbs.

A caller complains that her package hasn't arrived, which is a common complaint. Another asks if he can change his pickup time. A third caller is confused about signature: is he supposed to sign for the delivery or will the package just be dropped at his doorstep? The reps are confident and friendly. They welcome any question or complaint even if they have heard it a thousand times before. The words "I am sorry" are uttered often. They are careful not to give the customers a feeling of being rushed, but try to resolve complaints quickly. Time is money.

Several years ago FedEx installed software that reps at the call centers can use to provide faster service. Many of the callers are already registered in the company's database. One of the most frequent requests is to send a FedEx worker to pick up a package. Using the software, a rep can handle such a request in 20 seconds. All she needs to do is enter a name, which leads to a zip code, which in turn leads to a tracking number. That number uniquely identifies the package. Some complaints are more complex. For example, a FedEx driver misunderstood a note a caller had left for him and therefore misdelivered a package. A complaint like that takes no more than 10 minutes to resolve.

An experienced and efficient rep can handle about 10 callers in 45 minutes. Ideally, though, nobody would call. If FedEx had its way, at least six of the ten callers would use their computers to go to FedEx's Web site and solve their problem by themselves—because about 60 percent of FedEx's clients have a computer connected to the Internet. Like other companies, FedEx tries to save labor by directing callers to its Web site. Yet, many people prefer to use the phone and talk to a human helper.

Every time a customer decides to use the company's Web site instead of telephoning, the company saves up to $1.87. Efforts to divert callers to the site have been fruitful. In 2005, FedEx call centers received 470,000 calls per day, 83,000 fewer than in 2000. This difference in calls translates into a saving of $57.56 million per year. The company's Web site handles an average of 60 million requests to track packages per month. Operating the Web site does cost money. Each of these requests costs FedEx 3 cents, amounting to $21.6 million per year. However, if all these requests were made by phone, the cost would exceed $1.36 billion per year. As it is impossible to divert all callers to the Web site, the company must maintain call centers. The annual cost of these call centers is $326 million. This cost might decrease over the years, as more and more customers use the Web site, but there will probably always be call centers, because FedEx does not want to lose frustrated customers.

Many people are still uncomfortable doing business at a Web site. The cost of a customer who is frustrated by the company Web site is incalculable. Experience shows that people are willing to encounter one or two obstacles with the Web site, but then they stop trying.

Since its establishment in 1971 as Federal Express Corp., the company was keen on information technologies, but over the years it used an increasing number of disparate systems for different business purposes, such as air freight, ground freight, special logistic operations, and custom shipping of critical items. By 1999, customer information was scattered in computer systems implemented over 14 years. To periodically test service, executives pretend to be customers. They discovered that customers who used more than one FedEx business were not treated consistently. For example, when claiming damages a customer had to fill out 37 fields on a claim form, such as tracking number, ship date, pickup location, and destination, even though FedEx systems already held data for 33 of those fields. The official change of "Federal Express" to "FedEx" started an important move: all the company units were to share the same information systems.

Meanwhile, FedEx's customer service centers were redesigned around a PC-based software desktop. If reps could pull up historical data on customers whenever they called—not just their shipping histories, but their preferences and even images of their paper bills—FedEx could provide better, faster service, both to individual customers and to businesses that sold goods through catalogs.

In 2000, management purchased customer relationship management software called Clarify. A new policy was established: systems and customer service experts are equally responsible for the call centers. Using PCs, reps can pull up historical data on customers whenever customers call. Customer records that are immediately available to reps include shipping histories, preferences, and images of the paper bills. Customers are happier now than they were just a few years ago. So are the reps. Turnover of service reps has decreased 20 percent.

Productivity is important, but so is the reps' service quality. They must be polite, provide customers with correct appropriate information, and try not to give customers a reason to call again. Typically, callers are either determined to speak to a human or they know the help they need is too complex to be available at the company's Web site. Therefore, callers require more time than in the past. The company periodically evaluates the reps' performance based on clearly stated goals that take all these factors into consideration. Typically, 32 percent of the reps' performance rating is based on the quality of their response, and 17 percent on their efficiency. The other 51 percent is based on attendance, adherence to scheduled breaks, and compliance with regulations.

Interestingly, customers are not interested in friendliness, but in quick and accurate information. FedEx constantly follows customer reactions to different help styles. Managers discovered then when reps' time is not limited, they tend to speak with customers beyond the time required to solve the problem. Customers perceive them as too talkative, and they get a bad impression about FedEx. Thus, reps are encouraged to get off the phone as soon as the problem is resolved rather than try to be "nice."

The professionals who work for the vendor of Clarify, the CRM software, spent time with reps to see how well the software serves them. They discovered that reps often move quickly from one window of information to another, and that sometimes they take extra time to find a window that "disappeared." The software engineers decided to modify Clarify so it interacts with Java code. This enables the reps to switch between windows and different applications of Clarify quickly during a call without reentering customer data. For instance, if a customer needs directions to pick up

a package, the rep can click the tab of the mapping application. Relying on the customer's account data, the application picks up the customer's zip code. Combining it with the code of the pickup center, the software immediately produces directions, which the rep can read to the customer.

While great improvements have already been accomplished both in service speed and quality, FedEx executives continue to look for ways to improve. They refuse to discuss what their next step is because it might be copied immediately by competitors, but they do reveal that their goal is to bring call centers to the point where a rep never has to put a customer on hold.

Experts expect a single "nervous system" for all types of customer calls by 2010. Software will accept all customer calls from the customer's PC, phone, or handheld device. Special software involving artificial intelligence techniques will screen all incoming calls, evaluate the problem's complexity, and decide whether to direct the calls to other software for resolution or to invite a human rep to intervene.

Source: Gage, D., "FedEx: Personal Touch," *Baseline* (www.baselinemag.com), January 13, 2005; www.fedex.com/us, 2007.

Thinking About the Case

1. What is CRM in general? Give examples of *different* CRM applications.

2. Enumerate and explain the various ways in which the CRM application discussed here (Clarify) saves costs or helps in other ways.

3. Which metrics would you use to measure *before* and *after* performance regarding the information technologies implemented in this case? Consider cost, service quality, cycle time, and any other performance factor and provide a specific metric (i.e., ratio, product, or absolute value).

4. As a customer, would you prefer more or less mechanized service in lieu of human help?

5. As an executive for FedEx or a similar company, what else would you implement using software and the Internet?

Less Paper, Better Reforestation

By the 1920s many of Washington State's trees had been harvested for timber. The uncontrolled activity caused many regions to lay barren. State officials realized that unless reforestation was initiated, the state might lose one of its most important resources. Nurseries were established, but the supply of seedlings did not meet demand. In 1958 the Department of Natural Resources established the L.T. Mike Webster Forest Nursery south of Olympia, named after the new department's supervisor: 270 acres and 30,000 square feet of greenhouses. The nursery operates like a private business. It sells seedlings to companies and the general public, and receives no funding from the state.

To a traveler along the coast of the Pacific Northwest, all the trees on cliffs and mountains may seem the same. But for the staff of 18 people at Webster Forest Nursery, it is important to know the details of each of those trees. And there are millions of them. The nursery collects seeds throughout the state and cultivates them. Then it plants the seedlings on state property and sells them to the public.

Over its 48-year history, the nursery has accumulated a wealth of knowledge about which plants succeed in which environment, which do not, and what can be done to ensure successful growth. It now produces 8-10 million seedlings annually. Staff members know, for example, that a Sitka Spruce whose seed came from Mount Rainier may not grow well on the coast. However, for many years much of this experience was lost when staff members retired. Information that was recorded was kept on paper.

The nursery tracks each seedling's history from extraction of the seed to reforestation.

The staff collects data on each planted tree to ensure that it receives the proper care. Data recorded includes the seed origin, current location and its elevation, treatment history, and growth progress. This information on millions of trees categorized by 42 species is important to ensure proper growth. Cultivating, planting, and nurturing each seed until it can grow costs tens or hundreds of dollars. Inaccurate information could result in substantial financial losses.

For decades, tracking was done the same way: staff members in the field recorded data on a clipboard. Back in the office, they copied the data onto index cards. The "database" looked like an old library catalog. It was impossible to have an accurate count of the nursery seedlings, let alone counts of reforested areas. The totals were estimated by the number of acres. By 1996 workload increased, and the nursery estimated that unless a technological solution was implemented, the employees would soon have to increase their work time by 33 percent. Only one staff member in the office could locate records, and she was about to retire.

The nursery's manager decided to automate the system. He hired Rudeen & Associates, a small consulting firm. Rudeen installed personal computers with a database management system and equipped the workers with handheld computers. It named the system RIMS (Reforestation Information Management Systems). The database is Oracle Version 8, and the handheld units are rugged Husky computers. In 1996 few off-the-shelf wireless devices were available, let alone applications to connect them wirelessly to any system. Thus, Rudeen hired another company to develop the proper software.

The same technology still serves Webster today. Workers record tracking data into the handheld units, which transmit to the database in the office and update the seedling records. Since information is much less error-prone, the nursery can be assured it sells seedlings that can survive where they are planted and fulfill their specific purpose of reforestation. Improvements have been made over the years. In October 2006 sales data was incorporated into RIMS. Now it is easier to track sales by species and to forecast future revenues. Automation also freed the staff to devote more time to adding greenhouses. The nursery has decided to add software for analyzing employee performance. Improvements since the initial implementation of the system cost $750,000, not an insignificant amount for a small organization whose annual revenue is $2.5 million. However, nobody doubts that the investment yielded excellent results.

Source: Pettis, A., *eWeek*, January 8, 2007; Webster Forest Nursery (www3.wadnr.gov/dnrapp3/webster) February 2007.

Thinking About the Case

1. What were the main deficiencies of the paper system? What was wrong with the fact that a human remembered where a record could be found?

2. List and explain the benefits of the RIMS system.

3. What can be the benefits of analyzing employee performance by the system?

4. Consider the types of information systems discussed in the chapter. Which type of system could probably help both the nursery and the Department of Natural Resources?

2

TWO

Strategic Uses of Information Systems

LEARNING OBJECTIVES

Executives know that information technology is not merely a resource to support day-to-day operations. Clever use of IT can significantly change an organization's long-term strategic position. Often, innovative use of information systems radically changes the way a firm conducts its business. Some information systems even change a firm's product or service, such as when innovative software is integrated into a physical product or when a service is readily available on the Web. Therefore, information systems are now an integral part of strategic planning for nearly all organizations.

When you finish this chapter, you will be able to:

- Explain what business strategy and strategic moves are.

- Illustrate how information systems can give businesses a competitive advantage.

- Identify basic initiatives for gaining a competitive advantage.

- Explain what makes an information system a *strategic* information system.

- Identify fundamental requirements for developing strategic information systems.

- Explain circumstances and initiatives that make one IT strategy succeed and another fail.

GARDENERS+:
Using Information Strategically

The *Gardeners+* information system was successful so far: the business had been operating for a little more than a year, it was profitable, and Mary, Amanda, and Ed had begun to enjoy running it. Using the affiliation format, they offered services like ground preparation, installation of sprinkler systems, fence construction and maintenance, and a few other services that their affiliated workers knew how to do. During the winter, they offered custom snow removal services for driveways and sidewalks as well as services to protect vulnerable plantings from the cold. These services generated cash flow to help meet monthly loan payments during the off-season.

The three entrepreneurs were looking for ways to expand their business and increase their revenues. An opportunity presented itself at the local Chamber of Commerce meeting.

Looking at Expansion

Mary regularly attended Chamber of Commerce meetings to keep in touch with the local business community. She was always looking for new opportunities. After one meeting, the manager of a large mall asked about the possibility of contracting with *Gardeners+* for backup gardening support. He had seen the work that *Gardeners+* provided at City Hall during last year's major summer holiday events and was impressed. His current service was inflexible and often failed to provide required services during holidays—peak times at the mall. Furthermore, he mentioned that many mall stores were also interested in occasional indoor gardening services for their special events. Mary consulted with her partners, and they all agreed to expand operations into the commercial arena and offer their services to the mall and its stores.

New business is a good thing, but also presents new challenges. Mary, Amanda, and Ed needed to adapt their subscription and commission scheme to accommodate new commercial customers. They could no longer assume that their affiliated gardeners would be available as needed or have the appropriate equipment. The business would have to purchase its own gardening equipment and develop new contracts that require more durable commitments from both gardeners and commercial customers.

Purchasing their own equipment meant increasing their fixed costs. They began to worry that such a venture would not be sufficiently profitable during the slower winter months, when much of the equipment would sit idle. Large commercial clients like the mall also meant a high degree of dependence on fewer clients; what would happen if *Gardeners+* were to lose them?

A New Line of Business?

Very soon, *Gardeners+* went from being the mall's backup service to being the mall's main service provider. The mall shops also began to request *Gardeners+* services more frequently. New commercial clients surfaced, including several condominiums and the merchant's association for an important shopping boulevard. These commercial services were much different from the residential services *Gardeners+* had initially offered. They were strictly scheduled, with firm commitments, and required that gardeners be more aware of the public relations aspect of the job. The partners realized that they needed to separate the two types of business; they decided to develop a new line of products and services for their commercial clients.

Ed and Amanda offered to develop a list of products and services that would not only be profitable for *Gardeners+* (based on recent sales history) but also flexible, distinct, and carefully defined for their new and potential corporate clients. Because these would be long-term contracts, they had to be careful not to underestimate costs or overestimate staff and equipment availability. Such errors could lead to long-term losses or render the group unable to comply with commitments due to lack of resources. They also had to ensure that these new commitments would not affect the quality of service rendered to their existing residential customers—at this point the only ones proven to generate profit and business growth.

The *Gardeners+* partners decided to: (1) increase their fixed assets with a large purchase of gardening equipment and (2) establish long-term contracts with a select group of gardeners. The future seemed bright, but they were faced with the daunting task of determining how far to go with these long-term commitments while at the same time assuring business profitability during the slow months and potential downturns.

Charting a Strategy with Information Systems

Amanda and Ed investigated the costs of equipment, consumables, and capable gardening skills for commercial locations. They also performed a market analysis on consolidated gardening companies that serviced their area and other regions in order to get a better understanding of both the business and potential competitors. To do this research they used the Web, requested written proposals from some companies, and even posed as potential clients during direct phone enquiries.

After entering the costs, prices, and modalities into their spreadsheet, they discovered that some services would not be profitable for a company of the size and structure of *Gardeners+*. For example, they could not service gardening jobs in elevated places, like balconies or building facades, which required very expensive and specialized equipment. Demand for such services in their area did not justify the investment in this equipment. Other, less specialized services offered by competitors also did not provide sufficient potential profit. For example, to service sites that required certification from a major ecological ("green") organization, they would have to limit their overall use of fertilizers and pest control chemicals. They would also have to hire ecologically certified personnel who would command higher wages (or commissions,

depending on the type of contract). Nonetheless, since Amanda was an ecological activist, they decided to conduct a survey on the subject. The partners gathered a focus group consisting of Julian's colleagues, who pointed out that they often got proposal requests for organic gardening; but after receiving estimates, the clients tended to request the original chemically-based services. Furthermore, the gardeners contended that they often lose such clients, who hire a different provider of conventional service rather than admit reluctance to pay the higher price of the organic service. Amanda agreed to drop the idea for the time being.

New Competition on the Block

Soon after Mary, Amanda, and Ed had made these important decisions, they received bad news. Word of *Gardeners+*'s success had apparently spread. The company that used to service the mall came back with a new and more flexible service offer. It was clear that they planned on retaking the mall as well as some of the other commercial sites that *Gardeners+* were servicing. The three partners were worried about competing with a more established firm that had a larger financial and geographical base. This competitor learned from *Gardeners+* and decided to correct its mistakes; it was also expanding its business into the residential market.

To help retain customers, the partners decided to implement the "*Gardeners+* Loyal Customer Program": customers would get their tenth service free after they had paid for nine similar (or more expensive) services. Amanda was able to easily prepare a list of customers that had already received nine or more eligible services, and the next day she called to inform each of them that the next service would be free. This would be just one of the many innovations that would help them remain profitable in this increasingly competitive market. They knew that they needed to keep on their toes if they wanted to remain profitable and grow.

STRATEGY AND STRATEGIC MOVES

A survey of 291 IT executives by the journal *CIO Insight* revealed the ever-changing role of IT in corporations. One-third of the executives said that their role was to create business strategy. The other two-thirds said their role was executing strategy. Either way, IT was expected to contribute to business strategy. Half the executives said that contributing to development of strategy has become more important to their supervisors, usually the company president or chairperson of the board.

The word "strategy" originates from the Greek word *strategos*, meaning "general." In war, a strategy is a framework, or an approach, to obtaining an advantageous position. Other disciplines, especially business, have borrowed the term. As you know from media coverage, corporate executives often discuss actions in ways that make business competition sound like war. Businesspeople must devise decisive courses of action to win—just as generals do. In business, a strategy is an approach designed to help an organization outperform its competitors. Unlike battle plans, however, business strategy often takes the form of creating new opportunities rather than beating rivals.

Although many information systems are built to solve problems, many others are built to seize opportunities. And, as anyone in business can tell you, identifying a problem is easier than creating an opportunity. Why? Because a problem already exists; it is an obstacle to a desired mode of operation and, as such, calls attention to itself. An opportunity, on the other hand, is less tangible. It takes a certain amount of imagination, creativity, and vision to identify an opportunity, or to create one and act on it. Information systems that help seize opportunities are often called **strategic information systems (SISs)**. They can be developed from scratch, or they can evolve from an organization's existing ISs. They are not defined by their technical features per se, but by how they are used, that is, for strategic advantage.

In a free-market economy, it is difficult for a business to do well without some strategic planning. Although strategies vary, they tend to fall into some basic categories, such as developing a new product, identifying an unmet consumer need, changing a service to entice more customers or retain existing clients, or taking any other action that increases the organization's value through improved performance.

POINT OF INTEREST

IT as Strategic Tool

A 2006 survey of 408 chief information officers by *CIO Magazine* revealed an interesting fact. Fifty-two percent of the CIOs surveyed said that in their companies the IT unit was viewed as a strategic organization. The other 48 percent said that IT in their organization was regarded as a support or staff function. Interestingly, the smaller the organization (in terms of revenue), the more the IT staff are viewed as strategic.

Source: Alter, A. E., "August 2006 IT Organization Survey: The Wall Between IT and Business is Falling Down," *CIO Insight*, August 29, 2006.

Many strategies do not, and cannot, involve information systems. But increasingly, corporations are able to implement certain strategies—such as maximizing sales and lowering costs—thanks to the innovative use of information systems. A company achieves **strategic advantage** by using strategy to maximize its strengths, resulting in a **competitive advantage.** When a business uses a strategy with the intent to *create* a market for new products or services, it does not aim to compete with other organizations who make the same product, because that market does not yet exist. Therefore, a strategic move is not always a competitive move in terms of competing with similar products or services. However, in a free-enterprise society, a market rarely remains the domain of one organization for long; thus, competition ensues almost immediately. So, we often use the terms "competitive advantage" and "strategic advantage" interchangeably.

You might have heard statements about using the Web strategically. Business competition is no longer limited to a particular country or even a region of the world. To increase the sale of goods and services, companies must regard the entire world as their market. Because thousands of corporations and over a billion consumers have access to the Web, augmenting business via the Web has become a strategic necessity. Many companies that utilized the Web early on have enjoyed greater market shares, more experience with the Web as a business enabler, and larger revenues than latecomers. Some companies developed information systems, or features of information systems, that are unique, such as Amazon's "one-click" online purchasing and Priceline's "name your own price" auctioning. However, simply extending business to the Web can no longer guarantee a strategic advantage. Doing so in an innovative way can. Practically any Web-based system that gives a company competitive advantage is a strategic information system.

ACHIEVING A COMPETITIVE ADVANTAGE

Consider competitive advantage in terms of a for-profit company, whose major goal is to maximize profits by lowering costs and increasing revenue. A for-profit company achieves competitive advantage when its profits increase significantly, most commonly through increased market share. Figure 2.1 lists eight basic initiatives that can be used to gain competitive advantage, including offering a product or service that competitors cannot provide or providing the same product or service more attractively to customers. It is important to understand that the eight listed are the most common, but not the only, types of business strategy an organization can pursue. It is also important to understand that strategic moves often consist of a combination of two or more of these initiatives and other steps, and that sometimes accomplishing one type of advantage creates another. The essence of strategy is innovation, so competitive advantage is often gained when an organization tries a strategy that no one has tried before.

FIGURE 2.1
Eight basic ways to gain competitive advantage

Initiative	Benefit
Reduce costs	A company can gain advantage if it can sell more units at a lower price while providing quality and maintaining or increasing its profit margin.
Raise barriers to market entrants	A company can gain advantage if it deters potential entrants into the market, enjoying less competition and more market potential.
Establish high switching costs	A company can gain advantage if it creates high switching costs, making it economically infeasible for customers to buy from competitors.
Create new products or services	A company can gain advantage if it offers a unique product or service.
Differentiate products or services	A company can gain advantage if it can attract customers by convincing them its product differs from the competition's.
Enhance products or services	A company can gain advantage if its product or service is better than anyone else's.
Establish alliances	Companies from different industries can help each other gain advantage by offering combined packages of goods or services at special prices.
Lock in suppliers or buyers	A company can gain advantage if it can lock in either suppliers or buyers, making it economically impractical for suppliers or buyers to deal with competitors.

For example, Dell was the first PC manufacturer to use the Web to take customer orders. Competitors have long imitated the practice, but Dell, first to gain a Web audience, gained more experience than other PC makers on this e-commerce vehicle and still sells more computers via the Web than its competitors. Figure 2.2 indicates that a company can use many strategies together to gain competitive advantage.

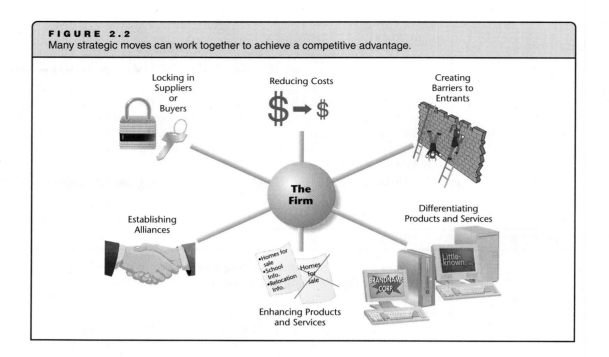

FIGURE 2.2
Many strategic moves can work together to achieve a competitive advantage.

Locking in Suppliers or Buyers

Reducing Costs

Creating Barriers to Entrants

The Firm

Establishing Alliances

Differentiating Products and Services

Enhancing Products and Services

Initiative #1: Reduce Costs

Customers like to pay as little as possible while still receiving the quality of service or product they need. One way to increase market share is to lower prices, and the best way to lower prices is to reduce costs. For instance, if carried out successfully, massive automation of any business process gives an organization competitive advantage. The reason is simple: automation makes an organization more productive, and any cost savings can be transferred to customers through lower prices. We saw this happen in the auto industry. In the 1970s, Japanese automakers brought robots to their production and assembly lines and reduced costs—and subsequently prices—quickly and dramatically. The robots weld, paint, and assemble parts at a far lower cost than manual labor. Until their competitors began to employ robots, the Japanese had a clear competitive advantage because they were able to sell high-quality cars for less than their competitors. A similar approach gave Intel, the computer microprocessor maker, a strategic advantage that it maintains to this day: much of the labor involved in making and testing microprocessors has been automated by information technology and robots. This enabled the company to substantially reduce the prices of its products.

In the service sector, the Web has created an opportunity to automate what until recently was considered an activity that only humans could perform: customer service. An enormous trend toward automating online customer service began with companies such as FedEx, which initially gave customers an opportunity to track their parcels' status by logging on to a dedicated, private network and database. The same approach is now implemented through the Web. Many sites today include answers to FAQs (frequently asked questions). Others have special programs that can respond to customer questions. Online service gives businesses two major benefits: it changes service from being labor intensive to technology intensive, which is much less expensive, and it provides customers easy access to a service 7 days a week, 24 hours a day. Any executives of companies that operate call centers will tell you that they work hard to shift callers off the phone and to their Web sites to receive the help they need. It not only cuts the costs of expensive human labor but also of telephone and mailing charges. Companies that are first to adopt advanced systems that reduce labor enjoy competitive advantage for as long as their competitors lag behind.

Understand the Notion of Strategic Information Systems

Although devising strategic moves is mainly the responsibility of senior management, let us remember Napoleon's words: "Every soldier carries a marshal's baton in his knapsack." To paraphrase: every junior worker is a potential senior executive. Thus, it is incumbent on every professional to try to think strategically for his or her organization. In fact, employees at the lowest levels have proposed some of the most brilliant strategic ideas. In today's highly competitive market, strategy might determine an organization's rise or fall.

An increasing number of strategic moves are possible only with the aid of ISs or by having ISs at the center of their strategy—that is, technology provides the product, service, or method that gains the organization strategic advantage. The potential for new business models on the Web is still great. Thus, professionals must understand how to use technology in strategic moves. Understanding how strategic information systems are conceived and implemented might help you suggest good ideas for such systems in your organization and facilitate your promotion up the organizational ladder.

Initiative #2: Raise Barriers to Market Entrants

The smaller the number of companies competing within an industry, the better off each company is. Therefore, an organization might gain competitive advantage by making it difficult, or impossible, for other organizations to produce the product or service it provides. Using expertise or technology that is unavailable to competitors or prohibitively expensive is one way to bar new entrants.

Companies **raise barriers to entrants** in a number of ways. Obtaining legal protection of intellectual property such as an invention or artistic work bars competitors from freely using it. Microsoft, IBM, and other software powerhouses have gained tremendous strategic advantages by copyrighting and patenting software. Numerous examples of such protection can be found on the Web. Priceline.com holds a patent for online reverse ("name your own price") auctioning, which has prevented competitors from entering its business space. Amazon.com secured a patent for one-click online purchasing, which enables customers to enter shipping and credit card information once and to place subsequent orders while skipping a verification Web page. Although the software is quite simple, Amazon obtained a patent for it in 1999 that won't expire until 2017. Amazon successfully sued Barnes & Noble (B&N) when it implemented the same technology on BN.com. Now B&N pays Amazon for its use. More recently, Amazon obtained a patent for the techniques it uses to guess what types of items a user might like to buy in the future. Exclusive use of the methods might give the company additional strategic advantage in online shopping. Protecting any invention, including hardware and software, with patents and copyrights provides an excellent barrier to potential entrants.

Another barrier to potential new market entrants is the high expense of entering the particular market. The pension fund management industry is a prime illustration. State Street Corporation is one of the industry's most successful examples. In the 1980s, State Street committed massive amounts of money to developing ISs that helped make the company a leader in managing pension funds and international bank accounts. The huge capital allocation required to build a system to compete successfully with State Street keeps potential entrants out of the market. Instead, other pension management corporations rent State Street's technology and expertise. In fact, State Street derives about 70 percent of its revenues from selling its IS services. This company is an interesting example of an entire business refocusing around its ISs.

Initiative #3: Establish High Switching Costs

Switching costs are expenses incurred when a customer stops buying a product or service from one business and starts buying it from another. Switching costs can be explicit (such as charges the seller levies on a customer for withdrawal from a contract) or implicit (such as the indirect costs in time and money spent adjusting to a new product that competes with the old).

Often, explicit switching costs are fixed, nonrecurring costs, such as a penalty a buyer must pay for terminating a deal early. In the cellular telephone service industry, you can usually get an attractive deal, but if you cancel the service before the one- or two-year contract ends, you have to pay a hefty penalty. So although another company's service might be more attractive, you might decide to wait out the full contract period because the penalty outweighs the benefits of the new company's service. When you do decide to switch, you might discover that the telephone is not suitable for service with any other telephone company. The cost of the telephone itself, then, is another disincentive to switch.

A perfect example of indirect switching expenses is the time and money required to learn new software. Once a company trains its personnel to use one word-processing or spreadsheet program, a competing software company must offer a very enticing deal to make switching worthwhile. The same principle holds for many other applications, such as database management systems, Web page editors, and graphical software. Consider Microsoft's popular MS Office suite; you can purchase the significantly less expensive Sun Microsystems' StarOffice, a software suite that is equivalent to MS Office. Better yet, you can download free of charge the entire suite of OpenOffice.org. Yet, few organizations or consumers who are accustomed to MS Office are willing to switch to StarOffice or OpenOffice.org.

Manufacturers of laser and ink-jet printers sell their printers at cost or below cost. However, once you purchase a printer, you must replace a depleted ink or toner cartridge with a costly cartridge that the printer manufacturer sells, or take a risk with other cartridges whose quality is often low. You face high costs if you consider switching to another printer brand. Thus, establishing high switching costs often locks in customers. Locking in customers by any means is a way to accomplish a strategic advantage, and is discussed later in this chapter.

High switching costs often apply when a company uses proprietary software, especially when the software is expensive, such as an ERP system. In addition to the initial price of the system, the client incurs other costs, some tangible and some not. Tangible costs include modification to suit the special needs of the client's unique business processes. Intangible costs include employees' learning the new system and the establishment of smooth working relations with the service unit of the software vendor.

Initiative #4: Create New Products or Services

Clearly, the ability to **create a new and unique product or service** that many organizations and individuals need gives an organization a great competitive advantage. Unfortunately, the advantage lasts only until other organizations in the industry start offering an identical or similar product or service for a comparable or lower price.

Examples of this scenario abound in the software industry. For instance, Lotus Development Corporation became the major player early on in the electronic spreadsheet market after it introduced its Lotus 1-2-3 program. When two competitors tried to market similar products, Lotus sued for copyright infringement and won the court case, sustaining its market dominance for several years. However, with time, Microsoft established its Excel spreadsheet application as the world leader, not only by aggressive marketing but also by including better features in its application.

Another example of a company creating a new service is eBay, the firm that dominates online auctions. The organization was first to offer this service, which became very popular within only a few months. While other firms now offer a similar service (e.g., Amazon.com and Yahoo! Auctions), the fact that eBay was the first to offer it gave the company a huge advantage. It quickly acquired a large number of sellers and bidders, a network that is so critical to creating a "mass" of clients, which in turn is the main draw for additional clients. It also gave eBay an advantage in experience and allowed it to open a gap that was difficult for competitors to close, even for giants such as Amazon.com. eBay is an example of an entire business that would be impossible without the Web and the information technologies that support the firm's service.

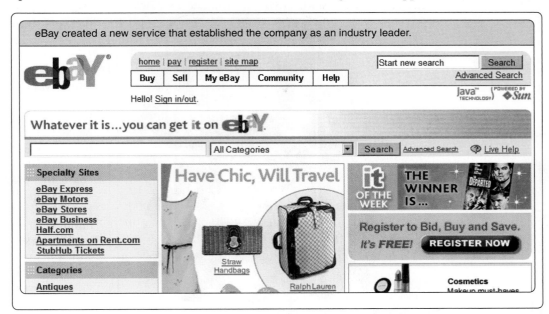

eBay's success demonstrates the strategic advantage of the **first mover**, an organization that is the first to offer a new product or service. By the time other organizations start offering the same product or service, the first mover has usually created some assets that cannot be held by the competitors: a superior brand name, a better technology or method for delivery, or a **critical mass**. A critical mass is a body of clients that is large enough to attract many other clients. In many cases, first movers simply enjoy longer experience, which in itself is an advantage over competitors.

XM and Sirius, satellite-based radio services, have changed radio broadcasting. Their broadcasts release radio services from the constraints of territorial boundaries and so far have avoided national content regulation. This is an example of a new service that is fast garnering an increasing client base. Some observers predict that in a decade or so, the number of listeners to this type of broadcast will surpass the number of listeners to traditional radio stations. Many radio personalities and radio stations now offer programs on satellite radio, hoping to participate in its strategic advantage. The two pioneers in this market, XM and Sirius, are reaping the rewards of first movers. Their combined subscribers totaled close to 14 million in 2006. In this case, however, both companies were first movers and therefore competition did not allow them to profit, so they agreed to merge. (At the time this book went to press the companies were waiting for approval of the merger from the U.S. Senate and Federal Communications Commission.)

A good example of a new product is Apple Computer's iPhone. Handheld devices that combine telephony and computers had been around for many years, but the iPhone introduced a new concept: no physical keys. All the functions are activated by using only touch-screen keys, and therefore can be more intuitively operated and offer more options. To compete with Apple, rivals will have to introduce a higher-quality device at a similar or lower price.

Some Web sites were the first to offer certain services that soon attracted millions of visitors per day. The high traffic they have created gives them a significant strategic asset in the form of advertising potential. YouTube enables individuals and corporations to place video clips. The site streams more than 100 million videos per day. Its popularity became so great that it was acquired for $1.65 billion by Google. MySpace.com and Facebook.com became the most popular social networking Web sites, making each of them a great potential for online advertising. This was the reason why News Corp., the large media conglomerate, purchased InterMix Media, the owner of MySpace.com, for $580 million.

iPhone was an innovative, attractive product when it was introduced in early 2007.

MCT/Landov

Being a first mover is not always a guarantee of long-term success, however. One example of how a first-mover strategic advantage can be lost within just a few months is in the Web browser arena. Netscape Corporation (now part of AOL) dominated the Web browser market, which was new in 1994. By allowing individual users to download its browser for free, it cornered up to 95 percent of the market. The wide use of the browser by individuals moved commercial organizations to purchase the product and other software compatible with the browser. Netscape's dominance quickly diminished when Microsoft aggressively marketed its own browser, which many perceived as at least as good as Netscape's. Microsoft provided Internet Explorer free of charge to anyone and then bundled it into the Microsoft Windows operating system software distributed with almost all PCs. Even after the court-ordered unbundling, its browser still dominated.

Other first movers have lost market share because they neglected to improve the service they pioneered. Few Web surfers remember Infoseek, the first commercial search engine. Google, which entered the search engine arena in 1998, improved the quality and speed of Web searches, offering a clutter-free home page. The strategy of its two young entrepreneurs was simple: provide the best search engine, and refrain from commercializing it for a while. Over a period of about three years Google established itself as the best search engine. In time, it started to capitalize on this prominence by selling sponsored links (the right side of the results of a user's search, and later the top shaded results). Most importantly, the organization never stopped improving its search algorithms and periodically has offered new services. The strategy has succeeded so much that "google it" has become synonymous with "search for it on the Web."

Initiative #5: Differentiate Products or Services

A company can achieve a competitive advantage by persuading consumers that its product or service is better than its competitors'. Called product **differentiation**, this advantage is usually achieved through advertising and customer experience. Consider Skype. Although the software was not the first to offer free phone calls over the Internet, its quality was higher than similar applications. People noticed the difference, and millions have downloaded and use the application. When the user base was large, the company (which was acquired by eBay) added many features, including video connection. It makes money by selling features for pay and mobile devices.

Brand-name success is a perfect example of product differentiation. Think of Levi's jeans, Chanel and Lucky perfumes, and Gap clothes. The customer buys the brand-name product,

An innovative Web service such as YouTube attracts millions of visitors, creating an advertising strategic advantage.

Google did not offer an original service, but its service has grown superior to other Web search services.

perceiving it to be superior to similar products. In fact, some products *are* the same, but units sold under a prestigious brand name sell for higher prices. You often see this phenomenon in the food, clothing, drug, and cosmetics markets.

Initiative #6: Enhance Products or Services

Instead of differentiating a product or service, an organization might actually **enhance existing products or services**, that is, add to the product or service to increase its value to the consumer. For example, car manufacturers might entice customers by offering a longer warranty period for their cars, and real-estate agents might attract more business by providing useful financing information to potential buyers.

Since the Internet opened its portals to commercial enterprises in the early 1990s, an increasing number of companies have supplemented their products and services. Their Web sites provide up-to-date information that helps customers utilize their purchased products better or receive additional services. Companies that pioneered such Internet use reaped substantial rewards. For example, Charles Schwab gained a competitive advantage over other, older brokerage companies such as Merrill Lynch by opening a site for online stock transactions. Within months, half its revenue came from this site. All brokerage houses followed and allow customers to trade through a Web site.

Other companies use the Internet to maintain their competitive edge by continually adding to and enhancing their online services. The Progressive Groups, the third largest U.S. car insurance company, is a good example. The company enables insured drivers to place a claim and follow its progress at the company's site. The company has connected its information systems with those of car dealerships and financing institutions. When a car is totaled (i.e., fixing it would cost more than purchasing a new car), the owner can receive a check to purchase a new car. However, since the company knows that purchasing a new car may be a hassle, the insured owner can use, free of charge, the company's Total Loss Concierge service. The company developed special software that retrieves details about the totaled vehicle. The details are shared with a network of dealerships, and the concierge selects the best alternatives in terms of compatibility with the client's needs and the price. The agent accompanies the client in the contacts with the dealerships. If the client still owes money to a lender, the Progressive agent uses the system to retrieve the financing information and sends it to a network of financing firms. The agents send the client the best alternatives. In the auto insurance industry, the Total Loss Concierge service is an enhancement offered only by Progressive.

Progressive uses innovative information technology to enhance its services and maintain a competitive advantage in the auto insurance industry.

Initiative #7: Establish Alliances

Companies can gain competitive advantage by combining services to make them more attractive (and usually less expensive) than purchasing services separately. An alliance may also be created to enable customers to use the same technology for purchases from different companies. These **alliances** provide two draws for customers: combined service is cheaper, and one-stop shopping or using the same technology is more convenient. The travel industry is very aggressive in this area. For example, airlines collaborate with hotel chains and car-rental firms to offer travel and lodging packages. Credit-card companies offer frequent flier miles for every dollar spent, discounts on ticket purchases from particular airlines, or discounts on products of an allied manufacturer. In all these cases, alliances create competitive advantages.

As Figure 2.3 indicates, by creating an alliance, organizations enjoy synergy: the combined profit for the allies from the sale of a package of goods or services exceeds the profits earned when each acts individually. Sometimes, the alliances are formed by more than two organizations. Consider the benefits you receive when you agree to accept a major credit card: discounts from several hotel chains, restaurant chains, flower delivery chains, and other stores; free insurance when renting a car; and frequent flier miles, to name a few. Similarly, travel Web sites such as Orbitz offer you the opportunity to reserve lodging and car rental at discounts while you make your airline reservations. The company has also established alliances with hotel chains and car-rental companies.

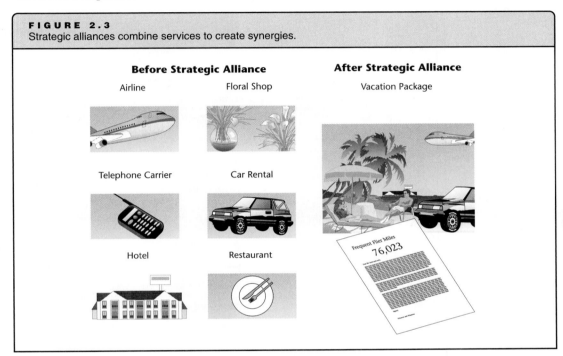

FIGURE 2.3
Strategic alliances combine services to create synergies.

Before Strategic Alliance

Airline Floral Shop

Telephone Carrier Car Rental

Hotel Restaurant

After Strategic Alliance

Vacation Package

Frequent Flier Miles
76,023

What is the common denominator among these companies? They each have an information system that tracks all these transactions and discounts. A package of attractive propositions entices clients who need these services (and most businesses do). Would this offer be feasible without an IS to track transactions and discounts? Probably not.

Growing Web use for e-commerce has pushed organizations to create alliances that would be unimaginable a few years ago. Consider the alliance between Hewlett-Packard and FedEx. HP is a leading manufacturer of computers and computer equipment. FedEx is a shipping company. HP maintains inventory of its products at FedEx facilities. When customers order items from HP via its Web site, HP routes the order, via the Web, to FedEx. FedEx packages the items and ships them to customers. This arrangement lets HP ship ordered items within hours rather than days. The alliance gives HP an advantage that other computer equipment makers do not share. Again, it is a clever IS that enables this strategy.

On the Web, an obvious example of alliances is an **affiliate program**. Anyone can place links to commercial sites on his or her personal Web site. When a visitor clicks through to a commercial site and makes a purchase, the first site's owner is paid a fee. Some online retailers have thousands of affiliates. The early adopters of these programs, such as Amazon.com, Buy.com, Priceline.com, and other large e-retailers, enjoyed a competitive advantage in gaining new customers. It is easy for any Web site holder to become an affiliate of Amazon.com.

Another example is the collaboration between Amazon.com and other retailers who leverage Amazon's technology. Target Corp. is one of America's largest retailers. To extend its operation to the Web, it formed a strategic alliance with the giant online retailer. If you go to Target's site, you will notice the words "Powered by Amazon.com." Amazon provides Target with its proprietary search engine, order-fulfillment and customer-service systems, and the patented one-click

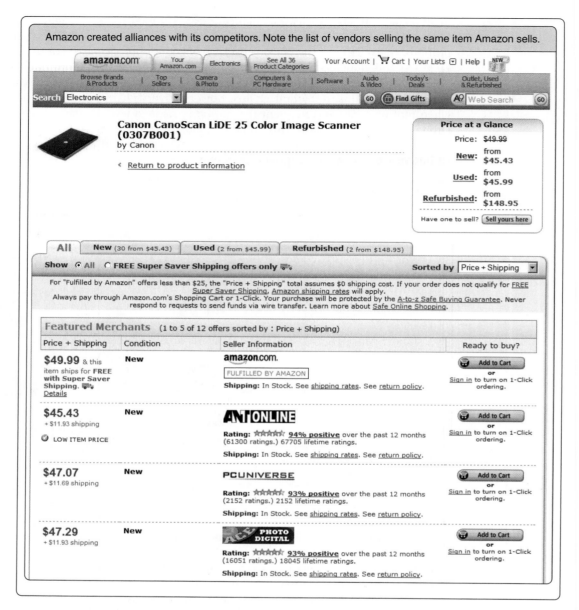

Amazon created alliances with its competitors. Note the list of vendors selling the same item Amazon sells.

shopping application, which lets customers pay for merchandise selected from the Target and Mervyns' sites from one electronic shopping cart (Mervyns is a Target subsidiary). In return, Amazon collects a percentage of all sales from Target's retail sites as well as annual fixed fees. Have we mentioned referrals? Next to the logos of Target and its subsidiaries, you also find Amazon's logo, which serves as a link to Amazon's site (where you also see the Target logo prominently displayed).

The Web has generated strategic alliances that would probably never be created offline. Can you imagine Wal-Mart inviting Sears to sell Sears' merchandise from Wal-Mart stores? This is exactly what Amazon does. Its site has links to products of other companies, and not just companies such as Target, with which it has a special relationship. When you search for an item on Amazon, you might find links not only to its own products but also to those of competitors, such as Circuit City, the consumer electronics chain. If this sounds strange, consider the rationale: Amazon wants customers to compare its price and its competitors' price for the same item and see that Amazon's is lower, mainly because Amazon manages its warehouses more efficiently than any other retailer in the world. Even if customers decide to purchase from the competitor through the Amazon site, Amazon receives a commission from the seller.

A growing number of companies use software to help analyze the vast amounts of data they collect. Some share the data and business intelligence with business partners because if their partners do better, so will they. For example, Marriott, the large hotel chain, provides online and traditional travel agencies with analyses about pricing, joint promotions, and inventory. The analytical results help the agencies optimize their operations, which results in more customers for Marriott.

Initiative #8: Lock in Suppliers or Buyers

Organizations can achieve competitive advantage if they are powerful enough to **lock in** suppliers to their mode of operation or buyers to their product. Possessing bargaining power—the leverage to influence buyers and suppliers—is the key to this approach. As such, companies so large that suppliers and buyers must listen to their demands use this tactic nearly exclusively.

A firm gains bargaining power with a supplier either when the firm has few competitors or when the firm is a major competitor in its industry. In the former case, the fewer the companies that make up a supplier's customer base, the more important each company is to the supplier. In the latter case, the more important a specific company is to a supplier's success, the greater bargaining power that company has over that supplier.

The most common leverage in bargaining is purchase volume. Companies that spend billions of dollars purchasing parts and services have the power to force their suppliers to conform to their methods of operation, and even to shift some costs onto suppliers as part of the business arrangement. Consider Wal-Mart, the world's largest retailer. Not only does the company use its substantial bargaining power to pressure suppliers to lower prices, but it also requires them to use information systems that are compatible with its own automated processes. The suppliers must use ISs that tell them when to ship products to Wal-Mart so that the giant retailer is never left understocked or overstocked. In recent years this power allowed the company to require its suppliers to use radio frequency identification (RFID) devices in packaging, to allow more accurate tracking of ordered, shelved, and sold items. This bargaining power and tight control of inventory enables Wal-Mart to enjoy considerable cost savings, which it passes on to customers, which keep growing in numbers thanks to the competitive prices. Many suppliers are locked in with Wal-Mart because of the sheer volume of business they have with the company: some sell a third to one-half of everything they produce to this single retailer, and some, such as the giant consumer products maker Procter & Gamble, have a "Vice President, Wal-Mart" as a member of the senior management.

One way to lock in *buyers* in a free market is to enjoy a situation in which customers fear high switching costs. In the software arena, enterprise applications are a good example. This type of software helps organizations manage a wide array of operations: purchasing, manufacturing, human resources, finance, and so forth. The software is expensive, costing millions of dollars. After a company purchases the software from a firm, it is locked in to that firm's services: training, implementation, updates, and so forth. Thus, companies that sell enterprise software, such as SAP, Oracle, and Infor Global Solutions, make great efforts to improve both their software and support services to maintain leadership in this market.

Another way to lock in clients is to **create a standard**. The software industry has pursued this strategy vigorously, especially in the Internet arena. For example, Microsoft's decision to give away its Web browser by letting both individuals and organizations download it free from its site was not altruistic. Microsoft executives knew that the greater the number of Internet Explorer (IE) users, the greater the user base. The greater the user base, the more likely organizations were to purchase Microsoft's proprietary software to help manage their Web sites. Also, once individual users committed to IE as their main browser, they were likely to purchase Microsoft software that enhanced the browser's capabilities.

Similarly, Adobe gives away its Acrobat Reader software, an application that lets Web surfers open and read documents created using different computers running different operating systems, such as various versions of Windows, the Mac operating system, and UNIX. When the Reader user base became large enough, organizations and individuals found it economically justifiable to purchase and use the full Acrobat application (the application used to create the documents) and related applications. Using this strategy put Adobe's PDF (portable data format) standard in an unrivaled position.

Another company, Macromedia Inc., now owned by Adobe, developed software called Flash to create Web page animations. It offers the Flash player for download free of charge but sells the development tool. Like PDF, Flash created a symbiotic situation to augment a market: the more individuals download the player, the more businesses are willing to purchase the development tool. The more companies engage Flash modules in their Web pages, the more individuals download the player, without which they cannot enjoy those animations.

The simplest way to lock in buyers is to create a physical or software limitation on using technology. This can be in the form of a company designing a socket for add-on plugs that takes only a specific size or form, or designing files so that they run only on its software. Apple Computer's iTunes is a classic example of the latter. The online music store is a popular site for purchasing music files. However, the files contain FairPlay DRM (digital rights management) software, which ensures the files run only on iPod, the company's music player. Digital music players made by competitors are locked out. Apple's decision had a significantly positive impact on its profits.

CREATING AND MAINTAINING STRATEGIC INFORMATION SYSTEMS

IT might offer many opportunities to accomplish a competitive edge, especially in industries that are using older software, such as the insurance industry. Insurance companies were among the early adopters of IT and have not changed much of their software. This is why some observers say the entire industry is inefficient. Once an insurance company adopts innovative software applications, it might gain competitive advantage. This might remind you of the airline industry. Many airlines still use antiquated hardware and software. As you'll learn later in the chapter, when JetBlue was established, it adopted the latest technologies, and this was a major reason for its great competitive advantage.

Companies can implement some of the strategic initiatives described in the previous section by using information systems. As we mentioned at the beginning of the chapter, a strategic information system (SIS) is any information system that can help an organization achieve a long-term competitive advantage. An SIS can be created from scratch, developed by modifying an existing system, or "discovered" by realizing that a system already in place can be used to strategic advantage. While companies continue to explore new ways of devising SISs, some successful SISs are the result of less lofty endeavors: the intention to improve mundane operations using IT has occasionally yielded a system with strategic qualities.

Strategic information systems combine ideas for making potentially winning business decisions and ideas for harnessing information technology to implement the decisions. For an information system to be an SIS, two conditions must exist. First, the information system must serve an organizational goal rather than simply provide information; and second, the organization's IS unit must work with managers of other functional units (including marketing, finance, purchasing, human resources, and so on) to pursue the organizational goal.

Creating an SIS

To create an SIS, top management must be involved from initial consideration through development and implementation. In other words, the SIS must be part of the overall organizational strategic plan. The danger always exists that a new SIS might be considered the IS unit's exclusive property. However, to succeed, the project must be a corporate effort, involving all managers who use the system.

Figure 2.4 presents questions that management should ask to determine whether to develop a new SIS. Executives meet to try to identify areas in which information can support a strategic goal. Only after completing the activities outlined in Figure 2.4 will management be able to conceptualize an SIS that seizes an opportunity.

1. **What would be the most effective way to gain an advantage?**
2. **Would more accessible or timely information to our employees, customers, or suppliers help establish a significant advantage? If so...**
3. **Can an information system be developed that provides more accessible and timely information?**
4. **Will the development effort be economically justified?**

◆ Can existing competitors afford to fund the development of a similar system?
    ◆ How long will it take the competitors to build their own, similar system?
    ◆ Can we make our system a moving target to the competition by constantly enhancing it, so that it always retains its superiority?

5. **What is the risk of not developing such a system?**
6. **Are alternative means of achieving the same goals available, and if so, how do they compare with the advantages and disadvantages of a new SIS?**

A word of caution regarding Question 4 in Figure 2.4, the issue of economic justification of an SIS: an increasing number of researchers and practitioners conclude that estimating the financial benefits of information systems is extremely difficult. This difficulty is especially true of SISs. The purpose of these systems is not simply to reduce costs or increase output per employee; many create an entirely new service or product. Some completely change the way an organization does business. Because so many fundamental business changes are involved, measuring the financial impact is difficult, if not impossible, even after implementation, let alone before. For example, if a bank is considering offering a full range of financial services via the Web, how can management know whether the move justifies the cost of the necessary software? It is difficult to estimate the success of such a bold approach in terms of how many new customers the bank would gain.

Yet, a great number of SISs are the unintended consequence of exploiting information technology to support activities that are not strategic. For example, in the 1990s, Owens & Minor, a distributor of hospital supplies, built a data warehouse from which to glean business intelligence. However, both its customers (mainly hospitals) and its suppliers (drug and medical instruments makers such as Johnson & Johnson) agreed to pay for mining the data warehouse to improve their decision making. In this case, the company did not plan to create an SIS, but the data warehouse and the tools that help mine it may become one, increasing Owens & Minor's profit in a business that has little to do with its original business.

Reengineering and Organizational Change

To implement an SIS and achieve competitive advantage, organizations sometimes must rethink the entire way they operate. While brainstorming about strategic plans, management should ask: "If we reestablished this business process from scratch, how would we do it?" The answer often leads to the decision to eliminate one set of operations and build others from the ground up. Changes such as these are called **reengineering**. Reengineering often involves adoption of new machinery and elimination of management layers. Frequently, information technology plays an important role in this process.

Reengineering's goal is not to gain small incremental cost savings, but to achieve great efficiency leaps—of 100 percent and even 1000 percent. With that degree of improvement, a company often gains competitive advantage. Interestingly, a company that undertakes reengineering along with implementing a new SIS cannot always tell whether the SIS was successful. The reengineering process makes it impossible to determine how much each change contributed to the organization's improved position.

Implementation of an SIS requires a business to revamp processes—to undergo organizational change—to gain an advantage. For example, when General Motors Corp. (GM) decided to manufacture a new car that would compete with Japanese cars, it chose a different production process from that of its other cars. Management first identified goals that could make the new car successful in terms of how to build it and also how to deliver and service it. Realizing that none of its existing divisions could meet these goals because of their organizational structures, their cultures, and their inadequate ISs, management established Saturn as an independent company with a completely separate operation.

Part of GM's initiative was to recognize the importance of Saturn dealerships in gaining competitive advantage. Through satellite communications, the new company gave dealers access to factory information. Clients could find out if, and exactly when, different cars with different features would be available.

Another feature of Saturn's SIS was improved customer service. Saturn embeds an electronic computer chip in the chassis of each car. The chip maintains a record of the car's technical details and the owner's name. When the car is serviced after the sale, new information is added to the chip. At their first service visit, many Saturn owners were surprised to be greeted by name as they rolled down their windows. While the quality of the car itself has been important to Saturn's success, the new SIS also played an important role. This technology was later copied by other automakers.

Interestingly, most reengineering of the 1990s and early 2000s failed and simply resulted in massive layoffs. Executives found it impossible to actually change many business processes. Business processes eventually did change in companies that adopted enterprise systems commonly called ERP systems. The reason: the new systems forced managers and employees to change their way of work.

POINT OF INTEREST

Abandoning Ship?

The U.S. Bureau of Labor Statistics reported that the number of women in IT positions has steadily declined over the past several years. About 984,000 women worked in eight IT categories in 2000, constituting 28.9 percent of all IT workers. In 2006, when the size of IT employment reached a record 3.74 million, only 908,000 women—26.2 percent of the total—were employed in IT positions. Thus, the decrease is both in relative and absolute terms. The reasons for abandoning or not choosing IT careers among women are unclear. A study by human resources consulting firm Sheila Creco Associates shows that overall leadership roles in IT receded in 2006 to the level of 2002. There was one point of light: the same study showed that the number of women holding the CIO position has increased 9 percent between 2000 and 2006.

Source: U.S. Bureau of Labor Statistics, June 2007; Cone, E., "Why Do Women Leave?" *CIO Insight*, June 7, 2007.

Competitive Advantage as a Moving Target

As you might have guessed, competitive advantage is often short-lived. In time, competitors imitate the leader, and the advantage diminishes. So, the quest for innovative strategies must be dynamic. Corporations must continuously contemplate new ways to use information technology to their advantage. In a way, companies' jockeying for the latest competitive advantage is a lot like an arms race. Side A develops an advanced weapon, then side B develops a similar weapon that terminates the advantage of side A, and so on.

In an environment where most information technology is available to all, SISs that are originally developed to create a strategic advantage quickly become an expected standard business practice. A prime example is the banking industry, where surveys indicate that increased IS expenditures did not yield long-range strategic advantages. The first banks to provide ATMs and online banking reaped some rewards in terms of labor savings and new customers, but the advantage disappeared because most banks now offer these services.

A system can only help a company sustain competitive advantage if the company continuously modifies and enhances the system, creating a moving target for competitors. American Airlines' Sabre—the online reservation system for travel agents—is a classic example. The system, which was designed in the 1950s, was redesigned in the late 1970s to sell travel agencies a new service, online airline reservations. But over the years, the company spun off an office automation package for travel agencies called Agency Data Systems. The reservation system now encompasses hotel reservations, car rentals, train schedules, theater tickets, and limousine rentals. When the Internet became accessible to businesses and consumers, the system was redesigned to let travelers use Sabre from their own computers. The system has been so successful that in some years American earned more from the technology than from its airline operations. The organizational unit that developed and operated the software became a separate IT powerhouse at AMR Corp., the parent company of American Airlines, and now operates as Sabre Holding Corporation, an independent company. It is the leading provider of technology for the travel industry. Travelocity, Inc., the popular Web-based travel site, is a subsidiary of Sabre, and, naturally, uses Sabre's software. Chances are you are using Sabre technology when you make airline reservations through other Web sites, as well.

We return to Amazon as an example of how ISs help companies maintain competitive advantage. Management believes that it must add new features to its Web site to attract buyers over and over again. The company continuously improves its Web pages' look and the online services it provides. Amazon has moved from merely selling books through the Web to providing best-seller lists, readers' reviews, and authors' interviews; selling almost any consumer product imaginable; and posting consumer wish lists, product reviews by customers, and other "cool stuff." The constant improvements help the company maintain its dominant position in online retailing.

However, all of these features have been imitated by competitors. Amazon now also offers Web hosting services and space for rent in its 10 million square feet of warehouses worldwide. It has also opened much of its software for developers to use.

Similarly, Google has offered access to some of its software. For example, it enables Webmasters and other Website owners to use software that finds out why Google's crawler—software that searches the Web and indexes new pages for search—has difficulties in indexing pages. The result: the site owners get their new pages indexed so the public can access them, and Google receives free labor in fixing problems its crawler faces. Organizations can integrate Google mapping software into their own intranets to map customer locations, track shipments, manage facilities, and perform other activities that are map-related. Amazon and Google have augmented the portfolio of services they provide to increase the circle of organizations and individuals who depend on them, thereby strengthening their strategic positions.

JETBLUE: A SUCCESS STORY

We usually expect entrepreneurs to enter a new and profitable industry, not an old, money-losing one. However, with the proper technology and management methods, it seems that some energetic people can gain strategic advantage where others have been hurting. The U.S. airline industry has seen mainly bad times since the industry's deregulation in the 1970s. The situation deteriorated as the 1990s drew to a close, and grew even worse after the terrible events of September 11, 2001. In 2001, the industry lost $7.7 billion, but JetBlue had a profit of $38.5 million on revenue of $320.4 million. It continued to be profitable in 2002, 2003, and 2004 along with only one other carrier, Southwest Airlines, while all other U.S. carriers had losses. JetBlue's revenues grew from $998.4 million in 2003 to $1.27 billion in 2004. It still enjoyed operating profit in 2005 and 2006. However, the ice storms of February 2007 caused the company to cancel many flights, which tarnished its reputation for a while.

JetBlue was established in February 2000 by David Neeleman, who serves as its CEO. Two decades earlier, in 1984, Neeleman cofounded Morris Air, a small airline in Salt Lake City, Utah, which was the first airline to offer ticketless travel, a program that was developed inside the company. Working with a college student he developed Open Skies, a computer program that integrates electronic ticketing, Internet reservations, and revenue management. Revenue management tools help an airline plan the most profitable routes and ticket pricing. Morris Air was sold to Southwest Airlines, which enthusiastically adopted the e-ticket idea. Neeleman became an executive at Southwest but left in frustration, because he believed that an airline could achieve much more efficiency with information technology. Now headquartered in Forest Hills, New York, JetBlue has gained a significant strategic advantage over larger and older airlines. The company's success is the result of understanding customers' priorities and gaining marked efficiencies through automating whatever IT can automate. Management also learned to break away from practices that inhibit efficiency and agility.

In a highly competitive industry that traditionally has had a narrow profit margin, JetBlue managed to gain strategic advantage by *reducing cost*, therefore reducing the price to the customer, and *improving a service*, especially in terms of on-time departures and arrivals.

For JetBlue, information technology is at least as important as fuel.

Richard Sheinwald/Bloomberg News/Landov

Massive Automation

We usually think of manufacturing organizations when mentioning automation, but benefits can also be gained by automating services. JetBlue uses Open Skies, the software that Neeleman developed. It is a combination reservation system and accounting system, and supports customer service and sales tracking. The company avoids travel agents. Booking a flight through a travel agent costs airlines $20 per ticket. JetBlue saves office space rent and electricity by using reservation agents who work from home (telecommuting is discussed in Chapter 6, "Business Networks and Telecommunications") and use VoIP (Voice over Internet Protocol, also discussed in Chapter 6) for telephoning. The company pays a flat fee of $25 per telephone line per month for these telecommuting agents. This reduces its handling cost per ticket to $4.50.

Because all tickets are electronic, there is no paper handling or related expense. JetBlue encourages customers to purchase their tickets online, and more than 79 percent of them do so, saving the company much labor. The cost of handling a ticket ordered via the Web is reduced to only 50 cents, as opposed to $4.50 paid to a reservation agent, and a far cry from the $20 when booking through a travel agent.

JetBlue automates other aspects of running an airline as well. Its maintenance workers use a maintenance information system from Dash Group to log all airplane parts and their time cycles, that is, when the parts must be replaced and where they can be found. The system reduces manual tracking costs.

Flight planning to maximize yield—the number of seats occupied on a flight—is executed on a flight-planning application from Bornemann Associates. It reduces planning costs and makes operations more efficient. JetBlue also uses an application that its team of 58 IT professionals developed in-house, called Blue Performance. It tracks operational data that is updated flight by flight. The company's intranet enables its 2,800 employees to access the performance data. Managers have up-to-the-minute metrics, so critical in airline operations, which enable them to respond immediately to problems.

When on the ground, employees use wireless devices to report and respond to any irregular event, from weather delays to passenger injuries. The response is quick, and the events are recorded in a database for later analysis.

When training pilots and other employees, no paper records are kept. An aviation training management system provides a database to track each employee's training record. It is easy to update and efficient for record retrieval.

Away from Tradition

JetBlue decided not to use the hub-and-spokes method of routing its airplanes, a practice used by all major airlines. Instead of having its airplanes land in one or two hubs and undergo maintenance there before taking off for the next leg of a route, it simply uses the most profitable routes between any two cities. All flights are point to point—no hubs, no spokes.

JetBlue was the first airline to establish paperless cockpits. The Federal Aviation Authority (FAA) mandates that pilots and other aircrew members have access to flight manuals. The manuals are the documents showing information about each flight, including route, weight, how the weight is spread on board, fuel quantity, and even details such as how many pets are on board. Other airlines update their manuals and then print them after every update. All JetBlue flight manuals are centrally maintained, and the pilots and first officers access and update the manuals on laptop computers that they carry into the cockpit. As soon as the data have been entered, employees have access to the information.

The laptops enable the pilots and first officers to calculate the weight and balance of their plane with a few keystrokes instead of relying on dispatchers at headquarters to do the calculations for them. JetBlue saves paper and time by having employees enter flight data. The company subscribes to SharePoint, a Web-based portal that enables electronic updates to flight manuals. This cuts 15 to 20 minutes from preflight preparations for every flight. The result is a savings of about 4,800 hours per year and planes that take off and land on time.

JetBlue continues to harness IT to maintain the strategic gap between the company and its competitors. Management planned a paperless frequent flier program, cockpit-monitoring cameras transmitting through satellites so that ground crews can monitor activity, and biometric applications in airport terminals. Biometrics use physical characteristics of people, such as fingerprints and retina scans, for authentication and access to physical places and online information systems. Biometrics are more secure than access codes. The IT team is also developing a new reservation system that will have features no other airline reservation system has.

Enhanced Service

Much of the technology that helps JetBlue employees provide better service is invisible to the customers, but it also has some more obvious winning features. JetBlue offers leather seats and individual real-time television on all its airplanes. Other airlines do not offer such seats on economy class, and offer only recorded television programs. The real-time TV service is offered under a contract with DirecTV.

Its use of IT technologies also placed the airline at the top of the list for on-schedule departures and arrivals, a service that is very important, especially to business travelers. Perhaps even better, JetBlue ranks at the top as having the fewest mishandled bags. Thanks to constant updates to the Open Skies system, the company has managed to maintain check-in time at less than one minute. When passengers arrive at JetBlue's terminal at JFK airport, they are directed by a large LCD display with a computer-generated voice telling them which window is available to serve them. Usually, checking baggage takes 45 seconds. When passengers arrive at their destination, they do not have to wait for their suitcases. Their electronically tagged suitcases wait for them at the baggage claim area.

Because of heightened security awareness, management decided to install hidden video cameras in the cabin and monitors in the cockpit. Technicians used the DirecTV wires to add the cameras and monitors. Customers are more comfortable knowing of this extra step to enhance their safety.

Impressive Performance

The most important metric in the airline industry is cost per available seat-mile (CASM), which is how much it costs to fly a passenger one mile of the journey. JetBlue has been able to maintain the lowest or next to lowest CASM in its first three years of operations. While its competitors' CASM is 11 cents or higher, JetBlue's CASM is less than 7 cents. While its competitors fill only 71 percent of seats, JetBlue fills 78 percent.

Late Mover Advantage

Some observers cite the fact that JetBlue is a late competitor as an important factor in its success. The company is not burdened with antiquated information systems, or as IT professionals like to call them, legacy systems. This allowed its CIO, Jeff Cohen (later succeeded by Duffy Mees), to implement the latest available technologies: fast databases, VoIP, a slick Web site, laptop computers with the latest algorithms for fast calculation of routes and loads in the cockpit, and other technologies. This situation illustrates the strategic advantage of the **late mover**.

JetBlue executives quip that while other airlines run on fuel, theirs runs on information technology. Cohen said that up to 40 percent of the software the company was using was beta or new software. Beta software is software that the developer gives to potential adopters for trial use. Talk about being on the cutting—and possibly bleeding—edge! Yet, competitors took notice. Delta Airlines established a subsidiary called Delta Song. The organization mimicked many of JetBlue's innovations, including live TV. It eventually was merged into Delta. Similarly, United Airlines created a nimble subsidiary airline called Ted to compete with JetBlue.

When ice storms wreaked havoc with airlines in February 2007, one of JetBlue's major problems was that crews who were supposed to be in a certain city were stranded in another, and therefore staffing of flights was affected. While the crisis was on, the IT team developed a special database and application to let crews call in their location and to replace it with the location still stored in the system. The development process took a mere 24 hours. The IT team also devised ways to communicate better with customers through broadcasting automated flight alerts via e-mail and mobile devices.

POINT OF INTEREST

The Cost of Success

Microsoft, the successful software giant, has faced many legal battles. In 2002, it settled an antitrust lawsuit with the U.S. government, but later faced similar issues with European Union prosecutors who refused to settle. Microsoft was fined $1 billion by the European Union over its ruling that the company must disclose the code of its operating systems to competing application developers. In 2006, Microsoft spent over $1.3 billion on its legal battle, about 10 percent of its net income.

Source: Downes, L., "A Tale of Two Microsofts ," *CIO Insight*, No. 77, January 2007, p. 27.

Size Matters

At what point do the public and the courts start to consider a successful strategy as a predatory, unfair business practice that makes competition from other businesses impossible, even if their products are better? For instance, should a firm that takes bold entrepreneurial steps to become a business leader be curbed when it succeeds in becoming powerful? Several court cases against Microsoft, the software industry leader, have focused on these questions. However, the questions are not simply legal issues. They are also important because they impact the economy and, as a result, society.

- **Historical Background.** In the 1970s, Microsoft was a small software company headed by its young president, Bill Gates, who established the company at age 19. The company was fortunate to find and buy an operating system from a small company in Seattle, Washington, for $50,000. An operating system (OS) is the software program that "mediates" between any computer program and the computer. Every application is developed with a particular operating system, or several operating systems, in mind. To a great extent, the operating system determines which applications a computer can run. Therefore, it is an extremely important program. We discuss operating systems and other types of software in Chapter 5, "Business Software.")

People who purchased a computer had to consider the OS to determine which applications they could run. After Microsoft bought the operating system, it entered into a contract with IBM, the most powerful computer manufacturer at that time. IBM needed an operating system for its new creation, the IBM PC, and they chose Microsoft's DOS (Disk Operating System). While Microsoft did not make much money on the IBM deal, its executives realized the strategic potential of contracting with "the big guy."

Indeed, the strategy paid off. Soon, Compaq (now part of Hewlett-Packard) and many other manufacturers started to market IBM PC clones, cheaper computers that performed as well as IBM PCs and that could run the same operating system and applications. Because Microsoft's contract with IBM allowed it to sell DOS to other parties, it made a fortune selling DOS to Compaq and others. Later, Microsoft developed Windows, an improved operating system, and the success story repeated itself. To this day, the majority of buyers of personal computers also buy a copy of some version of Windows.

One major key to gaining a decent share of the new Internet market was the widespread use of Web browsers. In the mid-1990s, more

than 80 percent of Web surfers used Netscape's browsers. Netscape (now part of AOL, a subsidiary of TimeWarner) was a young, entrepreneurial company selling innovative products. Microsoft decided to increase its own browser's market share of about 15 percent to a leading position. If a great number of people used its browser, Microsoft could expect hefty sales of related software, such as server management applications.

- **Controversial Practices.** No one would deny that Microsoft's attempt to compete in the browser market was legitimate. While Netscape gave its browsers away to individuals and educational institutions but charged for-profit organizations, Microsoft gave its browser to everyone free of charge. Also, the company took advantage of Windows dominance; it started bundling its browser with Windows, practically forcing any PC maker who wanted to sell the machines with the operating system installed to also install Internet Explorer (IE). The great majority of new PC owners used IE without even trying any other browser.

Within two years, a majority of Web surfers were using IE. But Netscape, the U.S. Department of Justice, and many individuals considered Microsoft's tactics unfair. Microsoft used its muscle in the operating system market to compel sellers of personal computers to include a copy of Internet Explorer with Windows. Furthermore, the browser was inseparable from newer Windows versions. Since sellers had to include Windows on every machine, and because it is practically the only operating system most buyers would accept, sellers had no choice but to succumb to the pressure. The U.S. Department of Justice and the Attorneys General of several states filed lawsuits claiming Microsoft violated fair trade practices. Subsequently, legal authorities in other countries, such as the European Union (EU) and Taiwan, also either probed the company or sued it. In 2004, the EU's antitrust office fined Microsoft 497 million euros ($665 million) for abusively wielding Windows' monopoly and for locking competitors out of the software market. Meanwhile, as competition in the digital audio and video media increased, Microsoft bundled its Media Player software with the Windows OS. In 2002, the U.S. Department of Justice settled with Microsoft on this issue, requiring the company only to enable users to hide Media Player and set another application as the default player. The EU demanded that Microsoft sell Windows without Media Player.

The EU also demanded that the company allow all software developers access to information about Windows, so that they could develop applications that would compete well with Microsoft's own applications. It claimed that developers of nonproprietary software (software that is not owned by anyone and can be used free of charge) were denied access to the Windows information altogether. In 2004, an EU court decided that Microsoft broke competition law and fined it $613 million. Due to Microsoft's refusal to comply with the court's demands, in 2006 the company was fined an additional $357 million. Microsoft appealed the decisions.

Contrary to public perception, the United States, the European Union, and many other countries do not outlaw monopolies. They only forbid unfair use of monopolistic power. Because anyone may compete in any market, it would be unfair to punish an entrepreneur for marketing unique products and mustering market power of any magnitude. Of concern in the eyes of U.S. law, for example, are two issues: (1) have any unfair practices helped the company gain monopolistic power, and (2) does the monopolistic situation serve customers well, or does it hurt them?

- **Up Side, Down Side.** Microsoft argues that although it could charge higher prices for Windows, it has not, because it wants to make Windows affordable to all. Microsoft also argues that, unlike typical monopolists, it invests huge amounts of money in research and development, which eventually benefit society in the form of better and less-expensive products. Microsoft's rivals in the software industry claim that Microsoft's practices stifle true competition. Both claims are difficult to measure. Some observers argue that allowing the same company to develop operating systems and many applications is good for consumers: the applications are compatible with each other; all use the same interface of menus and icons. Others suggest that Microsoft should be broken into two organizations, one that develops operating systems and another that develops only applications and competes fairly in that market. And some organizations and individuals simply fear the great power that a single person, Bill Gates, holds in an industry that so greatly impacts our economy and society. What is your opinion? What would you do about this issue?

FORD ON THE WEB: A FAILURE STORY

Sometimes what seems to be a great, forward-looking strategic move ends up as a colossal failure. It might be because of lack of attention to details or simply because the innovator could not predict the response of customers or business partners. Such was the great initiative of Jacques Nasser, the former CEO of Ford Motor Company, the second largest U.S. automaker.

The Ideas

When Nasser was appointed CEO of Ford in 1999, he regarded himself as an agent of change. He was eager to push the company into the Web, which was then at the height of its hype as a commercial vehicle. "We are now measuring speed in gigahertz, not horsepower," he said at the 2000 North American International Auto Show in Detroit. The concept cars sported, among other innovations, mobile Internet access. Ford Motor Co., he said, would put the Internet on wheels.

Ford launched Wingcast telematics, devices that would be installed in the company's vehicles and enable drivers and passengers to access the Web. To this end the company formed an alliance with Qualcomm, Inc., a telecommunications company, and Yahoo!

Ford created a joint venture with General Motors Corp. and DaimlerChrysler to establish Covisint, a Web site that served as an electronic market for parts suppliers who could bid online on requests for proposals posted by the automakers. Although not announced this way, the automakers' hope was that suppliers would fiercely compete in an open bidding process and cut their prices dramatically, so the auto companies could enjoy cost cuts. This was the business-to-business (B2B) part of Nassers's grand plan.

The business-to-consumer (B2C) idea was bolder: Ford wanted to push vehicle sales to the Web. Nasser wanted to bypass dealerships and retail the vehicles online directly to consumers. Consumers would go to the Web site, take a virtual test-drive, see images of a vehicle in all its available colors, order a vehicle, pay for it online, and then have it driven to their door. Ford would not only provide a great service but also save the dealer fees. The company called the site FordDirect.com. A special organizational unit, ConsumerConnect, was established to build the Web site and handle the direct sales.

Hitting the Wall

Apparently, buyers were not as enthusiastic about having Web access in their vehicles as Nasser predicted. In June 2001, Ford eliminated the Wingcast project. The B2B effort, Covisint, worked for a while, but not as expected. It was later sold to Compuware, a software development company. The B2C initiative failed.

The failure was not the result of faulty technology. There are excellent Web technologies that would support retail through the Web. There is no reason why a car cannot be selected, paid for, and delivered (with the help of companies that specialize in such delivery from the manufacturer to the buyer) via the Web. The company failed because it did not carefully consider state laws and its relationships with dealers.

Many state laws do not permit cutting an agent out of the sale. State franchising laws did not allow Ford to bypass its dealers. Also, since Ford would still rely on dealers to sell cars to people who do not have access to the Internet or who like to sit in a physical car and test-drive it, it could not cut the relationship all at once. Ford still needed the collaboration of the dealers, if it could overcome the legal hurdles, in order for direct sales to take off.

The Retreat

The circumstances convinced Ford to abandon its plan to sell directly to consumers. The ConsumerConnect unit was disbanded. FordDirect.com is now operated jointly by Ford and its 3,900 Ford and Lincoln Mercury dealerships. The site helps consumers find the vehicles they want, but they then have to find a dealer close to their homes who can deliver the vehicle. Like any car dealer, the site also offers used cars for sale, which is not what Ford would like to do. The price tag of this failed experiment was reported to be a hefty portion of the $1 billion Ford spent on its Internet initiative under Nasser's leadership.

Ford's management can find some solace in the continued operation of FordDirect.com. Although the grand plan did not materialize, the site is the origin point of 10,000 sales transactions per month. Ford reported that it sold 250,000 vehicles through the Web site in 2005. The site saves dealers marketing cost. A FordDirect sale costs dealers only $100, about one-fourth the cost per vehicle sold with traditional marketing. Ford also says the site helps it predict sales.

Some observers say that Ford's focus on the Internet was at times greater than on making automobiles. While other automakers were making modest profits in the period from 2000 to 2001, Ford posted losses. Nasser was forced to leave the company.

THE BLEEDING EDGE

As you might often hear, huge rewards go to whomever first implements a new idea. Innovators might enjoy a strategic advantage until competitors discover the benefits of a new business idea or a new technology. However, taking such steps before competitors have tested a system involves great risk. In some cases, failure results from rushing implementation without adequately testing a market. But even with careful planning, pioneers sometimes get burned.

For example, several supermarket chains tried self-checkout stations in the mid-1990s. Consumers were expected to ring up their own purchases. By and large, investment in such devices failed not because the technology was bad, but because many consumers either preferred the human touch, or because they did not want to learn how to correct mistakes when the devices did not pick up the price of an item or picked it up twice. Recently, machines that are more user-friendly and less error-prone have been installed by several chains, and consumers have been more willing to use them.

While it is tempting to take the lead, the risk of business failure is quite high. Several organizations have experienced disasters with new business ideas, which are only magnified when implementing new technology. When failure occurs because an organization tries to be on the technological leading edge, observers call it the **bleeding edge**. The pioneering organization "bleeds" cash on a technology that increases costs instead of profits. Adopting a new technology involves great risk: there is no experience from which to learn, no guarantees that the technology will work well, and no certainty that customers, employees, or business partners will welcome it.

Being on the bleeding edge often means that implementation costs are significantly higher than anticipated, that the new technology does not work as well as expected, or that the parties who were supposed to benefit—employees, customers, or suppliers—do not like using it. Thus, instead of leading, the organization ends up bleeding, that is, suffering from high cost and lost market share. For this reason, some organizations decide to let competitors test new technology before they adopt it. They risk losing the initial rewards they might reap, but if a competitor succeeds, they can quickly adopt the technology and even try to use it better than the pioneering organization.

Microsoft generally takes this approach. It seizes an existing idea, improves it, and promotes the result with its great marketing power. For instance, the company did not invent word processing, but Word is the most popular word-processing application today. The company did not invent the electronic spreadsheet, but Excel is the most popular spreadsheet application. And Microsoft was not the first to introduce a PC database management application, but it sells the highly popular Access. The company joined the Internet rush late, but it developed and gave away Internet Explorer, the market leader in Web browsers. You might call this approach competing by emulating and improving, rather than competing by being on the leading edge.

Sometimes, companies wait quite a long time to ensure that a technology has matured before they start using it, even at the risk of diminishing their strategic position. Although data warehousing—the organization and summarization of huge amounts of transactional records for later analysis—has been around since the mid-1990s, The Home Depot, Inc., decided only in 2002 to build a data warehouse. Home Depot is the world's largest home improvement retailer. It started the project years after its main rival in the United States, Lowe's, had implemented a well-functioning data warehouse, which it used effectively for strategic decision making.

■ Some ISs have become strategic tools as a result of strategic planning; others have evolved into strategic tools. To compete better, executives need to define strategic goals and determine how new or improved ISs can support these goals. Rather than waiting complacently until a problem occurs, businesses actively look for opportunities to improve their position with information systems.

■ An IS that helps gain strategic advantage is called a strategic information system (SIS). To assure optimal utilization of IT for competitive advantage, executives must participate in generating ideas and champion new, innovative uses of information systems. In recent years, many of these ideas involved using the Internet.

■ A company achieves strategic advantage by using strategy to maximize its strengths, resulting in a competitive advantage.

■ Strategic advantage is often achieved by one or a combination of the following initiatives. Cost reduction enables a business to sell more units of its products or services while maintaining or increasing its profit margin. Raising barriers to potential entrants to the industry lets an organization maintain a sizable market share by developing systems that are prohibitively expensive for competitors to emulate. By establishing high switching costs, a business can make buying from competitors unattractive to clients. Developing totally new products and services can create an entirely new market for an organization, which can also enjoy the advantage of being a first mover for that product and market. And if the organization cannot create new products or services, it can still enjoy competitive advantage by differentiating its products so that customers view them as better than a competitor's products. Organizations also attain advantage by enhancing existing products or services. Many new services are the fruits of alliances between companies: each contributes its own expertise to package services that entice customers with an overall value greater than that offered by the separate services individually. Locking in clients or suppliers, that is, creating conditions that make dealing with competitors infeasible, is a powerful strategy to gain advantage.

■ In the software industry, creating standards often creates strategic advantage. A standard is an application used by a significant share of the users. To this end, many companies go as far as giving software away. When the standard has been established, the company enjoys a large sales volume of compatible and add-on software. Microsoft, the software giant, has been found guilty of using unfair trade practices in trying to establish standards and squash competitors.

■ Reengineering is the process of redesigning a business process from scratch to save hundreds of percentage points in costs. Almost always, reengineering involves implementing new ISs.

■ Strategic advantages from information systems are often short-lived, because competitors quickly emulate the systems for their own benefit. Therefore, looking for new opportunities must be an ongoing process. Companies can maintain the strategic advantage gained through an IS by continuously augmenting the services they provide.

■ To maintain a strategic advantage, organizations must develop new features to keep the system on the leading edge. But they must be mindful of the bleeding edge, the undesirable results (such as huge ongoing costs and loss of customers) of being the first to use new technology with the hope of establishing a competitive advantage. Early adopters find themselves on the bleeding edge when the new technology is not yet fully reliable or when customers are uncomfortable with it.

GARDENERS+ REVISITED

The three *Gardeners+* entrepreneurs have gained experience, used information systems to research options, and instituted changes to remain profitable and expand their business. They also face some new opportunities and questions about which strategic direction their business should take. The next section explores some of their strategic initiatives to see whether you think they can make improvements.

What Would You Do?

1. Their contract with the shopping mall enabled the owners of *Gardeners+* to learn about a new market and to add new services. Can you identify the strategic moves that they have already made to help them compete? Have any of their partners operated strategically? How? Be sure to consider the following ways of gaining a competitive advantage:

   - Reduce costs
   - Raise barriers to entrants
   - Establish high switching costs
   - Create new products and services
   - Differentiate products and services
   - Enhance products or services
   - Establish alliances
   - Lock in suppliers or buyers

2. Review the decision that Mary, Amanda, and Ed made to not pursue the ecological certification for their gardening services. In your opinion, was this decision correct? What additional information could they use to monitor the ecological gardening market in the future?

New Perspectives

1. Generating long-term contracts and purchasing more and larger gardening equipment provided Mary, Amanda, and Ed the opportunity to completely rethink the hiring process and their mode of providing services—to reengineer. Think of some additional ways in which they could further redesign their contracts and services. Consider whether separating the two lines (commercial and residential) was a good decision or not. Look for additional changes that they could make to compete more effectively.

2. With the mall's former landscape services provider returning with new and more flexible services, how can Mary, Amanda, and Ed continue to monitor costs and profits closely so that they can make timely and appropriate changes that will allow them to remain competitive? They already have loyal customers and affiliated gardeners. How can they use their existing information systems to compete effectively against this renewed and highly aggressive competitor? Suggest at least three ways to help them compete. Could a Web site help them? Why or why not?

KEY TERMS

1. In what respect does business strategy resemble military strategy?

2. Refer to Chapter 1's discussion of different types of information systems. Which types of ISs can gain strategic advantage and which cannot? Why?

3. What should an information system achieve for an organization in order to be considered a strategic information system?

4. What strategic goal can an IS attain that does not involve wresting market share from competitors?

5. What conditions must exist in an organization planning an SIS?

6. Sometimes it is difficult to convince top management to commit funds to develop and implement an SIS. Why?

7. An SIS often offers a corporation short-lived advantages. How so?

8. What is reengineering? Why is reengineering often mentioned along with IT?

9. Why have most reengineering projects failed? What has eventually affected reengineering in some companies?

10. Software developers have made great efforts to "create a standard." What does creating a standard mean in the software industry, and why are companies doing it?

11. What should an organization do to sustain the strategic benefits of an IS?

12. Adobe encourages PC users to download its Acrobat Reader and Flash Player free of charge. How does this eventually help Adobe strategically? If they give the application away, how does their generosity help them make money?

13. Referring to the list of strategic moves (see Figure 2.2), classify the initiatives of JetBlue.

14. What were the reasons for the failure of the original purpose of FordDirect.com? Who eventually gained from the system and what were the gains?

15. The executives of well-established airlines are not less smart than those at JetBlue, and yet, their larger airlines have not done what JetBlue has done. Why?

16. What does the term "first mover" mean?

17. Can a *late mover* have any strategic advantage with IT? What is the risk that a late mover takes?

18. What does the term "bleeding edge" mean?

19. Can an off-the-shelf computer program be used as an SIS? Why or why not?

20. The organizations that eventually use the systems, not consulting firms that try to help organizations, develop more successful SISs. What might be the reasons for this?

21. You head a small company. You have an idea for software that can give your company an advantage over competitors. Since you do not have a staff that can develop and implement the software, you decide to approach a software company. Other than the company's technical offerings, what additional company aspects are desirable?

22. Some argue that an SIS gives a company an unfair advantage and might even cause the demise of smaller, weaker companies that cannot afford to build similar systems. Is this good or bad for customers? Explain your opinion.

23. Why has the Web been the arena of so much competition in recent years?

24. Information systems play a major role in almost every reengineering project. Why?

25. Accounting and payroll ISs have never become SISs. Why? What other types of ISs are unlikely to ever provide strategic advantage for their owners?

26. Ford's CEO envisioned a future in which consumers log on to an automaker's Web site, design their cars online, wait for the cars to be manufactured (design transformed into electronic blueprints), and have the car delivered to

their door. Do you think we will see this in practice within the next decade? Why, or why not?

27. Give two examples of other products or services whose delivery time could be cut from days to minutes with the aid of IT.

28. What is the role of ISs in alliances such as airlines and credit-card issuers? Why would such alliances be practically infeasible without IT?

29. JetBlue used new software that had not been tested by other companies. If you were a CIO, would you use software that is still in beta (untested with live data) in your organization?

30. You are an executive for a large organization that provides services to state and federal agencies. A software development firm approached you with an offer to implement new software that might give your organization a strategic advantage by reducing the service delivery cycle by several days. What would you do to avoid putting your organization on the "bleeding edge" while still considering the new software?

31. When a software developer creates a *de facto* standard (i.e., not the official standard, but something so widely used that it becomes a standard), it has monopolistic power. Should governments intervene to prevent this practice? Explain your opinion.

32. Suppose you are a venture capitalist considering a proposal to invest millions of dollars in a new online business. What questions would you ask the enthusiastic young people who have approached you for funds?

33. What are the potential risks of a single organization controlling much of the market for essential software?

34. Although Apple Computer Inc. introduced a personal computer and software that was superior to those produced by IBM and other companies, it failed to capture the lion's share of the PC market. However, it did capture a large share of the digital music player market. Do a little research. What was the difference in the company's approach to the two types of products? What is your conclusion regarding the proper approach when developing a new digital product?

35. Prepare a brief essay that includes an example of each of the following strategic moves: raising barriers to entrants (*Hint*: intellectual property), establishing high switching costs, creating a new product or service (*Hint*: the Web), and establishing alliances. The examples do not necessarily have to involve IT. Do not use examples already presented in the text. You may use examples from actual events or your own suggestions, but the examples must be practical.

36. A publishing company wants to publish electronic books on small CDs. To read the discs, users will need a device called an electronic book reader. At least two firms have developed e-book technologies that the publisher can adopt. The publisher hires you as a strategic consultant. Write a report explaining the strategic moves you suggest. What would you advise the company to do: try to develop its own e-book reader or purchase a license for existing technology? Who should be the initial target audience for the product? What should be the company's major goal in the first two or three years: profit, market share, user base, technological improvement, or perhaps having the largest sales force in this industry? Should the company give anything away? Prepare a detailed report enumerating and explaining your suggestions.

37. You are a software-marketing expert. A new software development firm has hired you to advise it on pricing and marketing strategies of its new application. After some research, you conclude that the firm can be successful either by selling at a high unit price (in which case, probably only businesses would purchase licenses to use the application), or at a very low price, which would be attractive to many individuals and companies. You estimate that by the end of the sixth year of the marketing effort competing software will be offered, which will bring the number of units sold to zero. For alternative A, the price would be $400 per license, and you expect 500 adopters in the first year and an annual growth of adopters of 70 percent. For alternative B, the price would be $30, and you expect 600,000 adopters in the first year and an annual growth of adopters of 4 percent. Use a spreadsheet application to calculate revenue, and tell the firm which strategy is expected to bring in greater revenue. Enter the prices and number of first-year adopters for each alternative only once, each in a single cell, and use absolute referencing to those cells.

38. Use PowerPoint or other presentation software to present the ideas you generated in Question 1 or 2 of "Applying Concepts." Use the program's features to make a convincing and visually pleasing presentation.

39. Do a library or Web search of business journals and magazines such as the *Wall Street Journal, BusinessWeek , Forbes,* or *Fortune.* Find a story on a business's strategic use of data, information, or information systems. (*Note*: The writer might not have identified the strategic use, but you might find that the use served strategic goals.) Prepare a report explaining the opportunity seized. Did the organization create a new product or service, improve one, or manage to capture a significantly greater market share of an existing product or service? How did the data, information, or information system play a major role in the strategic move?

40. Consider the information provided in the "Ethical & Societal Issues" box of this chapter. Prepare extensive lists of pros and cons. The pros should aim to convince an audience why Microsoft, or a similar company, should be left alone to practice its business maneuvers. The cons should aim to convince an audience why governments should intervene in how corporations such as Microsoft behave and explain what such interventions are meant to accomplish.

41. Brainstorm with your team to answer the question: "Which information technology over the past two years has epitomized a unique product or service that was 'ahead of the curve' for a significant amount of time?" This might be a physical product using IT or an online service that was, or still is, unique. List the reasons each of the team members liked this product or service.

42. Some information technologies had a certain original purpose but were creatively used to serve additional purposes. For example, companies have used caller ID to retrieve customer records as soon as a customer telephones. This saves labor and increases service quality. You and your teammates are consultants who work with many businesses. Offering your clients original ideas will increase your success. Select an information technology or an IT feature that can be leveraged in ways not originally conceived. How can your clients (in manufacturing, service, or any other business sector) use this feature to gain strategic advantage? Prepare a rationale.

FROM IDEAS TO APPLICATION: REAL CASES

Knight in Shining Trucks

The trucking business is highly competitive. In recent years many trucking companies have adopted information technologies to improve performance in what seems like a conservative industry. Those who innovate can reap great rewards. Trucking is still the least expensive and fastest way to move goods from one location to another in the United States; 80 percent of all goods transported in this country are moved on trucks.

Knight Transportation, Inc. was established in 1989 and is headquartered in Phoenix, Arizona. The company has experienced fast growth. At the start of 2002 it operated 1,087 trucks ("tractors" in industry lingo) and 4,834 trailers. In mid-2007 it operated 3,400 trucks and 7,900 trailers, of which 400 were refrigerated.

The company transports a great variety of items: consumer staples, paper products, packaging and plastic materials, manufactured goods, and imported and exported goods. Its refrigerated trailers transport perishable foods. With the growth of its fleet and variety of items handled, the company felt it was necessary to know more about its truck and trailer locations and the contents of trailers.

These needs were in addition to the old challenge of locating stolen trucks. Thieves often follow a truck from the time it is loaded until the driver stops for the night. In one case, a ring of criminals stole items with a combined value of $2.2 million from several trucks. The ring, which was arrested near Chicago in 2006, stole prescription drugs, liquor, copy machines, cookies, and car parts. The director of communications for Knight Transportation estimates that truck cargo theft is more pervasive than bank robbery.

Trucking is still largely an inefficient industry, mainly because timing is highly dependent on clients: the senders and receivers. Although handling perishable foods is fast, the story with other items is different. A truck may wait hours or even days until a store has the room or personnel to off-load the cargo.

Communication between truckers and their delivery sites is poor. The company often does not know which of its trailers is loaded and which is not, let alone the contents. It often sends trucks with empty trailers long distances to pick up cargos, while there is an available trailer much closer to the location. This increases the need for more trailers than is actually necessary. In the past, Knight maintained a ratio of more than four trailers per truck. These are costs that could be saved with more accurate and timely information.

The company explored ways to use global positioning systems (GPS). Many of the companies specializing in the field offer systems that locate drivers to ensure they do not stop excessively or do not take longer-than-necessary routes. Knight wanted more information than this.

In 2001, management decided to engage Terion, Inc. of Plano, Texas. At that time, Terion's product, FleetView, could only locate trucks through its GPS devices and communicate the location of trailers. In 2005, Terion introduced an adjunct product called Cargo Sensor, which is mounted inside a trailer to detect the presence of cargo. The sensor uses ultrasonic, high-frequency sound waves to determine if an environment is empty and reports to the fleet operator. Cargo Sensor is adaptable to trailer lengths and interiors, such as metal or plywood, and can record the exact time a trailer is emptied. The sensor detects exactly when the trailer doors are open and closed. Both this information and the trailer location are transmitted over the Verizon cellular network. The FleetView device costs $500 and can be installed in an hour. Terion charged Knight $125 per installation. Later, Knight decided to have the device installed at the trailer manufacturing site, which lowered the installation price to $100.

Fleet operators can also save money by timely maintenance of their trucks. FleetView sensors are also able to relay information about fuel level and tire air pressure. Drivers may receive calls to take care of the truck accordingly, or bring in trucks for routine maintenance at the proper time. Knight decided to install the sensors in all its trailers.

The information transmitted from the road gives Knight information about trailer contents, location, and times. Special software aggregates the information for staff at headquarters, and the workers receive a complete picture of vehicle activities. They can immediately detect when a trailer is hitched to an unknown or unauthorized truck, which could mean a theft is in progress.

Before the Terion device was implemented, drivers could drop off a trailer at a retailer such as Wal-Mart and then sit idle. Now, managers are able to go onto the company's Web site, bring up the current record of each trailer, and have an instant view of where the trailers are and where they are moving.

Knight management has not computed all the benefits from the new system. However, it estimates that it has saved at least $1 million dollars in fuel alone since installation. The carrier is able to dispatch its trucks on more efficient routes and reduce the need to drive around in search of an empty trailer. Management also estimates that the company's fleet operators spend one hour less daily trying to locate trailers.

For many years the trucking industry could not reduce the ratio of trailers to trucks from 3 to 1. In 2002, Knight's ratio was 4.4 to 1. Although its fleet grew, the ratio now is only 2.3 to 1. In this industry, this is a significant indicator of efficiency. Perhaps this is why the company is profitable. In 2006, Knight enjoyed an 11 percent profit.

Sources: Pettis, A., "Knight Gets a Handle on Trucks," *eWeek*, July 31, 2006; U.S. Department of Transportation, 2002; finance. yahoo.com, 2007; www.knighttransportation.com, 2007.

Thinking About the Case

1. What were the inefficiencies at Knight before Fleet-View was installed?

2. What information do managers have now that they did not have before?

3. What are the indicators for greater efficiency at Knight Transportation now?

4. Of the approaches to gaining strategic advantage discussed in this chapter, which one applies to this case?

5. Considering the devices and software that Knight uses, can it keep an advantage over competitors for long? Explain why or why not.

As Vast as the Amazon

In 1995, when Jeff Bezos decided to establish the largest bookstore on earth, he wanted to call it Cadabra, as in Abracadabra, to invoke a sense of magic for people who would buy books online. His attorney thought that the word might remind of cadaver, and suggested the alternative name: Amazon. Since then, Amazon.com has expanded by leaps and bounds and is anything but a cadaver.

In 2006, the company came out with a new initiative. It offers its own network of Web-based storage devices to businesses and individuals, who pay only for the storage they use. Clients use the Web to store as much data as they want. Amazon says: "Pay only for what you use. There is no minimum fee, and no start-up cost." Clients pay 20 cents per gigabyte (1 billion bytes) of data when it is transferred, and 15 cents per GB per month of storage space used. The company calls its storage business S3, for Simple Storage Service.

Amazon promises scalable, reliable, fast, inexpensive, and simple storage. Scalable means that a client can increase the amount of stored data whenever the need arises without hassle or additional costs. Reliable refers to the percentage of uptime, the percentage of time that the service is available. Amazon promises an uptime of 99.99 percent. That is, there is a chance of .01 percent that a client will not be able to upload or download stored data. When you purchase your own hard disks, you pay about 40 cents per 1 GB. Thus, Amazon's rate of 15 cents per GB is, indeed, inexpensive. Simplicity helps clients to avoid the cost and time in establishing their own network of storage devices. Amazon offers its own storage network, one that is distributed around the globe and on which data elements are replicated, so that if one server is down, another can still provide the data.

S3 is attractive to companies that do business online and need to store vast amounts of data that must be accessible from any part of the world. Reliability and speed of access are then of utmost importance. As an example, consider one client of this service: SmugMug, an online photo sharing company.

Photo sharing is popular with the public. Many online companies allow customers to place photos for social networking, share photos with family and friends, and post photos so they can order paper prints. SmugMug says its site is "Like Fort Knox for your photos." It offers many editing features and the ability to organize personal photos in online galleries. As the popularity of its site is booming, the need for storage space and the facilities to access the photos grows. One alternative for SmugMug was to augment its own facilities and maintain them. The other was to turn to S3.

SmugMug has only 15 employees. It saves over 70 million photos for more than 150,000 paying customers. It enjoys rapid growth but has a limited staff and data-center space. The single largest expense is data storage. It uses its own disks for one copy of every photo, and S3 as a backup. S3 enables the small company to compete with the larger online photo service firms without increasing staff or spending money on backup hardware. If its own data center or an Amazon.com data storage location fails, another S3 data center provides a backup, and subscribers do not even know of the incident.

Because each original photo has six display copies, the company maintains about half a billion images. Therefore, SmugMug increases its use of S3 by 10 terabytes (10,000 GB) each month. Before it subscribed to S3, SmugMug was dependent on its own redundant storage and on several data centers. This entailed high costs. With S3, the company expects to save $500,000 in expenditures on disk drives, and another $500,000 on the backup redundant disk drives.

Companies subscribing to the service need to develop the proper code to link their site to Amazon's storage service. This took SmugMug only a week. Since the company started to use the service, its customers have experienced no downtime. At competitors, when the site's service goes down, subscribers typically experience 3-4 hours of downtime. SmugMug's site did go down several times since it subscribed to S3, but the failover system kept access available to SmugMug customers. SmugMug's CEO asserts that his customers' photos are safer thanks to storage by two companies rather than one, and having backup data centers located in at least three states.

Interestingly, Amazon.com does not expect to make money on S3 in the near future. However, it certainly uses its technological muscles in a new strategic direction.

Source: Amazon.com, 2007; SmugMug.com, 2007; Cone, E., "Amazon at Your Service," *www.cioinsight.com*, January 7, 2007.

Thinking About the Case

1. Amazon.com tries to take strategic advantage of its resources. Of what physical resources does Amazon.com take advantage?

2. Does Amazon.com strategically leverage anything else in addition to physical assets?

3. What are the benefits of S3 for small companies?

4. At its current rates for S3 services, Amazon.com does not make a great profit (or any at all). Why, then, do you think Amazon.com offers S3?

IT Makes Cents

Does Avis Walton mind receiving orders from a machine when he works? No, he actually thinks it is "cool." Avis works as a "picker" for 99 Cents Only Stores. Walton spends his workday in a 750,000 square-foot distribution center in Katy, near Houston, Texas, riding an electric vehicle. He wears an earbud that streams instructions from a central information system. The female voice gives him a row number, then a section number, and then a bin number. He scans the tag on the bin's front with a wireless hand-held computer to confirm that he is at the right bin. The voice then orders him to pick so many cases. He gets off the vehicle, picks up the boxes, and places them on a pallet. He confirms the pick into a microphone. The voice now sends him to his next assignment.

He and his 15 fellow pickers are used to the electronic voice. It is generated by a computer that runs the distribution center's warehouse-management software. It instructs them which items to pick for individual stores. It also calculates the most efficient routes while ensuring that the carts do not crash into each other. The "lady" tells the pickers which bins need to be replenished and where to find the items to replenish those bins. Pickers place the boxes on a three-story conveyor. Laser scanners quickly scan box tags and route the boxes to 20 different lanes, ensuring that each box is on the proper path to a pallet waiting below for specific stores. The system also plans loading to utilize maximum space on each truck.

99 Cents Only Stores is America's oldest chain of one-price stores. The chain consists of 251 stores in California, Nevada, Arizona, and Texas. The business was started as a single store in Los Angeles in 1984 by David Gold. Now age 73, he still comes to the office daily at 4 a.m. The company never had a year in which it lost money. Between 1996, when 99 Cents went public, and 2003, the company's stock price climbed from $3.12 to $36.22. The U.S. retail market includes several other chains of fixed-price stores, and competition is fierce. The chain does better than its competitors in every measure important in the retail industry: sales per square foot and net profit margin on revenue. In 2003, profit margin was 8.3 percent while profit at Wal-Mart was 3.1 percent and at Kroger Co., the supermarket chain, a mere 2.1 percent (but typical for supermarkets). Gold, who in 2004 stepped down as the company's CEO, remains active as the Chairman of the Board, and his two sons and son-in-law run the company. The Gold family owns about 35 percent of the company.

Despite sales revenues of over 1.1 billion in the year ended March 31, 2007, the amount of spending on IT is relatively small, only $5 million in 2003, and about the same small proportion in 2006. However, Robert Adams, vice president of IS, selects IT projects carefully. Each store has a wireless local area network (WLAN) and connection to the Internet. All district managers carry cell phones, which they can also use as walkie-talkies. When Adams moved from another company to work for the chain, he was afraid he would not get the budget he might need for new systems because the management would not see the need to invest in technology. The contrary happened. Because the company is family-run, decisions are made quickly. He does not need to go through formal meetings. Therefore, the time between request and implementation is very short.

The fixed-price-store industry, popularly known as dollar stores, has been slow to adopt state-of-the-art technology. Only recently have such chains started adopting modern systems, and 99 Cents seems to be ahead of them. Some software companies, such as HighJump Software, design systems that can specifically support the operations of these chains. IT has enabled 99 Cents to differentiate itself from similar chains. The floor space of competitors is typically 4,000 to 6,000 square feet, and each store has annual revenue of $1 million. A 99 Cents store is 22,000 square feet and has annual revenue of $4.3 million. The targeted audience, too, is different. While other stores target neighborhoods with low-to-medium incomes, David Gold observed that rich people, too, like to save money. His company's most profitable store is located close to Beverly Hills, has an area of 18,000 square feet, and earns an average of $10 million annually.

If you have shopped more than once at the same dollar store, you probably noticed that an item you purchased the first time is no longer available on a subsequent visit. This is typical, because dollar stores purchase not by item but by price. When purchasing officers spot an opportunity to buy a lot of a discontinued product, they offer a very low price and purchase it. It is difficult for these chains to reorder the same items at the same low price. 99 Cents succeeds in reordering 60 percent of its inventory. The rest are one-time-only close-outs.

Gold and his executives have a simple goal, which is to establish the shortest path between an inexpensive item and a paying customer. This drives all the decisions on which IT pursues. And IT plays a major role in identifying suitable merchandise, efficiently receiving it at the distribution centers, and then distributing it to the stores while avoiding overstocks or understocks. Interestingly, Gold is not fond of computers. He rarely uses his own office PC. He does not have anything against IT, he says, he just dislikes big spending on IT if the information it produces is not used. He is also annoyed by the average IT professional, who keeps himself above the nontechnical masses. Adams, he says, is different. Adams is personable, a perfect choice for the company, Gold says.

Adams has an 18-person IT team to which he delegates much authority. However, he is a demanding boss who leads by example. David Gold was impressed when Adams wrote the entire code for the company's point-of-sale systems. He and his team write code whenever it is cheaper to purchase ready-made software and modify it than spend the resources to develop the software from scratch in-house. Since 40 percent of its merchandise consists of one-time-only inventory that will never be purchased again, 99 Cents requires systems that can accept new items on the fly. Adams' team ensures that the ISs are flexible. If the

decision is to develop software in-house, Adams spends much time with the project team. He still regards himself as a software developer and refuses to pay another company much money for modifications or for new software.

Until 2004, the company had a single distribution center in Commerce City, California. In mid-2004, management decided to expand to Texas and build another distribution center there. It purchased facilities and equipment for $23 million from supermarket giant Albertson's, Inc., which had invested $80 million in the facility in 1995. Adams had only four months to equip the warehouse with the proper IT so it could start operations. This time it did not make business sense to develop code in-house. Adams contacted HighJump Software, a subsidiary of 3M, which sells warehouse management software.

HighJump's software, called Warehouse Advantage, supports all the activities that occur from the time products enter the warehouse to the moment they leave. A Voxware computer receives the picking profile from Warehouse Advantage and tells workers what to pick and where to find it. At the retail stores, employees can use a Web-based system to access information about the status of incoming shipments. Management uses Advantage Dashboard for a high-level view of facility and worker performance expressed as metrics and graphs. Managers receive real-time inventory levels and order volumes of various products. The new systems are proving themselves. Picking accuracy, that is, picking and shipping the right item, is 90 percent at the California distribution center. At the Texas center it is 99 percent. Picking speed at Texas is 20 percent greater than at the California center. The system works so well that Adams decided to implement it in California.

With all his enthusiasm for IT, Adams avoids implementing cutting-edge technologies. He says the company is too small and traditional to sustain "bleeding edge" technologies. The strategic advantage he believes 99 Cents has is in the business intelligence with which the company integrates proven technologies into its operations. He says he prioritizes IT projects by how much obvious return on investment he sees in them. When it is obvious a certain technology will gain his company efficiency, he implements it. Often, his team completes only a part of a project, so it can start a new project that helps the company more. Adams says reprioritizing allows the company to get the greatest benefits from all IT projects. What is not completed now can be completed after the other, more important project is completed.

All dollar store customers like bargains, but the customers of 99 Cents Only visit their favorite stores more often and buy more. And they probably do not know that ever-better IT ensures that they can find those great, inexpensive items on the shelves almost as soon as 99 Cents Only can find them.

Source: Rae-Supree, J., "99 Cents Only Stores' Efficient IT Infrastructure," *CIO Insight*, January 1, 2004; www.99only.com, 2007.

Thinking About the Case

1. Is 99 Cents Only on the leading edge of IT? Is it on the bleeding edge?

2. What characteristics of the dollar store industry make it so important to increase efficiency?

3. The company has performed better than its competitors. In terms of the eight initiatives discussed in this chapter, which initiative or initiatives has gained it the competitive advantage?

4. 99 Cents Only must modify its information systems frequently. Why?

5. Often, CIOs are frustrated with the time it takes senior management to support their strategic initiatives and with the difficulty of earmarking funds for such initiatives. How is 99 Cents Only different in this respect?

3

THREE

Business Functions and Supply Chains

LEARNING OBJECTIVES

In an economy that produces and consumes so much information, professionals must know how to use information systems in virtually every business activity. Managers must have an overall understanding of all elements of a system, so that they know what options are available to control quality, costs, and resources. Modern information systems encompass entire business cycles, often called supply chains.

When you finish this chapter, you will be able to:

- Identify various business functions and the role of ISs in these functions.

- Explain how ISs in the basic business functions relate to each other.

- Articulate what supply chains are and how information technology supports management of supply chains.

- Enumerate the purposes of customer relationship management systems.

- Explain enterprise resource planning systems.

GARDENERS+:
Continued Growth and Specialization

Something had to give: *Gardeners+*'s expansion into commercial services and its investment in additional gardening equipment worked so well that the back-office became overloaded. The transcription of paper proposals and contracts for customers and affiliated gardeners was too much for the current system. Both the number of transactions and the complexity of each one made the process overwhelming. In terms of the back-office workload, one commercial customer was equivalent to dozens of residential customers, and one data entry mistake in such transactions could be very costly. Furthermore, the residential customer transactions had also become more complicated, because the list of available services was growing larger. The loyal customer program and other marketing initiatives increased sales as well as the back-office workload. Amanda and Ed were so overwhelmed with data entry they were unable to handle their other business responsibilities. Fortunately, Amanda had already integrated the transaction processing and accounting systems, thus eliminating the additional task of moving the information from one system to the other. Ed had hired a part-time assistant to help him with the transcription of paper forms, but they were still behind. If they were to keep up with the workload and remain profitable, something had to be done to make their processes less labor-intensive.

The bulk of the information on the paper forms was filled in by customers when orders were taken on-site. Over the phone, Mary and a new employee filled forms in by hand when the PC was not available, and it rarely was. The remaining paper forms were completed by the gardeners, who turned them in when they visited the office.

Amanda proposed that they use electronic forms that could be filled in on a notebook computer or even a PDA and then downloaded into the business's main system, minimizing tedious and error-prone transcription. She also suggested that they develop a Web site where customers could enter requests directly into the system. Finally, she proposed that they acquire a PC with a touch screen that the gardeners could use to enter data into the computer system themselves, after some initial training.

This automation generated costs for new equipment and additional programming; nonetheless, it was worthwhile because the time saved by skipping the paper stage and directly entering the sales transactions allowed the partners to concentrate on the bigger issues: tracking sales, costs, and profitability.

Ed was able to go back to his ongoing overall analysis of the business. When he printed sales reports segregated by service type, he noticed that revenues from the mall and other commercial customers were dropping. Their competitor was gaining ground by convincing its former clients to return. Ed also noticed that business was even worse during colder months. *Gardeners+* needed to turn the situation around.

A New Opportunity

To better compete with their large, high-profile competitor, Mary, Amanda, and Ed decided to increase their gardeners' payroll with more stable and long-term contracts. They also decided to rent commercial space to acquire a more corporate look than their garage home office.

During a dinner at the Chamber of Commerce, Mary learned that the mall had a few spaces for rent. There was a lower rate for current mall tenants, and although *Gardeners+* was not a tenant, the manager was willing to extend the discount to a company they had been doing business with. This storefront at the mall would make the business more visible and allow them to capture additional residential customers.

To handle the increased workload, the three entrepreneurs hired a full-time office assistant.

Amanda, with the help of several suppliers, developed the required functionality for the transaction processing system and planned the telecommunications between the mall storefront and their home office in the garage.

Advertising Needs and Promotions

To announce the opening of their new storefront, Mary used a desktop publishing program to create flyers. The flyers were handed out to customers and potential customers by the gardeners during their jobs and by a few hired teenagers who placed them on the windshields of cars parked at the mall.

The flyers included a feedback form; those who returned the forms would receive discount coupons on their next gardening job. They hoped these special offers would attract new residential customers, and therefore help defray the costs of this new rented space. Mary also suggested that they produce radio commercials or other low-cost mass advertisements, but none of the partners had any marketing experience and were uncertain about which type of media they should use. They needed professional help.

Moving Forward

Gardeners+ had come a long way since its start, but the three partners still had decisions to make and changes to undergo. With the opening of the mall storefront, they needed to revamp their information systems to integrate the garage and the storefront. And they needed to expand their payment methods to include credit cards.

Gardeners+ also needed to automate its payment systems. Ed had to continuously double-check payments using a calculator, and then manually revise the information on the system. Some checks had to be written by hand, since making corrections on the system was slow and cumbersome. When the business was small and they knew all the gardeners well, errors or delays were less important, but as the business grew they realized that payments needed to be prompt and accurate.

Furthermore, tracking the growing amount of gardening equipment became very unwieldy. Amanda had always tracked the inventory closely, but with the expansion it had become too difficult to keep current. Mary and Ed agreed that an inventory control system was needed, so they invested in QuickBooks®, a software program that could integrate other functions such as online credit verification, sales and expense tracking, payroll and accounting management, sales and payroll tax calculation, invoicing, and check printing. The system could not handle contracts or the management of the affiliate gardeners, but Amanda, with the help of an expert QuickBooks® consultant, developed software to interconnect both systems in an acceptable (although not ideal) manner.

This more comprehensive system was a great step forward. It was clear to the partners that a well-designed and well-run information system was essential, and that they needed to make sure that their technology kept up with the business in the future.

EFFECTIVENESS AND EFFICIENCY

The telephones at the offices of Capital One Financial Corp., a leading credit-card issuer and a Fortune 500 company, ring more than a million times per week. Cardholders call to ask about their balance or to ensure that the company received their recent payment. While callers almost immediately hear a human voice at the other end, computers actually do the initial work. The computers use the caller's telephone number to search the company's huge databases. Inferring from previous calls and numerous recorded credit-card transactions of the caller, the computers predict the reason for calling. Based on the assumed reason, the computers channel the call to one of 50 employees who can best handle the situation. Important information about the caller is brought up on the employee's computer monitor. Although callers usually do not contact the company to make purchases, the computer also brings up information about what the caller might want to purchase. As soon as the customer service representative provides the caller with satisfactory answers, he or she also offers the cardholder special sales. Many callers do indeed purchase the offered merchandise. All of these steps—accepting the call, reviewing and analyzing the data, routing the call, and recommending merchandise—take the computers a mere tenth of a second. Effective operations and efficient response are what made Capital One an industry leader.

It is often said that the use of information technology makes our work more effective, more efficient, or both. What do these terms mean? **Effectiveness** defines the degree to which a goal is achieved. Thus, a system is more or less effective depending on (1) how much of a particular goal it achieves, and (2) the degree to which it achieves better outcomes than other systems do.

Efficiency is determined by the relationship between resources expended and the benefits gained in achieving a goal. Expressed mathematically,

$$\text{Efficiency} \quad = \quad \frac{\text{Benefits}}{\text{Costs}}$$

One system is more efficient than another if its operating costs are lower for the same or better quality product, or if the product's quality is greater for the same or lower costs. The term "productivity" is commonly used as a synonym for efficiency. However, **productivity** specifically refers to the efficiency of *human* resources. Productivity improves when fewer workers are required to produce the same amount of output, or, alternatively, when the same number of workers produce a greater amount of output. This is why IT professionals often speak of "productivity tools," which are software applications that help workers produce more in less time. The closer the result of an effort is to the ultimate goal, the more effective the effort. The fewer resources spent on achieving a goal, the more efficient the effort.

Suppose your goal is to design a new car with fuel economy of 60 miles per gallon. If you manage to build it, then you produce the product effectively. If the car does not meet the requirement, your effort is ineffective. If your competitor makes a car with the same features and performance, but uses fewer people and resources, then your competitor is not only as effective as you but also more efficient. ISs contribute to both the effectiveness and efficiency of businesses, especially when serving specific business functions, such as accounting, finance, and engineering, and when used to help companies achieve their goals more quickly by facilitating collaborative work.

One way to look at business functions and their supporting systems is to follow typical business cycles, which often begin with marketing and sales activities (see Figure 3.1). Serving customers better and faster, as well as learning more about their experiences and preferences, is facilitated by **customer relationship management (CRM)** systems. When customers place orders, the orders are executed in the supply chain. Often, information about the customer is collected as orders are taken. This information may be useful down the road. Customer relationship management continues after delivery of the ordered goods in the forms of customer service and more marketing. When an organization enjoys the support of CRM and supply chain management (SCM) systems, it can plan its resources well. Combined, these systems are often referred to as enterprise resource planning (ERP) systems.

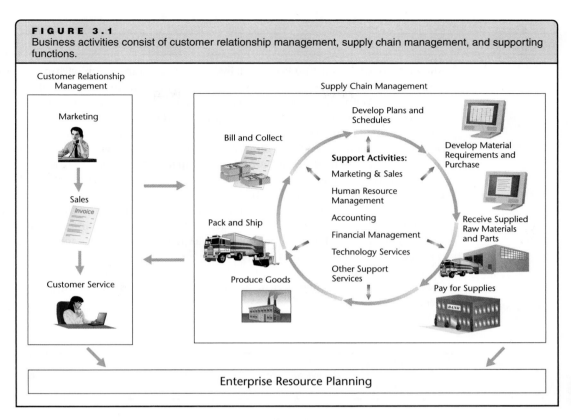

FIGURE 3.1
Business activities consist of customer relationship management, supply chain management, and supporting functions.

Figure 3.2 shows some of the most common business activities and their interdependence. For example, cost accounting systems are linked to payroll, benefits, and purchasing systems to accumulate the cost of products manufactured by a company; and information from purchasing systems flows to both cost accounting and financial reporting systems. The following discussion addresses the role of information systems, one business function at a time.

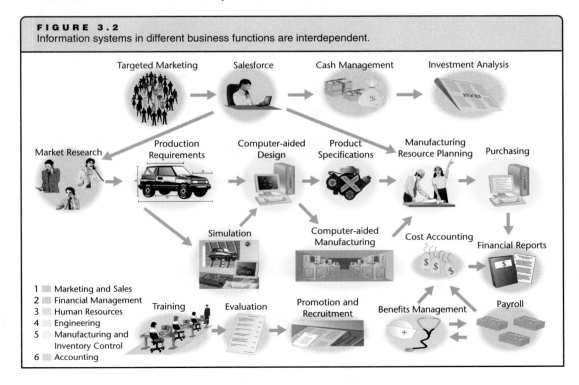

FIGURE 3.2
Information systems in different business functions are interdependent.

The purpose of accounting is to track every financial transaction within a company—from a few cents expenditure to a multimillion dollar purchase, from salaries and benefits to the sale of every item. Without tracking the costs of labor, materials, and purchased services using a cost-accounting system, a company might discover too late that it sells products below what it costs to make them. Without a system of accounts receivable, managers might not know who owes the company how much money and when it is due. Without an accounts payable system, they cannot know how much money the company owes suppliers and when payments are due. Without a system that records and helps plan cash flow, managers cannot keep enough cash in the bank to make payments on schedule. At the year's end, the company cannot present a picture of its financial situation—called a balance sheet—and a profit-and-loss report, unless it maintains a general ledger to record every transaction with a financial impact. Accounting systems are required by law and for proper management. General ledger, accounts receivable, accounts payable, and cash-flow books conveniently lend themselves to computerization and can easily generate balance sheets and profit-and-loss statements from records (see Figure 3.3). The word "books" in this reference is now a relic of former times. Accounting ISs are, of course, fully electronic.

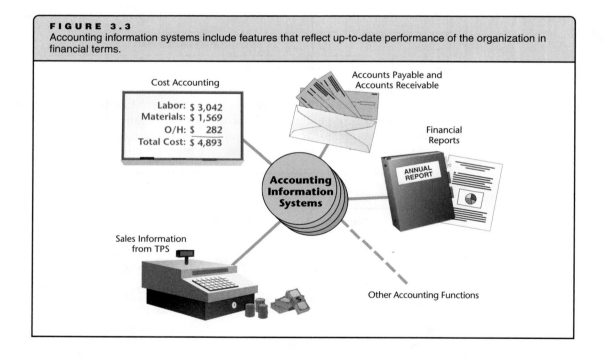

FIGURE 3.3
Accounting information systems include features that reflect up-to-date performance of the organization in financial terms.

Typically, accounting ISs receive records of routine business transactions—such as the purchase of raw materials or services, or the sale of manufactured goods—from transaction processing systems (TPSs), which include point-of-sale (POS) machines. Such a system automatically routes every purchase of raw materials or services to the accounts payable system, which uses it to produce checks or transfer funds to a vendor's bank account. Whenever a sale is recorded, the transaction is routed to the accounts receivable system (which generates invoices) and other destinations. Totals of accounts receivable and accounts payable can be automatically transferred to a balance sheet. Data from the general ledger can be automatically compiled to generate a cash-flow report or a profit-and-loss report for the past month, quarter, or year. Accounting ISs can generate any of these reports on demand, as well as at scheduled times.

Today's professionals are expected to be knowledgeable not only in their specific line of work but also in other areas. And since practically every business process involves information technology, new hires are expected to know, or quickly learn, how to use the proper ISs in their respective positions. Many employers look for generalists rather than specialists and focus on the techno-manager, a manager well-versed in information technology as it relates to the entire supply chain.

Because many ISs serve multiple functions and interface with other systems, it is extremely important for a professional to be familiar with the way ISs facilitate work in areas outside his or her expertise. If you work for a commercial organization, you are bound to be part of a supply chain or work for a unit that supports a supply chain. Knowledge of systems in different business areas helps you cooperate with your peers and coordinate efforts that cross departmental boundaries. Because professionals often have opportunities to be promoted to positions in other disciplines, the more you know, the better your chances of being "cross-promoted."

When a company develops and manufactures a new product that has never been available on the market, how can it determine a price that covers costs and generates a decent profit? It must maintain a system that tracks the costs of labor, materials, consulting fees, and every other expense related to the product's development and manufacture. Cost-accounting systems, used to accumulate data about costs involved in producing specific products, make excellent use of IT to compile pricing data. ISs also help allocate costs to specific work orders. A **work order** is an authorization to perform work for a specific purpose, such as constructing a part of an airplane. When interfaced with payroll and purchasing ISs, a cost-accounting system automatically captures records of every penny spent (and originally recorded in the payroll and purchasing systems) and routes expenses to the appropriate work order. Because work orders are associated with specific products and services, the company now knows how much each product or service costs, or how much making a part of a final product costs. This can help the company in future pricing of products or services.

Accounting ISs are also used extensively for managerial purposes, assisting in organizing quarterly and annual budgets for departments, divisions, and entire corporations. The same systems help managers control their budgets by tracking income and expense in real time and comparing them with the amounts predicted in the budget. Budget applications are designed with proper controls, so that the system does not allow spending funds for a specific purpose beyond the amount that was budgeted.

FINANCE

A firm's health is often measured by its finances, and ISs can significantly improve financial management (see Figure 3.4). The goal of financial managers, including controllers and treasurers, is to manage an organization's money as efficiently as possible. They achieve this goal by (1) collecting payables as soon as possible, (2) making payments at the latest time allowed by contract or law, (3) ensuring that sufficient funds are available for day-to-day operations, and (4) taking advantage of opportunities to accrue the highest yield on funds not used for current activities. These goals can be best met by careful cash management and investment analysis.

Cash Management

Financial information systems help managers track a company's finances. These systems record every payment and cash receipt to reflect cash movement, employ budgeting software to track plans for company finances, and include capital investment systems to manage investments,

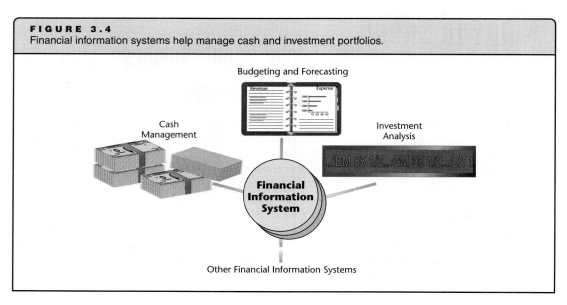

FIGURE 3.4
Financial information systems help manage cash and investment portfolios.

Budgeting and Forecasting

Cash Management

Investment Analysis

Financial Information System

Other Financial Information Systems

thus balancing the need to accrue interest on idle money against the need to have cash available. The information on expected cash receipts comes from sales contracts, and the information on cash outlays comes from purchasing contracts as well as payroll and benefits schedules. Systems that deal specifically with cash are often called **cash management systems (CMSs)**. One common use for a CMS is to execute cash transactions in which financial institutions transfer huge amounts of money using **electronic funds transfer (EFT)**. EFT is the electronic transfer of cash from an account in one bank to an account in another bank. More than 80 percent of all payments of the U.S. government are made using EFT systems.

Investment Analysis and Service

Every investor's goal is to buy an asset and later sell it for more than it cost. When investing in securities, such as stocks and bonds, it is important to know the prices of securities in real time, that is, *right now*. The ability of financial ISs to record millions of securities prices and their changes over long time periods, coupled with the ability to manipulate numbers using software, puts powerful analysis tools in investment managers' hands. Within seconds, an investment analyst can use a financial IS to chart prices of a specific stock or bond over a given period, and then build or use preprogrammed models to estimate what might happen to securities prices in the future.

Even the smallest investment firm can provide clients with an inexpensive online service for buying and selling securities, providing on-demand statements listing the stocks they own (called a portfolio), periodic yield, and the portfolio's current value. Clients serve themselves through the Web sites of brokerage firms to place, buy, and sell orders. Execution of orders takes only a few seconds.

Nearly instantaneously, ISs provide subscriber brokers and their clients with financial news, stock prices, commodity prices, and currency exchange rates from multiple locations across the world. Consider what happens when a foreign currency's exchange rate fluctuates a fraction of a percent. A brokerage house can make a profit of several thousand dollars within two minutes of buying and selling several million dollars' worth of the foreign currency.

Financial managers need to consider many factors before they invest in a security. Some of the most important factors to consider are (1) risk, measured as the variability (degree of change) of the security's past yield; (2) expected return; and (3) liquidity, a measure of how fast an investment can be turned into cash. Special programs help calculate these factors and present the results either in tables or graphs to allow timely decision making.

I Want to Hold It

In 2006, 80 percent of U.S. federal benefit recipients received their payments electronically via direct deposit into their bank accounts. In a May 2006 report, the U.S. Treasury stated that if the other 20 percent—12 million recipients—agreed to direct deposit, the government could save tax payers over $120 million by eliminating the 150 million checks it mails each year. While it costs the government 83 cents to issue a check, it costs only 8 cents to issue a direct deposit. Direct deposits are also safer and more reliable than mailing checks. In a study conducted for the U.S. Treasury by the Federal Reserve office of St. Louis, recipients who insisted on paper checks explained, among other reasons, that "I want to be sure it's there" and "I want to hold it in my hands." The good news: every year about one percent of the check recipients agree to switch to electronic funds transfer (EFT). The not-so-good news: 4.5 million Americans who still receive checks do not have bank accounts.

Source: Federal Reserve Bank of St. Louis for the U.S. Treasury's Financial Management Service, "Understanding the Dependence on Paper Checks," 2004; "Go Direct," Philadelphia Financial Center, May 2006; "10 Fast Facts About Direct Deposit," *Countdown to Retirement*, (www.godirect.org), July 2007.

ENGINEERING

The time between generating an idea for a product and completing a prototype that can be mass-manufactured is known as engineering lead time, or **time to market**. Engineering includes **brainstorming** (the process of a group of colleagues meeting and working collaboratively to generate creative solutions and new ideas), developing a concept, creating mock-ups, building prototypes, testing, and other activities that require investments of time, labor, and money. Minimizing lead time is key to maintaining a competitive edge: it leaves competitors insufficient time to introduce their own products first. ISs can contribute significantly to this effort. Over the past two decades, automakers have used engineering and other ISs to reduce the time from product concept to market from 7 years to 18 months.

IT's greatest contribution to engineering is in the area of **computer-aided design (CAD)** and **rapid prototyping** (creating one-of-a-kind products to test design in three dimensions). Engineers can use computers to modify designs quickly and store drawings electronically. With collaborative software, they perform much of this process over the Internet: engineers can conduct remote conferences while viewing and developing plans and drawings together. The electronic drawings are then available to make rapid prototypes.

Rapid prototyping allows a model of a product to be produced within hours, rather than days or weeks. The model required is often a mock-up to show only the physical look and dimensions of a product, without the electronics or other components that are part of the full product. First, an image of the object is created on a computer. The computer is connected to a special machine that creates a physical, three-dimensional model by laying down hundreds or thousands of thin layers of liquid plastic or special resin. The model can be examined by engineers and marketing managers in the organization, or shown to clients.

When the prototypes are satisfactory, the electronic drawings and material specifications can be transferred from the CAD systems to **computer-aided manufacturing (CAM)** systems. CAM systems process the data to instruct machines, including robots, how to manufacture the parts and assemble the product (see Figure 3.5).

As we mentioned, automakers needed years to turn a concept into actual vehicles rolling out for sale. Now, thanks to CAD, CAM, rapid prototyping, and collaborative engineering software, the lead time has been reduced to months. The digital design of vehicles saves not only time but

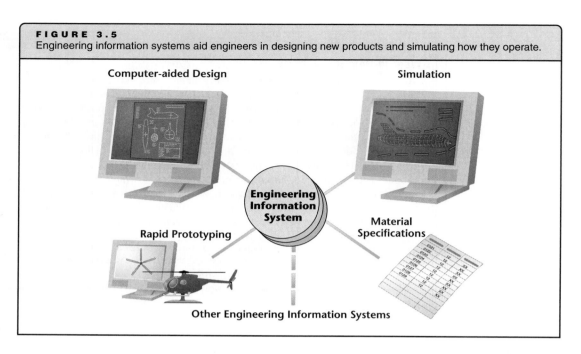

FIGURE 3.5
Engineering information systems aid engineers in designing new products and simulating how they operate.

Computer-aided Design

Simulation

Engineering Information System

Rapid Prototyping

Material Specifications

Other Engineering Information Systems

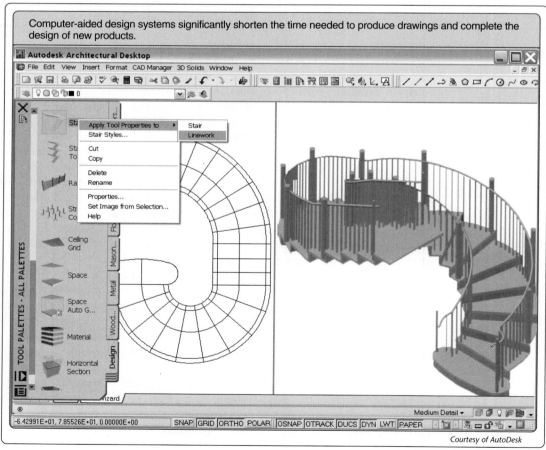

Computer-aided design systems significantly shorten the time needed to produce drawings and complete the design of new products.

Courtesy of AutoDesk

also the cost of cars crashed in tests; many of the tests can be performed with sophisticated software rather than with real cars. Similar benefits have been accomplished in aerospace and many other engineering and manufacturing industries.

Rapid prototyping shortens time to market. The white plastic model here is for an automotive differential housing.

Courtesy of Stratasys, Inc.

SUPPLY CHAIN MANAGEMENT

In its fundamental form, a **supply chain** consists of three major phases: procurement of raw materials, processing the materials into intermediate and finished goods, and delivery of the goods to customers. Processing raw materials into goods is manufacturing. **Supply chain management (SCM)** consists of monitoring, controlling, and facilitating supply chains, as depicted in the right side of Figure 3.1. Supply chain management (SCM) systems are information technologies that support SCM. SCM systems have been instrumental in reducing manufacturing costs, including the costs of managing resources and controlling inventory (see Figure 3.6). In retail, the manufacturing phase does not exist, so the term "supply chain" refers only to purchasing of finished goods and the delivery to customers of those goods. In the service industries the term "manufacturing" is practically meaningless, because no raw materials are purchased and processed.

As is clear from the previous discussion, much of the data required for manufacturing processes can flow directly from CAD systems to CAM systems as well as to inventory control systems and other systems that support planning and execution of manufacturing. While CAM systems participate in physical activities such as cutting and welding, other information systems help to plan and monitor manufacturing.

Information technology helps in the following manufacturing activities:

- Scheduling plant activities while optimizing the combined use of all resources—machines, personnel, tooling, and raw and interim materials.
- Planning material requirements based on current and forecasted demand.

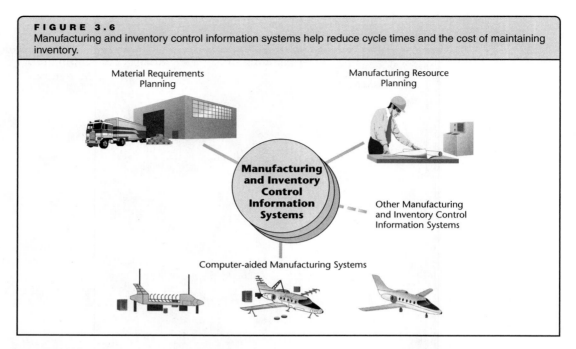

FIGURE 3.6
Manufacturing and inventory control information systems help reduce cycle times and the cost of maintaining inventory.

Material Requirements Planning

Manufacturing Resource Planning

Manufacturing and Inventory Control Information Systems

Other Manufacturing and Inventory Control Information Systems

Computer-aided Manufacturing Systems

- Reallocating materials rapidly from one order to another to satisfy due dates.
- Letting users manage inventories in real time, taking into consideration demand and the responsiveness of all work centers.
- Grouping work orders by characteristics of items ordered, such as color and width of products.
- Considering the qualifications of each resource (such as qualified labor, set-up crews, and specialized tools) to accomplish its task. For instance, people and raw materials can be moved from one assembly line to another to respond to machine breakdown or customer emergency, and design changes can be implemented quickly to respond to changes in customer wishes.

Material Requirements Planning and Purchasing

One area of manufacturing that has experienced the greatest improvement from IS is inventory control, or **material requirements planning (MRP)**. Traditional inventory-control techniques operated according to the basic principle that future inventory needs are based on past use: once used up, inventory was replaced. By contrast, replenishment in MRP is based on *future* need, calculated by MRP software from demand forecasts. MRP programs take customer demand as their initial input. The main input to MRP programs is the number of product units needed and the time at which they are needed; the programs then work back to calculate the amounts of resources required to produce subparts and assemblies. The programs use long-range forecasts to put long-lead material on order.

Other important input to MRP applications includes a list of all raw materials and subcomponent demands (called the **bill of materials**, or **BOM**) and the economic order quantity of different raw materials. The **economic order quantity (EOQ)** of a specific raw material is the optimal quantity that allows a business to minimize overstocking and save cost, without risking understocking and missing production deadlines. A special program calculates EOQ for each item. It considers several factors: the item's cost, the discount schedule for large quantities, the cost of warehousing ordered parts, the cost of alternative uses of the money (such as the interest the money could earn had it not been spent on inventory), and other factors affecting the cost of ordering the item. Some MRP applications are tied to a purchasing IS, to produce purchase orders automatically when the quantity on hand reaches a reorder level. The purchase order includes the economic order quantity.

MRP systems help reduce inventory cost while ensuring availability.

AP Images

Computer-aided manufacturing systems control robots.

© REUTERS/Rebecca Cook/Landov

Manufacturing Resource Planning

Manufacturing resource planning (MRP II) combines material requirements planning (MRP) with other manufacturing-related activities to plan the entire manufacturing process, not just inventory. (The "II" in MRP II is simply to distinguish this term from material requirements planning, another term with the same acronym.) MRP II systems can quickly modify schedules to accommodate orders, track production in real time, and fix quality slippage. The most important input of MRP II systems is the **master production schedule (MPS)**, which specifies how production capacity is to be used to meet customer demands and maintain inventories. Virtually every report generated by an MRP II package starts with, or is based on, the MPS. Purchases of materials and internal control of manufacturing work flow, for example, start with the MPS, so the MPS directly affects operational costs and asset use.

MRP II systems help balance production economies, customer demands, manufacturing capacity, and inventory levels over a planning horizon of several months. Successful MRP II systems have made a significant contribution to **just-in-time (JIT)** manufacturing, where suppliers ship parts directly to assembly lines, saving the cost of warehousing raw materials, parts, and subassemblies.

MRP and MRP II systems gave ERP systems their name. MRP II modules are now integrated into ERP systems. While MRP and MRP II were, indeed, used mainly for planning, the "P" in ERP is somewhat misleading, because ERP systems are used mainly for daily operations, in addition to planning.

Ideally, the ISs of manufacturing organizations and their suppliers would be linked in a way that makes them subsystems of one large system. The MRP II application of an organization that manufactures a final product would plan and determine the items required, their quantities, and the exact times they are needed at the assembly lines. Suppliers would ship items directly to assembly lines just before they are incorporated into the final product (hence the term *just-in-time manufacturing*). Manufacturing organizations have not yet reached the point where JIT is accomplished with every product, but they have made great progress toward this ideal.

The Internet facilitates such system linking. Companies that were quick to link their systems to their suppliers' systems attained strategic advantages. One such company is Cisco Systems, a world leader in design and manufacturing of telecommunications devices. The company used to maintain many manufacturing plants. In 2001, it had sold all but two. The company's ISs are linked through the Internet to the ISs of its suppliers, some of whom purchased the very plants that Cisco sold. Managers can track orders using these systems. They can tell Cisco clients the exact status of their orders and the time of delivery. Cisco managers keep track of the products they order and know at what phase of manufacturing and delivery each item is—as if *they* were running the manufacturing plants. More than 80 percent of what Cisco orders never passes through the company's facilities; the manufacturers ship the products directly to Cisco's clients.

Monitoring and Control

Information systems have been designed to control manufacturing processes, not just monitor them. Controlling processes is important to ensure quality. For example, Ford Motor Company implemented software that it calls Project Execution, which combines bar-coding and wireless technology to ensure quality. Since each vehicle is assembled on a chassis, each chassis is tagged with a unique bar code. A bar-code sensor is installed in each stop of the assembly line. The sensor transmits wireless signals to computers and electronically controlled gates. The "gates" are not physical ones, but points where the vehicle is checked. The purpose of the system is to ensure that no assembly steps are skipped, and that each vehicle passes a series of performance and quality tests along the way. If a step is missed, the gate does not let the vehicle leave the plant.

Shipping

When the process of manufacturing products is complete, the next link in the supply chain is shipping. Shipping is performed either by the manufacturer or by a hired shipping company. The variables that affect the cost and speed of shipping are numerous: length of routes, sequence of loading and unloading, type of shipped materials (e.g., perishable, hazardous, or fragile), fuel prices, road tolls, terrain and restricted roads, and many more. Therefore, the use of sophisticated software to optimize shipping time and the cost of labor, equipment use, and maintenance helps companies stay competitive. Figure 3.7 shows an example of such software.

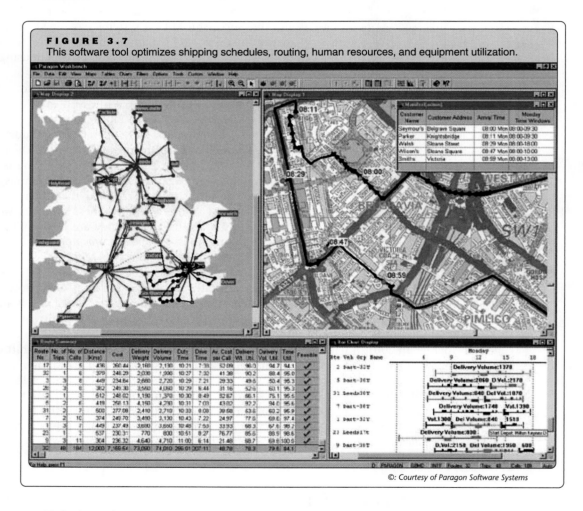

FIGURE 3.7
This software tool optimizes shipping schedules, routing, human resources, and equipment utilization.

©: Courtesy of Paragon Software Systems

Today's trucks are equipped with computers, global positioning systems (GPS), and satellite communication devices. You might have seen small antennas on trucks. The antenna receives real-time orders from a central shipping office, especially when routing changes are necessary, and transmits information about the truck, such as current location, the previous point of loading or unloading, and the next point of loading or unloading. Truckers rarely visit shipping offices. These systems allow them to be on the road doing productive work all the time, thanks to constant communication with the office.

Supply chain management software in transportation helps load trucks, ships, and airplanes in an optimal manner both in terms of space utilization and sequence of unloading. Figure 3.8 provides a visual description of an optimal loading of boxes on a truck before its dispatch. Figure 3.9 illustrates how information is communicated between a truck and a shipper's office.

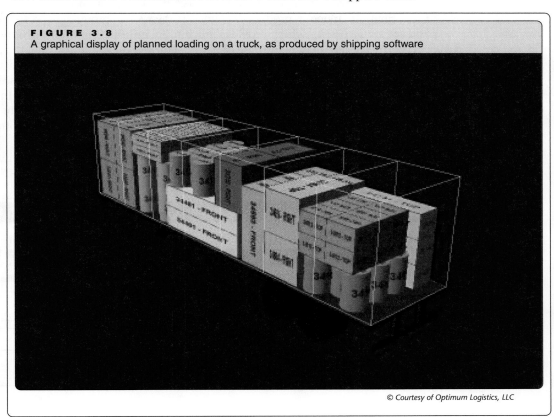

FIGURE 3.8
A graphical display of planned loading on a truck, as produced by shipping software

© Courtesy of Optimum Logistics, LLC

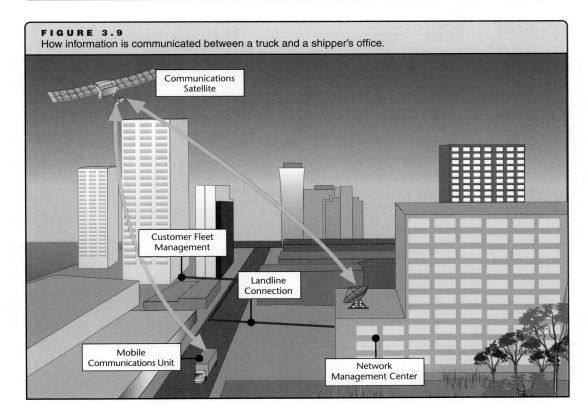

FIGURE 3.9
How information is communicated between a truck and a shipper's office.

Communications Satellite

Customer Fleet Management

Landline Connection

Mobile Communications Unit

Network Management Center

RFID in SCM

The most important development in hardware to support SCM has been a technology called **radio frequency identification (RFID)**. We discuss the technology itself in Chapter 6, "Business Networks and Telecommunications." RFID tags contain circuitry that allows recording of information about a product. When attached to a product, it contains an **electronic product code (EPC)**, which provides much more information than the universal product code (UPC) . The tag can include the date of manufacturing, the plant in which the product was made, lot number, expiration date, destination, and many other details that help track its movement and sale. The information can be read and also revised by special RFID transceivers (transmitter-receiver devices). Figure 3.10 shows the EPC and other information stored in one type of RFID product tag. Figure 3.11 shows an example of how RFID is used in a supply chain. Items with rewritable tags can contain maintenance history of products, which helps optimize maintenance of the items.

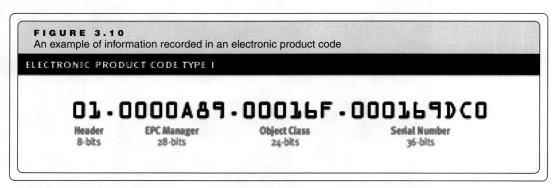

FIGURE 3.10
An example of information recorded in an electronic product code

ELECTRONIC PRODUCT CODE TYPE I

01·0000A89·00016F·000169DC0

| Header | EPC Manager | Object Class | Serial Number |
| 8-bits | 28-bits | 24-bits | 36-bits |

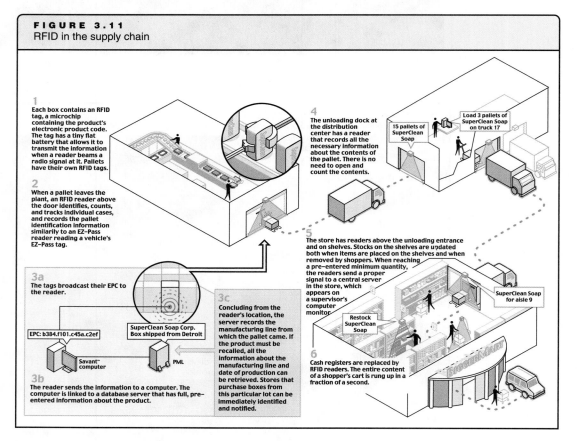

FIGURE 3.11
RFID in the supply chain

1 Each box contains an RFID tag, a microchip containing the product's electronic product code. The tag has a tiny flat battery that allows it to transmit the information when a reader beams a radio signal at it. Pallets have their own RFID tags.

2 When a pallet leaves the plant, an RFID reader above the door identifies, counts, and tracks individual cases, and records the pallet identification information similarily to an EZ–Pass reader reading a vehicle's EZ–Pass tag.

3a The tags broadcast their EPC to the reader.

EPC: b384.f101.c45a.c2ef

Savant™ computer

SuperClean Soap Corp. Box shipped from Detroit

PML

3b The reader sends the information to a computer. The computer is linked to a database server that has full, pre–entered information about the product.

3c Concluding from the reader's location, the server records the manufacturing line from which the pallet came. If the product must be recalled, all the information about the manufacturing line and date of production can be retrieved. Stores that purchase boxes from this particular lot can be immediately identified and notified.

4 The unloading dock at the distribution center has a reader that records all the necessary information about the contents of the pallet. There is no need to open and count the contents.

15 pallets of SuperClean Soap

Load 3 pallets of SuperClean Soap on truck 17

5 The store has readers above the unloading entrance and on shelves. Stocks on the shelves are updated both when items are placed on the shelves and when removed by shoppers. When reaching a pre–entered minimum quantity, the readers send a proper signal to a central server in the store, which appears on a supervisor's computer monitor.

SuperClean Soap for aisle 9

Restock SuperClean Soap

6 Cash registers are replaced by RFID readers. The entire content of a shopper's cart is rung up in a fraction of a second.

The same technology can also be used for other purposes, including detection of items that should be recalled because of hazardous components and accurate condemnation of expired items, such as drugs and auto parts. When a pattern of defects is discovered in a product, RFID helps pinpoint the plant at which it was produced and the particular lot from which it came. Only products from that lot are recalled and replaced or fixed. It does not take too long to determine the particular manufacturing phase in which the defect was caused. When the expiration date of an item arrives, a transceiver detects the fact and alerts personnel to remove the item from a shelf. Packaging of drugs and other items contain RFID tags with unique identifiers. Transceivers can detect whether the products are genuine.

CUSTOMER RELATIONSHIP MANAGEMENT

No commercial organization can survive without selling its products or services. Thus, businesses seek to provide products and services that consumers want—and to entice them to buy what the business produces. Businesses exert marketing efforts to pinpoint demographic groups that are most likely to buy products, to determine features that consumers desire most, and to provide the most efficient and effective ways to execute a sale when a consumer shows interest in the product or service. Because these efforts depend mainly on the analysis of huge amounts of data, ISs have become key tools to conceiving and executing marketing strategies. When marketing succeeds, ISs support the sales effort; to entice customers to continue to purchase, ISs support customer service (see Figure 3.12).

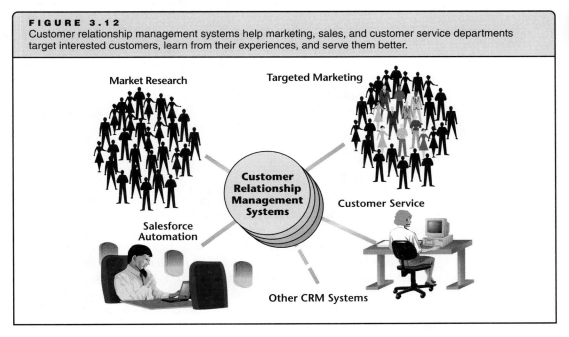

FIGURE 3.12
Customer relationship management systems help marketing, sales, and customer service departments target interested customers, learn from their experiences, and serve them better.

Customer relationship management (CRM) systems are designed to support any and all relationships with customers. Mostly, they support three areas: marketing, sales, and customer service. Modern CRM systems can help capture the entire customer experience with an organization, from response to an online advertisement to automatic replenishment of products to proactive service. With growing competition and so many options available to consumers, keeping customers satisfied is extremely important. Many executives will tell you that their companies do not make money (and might even lose money) on a first sale to a new customer because of the substantial investment in marketing. Thus, they constantly strive to improve customer service and periodically contact anyone who has ever purchased something from them to ensure repeat sales and to encourage customer loyalty. Any information technology that supports these efforts is considered a CRM system, but in recent years the effort has been to combine applications that support all three areas—marketing, sales, and customer service—to better understand what customers want, to be able to collect payment sooner, and to ensure timely shipping.

POINT OF INTEREST

What Happens in Vegas …

Casinos have eagerly adopted RFID technology. They embed tags in betting chips and antennas in betting tables to receive signals from the chips. The antennas transmit to computers that keep track of bets and prevent patrons from cheating. Combined with an optical detector positioned on the table, special software tracks each bettor's behavior and total gamble amounts, as well as how each hand was played.

This is not the only use of RFID in casinos. Bars in casinos and elsewhere in the U.S. suffer from "shrinkage" of $7 billion per year: bartenders overpour, undercharge, or fail to charge at all, and often steal full liquor bottles. To reduce these losses, bars install RFID devices, such as the Beverage Tracker. The Tracker is a spout equipped with an RFID tag and a measuring device. When a bartender pours a drink, the tipping of the bottle turns on the tag and the measuring device. The tag transmits to a computer the bottle's ID, the amount poured, the brand and size of the liquor bottle, and the time of the pour. Managers can better monitor and replenish inventory, and produce accurate bills to patrons organizing banquets.

Source: Jarvis, R., "Casinos Bet Big on RFID," *Business 2.0*, April 2005, p. 26; Swedbert, C., "Vegas Hotel-Casino Uses Tags to Keep Tabs on Liquor," *RFID Journal*, June 22, 2006.

CRM systems also provide an organization with an important element: all employees of the company who directly or indirectly serve a customer are "on the same page." Through their individual computers, everyone has immediate access to the status of an order for an item, a resolution of a buyer's complaint, or any other information that has to do with the customer. All who serve the customer are well-informed and receive the information from the same source. This is especially important in a long, complex sales cycle, because it minimizes response time and improves the quality of service for customers.

Market Research

Few organizations can sell their products and services without promotion; fewer still can promote successfully without market research. Market research systems help to find the populations and regions that are most likely to purchase a new product or service. They also help analyze how a new product fares in its first several months on the market.

Through interviews with consumers and retailers, market researchers collect information on what consumers like and dislike about products. When the researchers collect sufficient data, the marketing department uses statistical models to predict sales volumes of different products and of different designs of the same product. This critical information aids in planning manufacturing capacities and production lines. It is also extremely important for budgeting purposes. When questionnaires are involved, many companies offer Web-based forms instead of paper questionnaires. In some cases respondents use telephones to answer questions after a purchase, usually for a chance to win money prizes. The entered data are channeled into computer databases for future analysis.

Targeted Marketing

To save resources, businesses use IT to promote to people most likely to purchase their products. This activity is often referred to as **targeted marketing**. Great advances in database technology enable even the smallest low-budget business to use targeted marketing. The principle of targeted marketing is to define the prospective customer as accurately as possible, and then to direct

A retailer's sales receipt (right) invites the customer to participate in market research at its Web site with the online questionnaire (left).

OfficeMax

search   account login   cart subtotal: $0.00

Fast, FREE Delivery
on most orders over $50*

supplies   technology   furniture   ink & toner   order-by-number   favorites lists   print services

Stretch Your Tax Refund with **FREE Shipping!**

Concerning your shopping experience...:	Satisfied <					
1. Items are easy to find	●	○				
2. Advertised merchandise is in stock	●	○				
3. Regular merchandise is in stock	●	○				
4. Prices are competitive	●	○				
5. Cleanliness of the store	●	○				
6. Cleanliness of the restrooms	●	○				
7. Copy/Print order completed on time	●	○				
8. Copy/Print order completed correctly	●	○	○	○	○	○
9. Associates are visible on the sales floor and responsive	●	○	○	○	○	○
10. Associates are knowledgeable	●	○	○	○	○	○
11. Overall satisfaction with your shopping experience	●	○	○	○	○	○
12. Copy Center associates provided solutions or offered suggestions for my order	●	○	○	○	○	○
13. Checkout was fast and easy	●	○	○	○	○	○

OfficeMax®

OfficeMax #184
101 IRON LAKE BLVD.
EXTON,PA. 19341
(610) 594-3600

Tell us about your shopping experience
and enter to win 1 of 5 prizes at
www.officemax.com/store/survey
or to enter w/o purch., send a 3"x5"
card with name, address and phone # to
OfficeMax-Shopping Experience
Sweepstakes, 263 Shuman, Naperville, IL
60563. U.S. residents 18+ only. Void
where prohibited.

```
735290102058                    $29.99
TaxCut Premium Federal+Sta

SubTotal                        $29.99
Tax 6.000%                       $1.80
TOTAL                           $31.79

MasterCard                      $31.79
Card number:   vvvvvvvvvvvvvv7454
```

promotional dollars to those people most likely to purchase your product. Perhaps the best evidence of how much companies use ISs for targeted marketing is the use of the Internet for mass communication of unsolicited commercial e-mail, a practice called spamming. Many people loathe spamming, but it is certainly the least expensive method of advertising. Another controversial, but apparently effective, method is pop-up advertising, in which a small window pops up either in front of or behind a Web browser's window.

To define their target markets, businesses collect data everywhere they can: from sales transactions and warranty cards, or by purchasing databases with information about organizations and individuals. Using database management systems (DBMSs)—special programs to build and manipulate data pools—a company can sort and categorize consumers by age, gender, income, previous purchase of a related product, or any combination of these facts and other demographic information. The company then selects consumers whose characteristics match the company's customer profile and spends its promotional dollars to try to sell to those select customers.

The massive amount of personal information that corporations collect and purchase lets them prepare electronic dossiers on the interests, tastes, and buying habits of individuals. The information they possess lets them target "a market of one," namely, an individual rather than a group. Online purchase transactions and online product registrations by consumers provide a wealth of information to corporations. Vendors sort the information to send promotional material via ground mail or e-mail only to those customers whose profiles indicate potential interest.

Telemarketing (marketing over the telephone) makes extensive use of IT. The telemarketer uses a PC connected to a large database, which contains records of potential or existing customers. With a retrieved record displayed on the screen, a marketer dials the number by pressing a single key or clicking the mouse. The telemarketer speaks to the potential buyer while looking at that person's purchasing record with the organization or other organizations. Universities and charitable organizations use the same method to solicit donations.

Computer telephony integration (CTI) is a technique enabling a computer to use the digital signal coming through a telephone line as input in a computer system. It has been used often in marketing, sales, and customer service. For example, some mail-order firms use caller ID to better serve their customers. Caller ID was originally intended to identify the telephone number from which a person calls, but mail-order businesses quickly found a new use for the gadget. They connect it to their customer database. When you call to order, a simple program searches for your number, retrieves your record, and displays it on a PC monitor. You might be surprised when the person who receives your call greets you by name and later asks if you want to use the same credit-card number you used in your last purchase.

Techniques such as data mining take advantage of large data warehouses to find trends and shopping habits of various demographic groups. For example, the software discovers clusters of products that people tend to purchase together, and then the marketing experts promote the products as a combination, and might suggest displaying them together on store shelves. You will learn more about data mining in Chapter 11, "Business Intelligence and Knowledge Management."

With the proliferation of set-top boxes (devices that allow for personal programming and recording for digital televisions), several software companies, such as Visible World, Navic Networks, and OpenTV, have developed applications that may allow television networks to

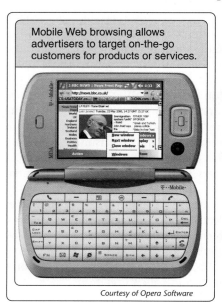

Mobile Web browsing allows advertisers to target on-the-go customers for products or services.

Courtesy of Opera Software

transition from the wasteful and expensive 30-second commercial to more personal advertising. Relying on information provided by households through these interactive boxes, they can select and transmit to each subscriber commercials only for products in which the subscriber is interested. For example, you will not receive commercials about pet food if you do not have pets, but you will receive commercials for gardening products and services if this is your hobby.

Use of information technology for targeted marketing has taken sophisticated forms on the Web. More than just targeting a certain demographic group, Web technologies enable retailers to *personalize* marketing when shopping and buying are conducted online. Special software used by online retailers tracks every visit consumers make and captures their "click streams" (the sequence of selections they make) and the amount of time they spend viewing each page. The retailer's software combines this information with data from online purchases to personalize the pages whenever consumers revisit the site. The reconstructed page introduces information about the products that the individual visitor is most likely to purchase. For example, two people with different purchasing records at Amazon.com who revisit the company's home page will find that they are looking at slightly different versions of the page. Amazon's software custom-composes the elements for each person according to his or her inferred interests in products. The ones that the software concludes might be of the highest interest are displayed or linked on the page.

CRM systems help customer service representatives support customers and learn more about their preferences.

© Manchan/Getty Images

Customer Service

Companies have saved millions of dollars per year by shifting customer service from employees to their Web site. Web-based customer service provides automated customer support 24 hours per day, 365 days per year. At the same time, it saves companies the cost of labor required when humans provide the same service. For example, letting customers pay their bills electronically not only provides convenience but also saves (both customers and companies) the cost of postage and paper and saves the company the time required for dealing with paper documents. Online billing costs only a small fraction of paper billing. The business research firm Gartner estimates the average invoice-to-payment cycle at 41 days, while online invoice and payment shortens the period by at least six days. Customers appreciate the discounts that many companies offer for accepting statements and paying bills online.

Some companies use Instant messaging to help customers online. However, totally self-help is by far the least expensive option for companies that are willing to install the proper software. Another research firm, Forrester Research, provided the following costs per customer contact (called "incident"): phone support session - $33; e-mail - $9.99; Instant messaging ("chat") - $7.80; self-service - $1.17. Considering that about 15 percent of all invoices are contested by customers, shifting service from human help to IT-support help provide significant savings for any company.

Online customer service applications have become increasingly sophisticated. They help track past purchases and payments, update online answers to frequently asked questions (FAQs) about products and services, and analyze customers' contacts with the company to maintain and update an electronic customer profile. The FAQ pages of many companies have been replaced with options for open-ended questions; instead of looking up a question that is similar to what you would ask, you can simply type in your question. Employing artificial intelligence software, the site will "understand" your question and provide a short list of links where you can find an answer.

Salesforce automation increases marketing and sales productivity.

© David Young-Wolff/Photo Edit

Salesforce Automation

Salesforce automation equips traveling salespeople with information technology to facilitate their productivity. Typically, salespeople are equipped with notebook computers that store promotional information for prospective customers, software for manipulating this information, and computerized forms. Many salespeople carry laptop computers or personal digital assistants (PDAs) that contain all the information they need, and which allow them to connect to their organizational information systems through the Internet. Salesforce automation can increase sales productivity significantly, making sales presentations more efficient and letting field representatives close deals on the spot, using preformatted contracts and forms.

Information technology lets salespeople present different options for products and services on the computer, rather than asking prospective customers to wait until the main office faxes or mails the information. At the end of the day or the week, salespeople can upload sales information to a computer at the main office, where it is raw input to the order-processing department, the manufacturing unit, or the shipping and invoicing departments.

Using PDAs that can establish a wireless connection to the Internet enables salespeople to check prices, confirm availability of the items in which a customer is interested, and place an order away from the office. The salespeople can then spend more time on the road, increasing direct contact with prospective customers.

Human resource management (HRM) has become more complex due to the fast growth in specialized occupations, the need to train and promote highly skilled employees, and the growing variety of benefits programs. Human resource management can be classified into five main activities: (1) employee record management, (2) promotion and recruitment, (3) training, (4) evaluation, and (5) compensation and benefits management (see Figure 3.13).

Employee Record Management

ISs facilitate employee record management. Human resource departments must keep personnel records to satisfy both external regulations (such as federal and state laws) and internal regulations, as well as for payroll and tax calculation and deposit, promotion consideration, and periodic reporting. Many HR ISs are now completely digitized (including employees' pictures), which dramatically reduces the space needed to store records, the time needed to retrieve them, and the costs of both.

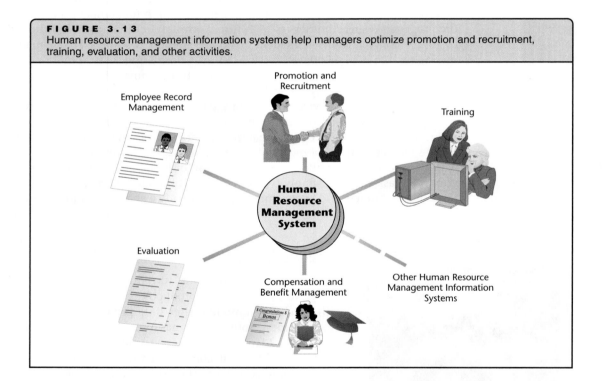

FIGURE 3.13
Human resource management information systems help managers optimize promotion and recruitment, training, evaluation, and other activities.

Promotion and Recruitment

To select the best-qualified person for a position, a human resource manager can search a database of applicants and existing employees' records for set criteria, such as type and length of education, particular experience, specific talents, and required licenses or certifications. Automating the selection process significantly minimizes time and money spent on recruitment, but it does require that a current database be maintained.

Intranets (intraorganizational networks that support Web applications) help HR managers post position vacancy announcements for employees to peruse and consider from their own PCs. This system is especially efficient in large organizations that employ thousands of workers, and even more so at multisite organizations.

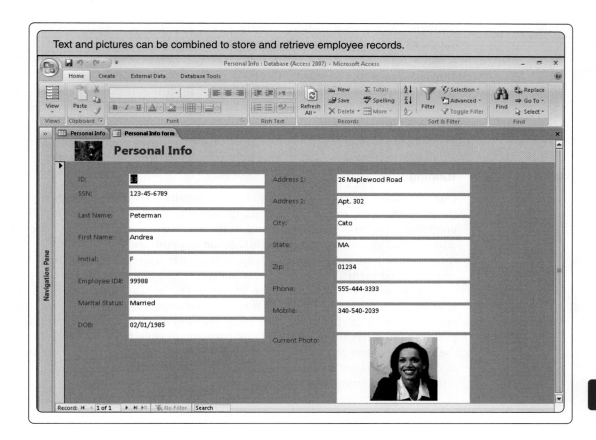

Text and pictures can be combined to store and retrieve employee records.

POINT OF INTEREST

Job Fishing? Start with Proper Baits.

A growing number of companies require an electronic copy of your résumé when you apply for a position. The résumé is added to a database. Human resource managers search the database by keywords. The search yields a list of résumés containing words that match the keywords. Therefore, you can enhance your chances of obtaining the position you want by including as many relevant keywords in your résumé as you can. One effective way to find these keywords is to examine online and print ads for the position you are seeking. Include these keywords—sometimes called "buzzwords"—and increase the chances of your résumé coming up in the recruiters' search.

With the growing number of job applicants, many companies refuse to receive paper applications and résumés. Consider the number of job applications the following companies received in 2006: Google – 1,245,000; Starbucks Coffee – 594,638; Nordstrom – 273,904; Genentech – 265,797. Therefore, it is no wonder that some companies may accept such documents via e-mail, but that others accept only forms that are filled out and submitted online. Using keywords, recruiting officers can then use special software to scour a database for the most-qualified candidates. HR consultants say that this process reduces the time spent on a typical search from several hours to several minutes. Some software companies sell automated recruiting and selection software to support such activities. For example, PeopleAdmin, Inc. offers software by the same name. HR managers save the cost of publishing help wanted ads and can start reviewing résumés as soon as applicants respond online instead of waiting the typical 6–8 days from traditional advertising.

Some companies use the entire Web as a database for their search, which means they include in the search many people who have never applied for a job with them but have posted their résumés. Consider Humana, Inc., a large health-care organization. The company uses software that searches the Web for résumés and then matches qualified candidates with job openings. The company then uses a tracking system to e-mail candidates while updating the corporate HR databases with their résumés. Adopting this approach cut the cost of processing qualifying résumés from $128 to a mere 6 cents. Overall, the new process saves Humana $8.3 million annually and more effectively forecasts the fit of candidates tracked through the recruiting system. Across industries, as companies move from traditional recruiting to recruiting through the Web, the cost per hire drops from $5,000–12,000 to $2,000–5,000, depending on the set of skills required and the level of the position.

Training

One important function of human resource departments is improving employee skills. In both the manufacturing and service sectors, multimedia software training is rapidly replacing training programs involving classrooms and teachers. Such applications include interactive, three-dimensional simulated environments. Some applications contain sophisticated virtual reality components. For example, one such application trains workers to handle wrought iron that must be hammered manually. The worker wears special goggles and holds a hammer in one hand and a piece of metal in the other, over an anvil. The worker "sees" the metal piece through the goggles, "hears" the hitting sound through earphones, and receives a programmed, realistic jolt every time he "hits" the metal. This safely prepares the worker for the dangerous work instead of putting him at risk for injury before he has enough experience to do the actual work. Although the initial investment in multimedia training systems might be high, human resource managers find the systems very effective. Surgeons train using similar systems to operate on virtual patients rather than risk injuries to human patients.

Training software emulates situations in which an employee must act and includes tests and modules to evaluate a trainee's performance. In addition to the savings in trainers' time, there are other benefits. The trainee is more comfortable because he or she controls the speed at which the sessions run. The software lets the trainee go back to a certain point in the session if a concept is missed. Also, the software can emulate hazardous situations, thereby testing employee performance in a safe environment. And if training in a real environment involves destruction of equipment or consumption of materials, virtual reality training applications accomplish the same results in skill enhancement without destruction or waste.

Developments in IT enable organizations to reduce training costs dramatically. Consider CVS, the largest U.S. drugstore chain. The company has more than 17,000 employees it calls technicians, whom it trains continually. The technicians must pass exams to be promoted. In 2000, the company installed PCs at its 400 training sites where technicians can take courses, review, and take exams during breaks or after work. The average cost per trainee was $50. Then, the company put the training materials on CD-ROMs. More than 80 percent of the trainees took the CDs home, so they could learn at their convenience. This approach reduced the average cost per employee to $15. The company also moved the training materials and exams to a central Web site so employees can personalize learning: using a Web browser, they can find the materials they need, bookmark selected Web pages, leave the training session when they wish, and come back to finish it later. When they do finish a training session, they can take certification tests. Their completed tests are then fed into a database at corporate headquarters so that managers can track who is ready to be promoted. The move to the Web reduced the average training cost per employee to $5.

Evaluation

Supervisors must periodically evaluate the technical ability, communication skills, professional conduct, and general behavior of employees. While objective factors are involved in evaluation—such as attendance rates and punctuality—employee evaluation is often very subjective. Assessing performance and effort levels, and their relative weights of importance, varies significantly, depending on who is evaluating. A supervisor might forget to include some factors altogether or might inappropriately weigh a particular aspect of performance. Subjectivity is particularly problematic when several employees are being considered for a promotion and their evaluations are compared to determine the strongest candidate. By helping to standardize the evaluation process across employees and departments, evaluation software adds a certain measure of objectivity and consistency.

In an evaluation, a supervisor provides feedback to an employee, records the evaluation for official records and future comparison, and accepts input from the employee. Software helps managers standardize their employee evaluations by providing step-by-step guides to writing performance reviews, a checklist of performance areas to include in the evaluation (with the option to add or remove topics), scales to indicate how strong the employee is in each area, and the ability to select the relative importance each factor should hold in the overall evaluation. Performance areas include written and oral communication, job knowledge, and management skills, with each topic broken down into basic elements to assist the supervisor in creating an accurate evaluation. A typical application guides the supervisor through all necessary factors and includes a help guide. When the evaluator finishes entering data, the application automatically computes a subtotal for each category and a weighted grade, which can then be electronically stored as part of the employee's record.

Compensation and Benefits Management

ISs help HR officers manage compensation (salaries, hourly pay, commissions, and bonuses) efficiently and effectively. Programs can easily calculate weekly, monthly, and hourly pay according to annual salaries and can include federal, state, and local tax tables to assist in complying with compensation regulations. This same system can also automatically generate paychecks or direct deposits, which are the electronic transfer of funds from the firm's bank account to the employee's.

Special software helps the HR department manage benefits, such as health insurance, life insurance, retirement plans, and sick and leave days, which are determined by seniority, amounts individuals pay into plans, and other factors. To optimize benefits, some companies use special software, incorporating expert systems (ISs that emulate human expertise) that determine the optimal health and retirement plans for each employee based on factors such as marital status, age, occupation, and other data.

Using intranets, many organizations allow their employees to access the benefits database directly and make changes to their preferences, such as selecting another health-care insurance program, or adding a family member as a beneficiary in a life insurance plan. When the company engages a third party for managing pension funds or other benefits, employees can go directly to the Web site of that company, not involving their own company resources at all. By making the changes directly from their PCs, employees reduce the amount of work of the HR staff and decrease the company's overhead costs.

Consumer Privacy

CRM systems, and IT in general, help businesses serve us better at lower costs. But in the process, we may lose some privacy. Consider the following scenario: you agree to give some financial information about yourself to a car dealership to finance the car you have just purchased. At a later date, you provide medical information when you purchase a prescription drug. Your credit-card company has enough information from your purchasing activity to know your culinary and fashion tastes better than you do. Whenever you interact with an organization online, information is recorded; and if you provide personally identifiable information – which is always the case when you make an online purchase – the information is added to your record held by this organization. It may also be recorded by a third party that contracts with the vendor. Finally, without your knowledge or consent, yet another organization gathers all this information and puts it in one big record that is practically a detailed personal dossier.

Organizations collect huge amounts of personal information. Every time you pay with your credit card you leave a personal record; the few details of your purchase are often used to update an already hefty dossier about your buying habits. Every time you provide personal information at a Web site, you either help open a new dossier with an organization or help other organizations update their dossier about you. In their zeal to market more effectively, businesses often violate consumer privacy.

- **What Is Privacy?** In the context of information, privacy is your right to control information about yourself. For example, you maintain your privacy if you keep to yourself your college grades, medical history, or the name of the organization with which you interviewed for a position. Someone who receives such information without your permission is violating your privacy.

- **Business Arguments.** Business leaders argue that they must collect and use personal data. Without personal data, they would have to waste time and money to target likely buyers. They need to know the repayment histories of individuals to help make prudent decisions on extending loans and credit. This ability to purchase and manipulate large amounts of consumer information makes the business world more democratic than it used to be. Small companies now have the same chances of targeting prospective buyers with good credit as

big companies, creating more opportunities and more competition, which eventually benefit consumers.

- **Consumer Arguments.** Consumers usually accept that they must divulge some private information to receive services, but many do not accept the mass violation of privacy. They resent unsolicited mail and e-mail sent by companies who know much about them although they have never provided personal details to these companies. They hate telephone calls from salespeople who obtained their records from companies that were supposed to keep their records confidential. And their greatest concern might be the "dossier phenomenon."

- **Losing Control.** In many cases, you volunteer information in return for some benefits, such as consumer loyalty points or participation in a sweepstakes. In others, you simply cannot receive the service or product unless you agree to provide certain personal details. In such cases, you give implicit or explicit informed consent to obtain information about yourself. However, once you provide information, you have little control over it. With some newer technology, such as RFID, you might not even be aware of who and when information is collected about you. You have just stepped out of a department store with your new clothing purchases. All are RFID-tagged. The store's systems recorded your visit and detailed what you purchased. Can you be sure that nobody else has the proper device to read and record what you purchased? Unless you removed the tag, the serial number embedded in the EPC of the sweater you purchased may uniquely identify you to a stalker whenever you wear it.

- **The Eight Commandments of Personal Data Collection and Maintenance.** In a free, market-oriented society, not allowing organizations to collect personal data is inconceivable. What can businesses do to help protect privacy? They can adhere to these rules to avoid misuse:

  Purpose. Companies should inform people who provide information of the specific, exclusive purpose for which the company maintains its data, and only use the data for another purpose with the subjects' consent. For example, this practice could protect people with a genetic proclivity for certain diseases from higher health insurance premiums.

Relevance. Companies should record and use only data necessary to fulfill their own purposes. For example, an applicant's credit record should not contain membership in political or religious organizations because that information is irrelevant in credit considerations.

Accuracy. Companies should ensure that the personal records they maintain are accurate. For example, many loan applicants have had terrible experiences because some of the data maintained by credit companies is erroneous. Careful data entry and periodic verification can enhance accuracy.

Currency. Companies should make sure that all data about an individual is current. If currency cannot be guaranteed, then data should be discarded periodically. Outdated information can create seriously negative repercussions. For example, a person who might have been unemployable due to past illness might not be able to get a job, even though he or she might be healthy now.

Security. Companies should limit access to data to only those who need to know. In addition to passwords, audit trails (which identify every employee who accesses a personal record and for what purpose) are also very effective tools for ensuring security. Extra caution must be practiced when personal data is accessible online by business partners.

Time Limitation. Companies should retain data only for the time period necessary. For example, there is no reason for a landlord to maintain your credit record after you move out.

Scrutiny. Companies should establish procedures to let individuals review their records and correct inaccuracies.

Sole Recording. When using a recording technology, a company should ensure that no other party can take advantage of the technology to record the same information. For example, if a store records an individual's purchases using RFID technology, it must ensure that the RFID tags embedded in the packaging or items are disabled as soon as the customer leaves the store.

Of course, many consumers will still feel that their privacy is invaded even if every business adopts these "commandments." How can you protect your privacy? Do not furnish your name, Social Security (or any other identifying) number, address, or any other private information if you do not know how it will be used. If you do provide detailed information, indicate that you do not wish the data to be shared with any other organization or individual. You can usually check a box to this effect on paper or Web forms. To avoid junk mail or junk e-mail, again check the proper box on Web forms. Do not fill out any online or paper forms with detailed data unless an opt-out option is available. Of course, many services we receive depend on our willingness to provide personal data, so at least some organizations must have personal information, but you can be selective. Always carefully weigh what you gain against the privacy you might lose.

SUPPLY CHAIN MANAGEMENT SYSTEMS

U.S. Department of Commerce statistics show two important patterns over the past two decades: the fluctuations in inventory as a percentage of gross domestic product (GDP) and the absolute ratio of inventory to GDP have steadily decreased. This means that the U.S. GDP is growing while less and less money is tied to inventory. The smaller the inventory, the more money can be spent on other resources. Much of this trend can be attributed to the use of ISs, especially SCM systems.

According to IBM, companies that implemented SCM systems have reduced inventory levels by 10–50 percent, improved the rate of accurate deliveries by 95–99 percent, reduced unscheduled work stoppages to 0–5 percent, reduced cycle time (from order to collection) by 10–20 percent, and reduced transportation costs by 10–15 percent.

Several enterprise applications, such as ERP systems, also serve as SCM systems. As Figure 3.14 illustrates, many such systems enable managers not only to monitor what goes on at their own units or organization but also to follow what goes on at the facilities of their suppliers and contractors. For example, at any given point in time managers can know the status of the following: an order now being handled by a contractor, by order number; the phase of manufacturing the produced units have reached; and the date of delivery, including any delays and their length. When purchasing parts, managers use the systems for issuing electronic purchase orders, and they can follow the fulfillment process at the supplier's facilities, such as when the parts were packed, when they were loaded on trucks, and when they are estimated to arrive at the managers' floor or the floor of another business partner who needed the parts. You, as a consumer, can get a sense of what SCM systems provide when you purchase a product on the Web and track its shipment status and delivery.

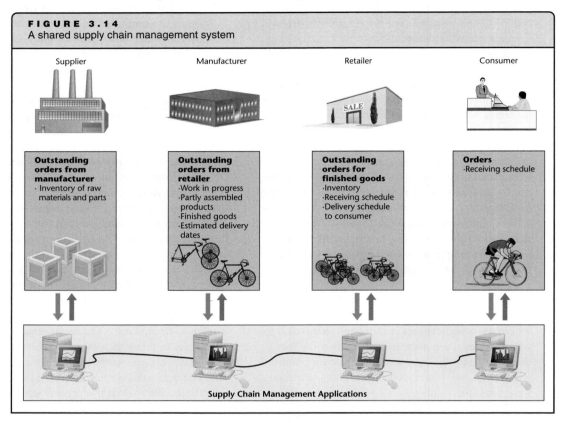

FIGURE 3.14
A shared supply chain management system

SCM applications streamline operations throughout the chain, from suppliers to customers, lowering inventories, decreasing production costs, and improving responsiveness to suppliers and clients. Harnessing the global network, managers can supervise an entire supply chain regardless of the location of the activity—at their own facilities or another organization's, at the same location or thousands of miles away. Older SCM systems connected two organizations. New ones connect several. For example, a distributor can reorder products from Organization A and simultaneously alert Organization B, the supplier of Organization A. The systems let all parties—suppliers, manufacturers, distributors, and customers—see the same information. A change made by any organization that affects an order can affect a corresponding change in scheduling and operations in the other organizations' activities.

Companies that have adopted SCM systems have seen improvement in three major areas: reduction in inventory, reduction in cycle time (the time it takes to complete a business process), and, as a result, reduction in production cost. Companies can reduce their inventory by communicating to their suppliers through a shared SCM system the exact number of units of each item they need and the exact time they need them. In ideal situations, they do not need to stockpile any inventory, saving warehouse costs. The management consulting firm Aberdeen Group estimates that companies using SCM systems through the Internet reduce purchase order processing cycles by 70–80 percent and pay 5–10 percent less for the items they purchase.

The Importance of Trust

SCM systems accomplish the greatest efficiencies when all businesses in the chain link their systems and share all the information that is pertinent to planning, production, and shipment. For example, Chevron, the giant gas and oil company, used to pump oil and deliver as much as it could to gas stations without accurate information about their future needs. Therefore, in many cases its clients, the gas stations, ran out of gas and had to wait for a delivery, or when the tanker delivered, it often had to leave some of the oil in the tank because the gas station tanks reached their full capacity. To avoid unplanned shortages, the company often had to purchase oil in the "spot market" and pay more for it than if it had pumped the oil from its own wells. Both situations cost the company millions of dollars. To avoid such situations, the company linked the gas stations' transaction information systems to its ISs, so that the company can plan drilling and refining based on the gas stations' demand. The increased efficiency saves Chevron much money.

However, not all organizations are willing to collaborate with their business partners. One reason is the fear that when Organization A purchases from Organization B and has access to Organization B's demand figures, it might disclose the information to competitors in an attempt to stir more competition and enjoy favorable prices. Another fear is that if Organization B realizes that at a certain point in time Organization A is in dire need of its raw materials, Organization B might take advantage of the situation and negotiate higher prices.

The first type of fear can be found in initial reluctance of suppliers to share information with large buyers such as Wal-Mart. Only the bargaining power of Wal-Mart and its insistence on sharing such information convinced suppliers to link their systems with those of Wal-Mart. The second type of fear still exists between General Motors and its main tire supplier, Goodyear. Goodyear could enjoy lower inventories if it had GM's demand schedule for tires. It could then calibrate its own order for raw materials and its manufacturing capacity to suit those of GM, save money, and pass at least some of the savings to its client in the form of cheaper products. It could always replenish its client's inventory of tires before GM ran out of them. Better yet, it could deliver the tires directly to the assembly lines just when they are needed, saving both GM and itself warehousing costs. Yet, GM is guarding its production schedule as confidential.

Thus, effective supply chain management between companies is not only a matter of appropriate technology but also a matter of trust and culture change. So far, most of the successful collaborations have been between a large company and its business partners, whereby the company uses its power to dictate collaboration. However, some large companies have tied their SCM systems out of mutual understanding that this would benefit both companies, even if the shared information reveals some unpleasant facts. For example, Procter & Gamble, Inc., the giant supplier of household products, has had its systems connected to those of Wal-Mart since 1987, when the term "supply chain management" was not even in use. By providing its retail information to P&G, Wal-Mart ensures that it never runs out of P&G products. A culture of sharing—you show me some of your information and I show you some of mine—is essential for the success of both companies and creates a sense of mutual dependence and true partnership.

SCM systems can be taken a step beyond the sale. The systems can be used for after-the-sale services. For example, Beckman Coulter, Inc., in Fullerton, California, makes blood analyzers and other medical devices. After it sells a machine, the company uses the Internet to link the machine from the client's facility to a computer in its Fullerton factory. Software on the computer runs 7 days per week, 24 hours per day to monitor the sold machine. When a problem occurs, the computer alerts a Beckman technician, who can repair the machine before it stops working. Beckman estimates that the system provides savings of $1 million annually, because malfunctions are captured at an early stage, which avoids the higher cost of fixing a more damaged machine. The added benefit is increased customer satisfaction. As business partners see the benefits of sharing data, trust has grown, and the fear of linking IS to those of other organizations is waning.

The Musical Chairs of Inventory

Recall the wonderful trend cited in the beginning of this section: the dollar value of inventory for U.S. businesses was growing at about 60 percent of the growth of GDP. However, much of the trend took place in the 1981–1991 period. In the 1992–2004 period, inventory as a percentage of GDP stayed fairly similar at about 3.5 percent. Apparently, while large corporations have the resources to install and run SCM systems to cut their own inventory, the ratio of inventory to revenue in small enterprises is growing because they do not use such systems. And sometimes, the companies that suffer the inventory ripple effect are not small. They might be powerful, and they even might manage their own SCM system, but their system might not be linked to their buyers' systems, so they cannot plan their production to reduce inventory.

For an example, let us return to the relationship between General Motors and Goodyear. The world's largest auto manufacturer improved "inventory turns" 55.2 percent between 1996 and 2001. Inventory turns is the number of times a business sells (or "turns over") its inventory per year. It is calculated by dividing the sales revenue by the average value of inventory. The greater the inventory turns number, the better. During the same period, Goodyear, GM's tire supplier, experienced a 21 percent decrease in inventory turns. The likely conclusion is that GM avoided purchasing tires from Goodyear until it needed them at the assembly line, but Goodyear did not have enough information on when, exactly, those tires would be required, and therefore kept overstocks. Had the SCM systems of the companies been linked, Goodyear could reduce inventory and see its inventory turns rise rather than fall. It is also reasonable to assume that, due to the cost savings, Goodyear would be able to sell tires to GM at a lower price. GM and other companies have created a situation where each company tries to "sit" with a lean inventory while, inadvertently, leaving another "standing" with an overstock. In order for all involved in a supply chain to enjoy efficiencies, the musical chairs, or "hot potato" situation, must stop.

Collaborative Logistics

The Web enables organizations from totally different industries to streamline operations through collaboration. In recent years an increasing number of businesses found a new way to cut shipping costs: they combine freight with other businesses, sharing their own trucks or the vehicles of trucking companies. The collaboration reduces partially empty trucks, or empty trucks between stops. To this end, the companies connect their SCM systems to the site of a company that specializes in optimization of logistics, such as Nistevo Corporation. The company manages the site and uses sophisticated software to calculate the shortest routes between departure and arrival points and the best combination of loads from two or more companies to share trucks and routes. The SCM systems of subscribing companies provide daily data into the shared system. The IS takes into consideration the type of freight to ensure safety and adherence to regulations. For example, the software is designed not to combine chemicals with food. Therefore, typical allies of food manufacturers have been paper manufacturers, for instance. The cost savings have been impressive.

The spice maker McCormick & Co., Inc. has reduced freight costs by 5–15 percent, while General Mills has realized savings of up to 7 percent of its overall logistics costs. Manufacturers of household paper products such as Georgia-Pacific and International Paper Co. share about 80 long-distance routes with General Mills on a regular basis, cutting freight costs for those shipments 5–20 percent. Because the success of collaborative shipping is so impressive, some experts expect competitors to share trucks, leaving competition to some other areas of their operations, such as development and manufacturing processes.

Another area where some companies have explored collaboration is warehousing. The principle here is the same: try to maximize the use of warehouse space, and if you cannot use all of it, allow other businesses to use the extra space. The way to accomplish this, again, is through the Web: a third party specializing in warehousing optimization combines warehousing needs and availability from member companies to offer optimal solutions.

ENTERPRISE RESOURCE PLANNING

A growing number of organizations elect to replace old, disparate ISs with enterprise applications that support all or most of the business activities we have described. As mentioned before, these systems are often referred to as **enterprise resource planning (ERP)** systems, although they are used not only for planning but also for managing daily operations. Designers of ERP systems take a systems approach to an enterprise. For example, the Manufacturing Resource Planning component of the system uses the information recorded on a sale to retrieve product specifications; the data is used to generate purchasing information such as items, quantities, and the timetable for suppliers to deliver for the purchasing department. As products are manufactured, the system tracks the stages of the work in progress. When items are ready to be shipped, the shipping department can retrieve information on the items for its operations. The system keeps shipping information such as content and destination, along with billing information, to help produce shipping and billing documentation. The system also records financial transactions involved in these activities, such as payment made from a bank account. The accounting component records the transactions. In addition, ERP systems also provide human resource modules for payroll, employee benefits management, and employee evaluation software. CRM components are also available and are tied to other components through orders applications and sales records. In terms of revenue, in 2007, the ERP market was divided among four vendors: SAP (36 percent), Oracle (24 percent), Microsoft (23 percent), and Sage Software (17 percent). SAP and Oracle have been the leaders in this field for several years. Oracle's market share has increased mainly through acquisitions of other ERP developers, such as J.D. Edwards and Siebel Systems.

Challenges and Disadvantages of ERP Systems

With successful ERP implementation, organizations can reap substantial rewards. However, ERP systems pose many challenges. The software packages are quite complex. Because they are not tailored to the needs of specific clients, they often require adjustment and fine-tuning for specific organizations. Therefore, their installation and testing involve experts who are usually employees of the software vendor or professionals who are certified for such work by the vendor.

Even with adjustments—often called "tweaking"—potential adopters must remember that the system was designed for an entire industry, not for the way an individual organization does business. If the organization has a competitive advantage thanks to a unique set of business processes, this advantage may diminish or disappear when the system is installed, because to a large degree the system dictates how business processes should be conducted. The system requirements are quite rigid, and therefore customization of ERP systems is limited.

ERP applications are expensive; modules cost millions of dollars. Buyers usually must allocate several more million dollars to pay for installation and modifications. Installation often takes many months to complete, and budget and time overruns are common.

The greatest advantage of ERP systems, the integration of many business processes, may become a challenge for the adopter. Because the operational lines between business units become blurred, there may be arguments over responsibility and accountability when something goes wrong. For example, the sales department may argue that the responsibility for an erroneous invoice is the accounting department's or even of a manufacturing unit that entered incorrect costs for the order. Also, a process that becomes a weak link in the supply chain may negatively affect other processes.

Implementation of ERP systems can fail because of formidable challenges: the gap between system capabilities and business needs, lack of expertise on the consultant's part, and mismanagement of the implementation project. The business research firm Standish Group found that only 10 percent of ERP implementation projects are completed as planned, on time, and within budget. Fifty-five percent are completed late or over budget (which usually means loss of business and revenue), and the other 35 percent of such projects are canceled because of difficulties. At Hewlett-Packard, one of the world's largest computer and IT equipment makers, a $400 million loss in the third quarter of 2004 was blamed on poorly managed migration to a new ERP system. Previous cases of difficult implementations of ERP systems are Hershey Foods and Nike. In both cases, the adopters blamed losses of hundreds of millions of dollars on late completions of ERP system installations. In one case, an industry leader was bankrupt as a result of unsuccessful implementation of an ERP system.

In 1995, FoxMeyer Health was the fourth largest pharmaceuticals distributor in the U.S., with annual sales of $5 billion. It was an early adopter of an SAP R/3 ERP system. It spent $100 million on the project. Counting on the new system to help increase the capacity to handle orders, it bid on and won a $1 billion per year contract with University HealthCare, a consortium of teaching hospitals. While the company's older system could process 420,000 orders per night, the new ERP system could process only 10,000. The new system caused other problems. The main distribution center in Ohio mishandled orders worth millions of dollars. FoxMeyer lost millions of dollars, partly because it could not collect payments on many orders. In 1996, the company filed for bankruptcy. In 1997, its assets were sold for $80 million to its archrival, McKesson.

Providing the Missing Reengineering

In our discussion of reengineering of business processes in the previous chapter, we noted that in the 1990s, most reengineering projects failed. Interestingly, in the late 1990s and early 2000s, ERP systems helped realize many of those reengineering ideas because the systems forced changes in processes. At the least, ERP systems integrated information from various organizational units, resulting in less labor, greater accuracy, and shorter cycles.

ERP systems also help organizations move away from the traditional silos of functional units to business processes, an approach that helps many of them operate better. Suppliers and customers do not care whose responsibility it is to take care of their orders and payments. Therefore, organizations are better off planning and managing processes rather than organizational units. Despite the risks and the high costs involved, a growing number of companies adopt ERP systems.

- Effectiveness is the degree to which a task is accomplished. The better a person performs a job, the more effective he or she is. Efficiency is measured as the ratio of output to input—the greater the ratio, the more efficient the process. ISs can help companies attain more effective and efficient business processes. Productivity is the measure of people's efficiency. When people use ISs, their productivity increases.

- ISs have been integrated into almost every functional business area. In accounting and payroll, because of the routine and structured nature of accounting tasks, the systems automatically post transactions in the books and automate the generation of reports for management and for legal requirements.

- Financial ISs help managers track cash available for transactions, while ensuring that available money is invested in short- or long-term programs to yield the highest interest possible. Investment analysis ISs help build portfolios based on historical performance and other characteristics of securities.

- Computer-aided design (CAD) systems help engineers design new products and save and modify drawings electronically. Computer-aided manufacturing (CAM) systems direct machines in manufacturing parts and assembling products.

- Supply chain management systems optimize workload, speed, and cost in the supply chains for procurement of raw materials, manufacturing, and shipping of goods. ISs, especially MRP and MRP II systems, facilitate production scheduling and material requirements planning, and shorten lead time between idea and product. Shipping ISs help speed up delivery and cut costs. RFID technology helps promote and operate supply chain management (SCM) systems. Radio frequency identification (RFID) tags carry product information that can be tracked and updated.

- Customer relationship management (CRM) includes the entire cycle of relationships with customers, from marketing through sales to customer service. CRM ISs collect information about shoppers and customers and help target the most likely buyers of a product or service. Online customer service systems help customers help themselves via the Web 24 hours per day, 7 days per week, and save the company labor and telephone expenses. Salesforce automation allows traveling salespeople to spend more time with customers and less time in the office.

- Human resource management systems expedite staff selection and record-keeping. An increasing amount of recruiting is done via the Web. Managers often use evaluation software to help assess their subordinates' performance. Employees can use expert systems to choose health care and other benefits programs that best suit their situation.

- Companies can link their SCM systems to monitor the status of orders at their own facilities but also at those of their business partners, usually their suppliers. Such cooperation can create further efficiencies, but it requires a high degree of trust between organizations.

- Rather than use disparate ISs for business functions, many organizations opt to install a single system that encompasses all their business processes, or at least the major ones. They employ enterprise resource planning (ERP) systems to support their supply chain management and customer relationship management. Installation of ERP systems is expensive and challenging, and often involves budget and time overruns.

GARDENERS+ REVISITED

Gardeners+ has grown in the last three years from a small office/home office (SOHO) operation to a business with a high profile store in the mall, a large variety of services, hundreds of employed and affiliated gardeners, and over 2000 clients. The three partners noticed that their activities were becoming more specialized as the business grew and that they need information systems to support those activities. Help them sort through their systems.

What Would You Do?

1. Using the classifications in this chapter, identify the business functions within *Gardeners+*. Which information systems do Mary, Amanda, and Ed use now to streamline their operations? What other applications could they use?

2. Do you think that *Gardeners+* should invest in an ERP system at this time? Why or why not? If not, which types of information systems mentioned in this chapter would be appropriate short of an ERP system?

New Perspectives

1. A large guild of plumbers learned about the *Gardeners+* model and systems and thought that it could well work for them. They offered to license the model and the system from *Gardeners+*. The changes in the system itself would be minor. What information could the three entrepreneurs use to help them decide whether this opportunity is worth pursuing? Suggest how they could obtain this information.

2. Mary thought that instead of licensing the system, the plumbers could join *Gardeners+*, which would then offer both types of service (gardening and plumbing). How can they compare these two options (licensing the system versus integrating plumbing services into the current organization)? What are the pros and cons of each option, from both a business and a technical perspective?

KEY TERMS

bill of materials (BOM), 86
brainstorming, 83
cash management system (CMS), 82
computer-aided design (CAD), 83
computer-aided manufacturing (CAM), 83
customer relationship management (CRM), 78
economic order quantity (EOQ), 86
effectiveness, 78

efficiency, 78
electronic funds transfer (EFT), 82
electronic product code (EPC), 90
enterprise resource planning (ERP), 104
just-in-time (JIT), 87
manufacturing resource planning (MRP II), 87
master production schedule (MPS), 87

material requirements planning (MRP), 86
productivity, 78
radio frequency identification (RFID), 90
rapid prototyping, 83
supply chain, 85
supply chain management (SCM), 85
targeted marketing, 92
time to market, 83
work order, 81

1. What is a supply chain? What is the purpose of supply chain management systems?

2. What is the purpose of cost accounting ISs?

3. What is the relationship between CAD and CAM systems?

4. What are the concerns in cash management, and how do cash management ISs help financial managers?

5. What is time to market? How have ISs affected time to market?

6. In brief, what is the purpose of customer relationship management systems?

7. What are the typical components of ERP systems?

8. Although technologically the full linking of the SCM systems of suppliers and buyers is feasible, many buyers are reluctant to do so. Why?

9. Why do the ERP installation and testing of systems require that experts be involved? Why does the implementation of so many ERP systems face severe challenges or totally fail?

10. What is EOQ? Which two problems do ISs that calculate EOQ help minimize?

11. What is JIT? How do MRP and MRP II systems help achieve JIT?

12. For the human resource managers of some organizations the entire Web is a database of job candidates. How so?

13. What information technologies play a crucial role in marketing?

14. Many sales reps have no offices, yet they have access to huge resources, and their productivity is great. Explain how that is possible.

15. What is RFID, and what role does it play in SCM?

16. In the supply chain, shipping software helps mainly in two ways. What are they?

17. You established a small shop that manufactures a single product that you sell by mail. You purchase raw materials from several vendors and employ five full-time employees. For which business functions would you certainly use software?

18. Which of the ISs you listed for Question 17 would you link to each other, and for what purpose?

19. Why is it so important to have a quick response of online investment ISs? Give two examples of how such systems are critical.

20. Some experts say that ISs have great potential in manufacturing. Explain why. (*Hint*: Consider business process reengineering.)

21. Over the past decade, banks and investment firms have offered many services that would be impossible without ISs. Describe three such services and explain how IT makes them possible.

22. CAD systems replace older, manual tools in engineering, but they also contribute by maintaining all information in electronic form. How does this facilitate the work of draftspeople and engineers? How do such systems help the transition from engineering a product to manufacturing it?

23. ISs in both the manufacturing and service sectors often help to *optimize*. Give two examples of what they optimize.

24. The Web has significantly cut the cost of collecting data about shoppers and buyers. Explain how.

25. Sellers of consumer products argue that targeted marketing serves not only them but also their consumers. How so?

26. If you had to evaluate your own subordinates, would you prefer to evaluate them in written, open-ended form, or would you prefer to use employee evaluation software? Why?

27. As an employee, would you prefer that your supervisor evaluate you with the aid of employee evaluation software or without it? Why?

28. Try to remember the last time you gave someone your personal data, such as an ID number, e-mail address, or a physical address. What was the reason for asking for the data? Do you know how the data will be used by the receiver?

29. Some consumer advocates argue that organizations should pay every individual whenever they sell data about him or her to another organization. (They suggest 5 or 10 cents per sale.) Do you agree? Why?

30. Examine the list of precautions suggested in "Ethical & Societal Issues" for ensuring minimum invasion of privacy when businesses use personal data. Which steps can be taken without, or with minimal, added cost? Which steps would impose financial burdens on businesses? Why?

31. RFID tags are increasingly embedded in almost every type of good, from soda six-packs to clothing items. Consumer advocates fear that the technology might cause massive violation of privacy. Describe at least two ways in which this can happen. What controls or limitations would you impose on RFID tags and use to minimize the fears of invasion of privacy?

APPLYING CONCEPTS

32. You are the CEO of a company that runs 2 plants, manufactures 12 different products, and sells them in 15 world regions. List all the items of information (totals, metrics, etc.) that you would like to know at least on a quarterly basis. State which information can or cannot be obtained through company operated ISs and why.

33. Choose three distinct but related business functions (e.g., inventory control, purchasing, payroll, accounting, etc.). Write a short paper describing how interfacing the information systems of these three functions can improve an organization's performance.

34. Select a business process (possibly at a local firm) not mentioned in this chapter. Write an essay explaining how IS technology could make the process (1) more efficient and (2) more effective.

35. Write a three-page essay titled "Factory of the Future." Your factory will not require anybody in the manufacturing organization to enter any data into information systems. All the necessary information will come from customers at one end and suppliers at the other end. There will also be no need to type in any data for payments and collections. Explain how all this will work.

HANDS-ON ACTIVITIES

36. Many companies use e-mail to advertise their products. Your company is trying to sell a new product and is advised to use e-mail. All the e-mail addresses are of people who have agreed to receive promotional e-mail about products such as the one you try to sell. The profit on each unit sold is $200. Developing the attractive e-mail message, use of 2,750,000 e-mail addresses, and sending the message would cost $25,000. Experience shows that 5 percent of the initial recipients forward such messages to friends and family. Experience also shows that 2 percent of all recipients actually click the Web address included in the message and visit the commercial site. Of these visitors, 0.5 percent end up purchasing the advertised item.

Using Microsoft Excel or another spreadsheet, answer the following questions: (1) Would you generate a profit if you used this advertising opportunity? (2) Would you profit if you could e-mail only 1,000,000 people?

37. Form a team and design an IS for a small business that sells manufactured parts to other businesses. The system must handle customer order processing, sales, salesperson commissions, billing, and accounts receivable. Prepare a report describing the system's different components and their points of interface. What files are necessary? How will the business use data in each file? If you have command of Microsoft Access, create the tables for the above objects, and populate each one with three to five records.

38. Assume that you and your teammates are about to start a Web-based business for sporting goods. You wish to e-mail information to potential customers. Determine the demographic characteristics of your target audience. Search the Web for companies that sell consumer data that can serve you. Prepare a report about three such companies: their names, services, and prices (if available).

FROM IDEAS TO APPLICATION: REAL CASES

Winning the Bet

International Game Technology (IGT) is a leading manufacturer of slot machines and lottery machines for casinos and government lotteries. Headquartered in Reno, Nevada, with sales headquarters in Las Vegas, the company also maintains sales, manufacturing, and service sites in Africa, Australia, Europe, and South America. Its Reno site alone produces 140,000 machines annually. It has been profitable for many years. In 2005, it had a profit of $437 million on revenue of $2.4 billion, apparently a situation that would lull executives of other companies to think "If it ain't broke, don't fix it." Not IGT managers.

Until 2002, each business function had its own information system. IGT had different systems for handling sales, customer orders, manufacturing, and accounting. When managers wanted to receive information about a specific customer order, they had to go to each functional unit to receive a different piece of the information: customer details from the sales department, status of the machines being manufactured from the manufacturing units, and payment status from accounting. The accounting department itself had several software applications that handled different books, such as accounts receivable, accounts payable, and the general ledger.

As business was growing, managers complained that they could not get comprehensive information on orders. The IT department developed interface software to connect the systems, but there were still complaints that information was not coherent. The IT specialists admitted that they were maintaining a mishmash of software. The loudest complaints came from the accountants. Every year it took them two weeks "to close the books."

The accounting department pressured management to purchase a new system that would make their work more efficient. The CIO understood their plea but was afraid that satisfying this department's request would trigger similar requests from other units, such as engineering and manufacturing. The result might be a better information system for each department, but disparate systems that still were not connected to each other. On the CIO's advice, IGT management decided to implement an ERP system.

A steering committee and project team were assembled. Their members focused on business functionality rather than the technology. After the first selection, systems from three companies were considered: SAP, Oracle, and J.D. Edwards (which was later acquired by Oracle). After further consideration,

SAP won the contract, and IGT embarked on a two-year effort. In 2003, the company switched to using the R/3 ERP system. IGT did not disclose the cost of the project, but analysts estimate it was well over $10 million.

When the system was ready, three functions were incorporated into one enterprise system: product development, manufacturing, and finance. Like other ERP systems, R/3 is highly structured even when modified for a particular customer. As often happened, the new system forced IGT to change some of its business processes. However, the company chose SAP's system because it found it less rigid than other ERP systems. This was important to IGT, because it builds machines to order.

The system afforded the company several benefits. Price proposals are made based on more accurate information and estimates. Managers on the manufacturing floor can view or print out manufacturing process sheets at their own PCs. Employees can no longer ignore specifications or "cut corners." The system does not allow a process to continue when an attempt such as this is made. The products are made more efficiently and with fewer errors. The system connects all of the company's sites around the globe. One of the system's modules is project management, which enables managers to monitor design changes and costs involved in new product development.

The new system replaced the old MRP (material requirements planning) system, but the company still uses its internally developed factory control system, which has been successfully integrated into the SAP system. The factory control system enables managers to know which machines are built at which plant.

IGT reduced the average period of order to shipping from 9–10 weeks to 7–8 weeks. When a rush order is entered, IGT can now fulfill it in four weeks instead of seven weeks. Between 2002 and 2005 the error rates in orders for raw materials decreased from 10 percent to almost 0. Inventory turn increased from 6.3 to 8.4 percent per year.

IGT's CIO admits that the implementation was challenging. The company makes a variety of machines, which meant that many bills of materials had to be entered into the system (and new ones will have to be entered for new products). Adapting some features to the way IGT operates was not easy. However, the implementation was successful. The CIO credits the success to strong support from senior management, the establishment of a steering committee with members from all affected units, a capable project management team, a training program to help employees understand how to use the new system, and the rigorous testing the system underwent before it was used.

Source: Bartholomew, D., "ERP: Gaming Company Hits Jackpot," *Baseline*, October 2, 2006; (www.igt.com), 2007.

Thinking About the Case

1. What problems did IGT face before the implementation of the ERP system?

2. How does the new system help control processes?

3. Compared to the situation in 2002, what are the benefits of the ERP system?

4. IGT decided to continue operating its older factory control system. Why do you think it did so?

Resort to CRM

Maintaining the same size of customer service staff while transaction volume triples is quite a challenge, especially when much of the business depends on direct contact with individual consumers. With the proper technology, ResortCom succeeded in doing just that.

ResortCom International of San Diego, California, provides services to companies that develop and own resort properties. Services include managing time-share payments, loans, and handling credit-card transactions. Time-share corporations build resort sites throughout the world for individuals who wish to use them for vacations. For example, a family may buy a week's stay at a site, or select a week at one of several sites the company manages. The contract is usually for many years. In addition, property owners sell customers ancillary services, such as car rentals and recreational activities. Prompt, courteous service is extremely important in this industry.

Instead of dealing directly with sales and billing, many of these corporations hire the services of companies like ResortCom to bill customers and answer their questions. Property owners outsource this work to ResortCom because the firm specializes in maximizing the revenue (often called yield) and the owners do not have to invest in information technology to handle sales, billing, and collection.

ResortCom serves more than one million traveling customers. Every year, between October and January, ResortCom mails about 100,000 bills to customers— actually, customers of its clients. Hundreds of customers respond with questions about the bills, and many contest them. Most of these responses are via e-mail. Since the number of ResortCom clients has increased, so has the number of these e-mail messages. For example, in the first week of January 2006 it received 450 e-mail messages. In the same week of 2007, the number was 750. Some of the issues raised can be resolved immediately by the call center, but many must be forwarded to another department. Many such messages were treated after too long a time or were simply lost in the shuffle.

By 2004, ResortCom consolidated two phone call centers. It still had to resolve the e-mail problem. The company's vice president of operations looked for a proper CRM software package. While browsing the Web he noticed the site of RightNow Technologies and tried the online demo. He liked the manner in which the system tracked customer inquiries from origination to resolution.

As in other industries, communication with customers is done in several channels: mail, telephone, fax, and e-mail. RightNow's CRM software provides a complete multi-channel contact center. Regardless of how customers contact ResortCom, their communication is channeled to a central database. Each entry receives an incident code, including scanned mail and electronically saved faxes.

Not only members of the call center staff have access to the incident center information, but all members of the firm, including units often referred to as "back-office." (Back-office staff are those who are not in direct contact with customers.) If a customer disputes a credit-card charge, the call center staff communicate the incident to the finance office The call center personnel can check at any time to see if the finance office finished researching the dispute. The system automatically sends a periodic reminder to the finance office—or any other unit responsible for researching an incident—until the incident is resolved.

The CRM system ensures that nobody in the organization "drops the ball" as incidents are transmitted from one unit to another. It enhances accountability. This accountability gives each unit an incentive to resolve matters faster, and therefore the service cycle has been shortened. Now, customer issues are resolved 75 percent faster without additional staff. Every action that has taken place in the resolution process is recorded, and managers can follow the process from beginning to end. ResortCom's internal e-mail decreased by 30 percent; instead of e-mailing other employees for status, managers access the CRM system directly.

Another benefit of using the CRM system is reducing paper by 90 percent. Since so much of the information is now saved electronically, there is no need to create or maintain paper documents. The scanned images of paper mail and fax are attached to incident records. In time, thanks to cumulative information about individual customers, that relationship with each customer becomes more intimate.

The CRM system maintains data that makes it amenable to further analysis. ResortCom can use RightNow marketing tools to gain insight into customers' needs and preferences. This helps ResortCom to launch effective target marketing campaigns. The company has made several presentations to their clients, the resort site owners, to demonstrate how efficient and effective the CRM system is. They demonstrate how customer loyalty increased, and how the CRM enables Resort-Com to up-sell and cross-sell. (Up-selling is moving customers to purchase more expensive services. Cross-selling is selling additional services.) The demonstrated success of using the CRM system encourages the site owners to continue doing business with ResortCom.

Source: (www.rightnow.com), 2007; Watson, B.P., "Getting Out of the In-Box," *Baseline*, March 8, 2007.

Thinking About the Case

1. What were the challenges ResortCom faced before adoption of the CRM system?

2. Compared to the situation before the system was implemented, what are the benefits RightNow afforded ResortCom?

3. How did the company reduce the amount of paper used by 90 percent?

4. In addition to using fewer paper documents, what are the advantages of using electronic records?

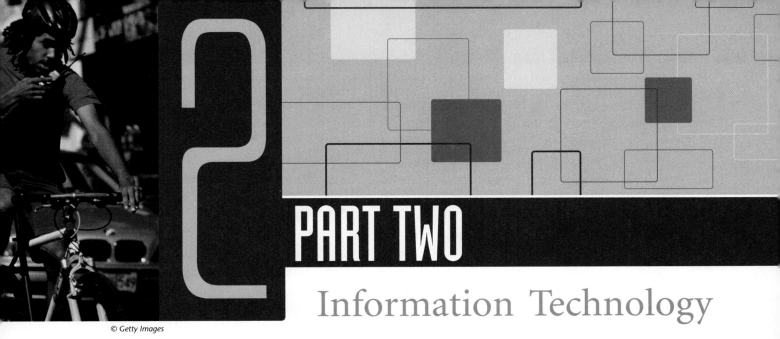

© Getty Images

PART TWO

Information Technology

CASE II: QUICKBIZ MESSENGERS

Andrew Langston looked out of his office window and smiled when he saw another of his bike messengers pedal in from a delivery. Had it really been a decade since he began QuickBiz Messengers? He'd come a long way from his early days in the business, when he got a phone call, hopped on his bike, and made the deliveries himself.

During college, Andrew competed in the cycling club's races. A friend told him that he worked part time for the local bicycle delivery service to keep in shape, so Andrew decided to sign up to earn a little extra cash. That was how he'd learned the ropes of the messenger delivery business. His employer had been operating for a long time in the city's central business district, and working there gave Andrew a taste of a different career option. After graduation, Andrew moved back to Seattle, his hometown, and started QuickBiz. It was the best way he could think to combine his love of cycling with the need to earn a living. Besides, at the time, Seattle had only a handful of small messenger services.

It was slow going at first. With such a small business and few funds, he had to watch every penny. But timing had helped him survive. With the business boom in the 1990s, the pace of business

transactions skyrocketed. Firms of all sizes needed additional services to carry out their day-to-day transactions, so deliveries needed to increase, too. "Instant service" became the watchwords of business in the Information Age. Meanwhile, traffic on Seattle's streets had grown heavier. Delays throughout the metropolitan area became a frustrating fact of life. Andrew found he could zip by the cars in downtown traffic as if they were parked—delivering his packages on time. He was proud that he'd built his business on a reputation for reliability. Now, here he was, president of a company with nearly 90 employees making deliveries by both bicycle and car. He'd met each challenge with the determination he'd had when he was racing. And as with his bicycle, he tried to keep his business running smoothly, although it didn't always run the way a well-oiled machine should.

Bumps in the Road

There was the time that a quickly opened door of a parked car had flattened one of his first messengers and landed him in the emergency room. With no way to communicate except a pager, Andrew didn't know where his messenger was until he regained consciousness and had the nurses call him. Andrew spent the afternoon worrying, calling local police stations, and trying to placate his customer about her missing delivery. Also, he remembered the time high winds had whipped up huge waves, which washed over the I-90 floating bridge. No traffic—including his car messengers—could get through for a day. QuickBiz had no system for traffic alerts then, so some messengers were stranded in the backup. The addition of cell phones and e-mailed delivery notices had certainly helped him maintain better contact in the field. Now if a messenger didn't arrive on time, he knew it sooner and could check the problem out directly.

Early Expansion and Growth

QuickBiz expanded rapidly over its first few years as demand grew for its services. Businesses found it cheaper to use a delivery service than to waste their employees' time running across town to make deliveries. The price for the service was another advantage—customers could get same-day delivery at prices much lower than the large package delivery services could offer.

As QuickBiz grew, Andrew gradually added staff to his payroll—both messengers and dispatchers—to handle repeat customers and routine route deliveries. The company served a variety of businesses: law firms needing contracts signed or papers filed, architects sending plans to their clients, medical and pharmaceutical suppliers who needed rush deliveries, public relations firms sending their copy to poster and sign suppliers, and other businesses needing quick deliveries to satellite offices, suppliers, or clients.

Andrew set up routes within the main business district to handle his regular customers' needs. He also accepted requests for special deliveries from drop-off or call-in business. Standard delivery was 2-hour service, with premium rates for faster service. If a business only needed same-day service,

then it could opt for the economy rate. QuickBiz made deliveries year-round, in any kind of weather, which in Seattle usually meant rain or occasional snow. Regular service operated Monday through Friday, from 7 a.m. to 7 p.m. During the high-tech boom, QuickBiz also added premium service delivery on Saturdays.

Moving Beyond Bikes

After a few years of building QuickBiz's clientele, Andrew noticed that revenues began to plateau. His competitors were offering the same type of service, and there was only so much business to go around. He needed to think of some way to separate his business from the pack—and soon.

In looking over the customer feedback his messengers entered into their report forms, Andrew saw patterns emerging. Messengers said several of his customers that had satellite offices outside downtown and in nearby towns in the Puget Sound area had requested expanded routes. He also had repeated inquiries to serve several art galleries in the area. Handling fragile art glass and other one-of-a-kind, irreplaceable items definitely called for a safer delivery method than bicycles. So Andrew investigated the feasibility of adding car and truck deliveries to his business and decided to make the move.

Maintaining a fleet and drivers took the business to an entirely new level, but it also allowed QuickBiz to deliver a wider and more profitable range of services—deliveries no longer had to fit in a backpack or bike basket. Ultimately, adding automobile service allowed QuickBiz to double in size. The company now made about 700 deliveries per day and generated revenue of roughly $1.5 million annually.

With the addition of auto service, Andrew needed to develop new pricing scales and schedules. He used his financial information system to calculate all the costs that went into a delivery—such as car and truck purchases and maintenance, fuel costs, and driver salaries. Then he added a profit margin. Next, he used a mapping system to compute delivery route mileage based on the zip codes of sending and receiving parties. To cover the new territories, he added more employees, especially to the central office staff to handle customer orders and other business functions. Finally, he set special rates for "white glove" service for the galleries and medical centers.

Customers Come First

Still, even with the expansion, the key to QuickBiz's success remained its service quality. Andrew insisted each of his employees provide the same on-time deliveries and courteous service that he had when he biked the routes himself. Messengers were on the front lines, and they represented the company to customers, so their attitudes and hard work were critical to QuickBiz. Over the years, he'd had some run-ins with messengers over slack work habits, and a few had quit or just didn't work out and were let go. Andrew had documented problems in employees' computerized personnel files when necessary. But overall, he considered his employees part of an extended family and valued their loyalty. Ongoing training for messengers and dispatchers was important to maintain service levels. Above all, he wanted all his employees to enjoy the work they did.

Increasing Reliance on Information Systems

Throughout his expansions, Andrew had turned to information systems to increase his efficiency and handle growing amounts of data. Information technology has helped him in many areas, including:

- Automating payroll and accounting services.
- Streamlining customer paperwork.
- Tracking equipment maintenance and supplies.
- Routing deliveries.
- Maintaining customer and messenger contact.

- Providing customized services on the Web.
- Handling customer and employee database files.

In fact, for a business that many considered low-tech, QuickBiz has relied on very high-tech computer hardware and software.

Handheld and in-dash computers with GPS mapping applications had rescued quite a few new messengers who became lost in Seattle's maze of streets. So, information technology was certainly critical to his employees. A couple of years ago the company even added a Web site offering online ordering to handle increased customer demands. Customers were pleased with the new option. For his own work, databases enabled Andrew to know his customers and their needs and to track his employees and their productivity. The company had certainly followed the digital wave. Looking back, he knew he wouldn't be able to sustain his business without these technologies.

Back to Business

Andrew's thoughts were interrupted by Leslie Chen, his administrative assistant, who was knocking at the door.

"Andrew? Sorry to bother you. Time for our meeting with the tire supplier. They want to discuss our upcoming needs for the year."

"Maybe we can get a volume price break on our fleet this year," noted Andrew. "We added two new trucks, you know." He had used the same tire supplier since the addition of the firm's first motor vehicle. His business relationship was strong and long lasting. He'd heard that the supplier had offered some quantity price breaks to other businesses, so he was going to pull the entire purchasing history of the supplier and use the information to squeeze out better discounts this year. Every dollar saved was a dollar he could put to use somewhere else.

BUSINESS CHALLENGES

Throughout his business's expansion, Andrew Langston has had to meet several challenges—not the least of which was selecting and using information systems to keep his business competitive. Information systems have played a critical role in QuickBiz's history. You explore how Andrew met those challenges in the chapters of Part Two:

- In Chapter 4, "Business Hardware," you learn how to evaluate QuickBiz's hardware needs and determine whether it has used hardware resources wisely.

- In Chapter 5, "Business Software," you learn how to determine the types of software QuickBiz needs as it grows, adds employees and customers, and streamlines its business processes.

- In Chapter 6, "Business Networks and Telecommunications," you learn about the strategies QuickBiz uses to remain in constant contact with its messengers and customers—with the goal of improving its services.

- In Chapter 7, "Databases and Data Warehouses," you learn the importance of one of business's most powerful tools—databases—and see how QuickBiz uses database technology throughout its business operations.

© Getty Images

4

FOUR

Business Hardware

LEARNING OBJECTIVES

At the core of any modern information system stands at least one computer. Few machines have changed human life as radically as the computer, and few such complex machines have become so affordable to so many businesses and individuals in such a short time. Because computers are central to information systems and to business, to successfully implement ISs, you need to understand them. Businesses have many hardware choices, ranging from types of computers and memory devices to input and output devices. Understanding the capabilities of hardware and the options available can save companies millions of dollars. This chapter provides you with the knowledge to make intelligent decisions about computer hardware in your professional career.

When you finish this chapter, you will be able to:

- List major hardware components of computers and explain their functions.

- Classify computers into major categories, and identify their strengths and weaknesses.

- Identify and evaluate key criteria for deciding what computers or related devices to purchase.

- Discuss the possible health hazards of computer use.

QUICKBIZ MESSENGERS
Hardware Streamlines Processes

When Andrew Langston opened QuickBiz and worked solo, he wrote every log sheet and customer slip by hand—he had no computers. As business picked up, Andrew hired Sarah Truesdale to be his bookkeeper and receptionist. Sarah organized the office and set up basic business applications—word-processing, spreadsheet, and database programs—on the company's first PC. So, when Andrew received a delivery request, Sarah typed it in her daily log sheet, and Andrew pedaled off.

To handle his growing customer base, Andrew hired college students as part-time bike messengers. The messengers carried cell phones with Bluetooth headsets so that the dispatcher could contact them with updated client or route information. This way, QuickBiz messengers were continually circulating through downtown, ready for the next order.

Tracking Delivery Data

The system for tracking delivery data evolved over time. In his first improvement, Andrew ordered no-carbon-required (NCR) forms. The couriers carried these forms in their backpacks and had customers fill them out with their delivery information. Customers kept a copy of the form, and the couriers took the originals back to the main office. Sarah input the customer information, such as order number, address, and type of service, along with the facts of the delivery—start and end times, courier name, and delivery address. From these inputs, Sarah would generate monthly hard-copy invoices to mail to customers.

The NCR system worked well enough for a time, but the handwriting on the forms was often hard to make out, and the forms tended to smear in wet weather. Sarah was constantly questioning the couriers about the delivery details. To say the least, inputting the data was tedious, but it became completely unmanageable when Andrew expanded his

services to car and truck deliveries. Too many orders flowed into the office for Sarah to input.

New Hardware, New Systems

Andrew and Sarah put their heads together to devise a new process for data input. Technology came to the rescue in the form of the increasingly popular handheld computers. Andrew and Sarah designed new forms containing two matching bar codes representing an order number; one bar code had adhesive and could be detached from the form. Regular clients were also issued their own bar codes representing their identification information. Messengers were equipped with handheld computers with bar-code readers. When a messenger arrived at a customer's site, he or she simply swiped the order bar code and then the client's bar code, instantly creating a new order and entering client data. Messengers then attached the removable order bar code to the customer's package. They pushed a button to record pickup date and time on the handheld and entered delivery site information. Once at the delivery site, the messenger again swiped the package's bar code and recorded delivery times. All of the data was immediately stored in the handheld computer's memory.

This delivery-process improvement required a corresponding upgrade in QuickBiz's central office computer system. Andrew selected a powerful personal computer as a server, with networked client computer terminals for the dispatchers and office staff. Leslie Chen, QuickBiz's new administrative assistant, was brought on board to assist Sarah with main office functions. Leslie downloaded the delivery information from the messengers' handheld devices into the system, instantly capturing data. As an added service, some clients also requested delivery confirmation for legal documents and medical supplies, and the information was noted in the downloads. If confirmation was needed, Leslie would e-mail or fax the client with the delivery facts.

Backing It Up

To safeguard all of its client and delivery data, QuickBiz needed to back up its hard drives. Andrew and Sarah decided that two backups would be stored off-site at their houses. At first, QuickBiz backed data up on magnetic tape drives, but retrieving information from them was time-consuming. So, as soon as rewritable CD drives came on the market, Andrew purchased one. Recently, he purchased a rewritable DVD drive. Sarah and Leslie can now directly access data by customer, delivery dates, or courier, making retrieval a breeze and providing much faster service to their customers.

COMPUTER HARDWARE COMPONENTS

eBay, the world's largest auction business, posts an average of 600 million items for sale every quarter and serves more than 204 million buyers and sellers. These activities require a huge amount of hardware. The company uses 15,000 servers. It adds disks with storage capacity of 10 terabytes every week to accommodate new listings and transactions. The company's computers are spread all over the world, and are connected through the Internet.

> ### POINT OF INTEREST
>
> **Squeezing More Bytes**
>
> IBM is developing a new storage technology called Millipede, which allows computers to store data at a density of a trillion bytes per square inch, about 20 times denser than magnetic disks available today. The process uses 4,000 very fine silicon tips that punch holes onto a thin film of plastic. The tiny holes represent bits. The technology is called nanotechnology, because it is at the level of atoms. A storage device the size of a postage stamp will hold more than 1 trillion bits, enough to store 600,000 digital camera pictures. The chip was successfully demonstrated by IBM in 2005, and was slated to be commercially available by 2008.

Hardware, in computer terms, refers to the physical components of computers and related electronic devices such as PDAs. (Software, covered in the next chapter, refers to the sets of instructions that direct the hardware to perform particular tasks.) In corporate decision making, managers should consider software first, not hardware. Businesses need to first identify the tasks they want to support and the decisions they want to make, and therefore the information they need to produce. This information will help them determine the appropriate software, and they can then purchase the best hardware to run the software. A new organization can often make software-related decisions first. However, in a great majority of cases, established organizations already have a significant investment in hardware and, therefore, must often consider adopting new software within the constraints of their existing hardware. Regardless of size, age, function, or capability, most computers have the same basic components (see Figure 4.1) and operate according to the same basic principles. A computer must handle four operations: (1) accept data, (2) store data and instructions, (3) process data, and (4) output data and/or information. In recent years, data communication over a network has become an essential aspect of input and output for almost every computer, whether stationary or portable.

In general, every computer has these components:

- **Input devices** receive signals from outside the computer and transfer them into the computer. The most common input devices are the computer keyboard and mouse, but some input devices accept voice, image, or other signals.

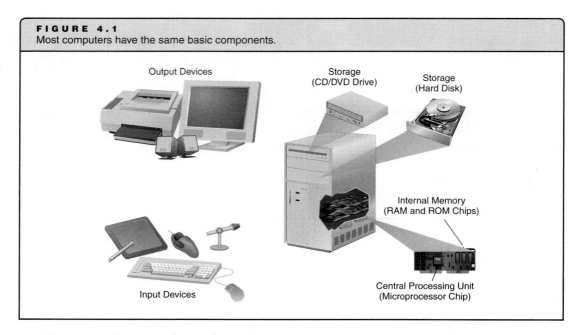

FIGURE 4.1
Most computers have the same basic components.

Output Devices

Storage
(CD/DVD Drive)

Storage
(Hard Disk)

Internal Memory
(RAM and ROM Chips)

Central Processing Unit
(Microprocessor Chip)

Input Devices

- The **central processing unit**, or **CPU**, is the most important part of any computer. The CPU accepts instructions and data, decodes and executes instructions, and stores results (output) in memory for later display. In technical terms, a CPU is a chip made of silicon, transistors, and numerous tiny soldered wires that form complex circuitry. The circuitry is built and programmed so that it can interpret electrical signals to run computers. Some computers have several CPUs. The increase in the power of computers and decrease in their prices have in large part been the result of engineers' ability to increase the number of transistors on these chips without increasing the chips' size.

- **Internal memory**, also called primary memory, is located near the CPU and stores data and instructions just before and immediately after the CPU processes them. This includes programs currently running on a machine, intermediate results of arithmetic operations, intermediate versions of documents being word processed, and data elements that represent the pictures displayed on a computer screen and the sounds played by the speakers. Most of a computer's internal memory is **RAM (random access memory)**, and a smaller amount is **ROM (read-only memory)**. RAM holds data and program instructions, and is volatile by design, that is, its contents are cleared when the computer is turned off or when a computer program is allowed to replace the data in it. ROM is nonvolatile. It contains data and instructions that do not change, mostly instructions the computer uses to load programs when it is powered on. The amount of RAM—often simply called memory—and the speed at which it operates are two of the properties that determine the power of a computer. The CPU and primary memory are usually plugged into a circuit board in the computer case called the **motherboard** or system board.

- **Storage** is different types of media—such as magnetic disks, magnetic tapes, optical discs, DVDs, and flash memory—that store data and information; however, unlike RAM, external memory allows for permanent storage. Thus, many external storage media are portable and can be moved from one computer to another.

- **Output devices**, most commonly computer monitors and printers, deliver information from the computer to a person. Additional output devices include speakers and digital audio players for audio output and specialized output devices such as Braille writers.

Recall the explanation of digital information in Chapter 1, "Business Information Systems: An Overview." Computers and other digital devices use two states to represent zeroes and ones. Representing only two states is easier than representing many states, and two states can be more accurately detected—that is, received—than many states.

Business majors and other non-IT professionals often ask: "Why do I have to study computer hardware?" The answer is threefold. You must know enough about hardware to be able to communicate your needs to IT professionals who can provide you with the devices you need for your work. If you are in a position to choose among various options and make a decision on certain hardware pieces, you must be sufficiently knowledgeable about hardware to make informed decisions. Finally, since you are or will be a professional, you will have to purchase hardware for your personal use. Keeping abreast of developments in hardware will make you an informed consumer, and you will be able to optimize your purchases.

In addition, knowledge of new technologies might give you ideas about how to develop new products and services to improve your organization's competitive position. Throughout history, necessity has been the mother of invention, but this is not so with information technology. Time and again inventions have been available long before business puts them to use. Professionals who realize that a certain development can give their companies an advantage will be rewarded for their vision.

The amount of data that computers process and store is measured in bits and bytes. A **bit** is a binary digit, a 0 or 1. A **byte** is a combination of eight bits. Most characters (except for those in complex languages) can be represented by a unique byte, because there are 256 (2^8) unique combinations, from 00000000 to 11111111. Therefore, when thinking of amounts of digital data, you can think of the number of bytes in terms of characters, such as letters, numerals, and special marks. Computer memory and storage capacity are measured in megabytes (MB, millions of bytes), gigabytes (GB, billions of bytes), and terabytes (TB, trillions of bytes) (see Figure 4.2).

FIGURE 4.2
Measuring amounts of digital data

1 KB (kilobyte) = 1,000 bytes
1 MB (megabyte) = 1,000,000 bytes
1 GB (gigabyte) = 1,000,000,000 bytes
1 TB (terabyte) = 1,000,000,000,000 bytes
1 PB (petabyte) = 1,000,000,000,000,000 bytes
1 EB (exabyte) = 1,000,000,000,000,000,000 bytes

CLASSIFICATION OF COMPUTERS

Computers come in a wide variety of classes, from supercomputers to handheld personal digital assistants. Computers are classified by their power, which is determined mainly by processing speed and memory size. However, the lines between the classes are not clear , and the class names have changed over the years. In general, the more powerful the computer, the higher its price.

Supercomputers

Supercomputers are the most powerful computers at any given time, but are built especially for assignments that require arithmetic speed. They would be overly expensive and impractical for most business situations. Usually, supercomputers are also the largest in physical size and the most expensive. Universities, research institutions, government agencies, and large corporations engaged in research and development are most likely to use them. Supercomputer manufacturers include IBM, Cray, Fujitsu, Hitachi, and NEC. Supercomputers' RAMs consist of billions of bytes, and their processing speed is billions of instructions per second. They usually cost at least $1 million.

Supercomputers are used predominantly by research institutions for complex computations.

Courtesy of NOAA

Supercomputers contain multiple processors that let them perform **parallel processing** and run at great speeds. For example, the Cray XT3 computer has 1,100 processors and a memory of 2.2 terabytes (TB). It can perform 5.9 trillion calculations per second. It solves in a few minutes problems that used to take several hours or days to solve. However, even this machine is slow in comparison to the Blue Gene used at Lawrence Livermore National Laboratory. This IBM computer has 131,000 processors (see Point of Interest) and was the world's fastest computer in 2007. Europe's fastest computer is an IBM machine located in a research center in Barcelona, Spain. It can make 40 trillion calculations per second. Its memory is equivalent to the combined memories of 20,000 PCs, and its hard disk storage has a capacity of 233 TB. Companies continue to increase the power of supercomputers. Europeans plan to build a computer that will surpass IBM's Blue Gene.

In parallel processing (sometimes called **multiprocessing**), several CPUs process different data at the same time. Uses of supercomputers include calculation of satellite orbits, weather forecasting, genetic decoding, optimization of oil exploration, and simulated testing of products that cannot otherwise be tested because of price or physical difficulty, as in the case of building a space station or the future transatmospheric plane, a commercial aircraft that will be capable of flying above the atmosphere to shorten flight time.

POINT OF INTEREST

The World's Most Powerful

The world's most powerful supercomputer, an IBM Blue Gene/P, was built for the Lawrence Livermore National Laboratory, a research institute in California. It employs 130,000 processors and takes the area of a half tennis court. The massive computer can perform 360 trillion calculations per second. A human charged with performing 360 trillion calculations would need about 90 million years. IBM also plans to build Blue Gene/Q, which will be able to perform 3 quadrillion calculations per second. That's 3,000,000,000,000,000 per second.

Until recently, only large engineering and life sciences businesses or governments could justify the cost of supercomputers. In 2005, IBM changed this by offering use of its Blue Gene computers over networks. Clients can log on via a secure Internet link to this supercomputer, which is located in Rochester, Minnesota. The machine can perform 5.7 trillion computations per second. Clients pay 50 to 90 cents per 1 million computing operations. One small company that needs such computing power, but could not afford the high purchasing cost, is QuantumBio, Inc. The small research company develops and tests new drugs for pharmaceutical companies. Having access to supercomputing allows the company to augment the variety of products it can offer.

In lieu of one large supercomputer, some organizations link a "cluster" of smaller computers via networks to create and enjoy similar computing power. Instead of a single machine with multiple processors, clustering uses the CPU power of multiple computers, with the same effect. This can be done with special software that links the CPUs of servers via a private or public network such as the Internet, all or part of the time.

Mainframe Computers

Mainframe computers are less powerful in computational speed and significantly less expensive than supercomputers. They cost several thousand to several hundred thousand dollars. Businesses that must handle business transactions and store large amounts of data in a central computer often use mainframes, which some IT professionals fondly call "big iron." These businesses include banks, insurance companies, large retail chains, and universities. Well-known mainframe manufacturers include IBM, Fujitsu, and Unisys. While the processing speed of mainframes is usually not higher than that of the fastest PCs, they often have multiple processors and their memories are significantly larger, measured in terabytes. By some estimates, 40–50 percent of the world's business data resides on mainframes. IBM, a major manufacturer of this class of computers, claims that about 60 percent of all data available on the Internet is stored and processed on mainframe computers. Like supercomputers, these computers are largely invisible to the public, although we access them often via the Internet.

POINT OF INTEREST

Super Detective

A supercomputer named XENON was developed for the Dutch internal revenue service and has also been adopted by the United Kingdom's HMRC—Her Majesty's Revenue and Customs service. XENON constantly sifts through thousands of Web-based transactions. The computer then cross-references the transactions against government taxation records. It checks to see if the selling business paid tax on each online transaction. The supercomputer is effective, saves many hours of labor, and helps businesses avoid the disruption of manual investigations.

Source: (www.bytestart.co.uk), February 22, 2007.

Midrange Computers

Midrange computers are smaller than mainframes and less powerful. They are usually used as a shared resource, serving hundreds of users that connect to the midrange computer from personal computers. Therefore, they act as servers, computers used to communicate to other computers and "serve" applications and data, both through the Internet and locally within organizations. The IBM AS/400, HP 9000, and HP Alpha families of computers are the best-known midrange computers. Like mainframe computers, midrange computers often use multiple processors. Classifying computers as midrange is becoming rare.

Microcomputers

Microcomputers is the collective name for all personal computers (PCs), notebook computers, and handheld computers. More powerful microcomputers are sometimes called **workstations**. Workstations are typically used for computer-aided design (CAD), computer-aided manufacturing (CAM), complex simulation, and scientific applications. As the performance of PCs steadily improves, computers that in the past were classified as midrange computers are now marketed as PCs, and the lines between computer categories continue to blur.

The power of microcomputers in terms of speed and memory capacity doubles about every two years. Most PCs now sold to individuals and businesses cost less than $1,000. However, a growing number of microcomputers are not PCs, but notebooks, handheld, and tablet computers. Many cell phones now also serve as handheld computers. Some global positioning system (GPS) devices double as navigation tools and handheld computers.

Handheld computers are popular devices for people who spend much time out of the office.

© Steve Lewis/Getty Images

Computers on the Go: Notebook, Handheld, and Tablet Computers

Computers are increasingly used outside the home, office, or school. Notebook or handheld computers are used to record and retrieve data for people on the go. The **notebook computer** (also called a laptop) is a compact, light, personal computer that can be powered by a rechargeable battery. These computers can operate for up to eight hours without recharging their batteries. Many notebooks have accessories that enable the user to communicate with other computers. All new notebook computers have internal circuitry that enables them to connect to networks and the Internet without

Tablet PCs are growing in popularity.

© Comstock Images

wires or cables. (Wireless technology is covered in Chapter 6, "Business Networks and Telecommunications.") Notebooks are quickly catching up to desktop PCs in terms of speed, memory, and hard disk capacity.

One highly popular class of computing machinery is the handheld computer, also known as the **personal digital assistant (PDA)**. Handheld computers appeared on the market in the early 1990s but became popular only toward the end of the decade. These devices are small enough to fit in the palm of your hand, and typically a **stylus** (a pen-like pointing and drawing device) is used to enter data through a touch screen, although some handhelds also have a small keyboard or can plug into a folding portable keyboard. With a special device called a projection keyboard, a virtual keyboard is projected on a surface and lets the users "type" as if they were using a full-size keyboard. A special sensor detects the location of each key and its "depression" by the user. Almost all new PDAs also serve as mobile phones.

Another microcomputer is the **tablet computer**, often called a tablet PC. It is a full-power PC in the form of a thick writing tablet. It looks like a notebook computer without a keyboard, although it can be connected to a keyboard and a mouse. Instead of a mouse, a stylus is used as the input device. The user can handwrite text, which automatically turns into typed text (as with some of the smaller handheld computers). The stylus is also used to click icons and select items from menus. The tablet PC is enthusiastically received among salespeople and hospital staffs. Forms now can be filled out directly on screen, eliminating hours of paperwork for sales representatives and nurses.

POINT OF INTEREST

Made In America?

Your notebook computer may have a Dell, HP, or Apple logo on its cover, but where was it really made? Very little of the labor involved in the making of these computers is performed in the United States. If your computer contains an Intel CPU, there is a 50 percent chance it was manufactured in Ireland or Israel. If it's an AMD CPU, it was made in Germany. The other components were most probably made in the following countries: graphics card—Taiwan; hard drive—Thailand; RAM—South Korea or Singapore; LCD—South Korea; battery—Japan; motherboard—China; case—Taiwan. In 2006, Taiwanese companies assumed the assembly work of close to 83 percent of the world's notebooks, but more than 85 percent of the assembly work was subcontracted to companies in China. An increasing proportion of notebook design is done in Taiwan. So, what's left for the U.S. "manufacturers"? Mainly design, advertising, shipping, and billing.

Source: Tweney, D., "What's Inside Your Laptop?", *PC Magazine*, April 10, 2007.

Converging Technologies

In recent years we have experienced an increasing trend of **technology convergence**, building several technologies into a single piece of hardware. This is true especially in handheld units. A unit might be called a cell phone or a digital camera, but it is also a computer and several other things. Consider the M-1, manufactured by Sanyo. It is a cell phone and a digital camera. It is also a television set, a digital sound recorder, and a stereo sound system that plays MP3 files, with an internal memory of 1 GB. A growing number of PDAs can also serve as GPS devices with speech directions. In homes, personal computers can be turned into entertainment centers that wirelessly transmit sound and television broadcasts to other computers or to sound systems and TV sets. Expect to see a growing convergence of digital technologies both in mobile units and in home devices.

A virtual keyboard affords users of handheld computers the comfort of a full-size keyboard by creating a "keyboard in the air." The device can be connected to any handheld computer.

AP Images

POINT OF INTEREST

Thanks, but No Thanks

In the United States, two-thirds of the mobile phones in use have the capability of accessing the Web. Yet, only 5 percent of these mobile handset owners ever connect to the Internet. The main reason is probably the high cost of such access. Other reasons are the small-size screens and too many menus.

Source: Burns, E., "Tech and Price Hinders Mobile Web Adoption," *ClickZ Stats* (www.clickz.com), March 19, 2007.

A PEEK INSIDE THE COMPUTER

It is not necessary to look under a car's hood to drive it, but it is important to know enough about how a car is built to know which car to buy. Similarly, professionals must know enough about the major components of a computer to understand what computing power and capabilities they buy or recommend for buying. The following discussion introduces the computer's most common parts and peripheral equipment and describes in some detail how these devices work.

The Central Processing Unit

The CPU is the computer's brain, where all processing takes place. The CPU consists of two units: the **control unit** and the **arithmetic logic unit (ALU)**. These units store and process data. The CPU is a silicon chip with multiple circuits. It carries signals that execute all processing within a computer. Because the chip is so small, it is often called a **microprocessor**, or simply a processor. Most modern computers use processors that combine two or more CPUs or "cores" on a single chip, called **multicore processors**. Multicore processors are capable of performing more than one task at a time (multitasking). For example, they can carry out a calculation in a spreadsheet and process a graphical design simultaneously. Processing more than one program, or processing several parts of a program, at the same time is often called **multithreading**, whereby each process is a thread.

Most modern computers contain multicore processors.

Courtesy of Intel Corporation

Microprocessors

Microprocessors are made of silicon embedded with transistors. A transistor is a semiconductor, a component that can serve as either a conductor or an insulator, depending on the voltage of electricity that tries to flow through it. This property is excellent for computer communications, because it provides a means to represent binary code's two states: a 1 (voltage conducted) or a 0 (voltage not conducted). Thus, transistors can sense binary signals that are actually encoded instructions telling the computer to conduct different operations.

The greater the number of transistors that can be embedded in the chip—which means the greater the number of circuits—the more powerful the microprocessor. Current processors can contain several hundred million circuits. Current technology enables chip makers to print circuits on silicon that is 0.1 micron thick, one thousand times thinner than a human hair. New processor-making technologies let engineers increase the processing speed of computers while enabling them to use less energy and give off less heat.

The Machine Cycle

When a program starts running in a computer, the CPU performs a routine sequence, illustrated in Figure 4.3 for a simple arithmetic function. First, the control unit, one of the two parts of the CPU, fetches an instruction from a program in primary memory and decodes it, that is, interprets what should be done. The control unit transmits this code to the other part of the CPU, the

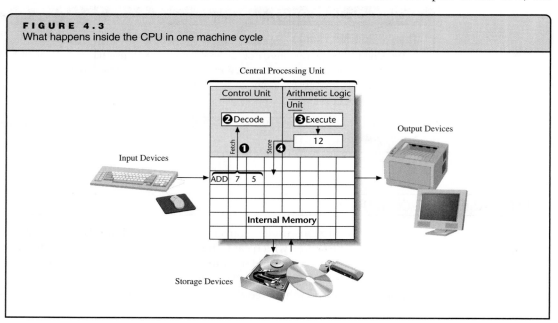

FIGURE 4.3
What happens inside the CPU in one machine cycle

arithmetic logic unit (ALU), which executes the instruction. Usually, the operation's result is needed for further operations. Therefore, the control unit takes the result and stores it in primary memory, or it leaves it in a memory location called register for a following instruction to use. The control unit then fetches the next instruction, decodes it, and "puts" it in the ALU, which executes the instruction. The control unit stores the result in primary memory, and so on, until the entire program is executed, or something happens that stops the cycle. Anything that stops the cycle is called an interrupt. It might be an instruction in the program itself, a power failure, or any other event that stops the CPU.

As you can see, the CPU performs four functions in every cycle: fetch, decode, execute, and store. Each cycle is called a **machine cycle**. CPUs can perform billions of machine cycles per second. The sequence of CPU operations must be paced so that different tasks do not collide. To this end, the control unit uses special circuitry called a **CPU clock**, which synchronizes all tasks. The clock is programmed to run operations at the maximum rate allowable. The number of pulses per second is called frequency, or **clock rate**. A machine cycle takes several clock pulses. CPU frequencies are measured in megahertz (MHz, millions of hertz), or gigahertz (GHz, billions of hertz). During the time it takes your eye to blink (about 0.2 second), a computer can execute hundreds of millions of instructions. Therefore, timing of computer operations is measured in very small fractions of a second (see Figure 4.4).

FIGURE 4.4
Computer time

1 millisecond = 1/1,000 (0.001) second
1 microsecond = 1/1,000,000 (0.000001) second
1 nanosecond = 1/1,000,000,000 (0.000000001) second
1 picosecond = 1/1,000,000,000,000 (0.000000000001) second

Interestingly, many computers now have a lower clock rate than computers of several years ago. This does not mean that such computers work more slowly. They have multicore processors, which are more efficient. They execute more instructions per machine cycle than the older single-core processors, and therefore are faster despite the lower clock rate. Therefore, both the cycles per second and instructions per cycle (IPC) should be considered when comparing speeds of processors.

The Word

The **data word** (or "word" for short) is the maximum number of bits that the control unit can fetch from primary memory in one machine cycle. The word's size is determined by the size of the CPU circuitry that holds information for processing. Obviously, the larger the word, the more instructions or data can be retrieved per second. Therefore, all other things being equal, the larger the word, the faster the computer. Current microcomputers have words of 32 and 64 bits.

The Arithmetic Logic Unit Operations

The ALU is the part of the CPU where all arithmetic and logic operations take place. Arithmetic operations include addition, subtraction, multiplication, division, exponentiation, logarithmic calculations, trigonometric computations, and other complex mathematical tasks. Logic operations compare numbers and strings of characters. For example, comparisons such as greater than, less than, and equal to are logic operations. The ALU also compares character strings that are not quantitative. For example, when you try to find a word in the text of a word-processing document, the ALU compares all words in the text to that specific word until it finds an identical word.

Computer Power

What makes one computer more powerful than another? The two major factors to consider are processing speed and memory capacity. A computer's speed is determined, among other factors, by the CPU clock rate (measured in MHz or GHz), and the amount of information the CPU can process per cycle (determined by the size of the data word and the capacity of internal data communication). However, the architecture of the various computer components also plays a significant role in determining processing speed. To mention one, consider the discussion of multicore CPUs. When two computers are built with the same components except the number of cores, the computer with the greater number of cores is faster.

All other things being equal, the greater the clock rate, the faster the machine, because it can fetch, decode, execute, and store more instructions per second. Similarly, the larger the data word, the faster the computer. A larger word means that in each trip to the primary memory, the control unit can retrieve more bits to process. Therefore, the CPU can execute a program faster.

You might have seen advertisements promoting a "64-bit computer." This means the data word's capacity is 64 bits. You must be cautious with regard to word size. A larger word does not always mean a faster computer, because the speed at which the bits move between the CPU and other components depends on the capacity of internal communication lines. The system bus—also called simply the **bus**—which is the electronic lines or traces used for communication inside the computer, might have a width of only 32 bits, while the word might contain 64 bits. The number of bits is also referred to as the width of the bus.

Buses have their own clock rate. The bus that computer makers usually mention in ads is the front side bus, which is the bus connecting the CPU to the memory. A typical front side bus clock rate is 800 MHz. The combination of bus width and clock rate determines throughput. **Throughput** is the number of bits per second that the bus can accommodate. Considering both factors, CPU clock rate (so many GHz) and bus throughput, enables you to compare properly the speeds of different computers.

Computer speed is also measured in **MIPS** (millions of instructions per second), which is not an accurate measure, because instructions have various levels of complexity. However, computer speed expressed in MIPS is often used to indicate overall processing speed because all factors that determine speed are considered: clock rate, data word size, and bus throughput, as well as other speed factors that we do not discuss here. Computer speeds expressed in MIPS have been used to indicate the dramatic reduction in the cost of computing; observers often divide the MIPS by the cost of a computer and marvel how the cost of computer power has decreased dramatically, from MIPS per dollar to MIPS per cent. In recent years, computer makers have also used the term "transactions per minute" (TPM), referring mainly to database transactions, but this ratio, too, is not an absolute measurement.

INPUT DEVICES

Computers must receive input to produce desired output. Input devices include all machines and other apparatuses used to enter instructions and data into the computer. Popular input devices include the keyboard, mouse, trackball, microphone, and various types of scanners. The most common input device is the keyboard.

Keyboard

The keyboard contains keys that users press to enter data into primary memory and instructions for programs to run. All keyboards include the basic letters of the alphabet, numbers, and punctuation marks—plus several function keys numbered F1, F2, and so on, that can be activated to execute preprogrammed functions, such as copying a highlighted sentence in a text file created with a word processor. With the growing use of the Web and use of computers to play music and video clips, keyboard manufacturers have added keys that facilitate Web browser commands such as Back and Forward, and music keys such as Volume and Play/Pause. On some keyboards you can bring up your e-mail application by pressing the Mail key or the calculator by pressing the Calculator key.

QWERTY and Dvorak Keyboards

The standard keyboard layout is called QWERTY, an acronym based on the top row of letter keys from left to right. Interestingly, the QWERTY keyboard was originally designed to slow down typing, because early mechanical typewriters jammed when users typed too fast. Today's electrical devices make this layout counterproductive. Other keyboard designs facilitate faster typing. On the Dvorak keyboard, the most frequently used keys are in the home, or central, row. Using this keyboard can increase typing speed by 95 percent. Some operating systems, such as Windows, let users map QWERTY keys into a Dvorak layout. Most computer users are reluctant to retrain themselves for the Dvorak map. In France and some other European countries, the A and Q keys are swapped, and the Z and W keys are swapped. These keyboards are known as AZERTY keyboards.

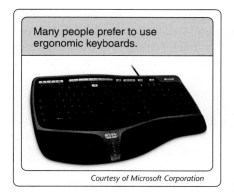

Many people prefer to use ergonomic keyboards.

Courtesy of Microsoft Corporation

Ergonomic Keyboards

One of the most prevalent computer-related work injuries is carpal tunnel syndrome, the pain or numbness caused by holding the forearms in an unnatural position for long periods. The repetitive motion of typing exacerbates this problem, causing repetitive-stress injuries (RSIs). In response, ergonomic keyboards are gaining popularity. **Ergonomics** is the study of the comfort and safety of human beings in their working environment. Ergonomic keyboards are split in the middle, and the two parts are twisted outward to better fit the natural position of the forearms.

Mouse, Trackball, and Trackpad

A **mouse** is an input device that controls an on-screen pointer to facilitate the point-and-click approach to executing different operations. It is most commonly used with a keyboard, although some programs use it exclusively. Mice have one to five buttons that let the user place the pointer anywhere on the screen, highlight portions of the screen, and select items from a menu.

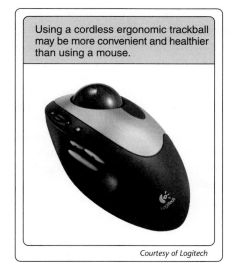

Using a cordless ergonomic trackball may be more convenient and healthier than using a mouse.

Courtesy of Logitech

When the user moves the mouse on the surface of a desk or a pad, the computer detects the movements, translates them into digital coordinates on the screen, and moves the pointer to imitate the mouse's movement. The buttons are used for clicking, locking, and dragging displayed information. A **trackball** is similar to a mouse, but the ball moves within the device, rather than over a surface. With a **trackpad**, a user controls the cursor by moving his or her finger along a touch-sensitive pad. Many notebook computers have built-in trackpads. Many mice and trackballs have a built-in wheel that scrolls pages displayed on the monitor.

Mice, trackballs, and keyboards are also available as wireless units that use infrared or radio technology. These units give users more flexibility, especially in software-based presentations, in which the presenter may move around with the mouse in his or her palm.

Touch Screen

Increasingly, interaction with computers is done through touch screens. This Microsoft "coffee table" facilitates many operations by touching and dragging with the fingers.

Courtesy of Microsoft Corporation

Sometimes a single device, such as a **touch screen**, may serve both as an input and output device. A touch screen lets the computer user choose operations by touching the options on the computer screen. Some common public applications use touch screens to provide advice to tourists, select lottery numbers, and ring in grocery items at self-serve supermarket checkouts. On handheld computers, the screen serves as both a display and input device. The user enters commands and data by touching a stylus on icons and menu items. With other touch screens, especially GPS units, you can execute commands by touching the screen with your fingers.

More and more, computers and other information devices are operated through touch screens. Global positioning systems (GPSs) have offered this convenience for some time. Some PDAs and mobile phones have touch screens. In the near future we will operate computers mainly or only through touch screens. For example, Microsoft's new "coffee table" touch screen computer allows interactions not only with human fingers at many points on the screen, but also with devices such as digital cameras. A picture can be taken, downloaded by placing the camera on the table, and manipulated—moved, enlarged, and more—by moving fingers on the screen.

Source Data Input Devices

In some businesses, the speed of data entry is a top priority. These businesses use machine reading devices, such as bar-code scanners, known as **source data input devices**. They copy data directly from the source, such as a bar code or magnetic-ink characters, without human intervention. They can also record data directly from other sources, including checks and credit cards. Source data input technologies are widely used in banking, credit-card processing, and shipping.

Source Data Technology

Mark-recognition devices are essential to successful source data entry. Special devices use *optical mark recognition* to detect the positions of marks on source documents, such as standardized test response forms. *Optical bar recognition* senses data encoded in the series of thick and thin black bars in bar codes.

A less accurate technology used for source data entry is *optical character recognition* (OCR). Unlike optical mark recognition, OCR technology is often used to try to interpret handwritten and printed texts not originally designed for source data entry. A special scanner scans the page and translates each character into a digitized representation. Software then tries to correlate the images with characters and stores interpreted text for further processing. Postal services around the world have experimented with OCR to replace human eyes and hands in the tedious job of mail sorting.

Note that OCR is not optical mark sensing. In optical mark sensing, the scanner senses a mark's *position*, not what the mark actually is. The mark's position determines the input. Because the mark's position rather than its shape determines the input data, mark sensing is far more accurate than OCR.

OCR has recently been integrated into mobile devices. For example, Samsung sells a cellular phone that can help save time entering information into the phone's address book. When you use the phone's digital camera to photograph a business card, the built-in character recognition software captures the information from the picture and enters it into the address book.

Banking

In the United States, commercial banks and the Federal Reserve Bank process about 200 million checks daily. Entering check data manually would make the process extremely expensive and slow. The bank identification number, account number, and check number are printed in special magnetic ink at the bottom of each check, as shown in Figure 4.5. A device called a magnetic-ink reader uses **magnetic-ink character recognition** (**MICR**, pronounced MIKE-er) to detect these numbers. A person at the bank enters the amount of the check, also in magnetic ink. The bank then records its check deposits by placing a large number of checks in a MICR device, which records check amounts and accounts from which the money is drawn.

Credit Cards

Credit cards, too, facilitate source data entry. Card number and holder information are coded on the magnetic strip on the card's back. When you charge a purchase with your credit card, the card is passed through the reader at the point of sale (POS) to record the account number and your name and address. The total amount charged is either keyed manually or recorded automatically from the cash register (often from a bar code on the item purchased).

Shipping and Inventory Control

You might have noticed that every package you receive through shipping companies such as UPS and FedEx has a bar code on it. Bar codes use the optical bar recognition techniques described earlier to represent information for both inventory control and shipment tracking. A package is scanned before it leaves the shipping facility, and the information is channeled into a computer

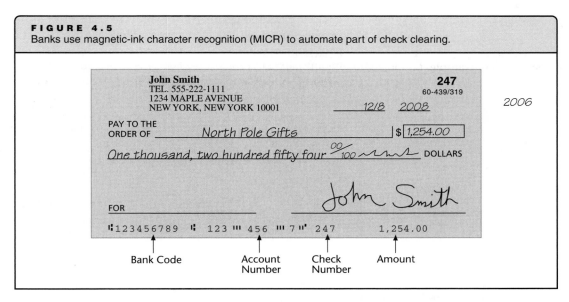

FIGURE 4.5
Banks use magnetic-ink character recognition (MICR) to automate part of check clearing.

that stores information such as the recipient's name and address. Whenever the item reaches a station, the bar code is scanned again. This information is combined with the identification information of the station. So, anyone with access to the shipping company's database can see exactly where the item has been and when, right up to the point of delivery. You can track an item by logging on to the shipping company's Web site and entering the item's tracking number. Since quick delivery is essential, source data input is extremely important in the shipping industry, because it is highly accurate and saves much labor and time. As discussed in Chapter 3, "Business Functions and Supply Chains," bar codes are being replaced with RFID tags for both shipping and inventory control.

Imaging

A growing number of organizations are **imaging**, or image processing, their documents. Doing so allows not only the storage of enormous amounts of data in less space than paper, but also much more efficient retrieval and filing. By scanning and indexing images, many companies have already reduced millions of paper documents to digitized pictures. They use the technology to store invoices, shipping documents, insurance policies and claims, personnel files, checks, and many other document types. The images are indexed and linked to relevant records in large databases, from which they can be retrieved and displayed on computer monitors. This technology is particularly useful when documents include signatures and graphics.

Once scanned, the original document can be destroyed because an exact copy can be generated on demand. Since it is in electronic form, it can be indexed. Indexing enables you to search a document by keywords and numbers. This reduces the average time of searching a document from several hours to about five seconds. In the United States, checking account holders receive one or two sheets of imaged canceled checks from their banks instead of a stack of their original checks. Customers who do their banking online can retrieve these images at any time. This system saves banks millions of dollars in paper, space, and handling costs. The images are often stored on DVDs. Because imaging reduces the amount of paper in organizations, some of the most enthusiastic adopters of imaging are companies in paper-intensive fields such as law, retail, insurance, banking, health care, and shipping.

Imaging technologies continue to progress. American Express, a financial services group with assets worth $232 billion, makes extensive use of imaging. For some years the group imaged documents at a rate of 25 pages per minute. Adopting new machines and software from the British company Captiva (which was later purchased by EMC), the group now images at a rate of 190 pages per minute. The indexing process was reduced from 45 seconds per document to

12 seconds per document. This allows American Express to record 3,000 client folders per hour. The number of employees involved in imaging was reduced from 95 to 45, and at an average salary of $30,000 per year, the immediate savings in the first year was about $1.5 million.

The French company Elior, Europe's third largest contract supplier of food, uses imaging to reduce the cycle of invoice validation and payment to its suppliers. Elior processes 4.5 million invoices annually. An electronic data interchange, which would eliminate paper invoices, could be implemented only with 12 percent of the suppliers. With most suppliers, who operate in various European counties, the company had to archive paper invoices to meet legal requirements. Because validating the invoice—ensuring that it is accurate—and physically filing and retrieving it was so time-consuming, the company often had to pay late payment fees. While a paperless solution was not possible, electronic processing of the invoices was. Elior installed an imaging system, provided by EMC Captiva, and connected it to the company's SAP ERP system. Elior now captures more than 900,000 invoices annually and validates each within one second. Erroneous payments decreased, and the company could speed payment and avoid late payment fees. The system cost Elior $200,000, but saves it $150,000 annually.

Speech Recognition

The way we communicate with computers is changing. We already mentioned touch screens. However, in some work environments, using manual input devices is either impossible or inconvenient. In other situations, such as customer service, using a computer to respond automatically to spoken customer queries can save labor costs. Instructing machines by speech can help in these instances. Consider the finding of Datamonitor, an IT-strategy consulting firm: the average call-center call costs the organization $5 if handled by an employee but only 50 cents when handled by self-service, speech-enabled systems. Speech recognition is fast becoming a staple of business. **Speech recognition**—also called voice recognition—is the process of translating human speech into computer-readable data and instructions. Although speech recognition systems vary in sophistication, all receive voice input from a microphone or telephone and process it with software.

Since help-desk labor is an area of great potential for reducing costs, several companies have developed speech recognition software. Nuance Communications, Inc. offers Dragon Naturally-Speaking for PCs for dictating text in word processors and e-mail. Pluggd, Inc. offers HearHere, software enabling voice-activated searches for specific sections of videos or podcasts. (We discuss podcasts in Chapter 8, "The Web-Enabled Enterprise.") Tellme's software is installed in automated 411 (telephone directory) services and call centers. TuVox's software is used in TiVo (the digital video recording device used with television), and by British Airways for the airline's call-routing and customer service applications. The Mac OS and Windows Vista operating systems include a voice recognition feature.

Soon, navigation systems will become a standard feature in vehicles. Toyota installs VoiceBox software in navigation systems of its new cars; IBM's Embedded ViaVoice is used in General Motor's OnStar and other dashboard command systems. So far a GPS system is capable only of providing voice directions; with the new system, you will be able to verbally ask the GPS system for directions, and it will speak them back.

Currently, the customer service departments of many companies use voice recognition of simple commands for telephone callers, who can utter answers to questions and receive recorded responses. However, customer complaints prod companies to employ more sophisticated voice recognition systems.

Some observers think speech-operated computers might increase already high noise levels in offices and add distraction. Imagine an office where everyone who currently types in their cubicles suddenly talked to computers. Also, speech recognition could become the source of pranks; people walking by could shout commands to other workers' computers.

Output devices include all electronic and electromechanical devices that deliver results of computer processing. We receive most information in visual form, either on screen or on paper. Therefore, this discussion focuses on the most popular output devices: monitors and printers. Output also includes audio signals, received through speakers and earphones, or downloaded to digital audio players. Soon we might also be able to enjoy smell output using digital technology.

LCD displays have largely replaced CRTs in organizations and households.

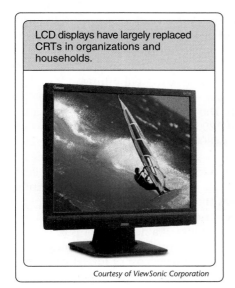

Courtesy of ViewSonic Corporation

Monitors

The most common output device is the computer monitor, which looks like and uses technology similar to a television screen. The two major types of monitors are cathode-ray tube (CRT) and flat-panel display. Images on a monitor are made up of small dots called **pixels** (*pi*cture *el*ements, with the addition of an *x* for easier pronunciation).

In a **CRT (cathode-ray tube)** monitor, the inner side of the screen has a layer of tiny phosphoric dots, which make up the pixels. These dots respond to electronic beams by displaying different colored light. An electron gun receives instructions from the computer and sweeps the rows of pixels, spraying a ray of electrons. When electrons hit a pixel, the pixel emits light for a limited time. The electronic gun bombards some pixels and skips others, creating a picture on the screen. Most new monitors are flat-panel. The only advantage of CRT monitors over flat-panel monitors is their speed of rendering a new picture. This is why people who often play computer games (popularly called "gamers") as well as artists who create digital video prefer CRT technology. However, the rendition speed gap between CRTs and flat monitors is closing fast. It is likely that within a few years we will rarely see CRT monitors in offices or homes.

Flat-panel monitors have gained popularity for personal computers and handheld computers, after years of use in notebook computers. The advantages of flat-panel monitors are their slim profile, sharper images, and lower power consumption. The most common type of flat-panel monitor is the **liquid crystal display (LCD)**. The price of LCD monitors has decreased sharply over the past several years, making them the most popular type of monitor. In LCD, a conductive, film-covered screen is filled with a liquid crystal, whose molecules can align in different planes when charged with a certain electrical voltage. The proper voltage applied to segments of the screen disrupts the crystal's regular structure in those areas, causing it to block light. Light continues to pass through the rest of the liquid. This combination of light and dark areas produces images of characters and pictures.

Any type of high definition television (HDTV) set can be connected to a computer (if it has the proper socket) and serve as a computer monitor.

The price of a monitor depends primarily on its size, measured as the diagonal length of the screen. Other price factors include brightness (the brighter the better), contrast ratio (the higher the better), and pixel pitch (how close the pixels are to each other; the closer the better).

The greater the number of pixels per unit area on the screen, the sharper the picture. Picture sharpness is called **resolution**. It is expressed as the number of pixels that fit the width and height of a complete screen image. Monitors come in various resolutions. Usually, the resolution required for clear text in edited documents is 640 × 350. If you multiply these numbers, you get the total number of pixels on the screen. Common resolutions are 1024 × 768, 1280 × 1024, 1600 × 1200, 1920 × 1200, and 2560 × 1600.

Good color monitors can display more than 16 million colors and hues. The number of colors and the overall quality of pictures also depends on the quality of the video card used inside the computer. The video card contains memory and circuitry to manipulate and display two- and three-dimensional images.

Printers

Printers can be classified into two basic types—nonimpact and impact—based on the technology they use to create images on paper.

Nonimpact Printers

The printer most commonly used today in businesses is the laser printer, which is a **nonimpact printer** because it creates images on a page without mechanically impacting the paper. Nonimpact printers include laser, ink-jet, electrostatic, and electrothermal printers. Laser printers are also page printers, because they print one whole page at a time. Laser and ink-jet printers produce very high-quality output, including color. Laser printing technology can create typeset quality equal to what you see in magazines and textbooks. Ink-jet printers can be used for photo-quality output, and therefore are often used to print pictures captured by digital cameras. All nonimpact printers have fewer moving parts than impact printers and are, therefore, significantly quieter. They are also much faster. The excellent quality of their output makes laser printers the choice of many individual and corporate users for desktop publishing.

Two qualities to check when purchasing a laser or ink-jet printer are speed, measured in pages per minute (PPM), and density, measured in dots per inch (DPI). The higher the density, the sharper the output. Desktop printers produce output at 300, 600, and 1200 DPI or more. Ink-jet printers are capable of producing output at much higher density, such as 4800 x 1200 DPI. The speed of desktop laser printers is 4 to 25 PPM. Color laser printing is somewhat slower due to the time it takes the printer to compose the image. Larger, commercial laser printers reach speeds of more than 400 PPM.

The low prices of laser and ink-jet printers might be misleading. Over the life of the printer, the buyer will spend much more money for the cartridges than for the printer. For example, a color laser printer that costs $200 typically requires four cartridges, each costing about $40. Just a single set of new cartridges costs almost as much as a new printer. If a new printer is to be used for high-volume printing, the initial larger expenditure on a laser printer makes business sense because the per-page cost of laser cartridges is lower than the per-page cost of ink-jet cartridges. However, ink-jet printers are more suitable for photo-quality prints because of their higher resolution, that is, a greater DPI density.

The latest ink-jet printer technology is Memjet. While in current ink-jet printers the printhead—the mechanism containing the ink cartridges—moves sideways while the page moves forward, in Memjet printers the printhead is page width, so it does not have to move. For example, in printers that use letter-size paper, the printhead contains 70,400 nozzles. This allows Memjet printers to reach speeds of 60 PPM.

Impact Printers

Printers are considered **impact printers** if they reproduce an image on a page using mechanical impact. Of this type, the only printers you might still encounter are dot-matrix printers. The printhead of **dot-matrix printers** consists of a matrix of little pins. When certain pins strike the ribbon against the paper, they mark the shape of a character or another form on the paper. Thus, each character or other image is made up of tiny dots. Dot-matrix printers produce low-quality output but are still in use in many businesses, because they can print multicopy forms.

STORAGE MEDIA

To maintain programs, data, and information for later use, data must be stored on a nonvolatile medium, that is, a medium that retains data even when not connected to electric power. Often, we also want to move stored data to a computer that is not part of a network, and we need to back up important programs and data as well. For these purposes, we use storage media. Although media are the materials on which information is stored, and the storage device is the media and the mechanism that stores and retrieves the information, the terms "storage media" and "storage devices" are often used interchangeably.

Storage devices come in different forms and use different materials, each with strengths and weaknesses. Cost, capacity, access speed, and access mode should all be considered when evaluating storage devices. Capacity is the amount of data the medium can hold, access speed is the amount of data that can be stored or retrieved per time unit, and access mode refers to the organization of data on the medium, either random or sequential.

Storage devices differ in the technology they use to maintain data (such as magnetic or optical) and in their physical structure (disks, tapes, or other forms). Physical structure might limit ways in which data can be organized on the medium. While disks allow any type of organization, tapes allow only sequential organization. This section discusses modes of access, looks at specific media and technologies, and considers the trade-offs that managers must consider when evaluating what type of storage media is best for a particular business.

Modes of Access

The two basic types of access modes for data storage are sequential and direct (random) access (see Figure 4.6). In **sequential storage**, data is organized one record after another. With sequential storage (the only option for magnetic or optical tapes), to read data from anywhere on the tape, you have to read through all the data before that point on the tape. Retrieving files from sequential devices is slower and less convenient than on devices that utilize direct access. In **direct access**, records are not organized sequentially, but by the physical address on the device, and can be accessed directly without going through other records. Devices that allow direct access storage are often called DASD (DAZ-dee), short for direct access storage device. They include magnetic and optical disks as well as **flash drives**, small storage devices that connect to a computer via a **universal serial bus (USB)** receptacle.

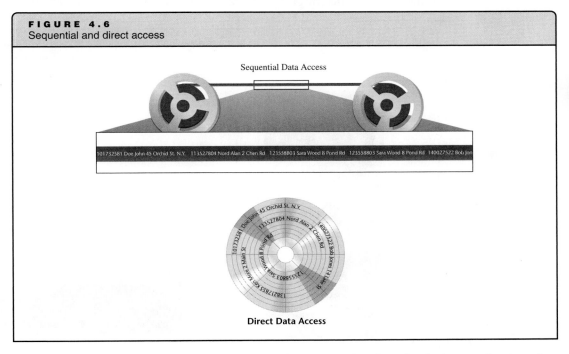

FIGURE 4.6
Sequential and direct access

Sequential Data Access

Direct Data Access

Storage and retrieval on sequential storage devices are slow but the devices are inexpensive. Therefore, tapes are suitable for backup purposes. Direct access storage media are the only practical way to organize and query databases.

Magnetic Tapes

Magnetic tapes similar to those used in tape recorders and VCRs are also used to store computer data. While some tape drives still use open reel tapes, most now use tape cartridges. Many of these cartridges look, in general, like the tapes used in audio tape players. One of the most popular types of tape cartridges is the Digital Linear Tape (DLT). In 2007, Quantum, a storage media manufacturer, offered tape cartridges with a capacity of 1.6 TB (terabytes) that access data at a rate of 120 MB per second. The cost of storage is measured in how much money is spent on each byte of storage capacity. Tapes provide the lowest cost in terms of cents per GB. The Quantum 1.6 TB tape costs six cents per GB.

DLT tape is an inexpensive way to back up data.

Courtesy of Hewlett-Packard Company

Backing up all or a designated part of data from its original storage medium needs to be done regularly. The entire hard disk of a PC can be backed up, or, in organizations, large amounts of data are backed up in case a hard disk crashes or an incident occurs that makes the original data irretrievable. Backing up can be done manually or automatically with the help of software. When the backup is done for an organization, often the organization makes use of a storage area network, a dedicated area where disk (and possibly tape) storage devices are connected through communication lines to organizational ISs for the sole purpose of data backup. Such networks are discussed later in this chapter. Backup and recovery procedures are discussed in Chapter 14, "Risks, Security, and Disaster Recovery."

Some organizations use magnetic tapes to automatically create two backups of all data. AOK, Germany's largest health insurance company with more than 25 million policyholders, uses 128 300-GB tape drives to store 44 TB of data. The amount of data grows at a rate of six percent per year. The data is backed up through a dispersed network of parallel tape drives. The company is well prepared for any incident that might destroy data.

For PCs, the most popular cartridges are connected to the computer via its USB ports. All PCs and other microcomputers are manufactured with several of these ports, which are used to connect many different peripheral devices, including external storage media.

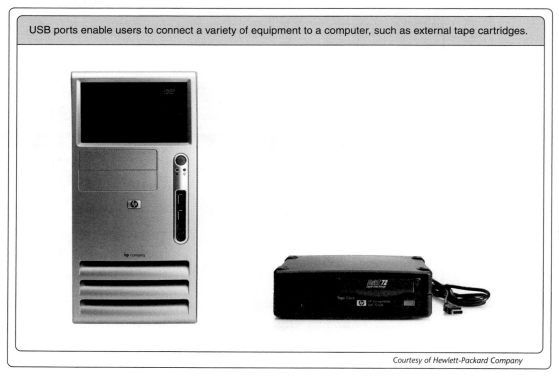

USB ports enable users to connect a variety of equipment to a computer, such as external tape cartridges.

Courtesy of Hewlett-Packard Company

Tapes are inexpensive but they have two major flaws. It takes a long time to copy from a tape. This is a serious concern when terabytes of data must be recopied to a disk from a tape. Tapes are also unreliable after about five years. To extend this period, a magnetic tape must be reeled back and forth every few months to maintain an even tension. Uneven tension, which always develops over time, may render some of the stored data unreadable.

Magnetic Disks

The most widely used storage medium is the **magnetic disk**. Magnetic disks include hard disks and floppy disks. As with information on magnetic tape, information on magnetic disks is coded in magnetized spots on the disk's surface.

Magnetic disks come in several forms, such as an external USB-connected disk and a microdrive that is installed in digital cameras.

Courtesy of LaCie;   Courtesy of IBM Corporation

PCs always come with at least one hard disk built in. (Hard disks are often mistakenly called hard drives. The disk is the storage medium itself; the drive is the mechanism that stores data to it and retrieves data from it. (However "hard disk," "hard drive," and "hard disk drive" are commonly used to mean the combination of the two, because the drive and disk are sold and installed as one unit.) A **hard disk** consists of one or more rigid platters installed in the same box that holds the CPU and other computer components, or attached externally to the computer, usually through a USB port. An external hard disk is portable; it easily can be connected to or disconnected from the computer without opening the computer box. External hard disks are usually more expensive than internal disks with the same capacity. Hard disks are capable of storing up to 1 TB of data. The cost of storing 1 GB has decreased to less than 40 cents.

Spending on storage devices accounts for about 30 percent of all IT expenditures in corporations. In recent years the most important impetus for acquisition of hard disks has been the construction of data warehouses, large databases that maintain mainly consumer purchase records. For example, Wal-Mart, the world's largest retailer, maintains close to 500 TB of consumer data.

The quickly decreasing cost of magnetic disks enables storage and streaming of thousands of video clips on the Web. In just two years of operations, YouTube (now part of Google) amassed a collection of video clips that required 45 terabytes of storage space. The company says it receives and stores 65,000 video clips per day. With an average size of 10 MB per clip, the company probably needs to add close to 20 TB of storage monthly.

Advanced DVDs can hold up to 50 GB of data.

Kyodo/Landov

Optical Discs

Optical discs are recorded by treating the disc surface so it reflects or does not reflect light. A special detecting device detects the reflections or nonreflections, which represent ones and zeroes of digital coding. The two basic categories of optical discs are **compact discs(CDs)** and **digital video discs (DVDs)** , also known as digital versatile discs. CDs come in several types: CD-ROM (Compact Disc, Read Only Memory), CD-R (recordable), and CD-RW (rewritable). Recordable DVDs come in a variety of recording options. The main advantage of optical discs is their storage capacity and portability. CDs and DVDs are also less expensive than hard disks in terms of bytes per dollar, although the cost gap is closing. Standard DVDs can store 4.7 GB per side for a total of 9.4 GB. More advanced DVDs, using techniques called blue laser and double storage, can reach capacities of 50 GB. However, the disadvantage of all optical discs is that the speed of storage and retrieval is currently slower than that of hard disks.

You might have noticed CD drive speeds listed in the form of 52X, 60X, or another X-number. Years ago, the original data retrieval (transfer) rate of CD drives was 150,000 bits per second, because this is the data transfer rate of audio playback. This number represents single speed, or "1X." Thus, 60X means 60 × 150,000 = 9,000,000 bits per second. The greater the data retrieval rate, the more desirable the drive. Note that writable CDs usually have different reading and writing speeds. Reading is often faster than writing. So, you might find that a CD drive reads at 60X but writes at only 24X.

Corporations use DVDs to store massive amounts of information, both for long-term storage and for operational use. They place manuals, drawings, and other large amounts of information that used to fill many books and file cabinets on a single or a few DVDs. Consider that the 32 volumes of *Encyclopaedia Britannica*—over 75,000 articles, including images and sounds—are stored on a single DVD, along with a dictionary and an atlas. In fact, the DVD contains both versions of the *Encyclopaedia*, the Student and Elementary editions.

Optical Tape

Optical tape uses the same technology as optical discs to store and retrieve data. The only difference is that the bits are organized sequentially, as they are on magnetic tape. Like magnetic tapes, optical tapes are made as reels or cassettes. Their storage capacity is enormous. A reel 14 inches in diameter stores over 1 terabyte (1 trillion bytes). A cassette stores about 9 gigabytes. Currently, the main use of optical tapes is in digital video camcorders; the technology is rarely used in corporations.

Flash Memory

Flash memory is becoming popular for both primary memory (memory inside the computer) and external storage. **Flash memory** is a memory chip that can be rewritten and hold its content without electric power. Flash memory consumes very little power and does not need a constant power supply to retain data when disconnected. It offers fast access times and is relatively immune to shock or vibration. These qualities make flash memory an excellent choice for portable devices such as MP3 players, digital cameras, and mobile phones, or as independent portable storage. Unlike other types of memory, erasing data can only be done in blocks of bytes, not individual bytes, and hence the name: a whole block of bytes is erased in a flash.

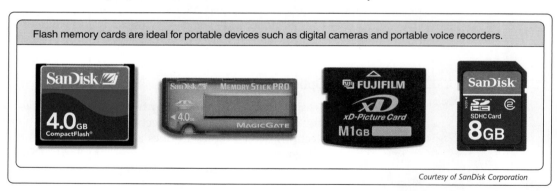

Flash memory cards are ideal for portable devices such as digital cameras and portable voice recorders.

Courtesy of SanDisk Corporation

Flash memory connects through USB ports to any computer. It often holds several gigabytes of data.

Courtesy of SanDisk Corporation

As an independent memory device, flash memory takes two main forms: as a memory card (often used in digital cameras and other portable devices), and as a **USB drive**, sometimes called a thumb drive or USB flash drive. Many computers and some monitors and printers include multiple built-in card readers that accommodate the most popular flash memory cards, such as SD (Secure Digital) and CF (Compact Flash). USB drives are about the size of an adult's thumb, and act as portable storage. (The name "drive" is a misnomer; there are no moving parts or disks in flash memory.) They plug into the computer through a USB port. As USB ports come standard in all microcomputers, it is easy to use a thumb drive to save data or transfer data between computers. There is usually no need to set up any software once the USB drive is plugged in. The device is recognized as an additional external storage device. USB drives come in storage capacities of up to tens of gigabytes, and their cost is decreasing rapidly.

Transfer rate (speed of storage and retrieval) of flash memory in USB flash drives and memory cards is usually indicated as a factor of X, similar to optical discs. A memory card of 133X is considered fast. Cards of the same storage capacity are significantly different in price due to transfer rate.

Flash memory is often called solid-state memory. In addition to its use in USB flash drives and memory cards, it is used in solid-state disks. A **solid-state disk (SSD)** is an alternative to magnetic disks. Again, the word "disk" is a misnomer, because this type of storage involves no disk. SSDs are attached to computers in a similar way to magnetic disks. The fact that there is no need to wait for a disk to rotate in order to locate data—a period of time called latency—makes SSDs up to 250 times faster than magnetic disks, especially if the SSD comes with its own CPU. The function of such CPUs is specifically to speed up data processing. SSDs are used by organizations to store frequently used software to prevent data processing "bottlenecks."

Many computers and some monitors have built-in USB ports and flash card slots. Two USB ports and four card readers are built into this LCD monitor.

DAS, NAS, and SAN

Organizations increasingly rely on storage systems that allow multiple users to share the same storage media over a network. In **direct-attached storage (DAS)**, the disk or array of disks is directly connected to a server. The storage devices might also be tapes, especially if the storage is for backup. Other computers on the network must access the server to use the disks or tapes. DAS is relatively easy to deploy and manage, and involves relatively low cost. However, speed of access to data might be compromised because the server also processes other software, such as e-mail and databases. Also, if the server is down, the other computers cannot access the storage devices. DAS might be suitable for localized file sharing, which is typical in small businesses. It is not easily scalable, because each additional server and its storage devices must be managed separately. Scalability is the ability to add more hardware or software to accommodate changing business needs.

Two other arrangements place the storage devices on the organization's network so that they can be accessed directly by all other computers. These approaches are known as network-attached storage (NAS) and storage area network (SAN).

Network-attached storage (NAS) is a device or "appliance" especially designed for networked storage. It comprises both the storage media, such as hard disks, and management software, which is fully dedicated to serving (accessing) files over the network. NAS relieves the server of handling storage, so the server can process other applications, such as e-mail and databases. Disks can store many terabytes of data in a small, centralized space, and managing such large storage in one place saves money. NAS is highly scalable. While in DAS each server runs its own operating system, NAS can communicate with servers running various operating systems, and therefore allow much flexibility when adding computers and other devices to the network.

Storage area network (SAN) is a network fully devoted to storage and transfer of data between servers and storage devices. The storage devices are part of this dedicated network, which is managed separately from the organization's local area network. (Networks are covered in Chapter 6, "Business Networks and Telecommunications.") A SAN may combine DAS and NAS devices. The communication lines in this network are high-speed optical fibers. The data transfer standards used in a SAN are different from those used by a NAS, and generally support higher speeds. NAS identifies data by files, or, as professionals say, at the file level. SAN identifies much larger quantities of data, called data blocks, and therefore can transfer and back up much larger amounts of data at a time. This is important when high speed of data transfer is important, such as in online business transactions that involve a large number of records in a stored database. A large number of users can simultaneously access data without delays. SANs are highly scalable. For these reasons, SANs are used by organizations that conduct business on the Web and require high-volume transaction processing. However, SANs are relatively expensive and their

Computers May Be Hazardous to Your Health

According to the U.S. National Institute of Occupational Safety and Health (NIOSH), about 75 million Americans—more than half the workforce—have jobs that require them to sit in front of a computer for many hours daily. An increasing number of studies show that working with computers threatens workers with a variety of hazards. These risks include repetitive-stress injuries (RSIs) due to long periods of repeated motions. According to the U.S. Bureau of Labor Statistics, RSIs cost American businesses an estimated $33 billion annually in workers' compensation claims. The U.S. Department of Labor estimates that about two-thirds of the reported injuries are due to working with computers. As computer-aided work has grown, RSIs have grown, too, to the extent that some scientists call these injuries an epidemic.

The most common computer-related type of RSI is carpal tunnel syndrome. It is the result of repetitive use of a keyboard. The injury causes pain in the forearms due to swelling and pressure on the median nerve passing through the wrist. Carpal tunnel syndrome may cause permanent disability. In rare cases workers lost their ability to return to work due to this injury.

Our eyes, too, are strained from computer work. Studies found that a programmer's eyes make as many as 30,000 movements in a workday. These are movements up, down, and to the sides, which strain the eye muscles. However, other studies found that while staring at a computer monitor people blink at one-sixth of the frequency that they blink normally. Blinking is important for moisturizing the eyeball, which helps kill harmful germs and eases eye strain. A study by NIOSH found that short breaks from work with computers that involve keyboards and video displays reduce eye soreness, visual blurring, and upper-body discomfort, while quantity and quality of work were not compromised. The agency estimates than more than half of those 75 million Americans who

stare at computer displays for long hours develop a health problem called computer vision syndrome (CVS), which is any combination of headaches, loss of focus, burning eyes, double vision, or blurred vision. The American Optometric Association reports that about 14 percent of patients schedule eye exams because of CVS.

The argument has been made that it is an employer's moral obligation to educate employees about such risks and to provide an environment that minimizes them. Both factors, the economic and ethical, have moved many employers to try to reduce the increasing "injuries of the Information Age." They do so by purchasing and installing ergonomic equipment, training employees how to use computers in a way that minimizes injuries, and enforcing periodic breaks from repetitive activities such as typing. The breaks help prevent both RSIs and eye strain. The Occupational Safety and Health Administration (OSHA), an arm of the U.S. Department of Labor, maintains a Web site, *www.osha.gov/SLTC/computerworkstation*, that provides useful tips on safe computer work. As a professional, it is likely you will spend much of your workday sitting in front of a computer. Read the tips and apply them to maintain your good health.

To minimize these risks, you can download and install on your computer one of several free programs. For example, Workrave—available at workrave.org— forces the user to take brief breaks ("micropauses") and less frequent but longer "rest breaks." It also limits the daily total time a worker can use a computer. When the computer is networked to other computers in the office, it does not allow the worker to use any of the others that are part of the network when a break or daily time limit is enforced. Working times of all workers whose computers are connected are recorded and tabulated on a server for review and analysis. The application also provides an animated exercise guide for the shoulders, arms, and eyes.

---

management is complex. In recent years, the technical differences between NAS and SAN have blurred.

DAS, NAS, and SAN often include **RAID** (redundant array of independent disks), whereby data is replicated on different disks to enhance processing speed and fault-tolerance. **Fault-tolerance** is the ability of the system to sustain failure of a disk, because the same data also appears on another disk.

Several companies specialize in NAS and SAN systems and the software that manages them, including Network Appliance, EMC, Hewlett-Packard, Hitachi, and IBM.

Business Considerations in Evaluating Storage Media

Before spending money on storage devices, professionals must consider several factors: the purpose of data storage, the amount of data to be stored, the required speed of data storage and retrieval, how portable the device needs to be, and, as always, cost.

Use of Stored Data

The first consideration before adopting storage media is how the data will be used, mainly, whether it will be used for current operations or as backup. If it is to be used for backup only, and not processing, magnetic tape, CDs, or DVDs would be a proper choice. Magnetic tape is less costly and holds more data per reel or cassette than a single CD; this should be a consideration, too. If the users need to access individual records quickly, then magnetic hard disks are the best choice. Thus, a business that allows customers to retrieve their records online should use fast magnetic disks. If the information is archival, such as encyclopedias or maps used by library patrons, the library should place the information on CDs or DVDs, because the user needs fast, direct retrieval of specific information (records), and might not tolerate sequential search on a tape. Archival information that should not be changed should be stored on write-once media.

Amount of Data Stored

When storage volume is the most important factor, professionals must first consider price per megabit or megabyte, that is, the ratio of dollars spent to storage capacity. If the medium is to be used solely for backup, their low cost makes magnetic tapes and DVDs an ideal choice. If the medium is to be used for fast retrieval, magnetic disks would be the best choice.

For some purposes, the capacity of the device is important. When a set of very large software applications and/or data must be stored on a single device, a device with a large capacity must be selected. For example, if a sales rep must be able to demonstrate applications totaling 4 GB, it might be more economical to store the data on five CDs, but this would be impractical because the rep would either have to first copy the content of all the CDs onto every PC where she makes a demonstration (which for security reasons might be prohibited by the hosting party), or she would have to swap the CDs throughout the demonstration. A small portable hard disk or USB flash drive of at least 4 GB would be a more practical option, albeit significantly more expensive.

Speed

The speed of magnetic disks (also called spindle speed) is often measured in rotations per minute (RPM). Current disks come with speeds of 5,400 to 15,000 RPM. For disks of the same size, a higher RPM means shorter data transfer time and usually better performance overall. While the great capacity and low cost of CDs and DVDs are appealing, the transfer rate of magnetic hard disks is still significantly better. If very high speed is required, SSD is currently the best choice, although its price is significantly higher than that of magnetic disks.

POINT OF INTEREST

Don't Place This Laptop on Your Lap

Defects in Sony batteries caused the largest recall in computer history. The batteries in several models of laptop computers overheated. Some exploded, and others burst into flames. Dell replaced the batteries in 4 million computers, Apple in 1.8 million, and Lenovo in 0.5 million. The Federal Consumer Safety Products Commission issued this helpful advice: "Do not use your computer on your lap."

Source: Horowitz, A., Jacobson, D., McNichol, T., Thomas, O., "101 Dumbest Moments in Business," *Business 2.0*, January/February 2007, p. 100.

Unit Space and Portability

Sometimes the cost of a gigabyte stored is not the most important consideration, but the physical size of the storage medium is. A portable hard disk drive might be economical and fast, but it is more practical for a traveling salesperson to carry a CD rather than an external hard disk. And even though a CD is significantly less expensive than a USB flash drive, the salesperson might find it more convenient to carry a 4 GB USB drive than carrying several CDs. CDs do not fit in shirt pockets, while a USB flash drive can be attached to a key chain or clipped to a shirt pocket. Even if storage cost is not as attractive as that of CDs, portability and the fact that USB ports are ubiquitous in PCs might push one toward selecting a USB flash drive.

Cost

Once professionals agree on the best type of data storage device for a particular business use, they need to consider cost. The approach is simple: obtain the greatest storage capacity for the smallest amount of money. In other words, for each proposed device, consider the ratio of cents per gigabyte of capacity. The lower the ratio, the more favorable the product. It is easy to find the ratio. If a 300 GB hard disk costs $120, the ratio is $120/300 GB, or 40 cents per gigabyte. If a 4 GB thumb drive costs $30, the cost per gigabyte is $30/4 GB, or $7.50. Thus, if the convenience and portability of a thumb drive is important to you, you will pay significantly more per GB of storage capacity.

Reliability and Life Expectancy

Although this is usually not the highest priority, businesses must also consider the storage medium's reliability and life expectancy. For instance, optical discs are more reliable and durable than magnetic disks. Magnetically stored data remains reliable for about 10 years, whereas CDs and DVDs are expected to store data reliably for 50 to 100 years (although they have not been around long enough to prove that).

Trade-Offs

As you can see, several factors must be considered when purchasing storage media, and often you must trade one quality of the device for another. For example, while USB drives are convenient and fast, they are also expensive and unacceptable for storing large amounts of transactional data, or even backing up large amounts of data, because of their relatively small capacity. Figure 4.7 summarizes characteristics of the most popular storage media. Obviously, terms such as "moderate cost" and "high capacity" are relative. Storage capacities and speeds of almost all storage media have increased over the years, and costs have decreased. Thus, the specific capacities, retrieval speeds, and costs change all the time. The table is presented for general comparison and reference, whereby "high" and "low" for each medium are relative to the other media.

FIGURE 4.7
Characteristics of storage media for business purposes

Medium		Capacity per Device Size	Recording and Retrieval Speed	Cost ($/GB)	Ideal for...	Capacity per Device	Limitations
Magnetic Hard Disk 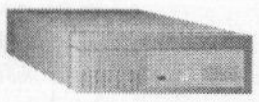		High	Very High	High	Immediate Transactions	Very High	Bulky, Heavy
Magnetic Tape 		High	Slow	Very Low	Backup	Very High	Not Suitable for Immediate Processing
Optical Tape 		Very High	High	Low	Backup	Very High	Limited Market
Recordable CD 		Very High	Medium	Very Low	Backup, Distribution of software	Low	Low Capacity per Device
Recordable DVD 		Very High	High	Very Low	Backup	Medium	Low Capacity per Device
Flash Memory 		High	High	High	Backup, Portability	Medium	Expensive

CONSIDERATIONS IN PURCHASING HARDWARE

Decisions about purchasing computers are usually made by an organization's IT professionals or with the help of a consulting firm. But surveys show an increasing trend of involving other employees in the decision-making process. More and more companies realize that effective use of computers depends on whether their employees are satisfied with the computers and other equipment installed in their workplace.

Before deciding what to purchase, consider the following variables:

- *The equipment's power:* Its speed, its memory size, and the capacity of its storage devices, such as the hard disk installed in the computer.

- *Expansion slots:* Computers should have enough slots to add circuitry cards for additional purposes, such as adding more powerful graphic cards and wireless cards on the motherboard (the board on which the CPU and other circuitry are installed). Additional memory cards increase the speed of processing by allowing more concurrent programs and data to run.

- *The number and type of external ports:* **Ports** are sockets used to connect a computer to external devices such as printers, hard disks, scanners, remote keyboards and pointers, and communication devices. More ports give more flexibility. Because so many external devices—hard disks, printers, scanners, thumb drives, digital cameras, presentation "clickers," and many others—connect to the computer through a USB port, the greater the number of USB ports, the more external devices can be added at the same time. Although USB hubs (devices that connect to a single port and provide several) can be used, this may cause inconvenience and increased costs. Built-in multiple card readers for flash memory make it convenient to read data from the cards instead of connecting the device that houses them, such as digital cameras.

- *The monitor type and resolution:* Higher resolution is more pleasing and less straining to the eyes. Larger monitors allow viewing the windows of many software applications simultaneously and require less scrolling.

- *Ergonomics:* Ergonomic equipment does not strain the back, arms, and eyes. For example, working with the keyboard must be comfortable. Traditional keyboards cause muscle pain when used for long sessions. Consider purchasing an ergonomic keyboard. Consider a trackball instead of a mouse; it requires only moving fingers rather than the forearm or the entire hand.

- *Compatibility:* IT managers must ensure that new devices will integrate with existing hardware, software, and networks. A new computer might have a different operating system or internal architecture. If it is to be used to host an important application, care must be taken to ensure that the application will run on the new machine. For example, commercial software vendors guarantee that their applications will run on a list of processors and operating systems. Professionals must consider **backward compatibility**, in which newer hardware is compatible with older hardware. (The same term applies to software.) For example, USB 2.0 devices are backward-compatible with USB 1.1 ports (although the communication speed then deteriorates to the speed of the older port). Compatibility between hardware and networks is also important. Newer handheld devices such as bar-code scanners might use an updated communication standard and no longer communicate with an existing warehouse network, because the new devices are not backward-compatible with the older standard transceivers.

- *The hardware footprint:* If space is scarce, you might want to consider the size of the computer and its peripheral equipment. The footprint is the area that a computer occupies. A smaller footprint leaves more desk space for other devices. This is one of the major reasons for adopting flat-panel monitors when they first appeared on the market.

- *The reliability of the vendor, the warranty policy, and the support given after the warranty expires:* Ask if the vendor provides a Web site and 24-hour help via telephone. Try to assess how soon the equipment will be obsolete, a difficult task given the reality of fast development in computer equipment.

- *Power consumption and noise:* Computers that consume less power help save money on electricity and usually also give off less heat. Computers use fans to cool down the circuitry. Quiet fans will make the work environment more pleasant.

- *Cost:* All of the preceding factors must be weighed against cost. Careful study might yield hardware with excellent performance for an affordable price. Perusing print and Web-based trade journals is helpful. Many periodicals provide tables evaluating comparable hardware, based on laboratory tests by impartial technicians. You do not have to be an IT professional to understand their evaluations.

Figure 4.8 summarizes the factors discussed in this chapter that you should consider when purchasing hardware. When comparing computers from different vendors, it is useful to establish a 10-point scale and score each category to indicate how well each computer addresses each important item. Your organization's or even your department's internal needs may require you to add some factors. The equipment receiving the highest score is the best in the evaluator's opinion.

Scalability and Updating Hardware

IT managers try to extend the productive life of hardware by ensuring that any equipment they buy is scalable. The principle of **scalability** implies that resources—in this case, hardware—can accommodate a growing amount of work either with or without upgrading. A scalable system can provide increased power as demands increase. For instance, many servers are designed to use multiple processors—4, 8, or 16 is not uncommon. If the server is initially installed with only a small number of processors, say two, then processors can be added over time to increase computing power. This way the machine will not have to be discarded too soon, and this helps protect the organization's initial investment. The same can be done for memory, storage, and other components.

However, some hardware is not scalable. Businesses tend to update their software, especially operating systems (such as Windows), when a new version is available, but many still maintain old hardware. While they avoid the cost of purchasing new hardware, this might actually cost the companies in lost productivity: newer software cannot run as fast or as reliably on the old machines. Often, excellent features of newer software are not available if it runs on older machines. For example, although Windows Vista offers greater security, faster file management, and superior visual effects over earlier Windows versions, most PCs at the time of its introduction were not powerful enough to run the new operating system.

Hardware should be disposed of and new hardware should be installed to avoid performance gaps between software and hardware. One rough formula to help determine when to replace hardware is the ratio of the average age of hardware pieces to the average age of the operating systems running on the machines. If the ratio is less than one, it might be time to replace some or all of the hardware.

If you are concerned that the equipment's useful life might be short because more powerful computers might be available within months, you can lease your system instead of buying it. Many vendors offer leasing programs. However, note that vendors are also aware of how quickly hardware becomes obsolete and price the leases accordingly; thus, you might find that the lease payment often covers the purchase price within a mere 18–24 months. Yet, many firms prefer leasing their PCs and notebook computers to purchasing them.

As you will see throughout this book, hardware components are combined in many different configurations to help businesses streamline operations and attain strategic goals. But hardware is rarely the first consideration in acquiring a new IS. When planning a new IS, managers should first determine their business needs and then consider which software can support those needs. Only then should they select the hardware that supports the software. The next chapter focuses on software.

FIGURE 4.8
Example of a hardware evaluation form

Factor	What to look for	Score
Power		
Speed	Greater frequency and word size	_____
RAM capacity	Larger	_____
Expandability	Greater number of board slots for additional devices and memory	_____
Ports	Greater number of ports for printer, external hard disk, communication devices, and other peripherals	_____
Ergonomics	Greater comfort and safety	_____
Compatibility		
with hardware	Compatibility with many other computers and peripheral devices from the same and other manufacturers	_____
with software	Compatibility with many software packages currently used and potentially to be used	_____
Footprint	Smaller area	_____
Support	Availability of telephone and online support for troubleshooting	_____
	Supply of information on new upgrades	_____
Warranty	Longer warranty period	_____
Cost	Lower cost	_____

- More professionals outside the IT field find themselves in the decision-making role regarding the purchase and use of computer hardware. Therefore, understanding hardware is important.

- For ease of reference, computers are classified into several categories according to their power. The most powerful are supercomputers, used mainly by research institutions for complex scientific calculations. Somewhat less powerful, but more suitable for business operations, are mainframe computers; many organizations still use them to process large databases and perform other tasks that require speed and large primary memory. Midrange computers are less powerful than mainframe computers and are often used as servers. Microcomputers include PCs and smaller computers, such as notebook, handheld, and tablet computers.

- Regardless of their size and power, all computers must have several components to function. The "brain" of every computer is its central processing unit (CPU), which consists of circuitry on a piece of silicon wafer and controls four basic operations: (1) it fetches instructions from memory, (2) it decodes them, (3) it executes them, and (4) it stores the results in memory.

- The rate at which the CPU does all this is the computer's clock rate.

- A computer's data word is the number of bits that can move through its CPU in one machine cycle.

- Speed, memory size, and the number of processor cores are among the determinants of a computer's power.

- The larger part of a computer's memory, RAM (random access memory), is volatile; that is, it keeps data only as long as electrical power is supplied. ROM (read-only memory) is nonvolatile.

- Unlike data in RAM, data stored in ROM stays in ROM when you turn the computer off. Similarly, all secondary storage media, such as magnetic disks, optical discs, and flash cards, are nonvolatile.

- Imaging devices help process large amounts of text and graphic data and have made the work of banks and other industries more productive.

- When evaluating storage media, factors to consider are capacity, transfer rate, portability, and the form of data organization that it allows. The latter determines the mode of access (sequential or direct).

- Data stored on tapes can only be organized and retrieved sequentially, therefore tapes are good for backup but not for transactions. Direct access storage devices, such as RAM, magnetic disks, and optical discs, allow random organization and retrieval. Direct organization provides faster storage and retrieval of records that must be accessed individually and quickly, such as records in airline reservation systems. Only direct-access devices are suitable for processing databases.

- When purchasing computers, professionals should consider computer power and other factors in addition to cost. Professionals should consider expandability of RAM, the availability of sockets (ports) for connecting peripheral equipment, and compatibility with existing hardware and software.

- Like many new technologies, information technology may pose health risks to users. The most common problems computer users experience are carpal tunnel syndrome and repetitive-stress injuries caused by the repetitive use of the keyboard over long time periods. Today, manufacturers of computer equipment pay more attention to health hazards and try to design devices ergonomically.

QUICKBIZ MESSENGERS REVISITED

QuickBiz's business has expanded from a one-person bicycle messenger service to a company with bicycles, cars, and trucks, as well as main office staff. As it expanded, the firm has upgraded its information systems to streamline its processes and handle its increasing customer load. Let's examine some of the changes it has made.

What Would You Do?

1. QuickBiz has used many different types of input and output devices throughout its history. How many can you identify? Create a two-column chart and list them under the headings Input and Output. Can you think of any other devices or technologies they haven't used yet that might help them?

2. Consider QuickBiz's change in storage media. It moved from magnetic tape backup to rewritable CD to DVDs. Go online to investigate the costs and capacities of current tape, CD, and DVD storage

systems. How do they compare? Do you think Andrew and Sarah were wise to change systems? Why or why not?

New Perspectives

1. Review Andrew and Sarah's decision to buy a server and handheld computers. What advantages does source data technology give to the messengers themselves? To the central office staff?

2. Seattle had a 6.8 magnitude earthquake. QuickBiz's main office suffered some damage during the quake. Its main information system was down for two days. Luckily, QuickBiz messengers could still make deliveries and save data on their handheld computers. But the crisis got Andrew thinking that his business needed additional safeguards. Discuss with your classmates and list some ways that QuickBiz can make sure its data and main information system can be backed up in case of a disaster.

KEY TERMS

REVIEW QUESTIONS

1. You have decided to buy parts and build your own personal computer. At the minimum, what are the components that you would need for this device to be considered a computer?

2. Modern CPUs contain cores. What is a core?

3. Multicore CPUs facilitate multithreading. What is multithreading?

4. Most people never get to see a supercomputer, let alone use one. Why? What are the most frequent uses of this type of computer?

5. Why are computers designed to work in binary form rather than by using multiple-value signals? Try to use the analogy of colors to explain your answer.

6. News about the death of mainframe computers has been greatly exaggerated. Explain.

7. IT professionals often speak of the merging of technologies. Think of handheld computers and cell phones. Give an example of such merging.

8. When a computer is offered for sale, one of its advertised characteristics is something such as "4 GHz." What does this mean, and what does it measure?

9. Why are computers said to be processing data digitally?

10. What is the difference between volatile and nonvolatile memory? Give one example of volatile memory and one example of nonvolatile memory.

11. What are the main qualities to look for in an LCD monitor?

12. Among the external storage devices discussed in this chapter, all but one store data on the surface of some material, and one in circuitry. Which one stores data in circuitry?

13. What is DVD technology? How does it differ from CD technology?

14. What does footprint mean in hardware? When is a footprint important in the office?

15. What are the most important features to consider before purchasing a PC?

16. On a continental tour, a traveling salesperson makes software-based presentations at every place he stops. He has ensured that there is a PC and projecting equipment at every site he visits. Occasionally, he needs to change the content of his presentation. He wants to carry as small a storage device as possible. What data storage device would you recommend he carry?

DISCUSSION QUESTIONS

17. Computers fail significantly less frequently than copy machines and printers. Why?

18. Comment on this statement: large computers, such as mainframes and supercomputers, have no future.

19. Because information technology advances so rapidly, professionals find it difficult to make informed decisions regarding computer and peripheral equipment purchases. What factors cause this difficulty?

20. End users' role in making hardware purchasing decisions is growing. Analyze the technological and operational reasons for this trend.

21. Would you replace a PC with a handheld computer for your studies or work? Why or why not?

22. Which storage medium would you use in each of the following situations: (1) airline reservations system, (2) information on employee benefits and professional conduct, and (3) online answers to customers' frequently asked questions (FAQs)? Explain your choices.

23. What health hazards are associated with computer use? What can be done to alleviate each type of health risk? Should the government pass laws to protect employees against such hazards?

24. The miniaturization and merging of technologies into highly portable devices has caused some annoyances. Give some examples.

25. Comment on the following statement: the useful life of a PC is about two years, therefore, it is not important whether the vendor is still in business in two or three years.

26. About 18–24 months into the life of a PC, a new PC becomes available that is twice as powerful. As a result, many IS managers opt to lease, rather than buy, PCs for employees. What factors would you consider in deciding whether to buy or to lease?

27. Thanks to DVD and other advanced technologies, a PC can combine the functions of a computer, telephone, fax machine, and television set. Would you give up your home telephone and television set if you could use your PC to make calls and watch television? Why or why not?

28. Sometimes useful information might be lost, not because the medium on which it was stored deteriorated or was damaged, but because no device was available to retrieve the information. How could that happen? Can you give examples?

29. You might have heard of the electronic book, a handheld device that allows readers to read a book from a CD. What are the advantages and disadvantages of such devices when compared with traditional books? Think in terms of portability, text clarity, searching for specific words or pages, and so on. What would you prefer: an electronic book or a paper book? Why?

30. Observers say that personal computers have become a commodity. What does the term "commodity" mean? How could this development impact businesses and homes?

31. A mechanic once recommended that the author of this book not purchase a car that has too many computer chips, because if those chips fail, they must be replaced; mechanics cannot fix them. Would you take the mechanic's advice? Why or why not?

32. Try to count how many hours per week you use a personal computer: at your home, in the PC lab, in the library, or elsewhere. Do you consider yourself "computer addicted"?

33. What do you expect will be the most popular storage devices for personal use in five years? What will be the most popular nonportable storage devices for corporate use in five years? Why?

34. Almost daily a new electronic device, often one that combines several technologies, is offered for sale. People sometimes refer to these devices as "gadgets," which hints that they might be nice to have but not really necessary or even useful. How do *you* delineate the difference between a gadget and a helpful device?

APPLYING CONCEPTS

35. Recommend one of the three hardware configurations described in the following table for each of the scenarios listed. Assume that all of the hardware configurations cost the same. Explain your choices.

    a. The employees of this firm do a lot of graphic design work. Graphics require large programs. Printouts must be high quality.

    b. This firm uses the computer mainly for word processing. The biggest application occupies 24 MB.

    c. Employees of this firm use scientific programs that run for many hours.

    d. It is imperative that employees be able to print reports quickly with reasonable print quality. They almost always print their reports from portable storage devices.

Features	Computer Configuration		
	X	Y	Z
RAM	1 GB	2 GB	1 GB
External storage			
Hard disk	200 GB	120 GB	60 GB
Thumb drive (USB 2.0)	256 MB	512 MB	256 MB
Speed (clock rate)	1.7 GHz	3.06 GHz	5 GHz
Printer	Laser	Ink-jet	Laser
	1200 DPI	600 DPI	600 DPI
	20 PPM	12 PPM	16 PPM

36. Assume you can choose among magnetic tapes, magnetic hard disks, recordable optical discs (CD-R: write once, read many), and flash memory USB drives. Consider each scenario independently of the others. For each of the following purposes, explain which one of the media you would choose and why. Start by saying which medium you have chosen. Then explain why.

a. You need to store thousands of employee records for several years. This is only a backup procedure. The information will never be processed from the backup medium.

b. The storage medium is used as part of an airline reservation system.

c. Your business sells machines that must be maintained well by your clients. You wish to provide them with a digital version of the maintenance manual. The manual includes an index (like one at the end of a book) with links to the proper pages.

d. You are a sales manager who travels often. You must store a large PowerPoint presentation that you show to prospective customers in their office. You do not carry a laptop computer, but there is a PC wherever you go. You do not want to carry CDs, because you found that the graphic-rich presentation moves too slowly from CDs.

e. You have a business on the Web. You maintain your own server and site. You provide much textual and graphical information from the site. Customers can search products and make purchases.

f. You want to store all the paintings of impressionist painters for use by your local library patrons. Patrons can search by artist name, artist nationality, or the painting's topic. The library would like multiple copies of what you store, and to be able to loan them to patrons for viewing at home.

g. You use the medium for a large database that your employees manipulate frequently.

h. You work for the IRS, and you need to archive the tax records of millions of taxpayers for several years. The archiving is done after all processing of tax filings are complete and after all refunds and payments have been made. IRS employees must occasionally go back and retrieve specific records from these files, and when they need a record, they want to access it directly.

37. Search the Web for remote-control devices to use with presentation applications such as PowerPoint. (Go to Web sites of online PC vendors such as cdw.com and pcconnection.com.) Examine the pictures of five different units. Summarize your thoughts about the ergonomics of these devices.

HANDS-ON ACTIVITIES

38. Your company is about to open a new branch. You were selected to equip the office with 20 personal computers, 10 notebook computers, and 5 laser printers. Management has asked that you purchase all the equipment from a single online vendor. Each PC must be purchased complete with a 19-inch LCD monitor. After interviewing employees about their typical computing needs, you developed the following scale:

PCs: Every 1 MHz of clock rate receives 1 point; every 1 MB of RAM receives 10 points; every 1 GB of hard disk storage receives 1 point. For CD-RW, each 1X of reading speed receives 1 point (writing and rewriting speeds are not essential, but the capabilities are required).

LCD monitors: Every 1:100 of contrast ratio gets 10 points. Other features are not essential.

Laptops: The same scoring as for PCs.

Printers: Every 1 PPM receives 100 points; every 1 DPI receives 1 point.

Research three online vendor sites for this equipment. Prepare a spreadsheet table with three columns, one for each vendor, and enter the information you found about each piece of equipment for each vendor. Enter a formula to add up the total number of points at the bottom of each column. Do not consider any factor that is not mentioned here. Find the vendor whose total points per dollar is the highest.

39. Try to forget the shapes of PCs, monitors, keyboards, and mice. Write a two-page description of your own ideas for an ergonomic workstation. Explain what about today's PCs and peripheral equipment does not fit human hands, eyes, and ears, and how you would like to change these devices' features and shapes for more comfortable and effective use. Be as revolutionary as your imagination allows.

40. Use a spreadsheet application to prepare a table that clearly shows (both in text and numbers) how to calculate the following. A music CD contains 750 million bytes. How long does it take to play all the music on it, assuming the disc plays at 1X? If the CD contains data, how long would it take to retrieve all the data from it into a computer's RAM, if you used a 60X CD drive?

TEAM ACTIVITIES

41. Your team has received $2,500 to purchase a computer system. Assume you have no equipment; everything needs to be purchased. Use the evaluation form in Figure 4.8. Visit the Web sites of three computer hardware vendors, and write down specifications of three sets of equipment. Include in each set a computer, a keyboard and mouse (or trackball), a compatible 19- or 21-inch LCD monitor, and a black-and-white laser printer. Your team should evaluate the features of each configuration, on a scale of 1 to 10 (1 = worst; 10 = best), and total the points. Which configuration (and, therefore, vendor) would you recommend to your fellow students? If you cannot spend your entire $2,500, any surplus should be considered a benefit. Be ready to explain your recommendation.

42. As in Activity 41, assume you have $2,500 available. You are to purchase your ideal PC, monitor, and printer, while utilizing all or almost all of your budget. Shop the Web for these devices, list them (item name, vendor, and capabilities) and their prices, and rationalize why this is the ideal system for your needs and desires.

FROM IDEAS TO APPLICATION: REAL CASES

Better Storage for Our Best Friends

"At Petco animals always come first," says the Web site of Petco Animal Supplies, Inc. Established in 1965, the company is a leading retailer of pet supplies, from cat collars to aquariums to pet food. It operates more than 850 stores in 49 states and the District of Columbia, and sells more than 10,000 pet-related products. To accommodate customers, it also operates an online store.

Because Petco maintains thousands of products in its warehouses, and because receiving and shipping takes place often, it must track in real time the location of each item. In its three main warehouses, workers use handheld devices equipped with both barcode and RFID capabilities. They are used to scan product barcodes, record inventory receipts, and track shipping instructions and execution. Because of the large number of stores and the huge variety of items, Petco is highly sensitive to disruptions and downtime of its information systems. Downtime may cause significant financial damage due to lost sales.

All this data must be recorded and backed up. Petco used magnetic tapes to back up data every few hours, but the approach was far from ideal. Tape backup is reliable, but the latency of a few hours posed a risk. If electric power was lost, so was several hours of data that could not be recorded. Also, recording on tape is labor-intensive, because tapes must be manually mounted and dismounted. Rewinding tapes is time-consuming, and therefore delays availability of new tapes for recording.

Another issue with data storage was that the company used the DAS (direct-attached storage) approach: each computer backed up to its own magnetic disk. The data could not be shared by all computers. This created two problems. Many of the disks were underutilized; much space—up to 50 percent—was never used. As the company grows, the total underutilized disk space grows as well. In addition, sharing the stored data was challenging.

To overcome these problems, Petco IT staff tried mirroring. In disk mirroring, the entire disk is automatically copied to a backup disk. While this reduces labor and makes data available immediately from the mirror disk, it also presents a problem. If the original disk is corrupt, such as infected with a virus, so is the mirror disk.

The IT staff examined SAN (storage area network) and NAS (network-attached storage) solutions. It found that SAN would require much maintenance, while NAS required much equipment to handle data communications. Petco opted for a system called iSCSI provided by Network Appliance, Inc., better known as NetApp. The system of backing up to DAS was replaced with backing up the Petco computers to NetApp servers over the Internet. iSCSI (pronounced "eye scuzzy") utilizes the existing Internet standards and network, and provides very fast data transfers. The adopter does not need to incur the typical expense of optical fiber networks associated with SAN. The magnetic disks and the software that manages them were implemented in the three main warehouses.

Compared to the DAS approach, using such a system reduces the total amount of required storage capacity, because much of the capacity of directly attached disks is never used. The new arrangement does not require as much storage planning as was required with DAS. The company can add storage capacity whenever data management needs require it. This eliminates wasted money spent on excess capacity. Thus, the storage system is scalable. It is easy and inexpensive to add more disks at any of the three warehouses.

Another benefit of using this technology was that the system could be installed without interruption to warehouse operations. In fact, warehouse workers did not notice the change. They left work on Friday, and when they returned on Monday morning everything looked the same to them.

Source: Pettis, A., "Petco's New Storage Gear is the Cat's Meow," *eWeek*, March 13, 2006; (www.petco.com), March 2007.

Thinking About the Case

1. What were the data backup problems when Petco used tapes?

2. What were the data backup problems when Petco used mirroring?

3. What are the disadvantages of using DAS, and how are these disadvantages compounded when a company grows?

4. What benefits did Petco acquire when adopting the current technology for backing up warehouse data?

Lean Times Require Innovative Systems

If necessity is the mother of invention, then Rock County, Wisconsin, has necessarily positioned itself on the leading edge of technological innovation. Supporting 29 different county departments, Rock County's IT team faces the same economic and budget realities confronting governments throughout the country. But Rock County officials are attempting to lower the cost of county operations through the use of technology. Cost-effectiveness is especially important in this jurisdiction of nearly 154,000 people located in the economically hard-hit southeastern section of Wisconsin. Rock County serves citizens scattered over 720 square miles on an annual budget of about $145 million.

"The pervasive theme of our county operation right now is driven by our budget realities," said Mickey Crittenden, director of IT for Rock County. "We can't necessarily seek the latest and greatest hardware and software—although we would like to—instead, we're focused on using technology to lower overall costs. And we've been quite successful in doing that."

Cost management has come in many forms for this large county that stretches along the meandering Rock River. But the overriding principles are basic: standardization and simplification. Consolidation has been a key issue—especially for servers. "We're trying to simplify our infrastructure by limiting the number of servers that we have installed throughout our data centers," Crittenden explained. "The big benefit of this is that it is easier to manage fewer servers, and there's also significant cost savings with licensing and maintenance of the equipment." Previously, all applications in use by the 29 departments were hosted on servers distributed over several sites. Varying hardware and systems spaced throughout the county made interagency communications nearly impossible and maintenance a costly nightmare.

In consolidating servers—including Hewlett-Packard (HP) NetServers, 9000 series, e3000 series, and HP ProLiant servers—Rock County centralized its processing resources. This provided the county with a centralized, highly adaptive system that is secure, scalable, and easily managed. The change saved time and money as IT staff members had to support fewer types of servers and continued to seek new and inventive ways to stretch shrinking budget dollars.

HP StorageWorks tape library enabled Rock County officials to quickly perform automatic backups each evening. "We're a lights-out operation at night, and we're using our tape library to back up all our servers and make sure all those servers are up and running in the morning without worrying that a backup procedure hasn't finished," Crittenden said.

Despite the consolidation initiative's extraordinary success, Rock County's IT workers manage a smaller number of servers overall. Uptime is maximized, and communication between county ISs is nearly seamless. IT professionals easily deploy new applications from a single source instead of traveling from department to department, or worse yet, from desktop to desktop. Once an application is placed on a server, it is available to all who might need it, remotely. "The result is a savings of tens of thousands of dollars for Rock County taxpayers," said Crittenden.

A storage area network (SAN) also helps maximize the use of county IT resources. Linking storage equipment on a single network allows Rock County to use the devices more efficiently. "Previously, we had servers with directly attached storage," explained Crittenden. "That meant some servers would be using 95 percent of their available disk space while others would be using only 15 percent. We just didn't have the storage where we needed it." HP SAN technology lets the county consolidate disk storage and allocate the capacity according to application requirements.

High-volume printing at county offices is also a cause for high costs. The IT Department had examined ways to cut these costs. "HP is helping us with the process of determining whether it is better to perhaps use one very capable high-end printer with a variety of features and functionality for a given department rather than having separate locally attached printers for employees," explained Crittenden. Currently, Rock County owns more than 500 heavily used HP laser printers. Only two of them are linked to a network through a server for centralized printing. Printer centralization allows agencies to better manage their printing and imaging needs, which can reduce overall expenses and increase user productivity. Crittenden believes the county's current centralization efforts will provide even greater IT efficiencies and cost savings.

In another cost-saving and efficiency-driving move, Crittenden has ventured onto the leading edge of technology by further empowering Rock County's current 1,000 desktop machines through implementation of a thin-client desktop strategy. A thin client is any personal computer that uses resources of a server rather than its own. Thus, the machine each employee uses will be much cheaper because it needs only a small hard disk. All applications and data will reside on the servers to which these personal computers are linked, and all files saved by county workers will be saved on the servers or SAN disks. If access to more software and files is required, the same desktops can still be used. Instead of upgrading them, the power of the servers can be upgraded, or the servers themselves, rather than 1,000 desktop computers, can be replaced.

Source: "Case Study: Streamlining Government," (www.hp.com), June 1, 2005.

Thinking About the Case

1. Why is centralization of resources—storage devices, application servers, and printers—so important in the particular case of Rock County?

2. A SAN enables Rock County to use storage devices more efficiently. Do further research on the difference in storage allocation between DAS and SAN. How does the SAN enable more efficient storage than the previous DAS arrangement at the county?

3. Centralized printing can increase productivity only if offices are not dispersed over a large area. Why?

4. What was the major change that reduced the cost of maintenance by the IT Department?

FIVE

Business Software

LEARNING OBJECTIVES

Hardware, as powerful as it might be, is useless without software. Software consists of instructions that tell the computer and its peripheral devices what to do and how to do it. These instructions are called programs or applications. Many IT professionals refer to computer programs as "systems" because they are composed of components working to achieve a common goal. As a professional, you must be able to make educated decisions regarding software selection. To do so, you need to understand the factors involved in developing, selecting, and using software.

When you finish this chapter, you will be able to:

- Explain the difference between application software and system software.

- Enumerate the different generations of programming languages and explain how they differ.

- Cite the latest major developments in application and system software.

- Identify and explain the roles of Web programming languages.

- Explain the types and uses of Web site design tools.

- Clarify the differences between proprietary software and open source software.

- List characteristics that are important in evaluating packaged software applications for business use.

- Understand the problem of software piracy and how it affects businesses and consumers.

QUICKBIZ MESSENGERS:
Software Steers a Path to Stability

Growth adds complexity. But the efficiency Andrew Langston achieved through information systems had helped him manage QuickBiz's complexity repeatedly through the years.

General Software Needs

When Andrew considered buying a new PC-based server system, he wanted to be sure that it could handle his needs. So, he listed the main business functions for which he needed software support:

- General word-processing software for letters and memos.
- Financial accounting and reporting software for tracking sales, invoicing, and paying taxes and license fees.
- Human resource information software to track full-time and part-time workers' time sheets and to generate their W-2 and 1099 income tax forms.
- Database management system software for recording employee and client information.
- Basic desktop publishing software for direct-mail pieces to send to prospective clients.

Andrew chose a software suite to handle most of the business functions because the pieces would work together well and share a common database. He also was able to purchase the financial, human resource, and desktop publishing software off the shelf.

Finding Efficient Routes

Andrew and his longtime messengers knew Seattle like the backs of their hands. They set their own best routes. Now that QuickBiz had more than 90 employees—some not native to the area—Andrew noticed that a few deliveries were delayed because messengers had taken the wrong route. Customers complained, and the problem needed to be solved to maintain QuickBiz's reputation.

Luckily, Andrew ran across an article in *InformationWeek* on a new routing program. The software could be loaded with a map and, given start and end points, it would generate the shortest route and logical delivery territories. He was surprised how well the software could organize the routes to save time, fuel, and—most importantly—money. It worked particularly well for the longer routes he'd added when service was extended beyond downtown.

The software was also tied into global positioning system (GPS) satellites so that messengers could get instant route information beamed to them as they worked. The software was installed on both the dispatchers' system and the messengers' handheld computers.

Staffing Challenges

Andrew also had trouble tracking his employees' availability for work. Sarah Truesdale and Leslie Chen had to make frequent manual changes to the schedule to ensure that routes had adequate coverage. Scheduling became increasingly complicated as the company grew and hired more part-time workers. Because many of these workers were college students, their availability changed from semester to semester, thus the entire schedule was revamped two or three times a year. Also, when someone called in sick, they had to scramble to line up a replacement. It was time to automate.

Sarah told Andrew about scheduling software that her friend, a nurse at a local hospital, had used at work. The employer simply input employees' available hours, and the program generated a schedule. Making changes was streamlined, too—the software could identify on-call employees or revise a worker's schedule quickly. Master schedules were posted at the end of the week for next week's work, and changes were generated as needed.

Using Financial Software for Assessing Performance

Andrew had always enjoyed the closeness of his small company. Employees worked hard to do their jobs well. To foster pride in efficiency, Andrew began a new program to track the number of deliveries and shortest delivery times for each messenger. The program also tracked any feedback he received—customer compliments and complaints or speeding tickets. He called the messengers together to alert them that beginning with next month's deliveries, he'd begin tracking their productivity under his new incentive program. At the end of the month, the two employees—a bicycle courier and an auto or truck courier—with the most deliveries, shortest delivery times per mile, fewest complaints, and most compliments would receive a bonus.

Andrew also evaluated the delivery territories to determine which were most profitable. He generated sales reports by region from the customer database. From the reports, he noticed that the Saturday delivery service in the downtown area wasn't generating enough revenue to cover its cost. Therefore, he decided to research this particular service further and see if its elimination could cause loss of regular services. He also adjusted the number of couriers to add more service to his most profitable routes. These changes would help boost the bottom line and keep QuickBiz rolling smoothly.

SOFTWARE: INSTRUCTIONS TO THE HARDWARE

At the 2007 Super Bowl, FBI agents carried mobile phones. Every phone also served as a video camera. The video was streamed to a central location, similar to the security centers with multiple screens. As the agents were patrolling the stadium, live video came in from their phones. The phones were not different from those carried by millions of people. So, how was this possible? The phones were equipped with special software called Reality Mobile. The same software will allow you and your friends to share video YouTube-style.

You use software all the time, not just when you use your computer. You use software when you drive a car, when you make a call from a mobile phone, and when you use the self-checkout station at a store. The purpose of much of the software used by organizations is to increase productivity.

When executives talk about productivity tools, they really mean computer programs, commonly known as software **applications**. Word processors, electronic spreadsheets, Web browsers, project management tools, collaborative work programs, and many other types of productivity tools are software that runs on computers and enables workers to produce more products and services in a given amount of time. This chapter discusses the differences between system software and application software, programming languages that are used to write software, and the types of software tools currently available.

Software is a series of instructions to a computer to execute any and all processes, such as displaying text, mathematically manipulating numbers, or copying or deleting documents. Computers only understand instructions made up of electrical signals alternating between two states, which eventually close or open tiny electrical circuits. Different sequences of signals represent different instructions to the computer. In the early days of computers, programming a computer meant actually changing the computer's wiring by opening and closing switches or moving plugs from one circuit to another. Because programs today consist of instructions that require no hardware reconfiguration, the skill of composing software programs is independent of building or directly manipulating hardware. As previously noted, software is executed not only on computers, but in every device that uses microprocessors, such as motor vehicles, digital cameras, and mobile phones. However, we will focus mainly on computer software that serves organizations.

The two major categories of software are application software and system software. **Application software** enables users to complete a particular application or task, such as word processing, investment analysis, data manipulation, or project management. **System software** enables application software to run on a computer, and manages the interaction between the CPU, memory, storage, input/output devices, and other computer components. Both types of software are discussed later in the chapter.

PROGRAMMING LANGUAGES AND SOFTWARE DEVELOPMENT TOOLS

Programs are needed for absolutely every operation a computer conducts. An operation can be as simple as adding 1 + 2, typing a word, or emitting a beep—or as involved as calculating the trajectory of a spacecraft bound for Mars. The process of writing programs is **programming**, also known as "writing code" and "software engineering."

Remember, the *only* language that computer hardware understands is a series of electrical signals that represent bits and bytes, which together provide computer hardware with instructions to carry out operations. But writing programs in this language—called **machine language**—requires a programmer to literally create long strings of ones and zeroes to represent different characters and symbols, work that is no longer required thanks to programming languages and other software development tools. **Assembly languages** made programming somewhat easier because they aggregated common commands into "words," although many of the "words" are not English-like. Higher-level **programming languages** enable the use of English-like statements to accomplish a goal, and these statements are translated by special software into the machine language.

Software development tools are even easier to use because they require practically no knowledge of programming languages to develop software. Programmers have at their disposal literally thousands of different programming languages, such as Visual Basic, Java, and C++. Programmers and nonprogrammers alike can use Web page development tools such as Adobe Dreamweaver or Microsoft FrontPage, which provide menus, icons, and palettes that the developer can select or click to create intricate Web pages, forms, animation, and links to organizational information systems. To develop the software development tools themselves, and to develop highly specialized software, programmers still have to write code in programming languages.

Figure 5.1 shows how programming languages have evolved dramatically over the years. Their different stages of development are known as generations. First-generation (machine language) and second-generation (assembly) languages were quite inefficient tools for code writing. They required lengthy written code for even the simplest procedures. In third- and fourth-generation languages, shorter, more human-friendly commands replaced lengthy code. Ultimately, it would be nice to be able to program using the daily grammar of your native language—English, Spanish, Hebrew, or any other language. But even then, the so-called natural language would have to be translated by another program into machine language.

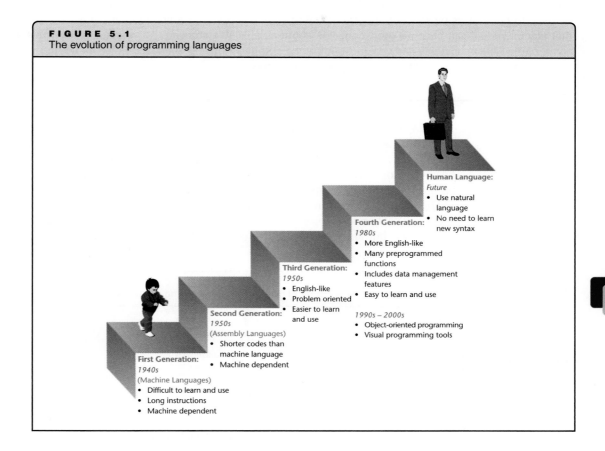

FIGURE 5.1
The evolution of programming languages

Third-generation languages (3GLs) are considered "procedural" because the programmer has to detail a logical procedure that solves the problem at hand. Third-generation languages reduced the programmer's time spent producing code. One 3GL statement is equivalent to 5–10 assembly language statements. Some common procedural languages include FORTRAN, COBOL, BASIC, RPG, Pascal, and C. Some of them, such as RPG and COBOL, are no longer in use or in limited use.

Fourth-generation languages (4GLs) make application development even easier. They are built around database management systems that allow the programmer to create database structures, populate them with data, and manipulate the data. Many routine procedures are preprogrammed and can be recalled by including a single word in the code. A single 4GL statement is equivalent to several 3GL statements, and therefore to dozens of assembly statements.

4GL commands are more English-like than commands in 3GL procedural languages. In fact, 4GLs are significantly less procedural than 3GLs. With 4GL commands, the programmer often only needs to type what is to be done, but doesn't need to specify how the procedure accomplishes the task. For example, if one column in a database is AGE, the programmer can simply use the preprogrammed command LIST AVERAGE(AGE) to display on the screen the average age, which is calculated from the age values in all the records. Similarly, preprogrammed functions are provided for total, standard deviation, count, median, and many more tasks. The list of preprogrammed functions in electronic spreadsheets such as Microsoft Excel has become so comprehensive that some people refer to them as 4GLs.

As a professional, you should regard software as a tool to further your productivity and education. Software can automate many processes that professionals must accomplish. Even simple software such as electronic spreadsheets can be used to build decision support applications. Software vendors offer a huge variety of programs. While it is doubtful that any individual can become knowledgeable about all available software, knowledge of the types of software and some particular applications lets you make informed comparisons and suggestions for improving your organization's software portfolio and your own library of personal software.

4GLs speed up the programming process. They are relatively easy to use by people who are not professional programmers, and therefore enable non-IT employees in many companies to produce applications on their own. The produced code is usually easy to change, which reduces the cost of software maintenance. Because 4GLs are very English-like, **debugging**—locating and fixing programming errors—is relatively easy.

Higher-level programming languages have their advantages, but also some disadvantages (see Figure 5.2). Therefore, programming languages are chosen based not only on programming productivity but also on the amount of control over the resulting software that is desired.

FIGURE 5.2
Advantages and disadvantages of higher-level programming languages

Advantages of Higher-Level Programming

◆ Ease of learning the language

◆ Ease of programming

◆ Significantly shorter code

◆ Ease of debugging

◆ Ease of maintenance (for example, modification of a procedure)

Disadvantages of Higher-Level Programming

◆ Less control over hardware

◆ Less efficient memory use

◆ Program runs more slowly

Visual Programming

To accelerate their work, programmers can use one of several **visual programming languages**, such as Microsoft Visual Basic, Borland Delphi, Micro Focus COBOL, ASNA Visual RPG, and Visual C++. These languages let programmers create field windows, scroll-down menus, click buttons, and other objects by simply choosing the proper icon from a palette. They can then use a flexible tool to shape and color these objects. (Note that here the term "object" is used loosely, not with its special meaning in the context of object-oriented languages, as discussed in the next section.) Seeing exactly and immediately how boxes and menus look on screen reduces the chance of bugs and helps programmers finish their jobs faster than if they had to write code. The appropriate code is written automatically for them when they click on elements. However, the programmer can always go back to the code and add or change statements for operations that cannot easily be accomplished by using the visual aids. Thus, knowledge of the programming language is still required.

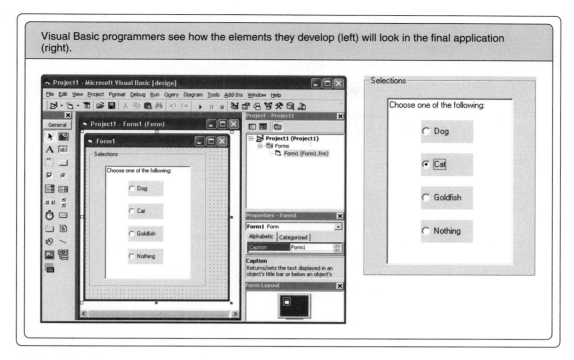

Visual Basic programmers see how the elements they develop (left) will look in the final application (right).

Object-Oriented Programming

An increasing amount of software is developed using **object-oriented programming (OOP) languages**. These languages use a modular approach, which offers two great advantages: ease of maintenance and efficiency in applications development (see Figure 5.3). In traditional programming, programmers receive specifications of how a program should process data and how it should interact with users, and then they write code. If business changes and the program must be modified, the programmer must change the code. In traditional programming, data and the operations to manipulate the data are kept separate from each other. In object-oriented programming, on the other hand, operations are linked to the data. For example, if the operation is to calculate an employee's gross pay, taxes, and net pay, selecting and clicking on the record triggers the calculation. Routine, frequent operations are kept with the data to be processed. Thus, OOP's primary emphasis is not on the procedure for performing a task, but on the objects involved in the task.

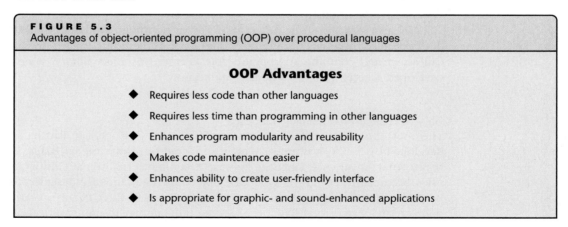

FIGURE 5.3
Advantages of object-oriented programming (OOP) over procedural languages

OOP Advantages

◆ Requires less code than other languages

◆ Requires less time than programming in other languages

◆ Enhances program modularity and reusability

◆ Makes code maintenance easier

◆ Enhances ability to create user-friendly interface

◆ Is appropriate for graphic- and sound-enhanced applications

What Is an Object in OOP?

Figure 5.4 illustrates how an object in OOP encapsulates a data set with the code used to operate on it. Data elements in the object are called "data members." They might be records, whole files, or another type of data structure. Data members have attributes that define the nature of the data, such as Social Security number, last name, and hourly rate. The code elements of the object are called "member functions" or "methods." These procedures operate on the data, such as calculating an employee's gross pay for the week. In object-oriented software, there is no direct access to data members of an object; they can be accessed only through the methods, which are part of the object. In our example, the object includes three methods: Weekly Pay, Overtime Pay, and Age. Weekly Pay calculates each employee's gross and net pay, Overtime Pay calculates each employee's overtime gross pay, and Age computes all employees' average age.

FIGURE 5.4
The object EMPLOYEE

E M P L O Y E E

Attributes

   Social Security Number

   Last Name

   First Name

   Address

   Date of Birth

   Hourly Rate

Methods

   Weekly Pay

   Overtime Pay

   Age

Ease of Maintenance and Development

Typically, about 80 percent of all work associated with software is spent on maintaining it. Maintenance primarily involves modifying programs to meet new business needs, but also debugging of errors that were not detected when testing the developed code. In object-oriented programming, software developers treat objects as parts, or standardized modules that work together and can be used and reused. Instead of creating large, complex, tightly intertwined programs, programmers create objects. Objects are developed in standard ways and have standard behaviors and interfaces. These modules enable software to be assembled rapidly rather than written laboriously.

OOP also makes creating programs easier for nonprogrammers. The inexperienced developer does not need to know *how* an object does what it does, only *what* it does. Thus, the developer can select and combine appropriate objects from an object library, which is a repository of developed objects, to build a desired application.

Object-Oriented Programming Languages

The most popular OOP languages are Smalltalk, C++, Object Pascal, and Java. Smalltalk, developed by Xerox, was an early object-oriented programming language. C++ has become the major commercial OOP language because it combines traditional C programming with object-oriented capabilities. Java is a popular object-oriented language designed to be platform independent, that is, to run on any computer regardless of the CPU or operating system. Another popular language, Visual Basic, enables the programmer to use graphical objects, but does not fulfill all the requirements of a true OOP language. For example, moving an icon to another application does not move the code associated with it. Some OOP languages are designed specifically for use in developing graphical user interfaces (GUIs). Elements of GUIs include windows, icons, scroll boxes, and other graphical images that help the user interact with the program with minimal effort. One of the earliest uses of Smalltalk was to develop GUIs.

Languages for the Web

Because an increasing amount of software is developed for Web sites and to link applications via the Internet, special software languages and tools have been developed for these tasks. Such programming languages include Java, JavaScript, J2EE (Java 2 Platform, Enterprise Edition), and PHP. The main advantage of Java, JavaScript, and J2EE is that the code produced—often called applets—can be executed well regardless of the operating system that the computer uses. Therefore, the same applet will be executed the same way on a computer running Windows or one running Mac OS X. This is a significant benefit, especially when the applets are developed to be posted at a Web site.

Consider Transportation Management System (TMS). This Web-based application helps FedEx Ground, the trucking segment of FedEx, to optimize the operations of its tractors, trailers, and dollies. The software, developed using the Java language, enables the organization's 29 hubs to share information via the Internet. More than 500 pickup/delivery terminals use the software. TMS helped save $100,000 per day in personnel and administrative costs and shortened by a day the time it takes to move shipments on many of the routes.

In recent years an increasing number of applications have been developed in Microsoft's .NET "environment." .NET is software that supports building and linking applications that can "talk to each other" on the Internet and enable Web browsers to invoke information resources, such as databases. In Microsoft's own words, .NET enables businesses to "quickly build, deploy, manage, and use connected, security-enhanced … Web services." Applications developed using .NET tools run on Microsoft operating systems, such as Windows Server 2003. We discuss operating systems later in this chapter.

LANGUAGE TRANSLATION: COMPILERS AND INTERPRETERS

Recall that computers understand only machine language. Just as assembly languages need assemblers, procedural languages need special programs to translate source code, which is the program as originally written, into object code, which is the same program in machine language. (Unfortunately, the word "object" is used for several different contexts. In the context of this section it has nothing to do with object-oriented languages.) The two types of programming language translators are compilers and interpreters. Compilers translate the higher-level code into an equivalent machine language code, but do not execute the code; the translated code must be run to check for programming errors. Interpreters translate each program statement and execute it.

A compiler (see Figure 5.5) scans the entire source code, looking for errors in the form (syntax) of the code. If it finds an error, it does not create the object code; instead, it generates an error message or a list of error messages. If the compiler finds no syntactic errors, it translates source code into object code, which the computer can execute. At this point, the programmer can save the object code. From now on, the user can simply run only the object code. This saves translation time.

An interpreter checks one statement at a time. If the first statement is free of syntactic errors, it interprets the statement into object code and makes the computer execute it. If the statement is erroneous, the interpreter issues an error message. In some environments, the programmer can immediately correct the statement. The computer then executes the corrected statement, and the interpreter moves on to check the next statement. Error-free statements are executed immediately.

Code written in interpreted programming languages can run only on machines whose disks store the interpreter. In contrast, compiled code is ready to run because it is in machine language and does not need to be translated. Most Visual Basic and Java translators are interpreters. Translators of FORTRAN, COBOL, C, C++, and most other 3GLs are compilers.

When you purchase an application, whether a computer game or a business program, you purchase a compiled version of the code, that is, the object code. There are three reasons for this. First, the application is executed immediately because there is no need to compile the code. Second, most users do not have the compiler for the source code. Third, the vendor does not wish

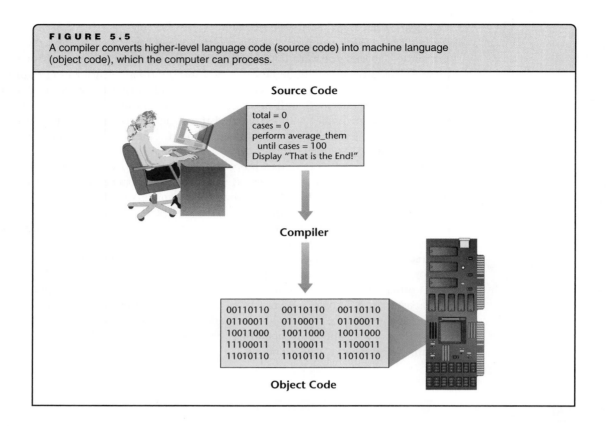

FIGURE 5.5
A compiler converts higher-level language code (source code) into machine language (object code), which the computer can process.

Source Code

```
total = 0
cases = 0
perform average_them
    until cases = 100
Display "That is the End!"
```

Compiler

Object Code

```
00110110   00110110   00110110
01100011   01100011   01100011
10011000   10011000   10011000
11100011   11100011   11100011
11010110   11010110   11010110
```

buyers to modify the code. If the program is sufficiently modified, the modified copies may be sold without violating intellectual property laws such as patents and copyrights. Source code can be modified by anyone who knows the programming language in which it was written; modifying object code is very difficult.

While testing code, programmers can use programming language translators to find syntactic errors. When they execute the program, they can find execution errors—also called runtime errors—such as division by zero or an excessive use of memory (memory leak). However, only the programmer can detect and prevent logical errors, because the logic relies, and should rely, solely on the way the programmer translated a way to produce a result into code.

POINT OF INTEREST

It's the Software, Stupid

In March 2007, Airbus had 167 orders for its A380 aircraft. The 239-foot (73-m) long plane is a double-decker with 555 seats when used for passengers. The total length of the wires in each passenger aircraft, 384 miles (618 km), means that the total wiring for the 167 ordered planes could circle the earth more than twice around the equator. Airbus is a British-French-German-Spanish venture. Both French and German engineers use computer-aided design (CAD) software called Catia, from Dassault Systems. However, while the French used Catia 5, the Germans used Catia 4. Catia 5 allows easy three-dimensional drawings. Catia 4 supports two-dimensional drawing (which can be converted into 3D only with special effort). The two teams were working on different parts of the aircraft. When technicians tried to install the wires, the wires were too short and could not be connected. Catia 4 and Catia 5 are incompatible. Training engineers on Catia 5 takes several months. A problem that could have been solved easily with forethought caused delays, order cancellations, and billions of dollars in losses.

Source: Duvall, M., Bartholomew, D., "PLM: Boeing's Dream, Airbus' Nightmare," (www.baselinemag.com) February 5, 2007; Schwartz, N.D., "Big Plane, Big Problems," *Fortune*, March 5, 2007.

APPLICATION SOFTWARE

As noted earlier, an application is a program developed to address a specific need. An application can also be software that lets nonprogrammers develop such programs. Most programs that professionals use are application programs, such as word-processing programs, spreadsheet programs, payroll programs, investment analysis programs, and work-scheduling and project management programs.

Programs designed to perform specific jobs, such as calculating and executing a company's payroll, are collectively called **application-specific software**. Programs that serve varied purposes, such as developing decision-making tools or creating documents, are called **general-purpose application software**. Spreadsheets and word processors are general-purpose applications.

General-purpose applications are available as **packaged software**; that is, they come ready to install from an external storage medium such as a CD or a file downloaded from a vendor's Web site. Application-specific software is not always so readily available. Managers must decide whether an off-the-shelf software package meets all of their needs. If it does, the company can simply purchase it. But if off-the-shelf or other ready-made software cannot address an organization's specified needs, managers must have a program developed, either within the organization or by another organization specializing in that type of software. We discuss alternative ways to acquire ready-made software in Chapter 13, "Choices in Systems Acquisition."

Office Productivity Applications

The purpose of *all* business software is to make the work of people more productive. However, applications that help employees in their routine office work often are called simply "productivity tools." They include word processors, spreadsheets, presentation tools, file and database management software, graphics programs, desktop publishing tools, and project management applications, as well as many others for more specialized purposes. Web browsers are also included in this group, because they help so many employees to find and communicate information in their daily work. Often, the tools are called desktop productivity tools, because they were developed to support home and office users on their personal computers.

While *word processors* are used mainly to type letters, articles, and other text documents, they also automate otherwise laborious tasks such as creating tables of contents and indexes. Some enable users to plan the binding and look of books up to the point of handing files to a high-quality printer for the production of the physical book. Examples of word processors include Microsoft Word, Corel WordPerfect, and Lotus WordPro.

Spreadsheets such as Microsoft Excel no longer limit users to entering numbers and performing basic arithmetic calculations. They include a long list of complex mathematical, statistical, financial, and other functions that users can integrate into analysis models. These functions are so powerful that statisticians often use them. Executives can build their own decision-support models with this robust tool. Spreadsheets also provide a large array of preformatted charts from which the user can select for presentation purposes.

Presentation tools such as Microsoft PowerPoint enable professionals and salespeople to quickly develop impressive presentations. One does not need to be a graphics expert, because the tools provide wide selections of font types and sizes and allow users to embed almost any art that they find (with permission!) or have created in graphics programs. Animations, sound, and video clips can be integrated into presentations and slide shows that can be posted to run on the Web as videocasts.

File management and data management tools enable the creation and manipulation of local or shared databases. Popular database management systems such as Microsoft Access are relatively easy to learn and create simple databases. They often include features that professional developers can use to create more complex databases.

Graphics programs make it easy to create intricate images and manipulate digital photographs. They are often used to create graphics to be placed on Web pages. The large selection of these tools includes Adobe's Illustrator and Photoshop, Corel Paint Shop, and MGI PhotoSuite, as well as the free IrfanView and Gimp.

Desktop publishing tools, such as Microsoft Publisher, Adobe FrameMaker, and Corel Ventura, enable both expert and novice to easily create professional looking pamphlets, newsletters, cards, calendars, and many other items for publication on paper or as Web pages. More professional tools, such as Quark, by a company of the same name, have significantly increased the productivity of the publishing industry.

Project management tools, such as Microsoft Project or the free Open Workbench, help managers of any type of project—such as building construction, product development, and software development—to plan projects and track their progress. Project managers enter information such as tasks and their expected completion dates, milestones, and resources required for each task: labor hours, materials, and services. The software alerts planners when they enter illogical information, such as scheduling a worker to work 120 hours in one week, and when tasks violate interdependencies. The latter happens when, for instance, planners schedule the start of Phase D before the completion of Phase C, though they had previously indicated that Phase D depends on the completion of Phase C.

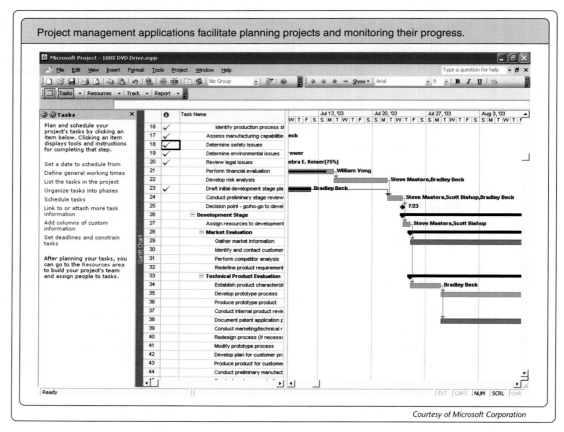

Project management applications facilitate planning projects and monitoring their progress.

Courtesy of Microsoft Corporation

Software developers often create **suites** of productivity tools. For example, most versions of Microsoft Office suite include a word processor (Word), spreadsheet (Excel), presentation application (PowerPoint), and an e-mail application (Outlook). Other examples of suites are IBM Lotus SmartSuite, and the free OpenOffice.org. When productivity tools are integrated into a software suite, the documents created can be interdependent using technologies such as object linking and embedding (OLE). You can create tables in a spreadsheet, copy them into a word-processed document or a presentation, and ensure that when you modify the tables in the spreadsheet they also change in the document or presentation. You can also embed links to Web sites in your documents. Linking among documents involves hypermedia technologies, and embedding information such as sound and video clips in documents uses multimedia technologies. These technologies are discussed in the next section.

A growing number of Web-based office applications are offered free of charge; all one needs is a Web browser. Typically, all the documents the user creates are saved at the application's server. This way, both the applications and documents can be accessed from anyplace with an Internet link.

For example, ThinkFree, offered by the company of the same name, is an online suite that includes a word processor, spreadsheet, graphical application, and a presentation application. It is promoted as a "free online alternative to Microsoft Office." Google offers Google Docs & Spreadsheets, free online word processor and spreadsheet applications. Documents are saved on Google's disks. TrimPath offers its free NumSum, a shareable spreadsheet, and several companies including Yahoo!, Google, and Kiko offer online shareable calendaring and scheduling applications for individuals and groups.

Object linking and embedding enables flexible and productive development of business reports and presentations.

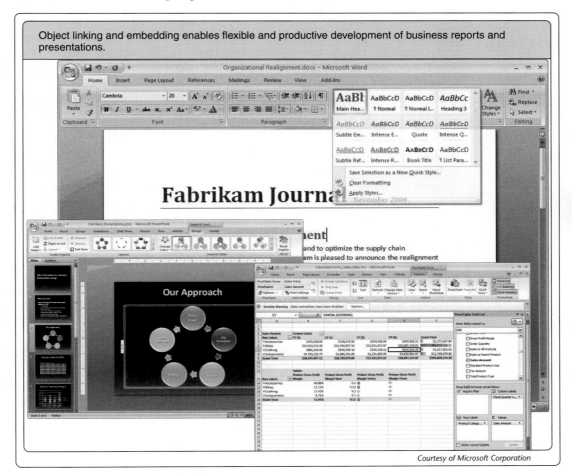

Courtesy of Microsoft Corporation

Hypermedia and Multimedia

Hypermedia is a feature that enables a user to access additional information by clicking on selected text or graphics. Hypermedia is the Web's most essential ingredient. When first conceived, the concept was limited to text and was called hypertext. Now, hypermedia is very common, used widely on software stored on CDs, and essential to Web-based documents as well as documents, charts, and presentations created using productivity tools. Any text or icon that can be clicked to jump to another place in a document or open a new document is called a link, whether on the Web or not. Often, we say that a word or icon is "clickable." Hypermedia enables linking text, pictures, sounds, animations, and video.

Hypermedia features are enabled by **Web page authoring tools**. They are also part of other applications, such as some word processors and presentation tools. You can easily create a PowerPoint presentation with marked text or an icon that calls up a picture, a sound, or an animation, or one that takes you to another slide. Programs that can handle many different types of data are called **multimedia software**. Multimedia is a powerful means of communicating, because it does not limit the method of communication. A natural extension of the computer's capabilities, it provides flexibility that lets people work the way they think, integrating all types and forms of information. Multimedia is tightly associated with hypermedia, because it often uses embedded links. These links are the essence of hypermedia and are used to communicate pictures, sounds, and video as part of the same message in a way that is similar to an educational lecture or a product manual. A few examples of the uses of multimedia are described in the following sections.

Multimedia in Education and Research

One of the most common uses of multimedia is in education. A student taking a multimedia-based lesson can view a scenario in one window and view text in another while listening to a recording of his or her professor. The student might then be asked to answer questions interactively, providing responses in another window on the screen. The same program might be designed to provide the student with feedback on her performance. With voice recognition software, multimedia programs used in language training can ask a student to pronounce certain words and evaluate the student's performance.

Another common use of multimedia is in compiling and integrating data from research. For instance, a researcher might use multimedia programs to view written articles and television news footage and to listen to radio clips.

Multimedia in Training

In many industries, multimedia is commonly used to simulate real-world situations for training exercises. For example, multimedia products that use video and voice and allow users to respond to questions about various situations have been used to teach workers for an electric utility company how to solve high-voltage wire problems. If they attempted to solve the same cases in the field, their lives would be jeopardized. Flight simulators use extensive multimedia software to simulate takeoff, landing, and other flight situations when training pilots before they fly real planes.

Multimedia in Business

Multimedia can be very useful in business situations as well. Consider this example: one manager writes a document that includes digitized photographs or video clips and possibly a "live" spreadsheet, which lets the user enter numbers and execute calculations. The manager sends the document to a colleague for review; the colleague tacks on a video and voice clip requesting clarification of a certain point. The compound document can be filed electronically, retrieved, altered, and communicated as appropriate, without ever being transformed into a paper document. In fact, multimedia by its very nature cannot be transferred to a paper document. Many Web sites include multimedia because of its interactive nature.

Mashups

Many companies, including Amazon.com, eBay, Flickr, Google, and Yahoo!, have opened their applications so that the applications, or some of their features, can be integrated with other software to create new useful applications. These integrated applications are called mashup applications, or simply **mashups**. For example, an amateur programmer can combine a mapping application from one Web site—such as Mapquest, Yahoo Maps, or Google Maps—with a local database of charity associations to show the locations and details of the associations on a map. The mapping application continues to provide its regular features, such as directions to and from the organizations' locations.

The programmer uses software elements from different applications and combines them, or some of their features, into a hybrid application. Since these software elements are constantly available on the Web, users of the mashup can enjoy it whenever their computers are connected to the Internet. The site mashupawards.com/winners provides an up-to-date list of creative mashups. For example, SignalMap uses Google Maps to show maps with areas of mobile phone dead spots. Users can avoid those areas to ensure connection quality. Users are invited to enter their zip code and mobile service provider as well as the quality of connection at different places from their own experience. The information is added to a database. Over time, a map of good spots and dead spots is built to the benefit of all users.

Web Site Design Tools

As a growing number of organizations established Web sites, and many needed to change the content of the Web pages daily or even hourly, the need for Web design tools grew. Popular Web page development packages include Microsoft FrontPage, SharePoint Designer and Expression Web, and Adobe Dreamweaver and GoLive. Many Internet service providers (ISPs) and Web site hosting companies also provide online tools for designing personal and commercial Web sites.

Web page development packages expedite development of Web pages. Like other visual tools, they provide menus, icons, and other features from which the developer can select. Therefore, developers have to write code only when a feature is not readily available. When using ready-made options, such as fill-in forms and animation effects, the code is automatically added. Since much of the code is in nonproprietary languages such as HTML and XML (which we discuss in Chapter 8, "The Web-Enabled Enterprise"), a programmer can start work with one development tool, such as FrontPage, and continue the work with another, such as Dreamweaver. Developers alternate if they find one tool is easier to use for quick development of icons, for example, whereas another offers a more appealing way to develop animations.

Many Internet service providers provide Web site design tools.

Groupware

Multimedia technologies are any applications that allow sharing of ideas and information resources among group members. Most of these applications are integrated with Web technologies. **Groupware** applications are programs that enable workers to collaborate in real time over the Web. They not only eliminate the need to travel and sit in the same physical room but also facilitate expression of ideas by demonstrating them through the combination of text, images, drawings, sound, animation, and video. Consider this example of a young company that cleverly uses groupware.

Kidrobot sells mainly vinyl toys. Unlike similar companies that spend thousands of dollars on special software, Kidrobot employees design new toys using an off-the-shelf application called Adobe Illustrator. They create six views of each new toy with exploded views of detailed areas such as eyelashes. The manufacturer of the newly designed toys is located in China. For a subscription fee of $100 per month the designers place the illustrations online using Basecamp, a Web-based project management application that helps people remotely collaborate on projects. The Chinese engineers use the files to create the vinyl dolls.

Virtual Reality

Virtual reality (VR) applications mimic sensory reality using software. They create the illusion of experiencing situations through simulated sight, hearing, and touch, such as flying in an airplane or forging a piece of hot metal. A user can sense virtual reality in several ways. The most sophisticated VR devices provide two important elements: immersion and interaction. They include goggles, gloves, earphones, and sometimes a moving base on which the user stands; all of these devices sense movement, respond to signals, and provide feedback to the user. In immersion, an individual senses that she or he is surrounded by the simulated environment. Interaction lets individuals simulate change in the environment by moving their hands or fingers. Users receive a three-dimensional visual sensation and hear stereophonic sound. With interactive gloves, the user can use hand motions to change the direction and "move" within the virtual environment. For instance, a VR system might be designed so the user experiences being a race car driver. In this case, when the user's hand makes a grabbing motion, sensors in the VR glove cause the hand in the VR image to "grab" the stick shift. The distinction between multimedia and VR can be hazy. Experts usually assert that only systems that include sensing helmets, gloves, and similar components, and which truly surround the user with a sense of a

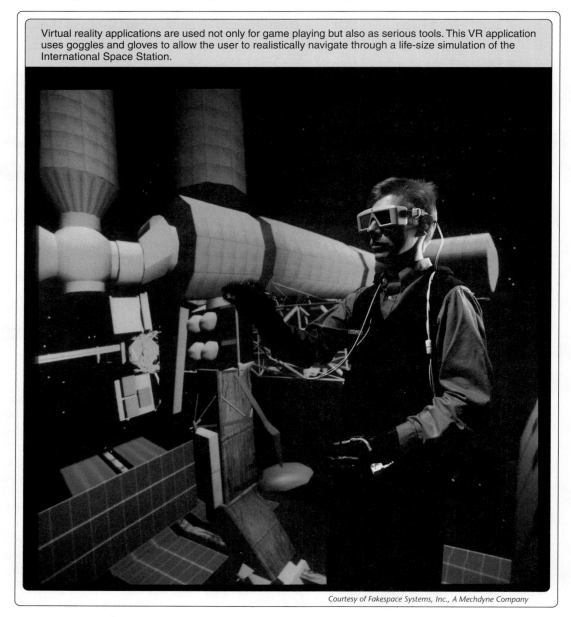

Virtual reality applications are used not only for game playing but also as serious tools. This VR application uses goggles and gloves to allow the user to realistically navigate through a life-size simulation of the International Space Station.

Courtesy of Fakespace Systems, Inc., A Mechdyne Company

real experience, are VR. However, many people refer to sophisticated multimedia applications that run on PCs as VR as well.

Business use of VR is growing. VR business applications can decrease the cost of planning buildings, machines, and vehicles. They already help marketing efforts to lure buyers to try new products. For instance, architects can use VR to let a potential buyer "tour" a house that has not yet been built. The buyer can then request changes in the floor plan and other features before construction begins. Volvo, the Swedish car and truck maker, invites prospective buyers to test-drive its latest models in VR. Companies such as Raytheon and Fluor Daniel use VR to help design new manufacturing plants.

Virtual reality has been implemented on the Web. Web-based VR has evolved from use on corporate intranets to public gatherings and other purposes. The best example is Second Life, an imaginary world where the representative figures of real people, called **avatars**, can meet and communicate. We discuss this concept in detail in Chapter 8, "The Web-Enabled Enterprise."

3-D Geographic Software

Similar to virtual reality but for somewhat different purposes, programmers develop three-dimensional models of geographic areas and whole cities. An increasing number of applications are being developed to create 3-D models of existing city blocks down to every hydrant and shrub. The raw materials are land and aerial photographs that cover the targeted area. The digital photographs are "sewn" together to allow a continuous "walk" or "travel" on a city street or university campus. This helps with navigation, whereby one can recognize buildings and landmarks by their similarity to the software images. This type of information can be delivered through the Web. When tied with a global positioning system (GPS), the software helps people who have never been to a place to navigate easily. In the near future, 3-D software such as this will help property rental companies manage their assets. For example, a manager will be able to click on an apartment on the 12th floor of a building and check information about the unit and let a potential renter have a view from the windows or balcony. Maintenance staffs will be able to virtually go into the walls and check pipes and electrical wiring, and fire companies will be able to navigate quickly and locate hydrants on their way to put out fires.

3-D city models help city planners maintain existing facilities and plan new blocks and parks.

Courtesy of GeoSim Systems Ltd.

GeoSim, a company that specializes in such software, has developed "virtual cities" including Philadelphia. One can virtually walk in the streets, drive a car, or fly above a city. Similar software was developed for university campuses, notably that of the University of Pennsylvania. Such software helps city planners and service agencies as well as tourism and travel agencies. Some of this software is demonstrated at www.geosimcities.com.

SYSTEM SOFTWARE

System software includes programs that are designed to carry out general routine operations, such as the interface between user and computer, loading a file, copying a file, or deleting a file, as well as managing memory resources and operating peripheral equipment such as monitors and printers. The purpose of system software is to manage computer resources and perform routine tasks that are not specific to any application. On one hand, system software is developed to work in partnership with as many applications as possible; on the other, applications can work with system software only if they are developed to be compatible with that software. The following discussion covers major types of system programs. Note that compilers and interpreters, which were discussed earlier, are also classified as system software.

Operating Systems

The **operating system (OS)** is the single most important program that runs on a computer and the most important type of system software. As Figure 5.6 illustrates, operating systems perform basic tasks, such as recognizing input from the keyboard and mouse, sending output to the computer display, keeping track of files and directories (groups of files) on disks, and sending documents to the printer. Without an operating system, no application can run on a computer. An operating system is developed for a certain microprocessor or multiple microprocessors. Programmers know which operations each microprocessor can perform and how it performs them. The OS must address technical details such as CPU circuitry and memory addresses.

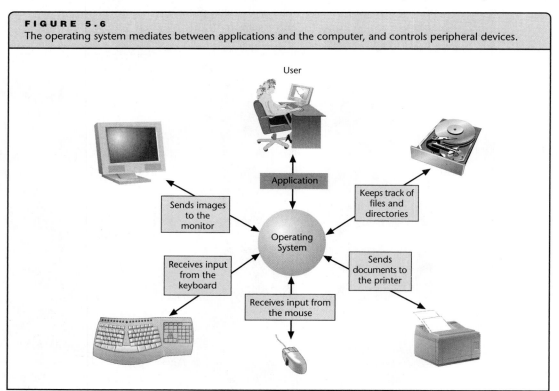

FIGURE 5.6
The operating system mediates between applications and the computer, and controls peripheral devices.

Therefore, OSs are usually developed with the aid of low-level programming languages, such as assembly languages, or with a language that can access low-level machine functions, such as C.

The OS is sometimes called the "traffic cop" or the "boss" of computer resources. Indeed, it is charged with control functions such as optimally allocating memory locations for an application program, copying the application from an external storage medium into memory, passing control to the CPU for execution of program instructions, and sending processing results to output devices. Operating systems are also often referred to as "platforms," because they are the platform on which all other applications "ride" when interacting with the hardware.

When application developers write code, they use the **application program interfaces (APIs)** for the operating system on which the application will run. APIs are software included in the operating system. A good API makes it easy to develop an application. Applications using the same API have similar interfaces.

From User to OS to CPU

Figure 5.7 shows the OS's position in the logical operation of a computer. The user interacts with the user interface using menus, icons, and commands provided by the application. The application converts some of the user's input into commands the OS understands, and the OS commands the CPU to carry out the operation. (Some commands are not delivered to the OS but directly from the application to the hardware.) The OS ensures that applications can use the CPU, memory, input and output devices, and the file system. The file system is software that stores, organizes, and retrieves files.

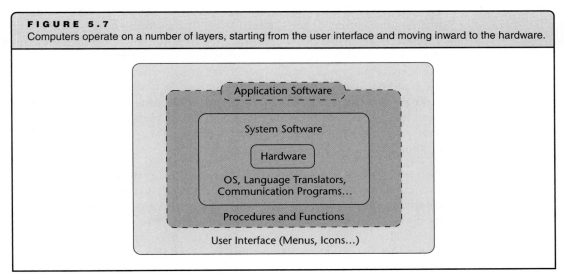

FIGURE 5.7
Computers operate on a number of layers, starting from the user interface and moving inward to the hardware.

For example, assume that you are using a word processor. You select a paragraph you wish to copy and paste. You select Copy from the menu. The word processor converts your choice into an appropriate command for the OS, which then instructs the CPU to copy the paragraph. A similar action takes place when you select Paste from the menu. Assume that you like a picture on a Web page and have permission to copy it. You right-click the picture and choose to copy it. The Web browser's menu might not look the same as the word processor's menu. However, when you select Copy Image, the operating system receives a command from the application that is identical to the one it received when you used the word processor. And when you paste, the Paste command that the OS receives from the browser is the same one it received from the word processor. Thus, developers of these two applications did not need to program the copy and paste operations; they only needed to know how their programs must call up these operations from the OS.

In addition to performing input and output services and controlling the CPU, many OSs perform accounting and statistical jobs, including recording times when a user logs on and logs off, the number of seconds the operator used the CPU in every session, and the number of pages a user printed. Some OSs also perform **utilities** such as hardware diagnostics, file comparison, file sorting, and the like. However, not all OSs provide all the utilities that might be necessary, in which case special utility programs must be used. Operating systems also include a number of security functions, such as the ability to set user passwords and restrict access to files and computer resources.

Operating System Functions

Operating systems provide several services, the most important of which is system management. System management refers to the efficient allocation of hardware resources to applications and includes tasks such as prompting the user for certain actions, allocating RAM locations for software and data, instructing the CPU to run or stop, allocating CPU time to different programs running at the same time, and instructing co-processors and peripheral equipment.

User Interface   An important part of the OS is the user interface. A graphical user interface (GUI) makes the use of the computer intuitive and easier to learn. The interface takes the form of easy-to-understand frames, icons, and menus. Users find it helpful to have most of the interface features identical regardless of the application they use, unless the application requires an interface element for a unique feature.

Memory Allocation   One of the most important functions of an operating system is memory management, especially RAM—the memory where data and program code must reside before being executed. Ideally, an entire application and all the data it processes reside in RAM until processing ends. However, when many applications are open concurrently, or when applications and data pools exceed the computer's RAM capacity, the operating system may use virtual memory. **Virtual memory** lets the user proceed as if significantly more RAM were available than really exists. Virtual memory uses the hard disk as an extension of RAM. A special module of the OS continually detects which parts of the application program are used frequently. The OS keeps these parts in RAM while leaving on the disk the least frequently used parts. Professionals call this activity "page swapping"—"pages" are program parts of equal size that the OS swaps between RAM and the disk, and the space on disk used as memory is called "swap space." Because hard disks are slower than RAM, opening many applications at the same time may reduce the speed at which programs are running. However, virtual memory enables concurrent use of many and large programs without need to purchase more RAM, which is significantly more expensive than disk memory.

Plug and Play   A good operating system should also facilitate fairly simple changes to hardware configuration. When a new device, such as an external hard disk, DVD burner, external communication device, or joystick, is attached to a computer, the operating system's job is to recognize the new attachment and its function. If the OS can do so (without your intervention) immediately after you attach the device, it is a **plug-and-play (PnP)** OS, and the device, too, is referred to as a plug-and-play device. To do so, the operating system must have access to the attached device's driver. A **driver** is the software that enables the OS to control a device, either one installed inside the computer box (such as a second video card) or an external device such as a flash memory drive. Thus, a true PnP OS, such as Windows XP, Windows Vista, or Mac OS, includes the drivers for many devices, or at least is fully compatible with a driver that is installed either from a disc or after it is downloaded from the Web. In recent years almost all external devices have been built to attach to a computer through a USB port. (USB ports were discussed in Chapter 4, "Business Hardware.")

Increasing Services from OSs   The trend in OS development is to incorporate more and more services that used to be provided by separate software. These services include database management, networking, and security. For instance, users now expect an OS to perform such security measures as tracking account numbers and passwords, controlling access to files and programs, and protecting the computer against viruses. OSs also check for access codes to ensure that only authorized users can access the computer. Modern operating systems provide networking functions previously handled by separate programs.

Current Operating Systems   As mentioned earlier, operating systems are designed to work with a particular microprocessor; consequently, different computers and types of microprocessors use different OSs. While specific operating systems exist for supercomputers, mainframe computers, midrange computers, and handheld computers, most people use a PC operating system. Popular operating systems for personal computers include Windows XP, Windows Vista, Linux, and Mac OS. Some operating systems are designed especially for computers used as servers in networks. For example, NetWare and Windows Server are popular network operating systems that are compatible with clients running DOS, all versions of Windows, and Mac OS. Figure 5.8 provides a list of popular operating systems. Some may come bundled with other applications. For example, i5/OS (formerly, OS/400) comes with powerful DB2 database management systems.

FIGURE 5.8
Popular operating systems

Typically run on...	Name	OS Developer
Mainframe	z/OS (zSeries)	IBM
	Clearpath/MCP, Clearpath IX	Unisys
Midrange	i5/OS	IBM
	Solaris	Sun Microsystems
PCs	Windows (XP, Vista)	Microsoft
	Linux (Ubuntu, Fedora, and many others)	Linus Torvalds and others
	Mac OS X	Apple Computer
	BSD	FreeBSD Foundation
	Solaris	Sun Microsystems
Servers	Windows Server 2003/2005/2008	Microsoft
	Unix	Bell Labs (originally), and later many Unix-like OSs by others
	NetWare	Novell
PDAs	Palm OS	Palm
	Windows CE, Windows Mobile	Microsoft

One OS that has grown in popularity is Linux, which can be obtained free of charge. Linux is based on UNIX, an operating system developed by AT&T Bell Labs in 1969 to run on midrange computers, and for 10 years was distributed free of charge. Different companies and individuals modified UNIX, developing variations of the OS such as Linux (developed by Linus Torvalds and others) and Solaris (developed by Sun Microsystems). Linux and other "open source" software is discussed in the next section.

One of the most important qualities of an OS is its *stability*. A stable OS does not cause the computer to freeze or produce error messages. It is expected to continue to function even if the user makes a mistake, in which case it should gracefully notify the user what happened and give an opportunity to resolve the problem, rather than stop functioning. Windows 95, Windows 98, and Windows Me were notoriously unstable. Although mishaps do occur with later versions of Windows, they are significantly more stable. OSs based on UNIX are known to be highly stable, and their stability is the main reason for their popularity, especially for running servers. Mac OS X versions are based on UNIX. Linux, too, is considered to be very stable.

Although Mac OS is installed on fewer than five percent of the world's computers, its share is growing. From the start, Mac operating systems have been more intuitive and user-friendly than Windows operating systems. In many respects they set the standards that Windows OSs later followed. Many current popular applications, such as Excel, were first designed for the Mac OS, and only later adapted for Windows, when the latter provided user interfaces similar to those of the Mac OS.

In 2007, Windows operating systems were used on 86.7 percent of the world's computers. This near monopoly is gradually but steadily changing as an increasing number of governments and commercial organizations adopt open source operating systems.

Other System Software

While operating systems are the most prevalent type of system software, other types of system programs include compilers and interpreters (discussed previously), communications software, and utilities. Some people also include in this class database management systems, which are discussed in Chapter 7, "Databases and Data Warehouses."

Communications software supports transmission and reception of data across computer networks. We discuss networking and telecommunications in Chapter 6, "Business Networks and Telecommunications." Utilities include programs that enhance the performance of computers, such as Symantec's Norton SystemWorks, which checks PCs for inefficiencies and fixes them. Utilities also include antivirus programs, firewalls, and other programs that detect and remove unwanted files and applications, such as cookies and spyware, or block data from being transmitted into or out of a networked computer. We discuss these topics in Chapter 8, "The Web-Enabled Enterprise," and Chapter 14, "Risks, Security, and Disaster Recovery."

OPEN SOURCE SOFTWARE

The great majority of business and individual software is proprietary, that is, software that is developed and sold for profit. The developers of **proprietary software** do not make the source code of their software public. The developer retains the rights to the software. In most cases you do not actually own the copies of applications that you purchase; you only purchase licenses to use those applications. In contrast to proprietary software, some programmers freely contribute to the development of a growing number of computer programs not for profit. The developers of **open source software** can obtain the source code free of charge, usually on the Web. Anyone who can contribute features or fix bugs is invited to do so. Anyone who wishes to download the latest version can do so free of charge. An open source program can be developed by a random group of programmers, rather than by a single company. Programmers share an application's basic code, find its weaknesses, debug it, and contribute new pieces. This process might yield better results than the traditional "closed" process of proprietary software, because so many talented programmers continuously try to show their prowess in improving the program. Some historians find the beginning of the open source "movement" in people such as

OpenOffice.org, an office suite; Firefox, a Web browser; and Thunderbird, an e-mail application, are just a few of hundreds of useful and popular open source applications.

Firefox

Thunderbird

Courtesy of The Mozilla Foundation

Courtesy of OpenOffice.org

Richard Stallman and his cohorts in the Free Software Foundation, who believe that software should be as free as the air we breathe and never sold for money.

The advantages of open source software over proprietary software are clear: the software has fewer bugs because thousands of independent programmers review the code, and it can offer more innovative features by incorporating ideas from a diverse set of experts from different countries and cultures who collaborate. The motive for developing and improving open source software is not monetary, but rather the satisfaction of solving programming problems and the recognition of one's contribution. Programmers who improve such software do it for fame and recognition by their peers the world over. They collaborate mainly via the Internet. They post patches of code that improve current code, or add extensions and plug-ins to enhance functionality of an application. These extensions are free for all to download and use. The major disadvantage is that development and support depend on the continued effort of an army of volunteers.

Tux the penguin is the unofficial trademark of Linux. Linux has made inroads into the corporate world, where it is used mainly to run servers.

Courtesy of Larry Ewing and The GIMP

Open source software includes hundreds of useful applications, such as the popular Web browser Mozilla Firefox, the e-mail application Thunderbird, the relational database management system MySQL, and the powerful programming language PERL (Practical Extraction and Report Language). The OpenOffice.org suite, which can be freely downloaded at www.openoffice.org, provides a complete alternative to Microsoft's Office suite of productivity applications.

Note that not all free software is also open source. For example, Microsoft's Internet Explorer browser can be downloaded and used free of charge, but the source code and documentation of the software are proprietary. On the other hand, the source code and documentation for Firefox is open: programmers from the world over can access the source code and improve it.

Linux is the best known open source operating system. A Finnish graduate student named Linus Torvalds developed it for his own use, but he has never claimed rights to the software. Hundreds of programmers have contributed code to Linux. Over time, Linux evolved into many different variants, some of which are free, such as Ubuntu, Red Hat Fedora, and Mandriva, while others such as Red Hat Enterprise Edition and SUSE charge for additional interface features and support services. Linux has become the OS of choice of many Internet service providers to run their Internet servers.

The major disadvantage of using Linux is the limited number of applications that can run on it, compared with the Windows platform.

POINT OF INTEREST

A Treasure Trove

You can find a free, open source application for virtually any purpose, from computer-aided design to instant messengers to games. Direct your Web browser to http://en.wikipedia.org/wiki/List_of_open-source_software_packages.

Reputable software companies including IBM, Intel, Hewlett-Packard (HP), and Dell have committed to supporting Linux by developing applications that run on it. A growing number of corporations have adopted it, including DaimlerChrysler, Tommy Hilfiger, and practically every major brokerage house on Wall Street. Linux is popular not only because it is stable, but because it is versatile: it can run on mainframe computers, PCs, handhelds, and electronic devices. The oil company Amerada Hess uses Linux on a supercomputer to help find oil and gas deposits. Pixar Animation Studios uses Linux machines to render digital animated characters. The brokerage firm Morgan Stanley replaced 4,000 servers that ran Microsoft Windows and Sun Solaris with cheaper machines that run Linux and saved $100 million. TiVo, the television recording device, runs on Linux. So do many game consoles and television sets that are connected to the Internet. A version of Linux, Linux Mobile, operates on some PDAs and mobile phones.

Many governments, both local and national, have decided to move to open source software. They do so mostly to save money but also to improve operations. Many adopters of Linux, for instance, have reported a tenfold improvement in the speed of their software-based operations when they moved from a commercial OS to Linux. Forty-two percent of Argentine companies use Linux. The governments of Brazil, Peru, and Chile mandated that all public administration agencies use only open source software when available. The Brazilian government switched more than 300,000 of its computers from Windows to Linux.

While many versions of Linux can be downloaded free of charge from the Web, most firms prefer to purchase a packaged version. Companies such as Novell, Red Hat, and VA Software sell the software and promise technical support. Usually, contracts also include software updates. Companies such as IBM and HP have made millions of dollars by bundling Linux with other system software and business applications, such as database management systems.

SOFTWARE LICENSING

The next time you "purchase" software, read carefully the "purchase" contract. You might be surprised to learn that you do not own the software you have just obtained. As noted earlier, most of the software that organizations and individuals obtain is not purchased; it is licensed. The client receives a software license, limited permission to use the software, either indefinitely or for a set time. When the use is time-limited, the client pays annual license fees. The only exceptions to this rule occur when an adopter uses its own employees to develop the software, when it hires the work of a software development firm, or when the adopter uses software developed by people who explicitly allow the user to change the software and sell the product.

POINT OF INTEREST

Frustrated with Software Licensing

A survey of 257 IT executives conducted by the research firm IDC found that they were frustrated with software licensing agreements. Their feeling was that even when software vendors offer substantial discounts, the clients end up paying much more than the initial price. Usually, the extra payments are for software maintenance or subscription fees. Subscription fees are often charged for support and upgrades. Interestingly, the executives believed that their companies used only 16 percent of the software they buy.

Source: Koch, C., "Do You Really Want Software as a Service?" *CIO* (www.cio.com), March 17, 2005.

Licensing of software comes in several models. The *permissive* model allows anyone to use, modify, and make the software into a product that can be sold or licensed for profit. The Berkeley Software Distribution (BSD) UNIX operating system is an example of software under this model. Another is the *General Public License* (GPL), which permits anyone to use, modify, and make applications with the code, but not to use it in proprietary products for sale or licensing. This is the approach taken by the Free Software Foundation. Much of the software we use is *proprietary*, which means the code is owned by someone who has the right to sell or license it to us.

Microsoft, SAP, Oracle, and all other for-profit organizations that develop software own their software and license it. Such licensing takes several forms, such as a fee per user per year, or a site license for a limited or unlimited use regardless of how many users use the software. The latter type of agreement is sometimes signed between a software vendor and a higher-education institution.

CONSIDERATIONS FOR PACKAGED SOFTWARE

When an application is developed especially for an organization, specific program goals and custom requirements are considered during the development process. Such requirements include business needs, organizational culture needs, the need to interface with other systems, and performance issues, such as response time. However, organizations find ways to satisfy many needs with ready-made software as well. Figure 5.9 summarizes important factors and outlines what you should look for when purchasing software.

The Los Angeles County Sheriff's Department purchased a license to install software offered by DataWall, Inc. The license allowed the department to install 3,700 copies of the software, but it installed 6,000 copies. The department claimed that the number of employees using the software concurrently could not exceed 3,700. The sheriff's department was sued. An appellate court did not accept the department's argument. The defendant ended up paying over $750,000 in fines and attorney fees.

Software piracy, the illegal copying of software, is probably one of the most pervasive crimes. Software piracy has several forms: making copies from a single paid copy of the software; using the Internet to download software from a Web site without paying for it, or copying software through use of peer-to-peer applications; using one licensed copy to install an application on multiple computers; taking advantage of upgrade offers without having paid for a legal copy of the updated version; using for commercial purposes copies that were acquired with discounts for home or educational use; and using at home a copy that was purchased by an employer under a license to use only on the employer's premises. The software industry established two organizations to protect software developers from piracy: Business Software Alliance (BSA) and the Software & Information Industry Association (SIIA). The two organizations were established by major software companies and are supported by the majority of the world's software development firms. Both organizations have Web sites that encourage everyone to report pirated software. Occasionally, the organizations sponsor studies that estimate the proportion and financial damage that piracy causes in various world regions.

As the amount of software sold on the market grows, so do the estimated losses that the software industry suffers from piracy. In the 1980s and 1990s, the global financial damage was estimated at $10–$12 billion annually. However, an annual study conducted jointly by BSA and the IT research company IDC (International Data Corporation) reported that the loss to the software industry reached $29 billion in 2003, $33 billion in 2004, and $34 billion in 2005. The studies found that about a third of the world's installed PC software was pirated. In 2005, in terms of absolute dollars lost, the top three countries where the losses occurred were the U.S. ($6.895 billion), China ($3.884 billion), and France ($3.191 billion). In terms of the piracy rate (proportion of pirated software units to total installed software units), the highest rates were in Vietnam (90 percent), Zimbabwe (90 percent), Indonesia (87 percent), and China (86 percent). The lowest rate was in the U.S. (21 percent). IDC estimated that over the period 2006-2010, the world would install $300 billion worth of PC software. Examining growth trends in software use in the various countries and their piracy rates, BSA and IDC expected $200 billion of that amount would consist of pirated software. A spokesperson for BSA explained that piracy deprives local governments of tax revenue, costs jobs in the technology supply chain (developers, distributors, retailers), and cripples local software companies.

Critics question the methods used to reach these estimates. Furthermore, they say that even if the estimates of pirated software are correct, the conclusions are exaggerated, because not all who pirated software would necessarily acquire it if they had to pay for it. Thus, if everybody was forced to pay for the software, the software companies would not collect the entire worth of installed software, but much less. Still, it is reasonable to assume that many pirates actually needed the software, and would pay for it had piracy not existed.

Laws in most countries treat software the same way as they do books, DVD movies, and other types of intellectual property: copies (except for one copy of the software for archival purposes) may not be made without permission of the copyright or patent holder. Yet, the crime is pervasive because it is easy to commit and rarely is punished.

While Figure 5.9 provides a general framework for evaluating ready-made software, each item might be augmented with further inquiry, depending on the program's main purpose. For example, potential buyers often test a word-processing program for features such as availability of different fonts, dictionary size, response time to search operations, ability to create tables of contents and indexes, and other features. Electronic spreadsheet programs are tested for speed of recalculation of formulas, charting, and other features typical of this type of software. Web page development applications are tested for ease of creating various layouts and graphical designs, as well as the ability to maintain desired template appearance and integrity of links among related

FIGURE 5.9
Sample software evaluation form

Factor	What to Look For	Score
Fitness for purpose	◆ Try to maximize the number of needs satisfied.	_____
Ease of learning to use	◆ The shorter the learning time, the better.	_____
Ease of use	◆ The easier a program is to use, the better. ◆ Try to minimize the number of commands that need to be memorized. ◆ The more intuitive the icons, the better.	_____
Compatibility with other software	◆ Try to maximize compatibility with related software and with other operating systems. ◆ Try to maximize portability of data and output to other programs.	_____
Reputation of vendor	◆ Use professional contacts and references to gather background information on the vendor. ◆ Be sure the vendor can deliver what it promises. ◆ Be sure the vendor stands by its pricing.	_____
Availability and quality of telephone and online support	◆ Ask references about their experience. ◆ Look for knowledgeable staff on Web and phone support.	_____
Networking	◆ Try to maximize ability of many computers to share the software.	_____
Cost	◆ Seek detailed pricing information. ◆ Seek the best price, while maintaining quality and performance. ◆ Consider the total cost of ownership: annual license fees, support cost, necessary hardware upgrades, and other costs associated with use of the software.	_____

pages and associated media. Many trade journals, such as *PC World* and *PC Magazine*, maintain labs in which they test competing applications. Experts test different applications on the same computer and report the results.

POINT OF INTEREST

Claytronics

Computer scientists are working on a revolutionary idea: Claytronics, a contraction of clay electronics. The clay will be made from millions of tiny microprocessors called catoms (clay-tronic atoms). A catom will be less than one millimeter in diameter. Catoms will be programmed to form various objects, including clones of humans. Using wireless communication, we will be able to reprogram an object to regroup its catoms to form a new object, for example, turn a laptop computer into a cell phone. We should expect to see the technology at work by 2017. If it works, people who wish to be "present" at a remote meeting will be able to create claytronic clones of themselves at the meeting's location instead of using teleconferencing.

Source: Yen, Y.W., "Forget Nanotech. Think Claytronics," *Business 2.0*, May 2007, p. 33.

The factors to be considered when purchasing large software packages such as ERP software are significantly more complex. The purchasing organization must consider not only the cost of the software, which is usually millions of dollars, but also the amount of time it will take to implement the software, the cost of interrupting ongoing operations, the difficulty and cost of modifying the software for the organization's specific needs, and many other issues.

- "Software" is the collective term for computer programs, which are sets of instructions to computer hardware.

- Software is classified into two general categories. System software manages computer resources, such as CPU time and memory allocation, and carries out routine operations, such as translation and data communication. Application software is a program developed specifically to satisfy some business need, such as payroll or market analysis. Application software can include programs that carry out narrowly focused tasks, or general-purpose applications, such as spreadsheets and word processors.

- To develop software, programmers use programming languages and software development tools. Third-generation languages (3GLs) are more English-like than machine language and assembly languages, and allow more productive programming, meaning that they require less time to develop the same code. Fourth-generation languages (4GLs) are even more English-like and provide many preprogrammed functions. Object-oriented programming (OOP) languages facilitate creation of reusable objects, which are data encapsulated along with the procedures that manipulate them. Visual programming languages help programmers develop code by using icons and other graphics while code is developed automatically by manipulating the graphics.

- As an increasing amount of software is linked to the Internet, many software tools have been created especially for development of Web pages and the software that links Web pages with organizational information resources, such as databases. They include programming languages such as Java, JavaScript, and PHP, and Web page development packages such as FrontPage, Dreamweaver, and GoLive. Java and other languages for the Web produce code that runs on various computers and therefore is very useful for the Web.

- All code written in a programming language other than machine language must be translated into machine language code by special programs, either compilers or interpreters. The translation creates object code from the source code. Software offered for sale is usually object code.

- Some application programs are custom-designed, but many are packaged. The majority of packaged applications are purchased off the shelf, although "off the shelf" might actually mean downloading the application through the Internet.

- Office productivity tools help workers accomplish more in less time. The most pervasive of these tools include word processors, spreadsheets, presentation tools, file and database management software, graphics programs, desktop publishing tools, and project management tools. Some of them are offered as suites.

- Hypermedia and multimedia technologies are useful tools for training, education, research, and business.

- Groupware combines hypermedia and multimedia with Web technologies to help people in separate locations collaborate in their work.

- Virtual reality tools help build software models of products and structures. Virtual reality applications help in training and help build models that are less costly than physical ones.

- Three-dimensional geographic software helps model city blocks and campuses. Combined with other information, it is useful in city service planning and real estate management.

- A growing number of applications are developed using Web programming languages and software tools such as those included in Microsoft .NET. The applications support Web services and access to information resources from Web browsers.

- The most important type of system software is operating systems, also referred to as "platforms." Operating systems carry out an ever-growing number of functions, and include networking and security features. System software also includes utility programs.

- Open source software is being adopted by a growing number of businesses and governments. The source code and its documentation are open to all to review and improve. Open source applications and system software can be downloaded from the Web. Programmers continually improve the code, not for monetary remuneration, but to prove their programming prowess and gain the appreciation of the users. This practice yielded the powerful operating system Linux as well as hundreds of useful applications.

- While some software is purchased, much of it is licensed. The user purchases the right to use the software for a limited time or indefinitely, but does not own the software.

- Businesses should follow a systematic evaluation to determine the suitability of ready-made software to their needs. Consideration of software includes many factors, among which are fitness for purpose, ease of learning to use, ease of use, reputation of the vendor, and expected quality of support from the vendor.

- While software prices have decreased over the years, software piracy is still a problem. About a third of the software used around the world has been illegally copied.

QUICKBIZ MESSENGERS REVISITED

QuickBiz has purchased quite a bit of software through the years. As the company has grown, it has used software to perform routine business functions, develop new routes, and generate employee schedules. It has also begun to use financial software to analyze route profitability and to motivate healthy competition among its messengers.

What Would You Do?

1. When Andrew Langston decided he needed a new information system, he started by listing the basic functions that he needed software to perform. Why would he start with software needs first? Would you do the same?

2. Some QuickBiz employees have taken the wrong routes to make their deliveries. In addition to purchasing the routing software, Andrew is considering implementing a training program to familiarize staff with the streets of Seattle and the communities surrounding Puget Sound. He's asked you to help him come up with some different training ideas. Develop a report on general types of software he might use to train his employees.

3. Explain the importance to QuickBiz of keeping up to date on software developments. Where did Andrew get his information on current software? What business systems did this improve? Where else could QuickBiz look for software news? List some sources.

New Perspectives

1. A couple of QuickBiz's messengers were skeptical of the new routing software's capabilities. They were discussing its usefulness at lunch. Debate the following statement from their discussion: "No software can do what a human can do. I can figure out my routes much better than it can."

2. Andrew's cousin works for a seafood company. His company uses a Linux-based system, and he is recommending that Andrew switch from his Windows-based system to Linux. List the pros and cons of this step for QuickBiz.

3. If you were Andrew, how would you use software to determine whether eliminating downtown Saturday deliveries could also have an effect on revenue from regular services?

applet, 165
application, 159
application program interface
    (API), 175
application software, 160
application-specific
    software, 167
assembly language, 160
avatar, 173
compiler, 165
debugging, 162
driver, 176
general-purpose application
    software, 167

groupware, 171
hypermedia, 169
interpreter, 165
machine language, 160
mashups, 170
multimedia software, 170
object code, 165
object-oriented programming
    (OOP) language, 163
open source software, 178
operating system (OS), 174
packaged software, 167
plug-and-play (PnP), 176
programming, 160

programming language, 160
programming language
    translators, 165
proprietary software, 178
software, 159
source code, 165
suite, 168
system software, 160
utilities, 175
virtual memory, 176
virtual reality (VR), 172
visual programming
    language, 162
Web page authoring tools, 169

REVIEW QUESTIONS

1. Why would any programmer today use a low-level programming language such as assembler rather than a higher-level language?

2. The use of 4GLs is said to contribute to programmer productivity. How so?

3. What is multimedia? Give five examples of how this technology can be used in training, customer service, and education.

4. With so many ready-made software packages available, why do some companies commission software development projects?

5. Office applications are often called productivity tools. Why?

6. Electronic spreadsheets are great tools for modeling. Give an example of a model that shows gradual growth of a phenomenon and describe how you would implement it in a spreadsheet.

7. Why can hypermedia not be implemented on paper? Give an example of what you can communicate with hypermedia that you would not be able to communicate on paper.

8. What are the different media in multimedia?

9. Immersion is an important element of virtual reality. What does it mean?

10. What is the importance of 3-D geographic software? For which types of organizations is it useful?

11. What is the difference between system software and application software?

12. System software is often written using low-level programming languages. Why?

13. Linux is a free and stable operating system, which is a great advantage. What are the disadvantages of adopting it?

14. What is the difference between an interpreter and a compiler?

15. To a compiler or interpreter any logic is legitimate, even if it results in a bad program. Why can't compilers and interpreters detect logic errors in a program?

16. What are the main elements to consider when purchasing ready-made software for an organization?

17. What is open source software? To what does the word "source" refer?

18. Give three reasons why Linux has become a popular server operating system.

19. Why has the trend been to purchase (often, license) software rather than have it tailor-made for organizations?

20. Think of a standard application such as a payroll system. What might drive an organization to develop its own payroll application rather than purchase a ready-made application?

21. Practically all operating systems that run on PCs have graphical user interfaces. What additional (or different) elements would you like to see in operating systems and applications to make them more intuitive?

22. A decision to adopt Linux or another open source operating system is not an easy one for IS managers. What are their concerns? (*Hint*: Think of the relationships between OSs and applications.)

23. Some companies sell open source software, such as Linux. Companies and many individuals buy the software rather than download it free of charge. Why? Would you buy such software or simply download it from the Web?

24. Widespread free application software, such as OpenOffice.org, that runs on a variety of OSs, as well as Web-based applications such as Google Docs & Spreadsheets, threatens to eat into Microsoft's potential revenue. Why?

25. The more an application takes advantage of a GUI, the more suitable it is for international use. How so?

26. Increasingly accurate voice recognition software and sophisticated software that can interpret commands in natural language are bringing us closer to the days of operating a computer by speaking to it. Would you rather speak to a computer than use a keyboard, mouse, or some other input device? Why or why not?

27. Why is software piracy so pervasive? What are your innovative ideas to reduce this problem?

28. Most pressure to legislate and enforce copyright laws for software has come from North America and Western Europe and not from other parts of the world. Why?

29. Do you think open source software will proliferate or disappear?

30. If you were so proficient in programming languages that you could improve open source code (such as the Linux operating system, the Firefox browser, or any of hundreds of applications), would you do it for no monetary compensation? Why or why not?

31. In what ways can young people who seek IT careers benefit by participating in improving open source software?

32. Some observers compare open source software to water. Both the software and water are free, but some companies manage to generate revenue from selling them. How?

33. HeadHunter, Inc., is a new personnel recruiting and placement company. The well-established and cash-rich management consulting company that founded HeadHunter is intent on providing adequate financial resources for the new firm to acquire information systems. Head-Hunter has opened offices in eight major U.S. cities and two European cities.

    Recruiting specialists exchange written correspondence with prospective clients, both managers looking for new positions and companies that might hire them. Records of both recruits and client companies must be kept and updated. All 10 branches should be able to exchange information in real time to maximize the potential markets on both continents. HeadHunter professionals will often travel to make presentations before human resource managers and other executives.

    The majority of HeadHunter's own personnel are college graduates who lack programming skills. HeadHunter management would like to adopt software that is easy to learn and use.

    a. List the types of software the firm needs, both system software and applications.

    b. Research trade journals. Suggest specific software packages for the firm.

Honest Abe and Cars R Us are two fiercely competitive car dealerships. Recently, both started to sell Sniper Hybrid, a new model from Green Motors. Dealers' cost of the car is $19,600. Green Motors pays a dealership $200 for each car sold, and the dealership also keeps whatever markup it adds to the cost. Both dealers start selling the car at the price of $20,600.

Immediately after the two dealerships started to offer the car, each decided to lower the price until the other dealership stopped selling the car. However, their price reduction policies differed. Honest Abe's policy is as follows: at the end of each day, the company sets the price for the next day at the competitor's price minus $50. Cars R Us's policy is the following: at the end of each day, the company sets the price for the next day at the competitor's price minus one percent.

Each dealership decided to stop selling the car as soon as it sells a car at a loss instead of a profit.

34. Using a spreadsheet application such as Microsoft Excel, enter the initial numbers and build a model that will help you answer the following questions:
    a. Which dealer will stop selling the car first?
    b. How many days after it starts to sell the car will this dealer stop selling it?
    c. How much money will this dealer lose per car on the first day it loses money on this car?

35. Team up with two students from your class. As a team, choose two operating systems that run on PCs. Research their features. If both operating systems are available at your school, try them. Write a comparison of their features. Conclude with recommendations about which system you would prefer to adopt for a small business, and why. Focus only on features, not cost.

36. Team up with another student. Log on to www.openoffice.org and prepare a report that covers the following points: (1) Who established this site, and for what purpose? (2) What type of application is the subject of this site? (3) Who contributed the original source code for this project? (4) Who is invited to participate in this project? (5) Would you recommend to a small business with little cash to download and use the software? (6) Would you recommend to a larger and richer organization to do so? Why or why not? In your assessment, address the issues of compatibility with other software, support, ease of training, and ease of use.

37. Team up with another student. Use three different free online video conferencing applications. Try to use all the features. Compose a comparative summary of the features: (1) intuitiveness of the icon and menus, (2) ease of learning and ease of use, (3) quality of the sound, and (4) quality of the video.

FROM IDEAS TO APPLICATION: REAL CASES

Less May Be More

Sometimes, the best is the worst enemy of the good. In software, too much sophistication may alienate customers instead of improve service. Managers at one successful business learned this the hard way.

Citizens National Bank of Texas has a long history. It was established in Waxahachie, Texas, in 1868, and is still privately held. The bank has 16 offices. It does not try to compete with big banks, but rather caters to small communities of 500-25,000, and emphasizes friendly customer service. The extra care has paid off: the bank enjoys two to two and a half cross-sales per customer, a ratio considered high in banking. This means that, on average, each customer has purchased more than two products, such as a checking account, savings account, or home loan. Banks' profitability is highly related to the number of services the same customer purchases. Citizens' best customers use six or seven products.

For many years the bank tracked customer contacts manually. Relationship bankers—as the bank's sales people are called—wrote down on paper the details of contacts. Every Monday, management received a sales report covering the previous week and the calls each banker was going to make that week. With 50,000 customers and many prospective ones, this information soon became reams of paper. The documents contained good information, but the information was difficult to glean and manage.

In 2001, management decided to install a customer relationship management (CRM) system. The product chosen was a CRM package from Siebel Systems (which was later acquired by Oracle). Citizens National hoped to enable the CEO and the 16 relationship bankers to improve tracking of prospective customers and increase the number of contacts the bankers made.

The bank hired a local consulting firm that specializes in installing software for small businesses. The cost of the new software was $150,000. Installing it and adapting it to the bankers' needs cost another $350,000. Siebel's CRM software is recognized as very good, but it was overkill for the bankers, who were typically old-fashioned sales people not keen on using technology. The system has many features that the bankers did not need, and lacked simple features that they did need. It was too sophisticated for Citizens' simple handling of customers. Much of the adaptation time was spent turning off unused features.

Large companies often use CRM systems to set up customer support cases, which are files with details of complaints and how they were resolved, from beginning to end. The bank did not need this function. When customers call to complain, the call center handles the complaint immediately, or channels it to the appropriate officer for immediate resolution. If a customer needs a new checkbook, the call center sends an e-mail message to the proper bank worker who handles checkbook orders. The request for the activity is scheduled, and the bank worker handles such requests in the order in which they come in. The new system did not support this simple way of operation.

The bankers found navigating the new system challenging. Moving from a window to another relevant window was not intuitive, and the bankers wasted much time. They expected to see the typical opportunities to sell more services to a customer listed in the record of the customer. For example, they expected to see an opportunity to offer the customer a business loan listed. However, such opportunities must be entered by the system's users when the customer record is set up.

Different relationships with the same customer might be on different screens. The bankers were confused, and could not get a good sense of all the relationships a customer might have with the bank (such as which services the customer used or a history of complaints the customer had). They had to flip screens constantly. The relationships were organized in a manner inconvenient to the bankers. A consultant that specializes in CRM systems observed that Siebel's system had everything, which is typically too much for small businesses. These clients, he said, usually lose the forest for the trees.

Another challenge was integrating Siebel's software with Citizens' banking software. Like many small and medium size banks, it uses Kirchman Bankway. The software helps process and track deposits, loans, and trust accounts. While the Kirchman software keeps customer last and first name in a single field in its database, Siebel's systems keeps them in two separate fields. This and other differences made integration of the systems time-consuming. The bankers did not expect the integration time to be so long.

After three years, Citizens' management decided to abandon Siebel's system. The bank's consulting firm was committed to automating CRM at the bank. It offered an alternative: QuickBase. QuickBase is offered by Intuit, the company whose fame comes from software such as the financial analysis application Quicken and the small business accounting application QuickBooks. QuickBase offers online tools for project management, sales and customer management, professional services, marketing, IT management, real estate management, and other tasks. It is flexible, and therefore can be used for many purposes. Clients can choose from over 50 ready-made applications or use tools to customize applications to suit their needs. One can maintain data in two different ways: database style and spreadsheet style. The package is available to customers through the Internet. Therefore, it is available immediately. Customers need to install no software on their own computers. Employees use the package through their Web browsers. It is easy to import spreadsheet and database data to a QuickBase spreadsheet or database.

QuickBase is not a CRM system, but provides the tools to build one. Many companies use it for group collaboration because of its accessibility through the Web. Clients of the service pay a one-time fee of $249 for the first 10 users, and $3 per additional user per month. Sales people can use the software in two modes: forms, in which customer details and the history of all contacts with the customer can be recorded; or views. Views are spreadsheets in which, for example, all the sales opportunities for each customer can be listed. Citizens' uses QuickBase to classify sales opportunities by phase: identifying potential customers, analyzing their needs, preparing a proposal, and the result of the effort: won or lost.

The consulting firm prepared interface software using XML (eXtensible Markup Language). The interface links the records in the Kirchman software with the customer records in the online QuickBase records. A bank employee can click on a customer file and view all the contacts made with the customer and what actions have been taken. This helps the bank learn more about customers and know on which ones to focus its sales efforts.

To mitigate the challenge of conservative bankers who dislike technology, management assigned each loan officer an administrative assistant. The loan officer dictates information to the assistant, and the assistant enters it into QuickBase. While the extra work for administrative assistants seems counterproductive, it is not. Management reasons that bank officers' time is better used to call prospects than to struggle with new technology.

Source: Bartholomew, D., "Why Citizens National Bank Threw Out Siebel in Favor of Intuit's QuickBase," *Baseline*, February 26, 2007; (www.cnbwax.com), April 2007; (www.quickbase.com), April 2007.

Thinking About the Case

1. What were the goals of installing CRM software?

2. A Siebel executive commented that the company's CRM system does not fit the needs of all clients. He noted that the clients need to decide if they want an application or a tool kit. Research the term "tool kit." Considering Siebel's CRM system and Intuit QuickBase, which is an application and which is a tool kit? Explain why.

3. The bank's president said that management learned a lesson, and that the $500,000 spent on the abandoned CRM system was tuition for that education. What would you have done in the first place to avoid this "tuition"?

Less Paper, More Efficiency

Whirlpool is the world's leading manufacturer of major home appliances, including refrigerators, washing machines, dryers, dishwashers, and more. It has more than 80,000 employees in over 60 manufacturing and technology research centers around the world, and enjoys annual sales of $19 billion. Whirlpool attributes its success to efficient engineering and manufacturing processes and meticulous attention to quality. Yet, it seemed that there was still room to improve efficiency in one area: order-related business processes.

Whirlpool's main clients are builders who purchase the appliances for new construction sites, and home appliance retail chains. Typically, these clients use low-tech processes. They send orders by mail, phone, fax, or e-mail attachment. This forced Whirlpool to handle orders manually. In addition to being labor-intensive, the process also entailed long order-process times and a high potential for errors, as well as misplaced and lost documents.

Whirlpool's order center is in Knoxville, Tennessee. Until 2006, order handling involved much paper. More than 2 billion paper documents were entered annually by the order entry department. In March of that year, Whirlpool acquired one of its major competitors, Maytag. Management realized that handling orders by telephone, fax, and paper would increase significantly. It decided to automate as much of the process as possible.

Whirlpool uses SAP's R/3 ERP system. Automating order processing had to be tied to the system. The company was already using Esker's DeliveryWare 4.0, software that supports document automation processing. Esker is a business partner of Dolphin IT

Project and Consulting Corporation, which specializes in providing SAP R/3 solutions for document and data archiving, business workflow, and content management. During its eight-year relationship with Esker, Whirlpool used part of its software—called Fax Server—to manage outgoing faxes. However, the software operated independently of SAP's ERP system, although DeliveryWare can extend SAP software to handle documents.

Whirlpool hired both companies to expand the integration of DeliveryWare to the SAP software. The software now provides optical character recognition (OCR, discussed in Chapter 4, "Business Hardware"), software that recognizes print and handwritten documents.

When an order comes in any form—mail, fax, or e-mail attachment—DeliveryWare scans the image to turn it into digital text, looks up to whom at Whirlpool the order should be routed, and sends the employee a notification. The employee who receives the notification accesses the order remotely via a Web browser. The electronic data is channeled into the SAP system for processing, which starts by creating an internal sales order. Employees are allowed to override the automatic process if there are some exceptions, such as price discrepancies.

The integrated software has helped in several ways. The time between receipt of order and order entry decreased from almost four days to one day. Much of the paper involved in the process has been eliminated, because after paper is scanned there is no need to archive it. E-mail-attached orders do not need to be printed out. The rate of errors decreased. Whirlpool hopes to further reduce the cycle time from one day to several minutes. It reduced the number of employees engaged in order processing by five percent.

The final phase of the software integration was to automate proof of delivery (POD). When appliances are ready to be shipped to a client, Whirlpool's invoices coming out of the SAP system are printed by the DeliveryWare application. The software finds the associated order and the bill of lading. The three documents are merged into one document, which is then sent with the appliances to the client. The automatic process reduces labor and shortens the time for cash collection. Keeping documents in digital forms and electronically related to each other supports communication between Whirlpool's marketing and warranty staffs. Thus, one application, DeliveryWare, well integrated into the organization's ERP system, has improved the efficiency of several processes.

Whirlpool's IT department, headquartered in Benton Harbor, Michigan, credits the success of this project to the cooperation and commitment of the three companies. The project team, which continues to work on automating additional processes, includes representatives from Whirlpool's IT department, Esker, and Dolphin.

Source: Haber, L., "Whirlpool Soothes Highs and Lows," *eWeek*, December 6, 2006; Bowen, G.N., "Outsourcing Simplifies the Paperwork for Appliance Manufacturer Whirlpool," (www.outsourcing-information-technology.com), February 2007; (www.whirlpoolcorp.com), April 2007; (www.dolphin-corp.com), April 2007; (www.esker.com), April 2007.

Thinking About the Case

1. What software packages are now used by Whirlpool?

2. Which aspects of order processing have been improved thanks to the software?

3. If you could convince all of Whirlpool's clients to adopt software that would further help Whirlpool, what type(s) of software would you recommend? (You may want to research business-to-business software before you answer this question.)

Stop! Wait! I Am Pulling Down a Menu!

San Jose, California, is considered one of the safest large cities in the United States. The city's 1,000 police officers serve 925,000 residents, making it the smallest officer-to-resident ratio in the country. From 1990 to 2004, the city's police department used a text-based mobile dispatch system. The system had been customized by its designer to meet the needs and preferences of the city's officers. Although there was some initial hesitation by officers to use the system, they eventually embraced it.

After more than a decade of reliable service, police and city officials decided to replace the system with new Windows-based touch-screen software. A new touch-screen computer was to be installed in every patrol car. It was designed to receive orders, send messages, write reports, receive maps of the city, and use GPS to let officers know where they are located and where other patrol cars are. San Jose government paid Intergraph, the company that developed the software, $4.7 million for the software, which was supposed to serve both the police and fire departments. However, the effort was plagued with problems from the start.

Even before the new system was installed, there were already grumblings at the department. Officers claimed nobody had ever sought their input about the design of the user interface. When they started using the new system, they were disappointed.

Tension had built up, but this was not the main concern of the San Jose Police Officers Association (SJPOA). The organization's leaders were not so much offended because they had not been asked about the system before it was developed. They were more concerned about the results of that failure to ask for their members' feedback. They were frustrated with the lack of training and error-infested software. Some people will inevitably complain when adapting to new technology, but when their lives and those of the public depend directly on the software's performance, the stakes are much higher.

Since its June 2004 operational debut, the system has had numerous major problems. The greatest concern is the increased difficulty in issuing the Code 99 command, the emergency contact when an officer is in danger and needs immediate help. Initially, officers had to strike one key to issue Code 99, but that resulted in too many false alarms. As a result, code entry for emergencies now requires a two-keystroke combination. Officers complain about having to find the right combination of touch-pad keys on a 12-inch screen while they are under fire or in hot pursuit of a suspect. One officer even crashed his squad car into a parked vehicle because he was so distracted by the information he had to enter using the touch-screen. Another problem was that with the new software it took patrol officers longer to find out whether a person they had stopped has a violent criminal record, which is vital information in a job that requires split-second decisions of life or death.

The police officers complained that they were not given sufficient training. However, the problems with the system had nothing to do with how the police officers used it; the software simply did not work. Two days after the system went live, it crashed. For the next few days, it was almost completely inaccessible. Its designers acknowledge that this was not a good way to build confidence with the officers. Yet, even after the system was modified to fix these problems, several more errors were discovered by the president of a user-interface design consulting firm that was hired by the SJPOA to review the software.

The mapping and GPS location tracking were supposed to be assist officers. Yet, the system's map information had some significant inaccuracies. Additionally, unneeded information took up screen space, and display fonts were hard to read. Even a simple task such as checking a driver's license plate was difficult to perform after the system had already been treated for bugs.

Every new technology has a learning curve that can last weeks or months until users feel sufficiently comfortable with it, but with this software the difficulties were not only a matter of a learning curve. Even tolerant and receptive officers have faced obstacles in trying to adapt. Intergraph's specialists spent weeks in San Jose to fix bugs and streamline procedures for the most basic patrol tasks, like the license plate verification.

Officers complained about receiving only three hours of training on software that is supposed to ensure their safety. In response, the department has offered more training sessions. The software runs on the Windows operating system, a fact that complicated matters for many of the police officers. Older officers were not comfortable with pull-down menus and other features of the interface. As a result, they have been more resistant to the new software than their younger, more computer-literate colleagues.

Observing police work, the consultants brought in by the SJPOA noted that choosing a Windows GUI with complex menu hierarchies does not make sense for anyone who has to use the system while driving a car. In addition, officers were trained on desktop computers with trackpads on keyboards instead of touch screens they actually have to use in the squad cars.

Dispatchers, too, have expressed dissatisfaction with the Intergraph system, especially because of risky delays in task execution. With the new software, officers have to wait longer to access information about any previous arrests for a detained suspect. Dispatchers also note the same concern expressed by their comrades on patrol: the new software cannot perform multiple tasks simultaneously. Like the officers, the dispatchers feel they should have been consulted about the software during the interface design stage. San Jose's police chief admits that in hindsight, incorporating more end-user input during the planning phase would have eased the introduction and implementation of the new system.

The Chicago Police Department had a similarly painful experience with a major dispatch system overhaul in 1999. Just as in San Jose's case, patrolling police officers were not asked for input before the software was developed, and the results left bad feelings across the department. Chicago eventually replaced the software with a newer system. This time, patrol officers were consulted, and their suggestions were considered before the programmers developed the applications. Unfortunately, San Jose's police department did not learn the lesson from the Windy City's experience.

Police departments in two Canadian cities, Calgary and Winnipeg, had similar disappointing experiences with the Intergraph system. Officials in other cities also have been frustrated, and some planned to scrap the system.

Perhaps San Jose might not have to replace the Intergraph software after all. The San Diego Sheriff's Department has used Intergraph's touch-screen software for six years with eventual success. Initially, the system had bugs similar to those experienced in San Jose, but Intergraph eventually fixed them. Also, San Diego officials conducted basic Windows training sessions for their sheriff's deputies, because some of them had no previous computer experience whatsoever. The sheriff's department also experienced some resistance to the new software. But fixing the bugs and providing good training did the trick, and the deputies adapted.

Source: Hafner, K., "Wanted by the Police: A Good Interface," *New York Times*, Technology Section (www.nytimes.com), November 11, 2004; Zapler, M., "New S.J. Dispatch System Flawed," *Mercury News* (www.mercurynews.com), September 22, 2004.

Thinking About the Case

1. Are the problems encountered by the police officers due to hardware or software?

2. Whom do you think is at fault for the unsuccessful implementation of the new software? Why?

3. People, especially the "technologically challenged," are often not receptive of new technologies. Was this a major issue in this case?

4. If you were the CEO of Intergraph before it assumed the project for San Jose, what would you do differently?

6

SIX

Business Networks and Telecommunications

LEARNING OBJECTIVES

Modern telecommunications technology allows businesses to send and receive information in seconds. Except when a physical transfer of goods or performance of a local service is involved, geographical distances are becoming insignificant in business transactions. When using computers and other digital devices, people can now work together as if they were sitting next to each other, even when they are thousands of miles apart. Financial transactions and information retrieval take seconds, and wireless technology enables us to perform these activities from almost anywhere and while on the go. Understanding the technology underlying telecommunications—its strengths, weaknesses, and available options—is essential in any professional career.

When you finish this chapter, you will be able to:

■ Describe business and home applications of digital telecommunications.

■ Identify the major media and devices used in telecommunications.

■ Explain the concept of network protocols.

■ Compare and contrast various networking and Internet services.

■ List networking technologies and trends that are likely to have an impact on businesses and information management in the near future.

■ Discuss the pros and cons of telecommuting.

QUICKBIZ MESSENGERS:
Communication Is Key

Mark Johnson, one of QuickBiz's longtime car messengers, was hopelessly stuck in traffic. An accident involving two semitrailer trucks had brought traffic on Interstate 5 to a dead stop. He desperately needed to contact his customer—a medical supply firm—to alert them that his delivery would be delayed. So he used his hands-free cellular phone to call the customer. His contact at the supply firm acknowledged his delay and told him that the supplies were a routine delivery to a hospital pharmacy and not to worry—as long as the hospital received the delivery sometime that day, they'd be fine. Mark apologized for the glitch and promised to get off at the next exit as soon as he could move again. Then he used his group e-mail program to warn other messengers to stay off I-5 for the time being. Maybe he could save somebody else a headache.

Using New Technology

When cellular phones with GPS (global positioning service) capability became affordable, Andrew Langston equipped each messenger with such a phone so headquarters would be able to locate and communicate with them instantly, and they would be able to communicate among themselves. Andrew also negotiated a good deal for text messaging. Text messaging was especially important in case cellular services deteriorated, because even in emergencies such as floods or earthquakes, text messaging has proven itself superior to cellular voice service. In addition, text message alerts could be broadcast to the entire delivery fleet. Now messengers could be rerouted around trouble spots. Of course, occasional delays for one or two messengers would still occur, but the problems now could be isolated.

As soon as the media began reporting a link between cell phone use and automobile accidents, Andrew decided to purchase hands-free Bluetooth car kits so that messengers could communicate safely with customers and the office. These devices also meant that his messengers didn't miss calls while they were fumbling for their phones.

Increasing Efficiency and Customer Satisfaction

Leslie Chen updated delivery information from the messengers' handheld computers into the database. As the business grew, however, Leslie spent more of her time uploading data from the handheld computers. A representative from the company's cell phone service provider told her about a wireless card that messengers could use to upload delivery information to the company's database. The messengers plug the card into a slot on their handheld computers and access the Internet through the cell phone providers' connection. Not only did this innovation save Leslie time, it also meant that messengers could update delivery information immediately upon delivery so that the company could provide the information to their customers right away. Delivery confirmations now could be sent via e-mail directly to the senders as soon as deliveries were made. Leslie no longer had to confirm special deliveries; messengers did so immediately and copied her on their transmittals, saving her time and the company money, all while increasing customer service.

Competitors Up the Ante

QuickBiz's competitors hadn't stood still either. A major competitor had improved its service by offering standard one-hour delivery time in nearby communities—half of QuickBiz's standard delivery time. So, Andrew responded by opening two satellite offices to get messengers to remote destinations more quickly. This allowed Andrew to match his competitor's new time frame and still make a profit on deliveries.

An additional benefit of the three-office configuration was enhanced data security. In 2004, when pipes burst and flooded the main office, QuickBiz did not have a recovery plan. Now, every time any data is recorded at one of the offices, it is automatically duplicated on disks at the other two offices via the Internet. Andrew felt much more secure knowing that important information would always be available when needed.

Choosing the Right Network Service Providers

To link its three offices, QuickBiz used an Internet service provider (ISP) offering digital subscriber line (DSL) service, and a company that specialized in installation of virtual private networks (VPNs). Andrew found the DSL service to be fast enough for his company's needs and very affordable, but the connection was not reliable enough for QuickBiz. So Andrew and Sarah Truesdale, the office manager, found themselves looking for an alternative. They considered cable and even a T1 line hookup, but the companies that offered those services would have to string their lines to the office sites, and they couldn't get to QuickBiz for six to eight weeks.

Some time earlier, Andrew noticed strange antennas popping up in the neighborhood. He remembered someone mentioning that a telecom company was establishing fixed wireless service in the area. Perhaps he could use that service. Indeed, the service was available at a fee comparable to the DSL service, and Andrew subscribed QuickBiz. The company that provided the VPN software made all the necessary arrangements to ensure that communication among the three offices remained private. Now, QuickBiz's three offices would have high-speed wireless Internet access as well as secure interoffice communications at a reasonable rate.

Intranets and Extranets

As the staff became more comfortable with Internet technology, they began to see its usefulness for other business functions. For example, the human resources manager set up an intranet to inform staff of the benefits program options and general company news. The information could be accessed from all three offices and through the cellular phones that messengers carried. He also set up a short orientation video to introduce new employees to the company.

Sarah and Leslie began to consider an extranet to expedite transactions with the firms that maintained their truck and bicycle fleets. They also explored the option to use the extranet of a national office supply superstore.

TELECOMMUNICATIONS IN BUSINESS

It is your first visit to Barcelona. You are standing at a bus stop, waiting for the bus that will take you on the next leg of your vacation tour. You pull out your mobile phone, send a short text message to a four-digit number, and receive a message with an accurate time when your bus will arrive. You then use the device to receive directions and maps describing how to get from one point to another. You use the time until the bus arrives to view a local TV program on your phone. On the bus, you use the device to check your e-mail. When you arrive at your destination you use the phone to find one of the many hotspots where you can connect to the Internet. You use the phone to e-mail and call home. Since the call uses the Internet, it is free.

POINT OF INTEREST

Growing E-Mail

The size of e-mail messages that people send and receive grows steadily. According to the Radicati Group, a technology research firm, in 2007 a typical corporate e-mail account generates about 18 MB of mail and attachments per business day per employee, or about 4.3GB of electronic data per user/per year. The number is expected to grow to 28 MB per day, or 6.7 GB per year by 2010.

Source: Preimesberger, C., "Firms Face Risks for Failing to Archive E-mails," *CIO Insight*, April 30, 2007.

Telecommunications, which is essential to smooth operations in today's business world, is the transmittal of data and information from one point to another. The Greek word *tele*, which means "distance," is part of such words as "telephone," "teleconference," and other words referring to technologies that allow communications over a distance. Thus, telecommunications is communications over a distance. Telephone, e-mail, the World Wide Web—none of these essential business services would be available without fast, reliable telecommunications. Telecommunications, made possible by networking technologies, has brought several improvements to business processes:

- *Better business communication.* When no physical objects need to be transferred from one place to another, telecommunications technology can make geographical distance irrelevant. E-mail, voice mail, instant messaging (IM), faxing, file transfer, mobile telephony, and teleconferencing enable detailed and instant communication, within and between organizations. Telecommunications can also be used by one person to monitor another person's performance in real time. The use of e-mail, IM, and voice mail has brought some secondary benefits to business communications by establishing a permanent written or electronic record of, and accountability for, ideas. Web-based instant messaging is used to support online shoppers in real time.The result is more accurate business communications and reduced need for manual recording.

- *Greater efficiency.* Telecommunications has made business processes more efficient. Any information that is recorded electronically can become immediately available to anyone involved in a business process, even when the business units are located far apart. For example, as soon as an order is placed, anyone in the organization who will be involved with it at any stage can view the order: from the marketing people, to purchasing officers, to manufacturing managers, to shipping workers, to billing and collection clerks. For example, if a store lacks a certain item, a clerk can check the entire chain's inventory and tell the customer the nearest store that has the item available. If a customer wishes to return an item, she can do so at any store of the chain because a sales associate can easily verify the purchase details. This may also help retail chains discover "serial returners."

- *Better distribution of data.* Organizations that can transmit vital data quickly from one computer to another can choose not to have centralized databases. Business units that need certain data frequently might store it locally, while others can access it remotely. Only fast, reliable transfer of data makes this efficient arrangement possible.

- *Instant transactions.* The availability of the Internet to millions of businesses and consumers has shifted a significant volume of business transactions to the Web. Both businesses and consumers can shop, purchase, and pay instantly online. Wireless technology has also made possible instant payment and data collection using small radio devices, such as electronic toll collection tags. In addition to commercial activities, people can use telecommunications for online education and entertainment.

- *Flexible and mobile workforce.* Employees do not have to come to the office to carry out their work as long as their jobs only involve the use and creation of information. They can telecommute using Internet connections. Salespeople, support personnel, and field workers are more mobile with wireless communication.

- *Alternative channels.* Services that used to be conducted through specialized dedicated channels can be conducted through alternative channels. For example, voice communication used to be conducted only through proprietary telephone networks but is now also conducted through the Internet, which decreased its cost. Radio and television broadcasts were conducted through radio frequencies and company-owned cables. Newer technologies enable organizations to broadcast over the Internet and provide telephone services over the Internet as well. Furthermore, Internet technologies allow individuals to broadcast text, sound, and video to subscribers' computers or to Web-capable mobile devices. (We discuss these technologies in Chapter 8, "The Web-Enabled Enterprise.")

At the same time you enjoy the opportunities created by telecommunications technology, you must recognize that it poses some risks. Once an organization connects its information systems to a public network, security becomes a challenge. Unauthorized access and data destruction are constant threats. Thus, organizations must establish proper security controls as preventive measures. We discuss the risks and security measures in Chapter 14, "Risks, Security, and Disaster Recovery."

TELECOMMUNICATIONS IN DAILY USE

We have grown so accustomed to telecommunications networks that we no longer think much about them in daily life; however, they are pervasive. The most widespread telecommunications uses are described in the following sections.

Cellular Phones

Cellular phones derive their name from the territories of service providers, which are divided into areas known as cells. Each cell has at its center a computerized transceiver (transmitter-receiver), which both transmits signals to another receiver and receives signals from another transmitter. When a call is placed on a cellular phone, the signal is first transmitted to the closest transceiver, which sends a signal through landlines that dial the desired phone number. If the receiving phone is also mobile, the call is communicated to the transceiver closest to the receiving phone. As the user moves from one area, or cell, to another, other transceivers pick up the transmission and receiving tasks.

Using cellular phone networks, people can transmit and receive calls almost anywhere, freeing them from a fixed office location. Cellular phones (often called mobile phones) can also be used for e-mail and faxing, and many are Web-enabled. Many mobile phones have been merged with digital cameras, PDAs, and GPS (global positioning system) circuitry. "My car is my office" is a reality for many professionals who spend much of their time traveling. As technology advances and more capabilities are squeezed into smaller casings, some professionals can say, "My pocket is my office."

The major advantage of cell phones is that they are attached to people, not offices. This is why, despite the higher cost of mobile phones over landline phones, some companies have decided to discard the latter and adopt the former for some or all of their employees. For example, in 2005, Ford Motor Company disconnected the landline phones of 8,000 employees and equipped them with mobile phones. The purpose is to make engineers more available to each other.

Some companies make the switch to mobile phones when they move their offices. Moving electronic switchboards and telephone lines to its new offices in Hawaii would have cost NovaSol, a scientific research firm, $30,000. The company decided to equip its 80 employees with cell phones. Other companies make the switch because so many employees already have both landline phones in the office and a cell phone for their time with customers or on manufacturing lines. For this reason, Dana Corp., a manufacturer of auto parts, removed most of the phones from its offices in Auburn Hills, Michigan. The lines left are used mainly for teleconferencing.

Videoconferencing saves time and travel expenses and reduces air pollution.

Courtesy of Polycom, Inc.

Videoconferencing

People sitting in conference rooms thousands of miles apart are brought together by their transmitted images and speech in what is called **videoconferencing**. Businesses use videoconferencing to save on travel costs and lodging, car fleets, and the time of highly salaried employees, whether they work in different organizations or at different sites of the same organization. From national and global perspectives, videoconferencing also reduces traffic congestion and air pollution. The increasing speed of Internet connections makes it easy for anyone with a high-speed link to establish videoconferences by using a peer-to-peer link or the services of a third party, a company that specializes in maintaining videoconferencing hardware and software. In the latter case, businesses pay a monthly fee for unlimited conferences or pay a per-use fee.

Wireless Payments and Warehousing

Radio frequency identification (RFID) technology, mentioned in Chapter 3, "Business Functions and Supply Chains," and covered in more detail later in this chapter, enables us to conduct transactions and to make payments quickly. An increasing number of drivers never approach a cash register or swipe credit cards when paying for fuel at gasoline stations. If you use a speed payment device such as ExxonMobil's Speedpass™, an RFID tag communicates with a device on the pump to record the details of the transaction. An antenna dish on the rooftop of the gas station communicates these details and checks your credit through a link to a large database located hundreds or even thousands of miles away and operated by the bank authorizing the charge. In this transaction, you use telecommunications twice: once between the device and the pump, and once between the gas station's antenna and the database. Wireless toll payment systems use a similar technology. A special transceiver installed at the toll plaza sends a signal that prompts the tag installed in your car to send back its own signal, including the unique owner's code, entry location, and time the vehicle passes by. The information is used to charge the account associated with the owner's number, and the information captured is transmitted to a large database of account information.

RFID technology is also used in warehouses where employees can use handheld units to check a central system for availability and location of items to be picked up from and stored in shelves or bins. When storing, the handhelds are used to update inventory databases. Such systems have made the work of "untethered employees" more efficient compared with older systems that require physical access to a computer. Wireless communications have many other uses, some of which are discussed in detail later in the chapter.

Why You Should Understand Telecommunications

As a professional, you will be responsible for ensuring that your organization maximizes its benefits from fast and reliable telecommunications. To do so, you might be involved in selecting from networking alternatives. To be a creative and productive contributor to these decisions, it is essential that you understand the fundamental promises and limitations of networking and telecommunications.

Many tasks that used to be in the sole domain of highly paid specialists are being performed by professionals whose main occupation is not IT. For example, creating small networks in businesses and homes used to be the responsibility of technicians. Now any professional is expected to know how to create hotspots and how to use a plethora of networks: wired, wireless, cellular, and Internet-based.

Peer-to-Peer File Sharing

One of the most exciting features in worldwide telecommunications is **peer-to-peer (P2P) file sharing** through the Internet: anyone with access to the Internet can download one of several free applications that help locate and download files from any online computer. You might have heard of some of these applications, such as LimeWire, BearShare, Morpheus, and KaZaA. While the concept has effectively served scientists who share scientific text files and application developers who exchange code, the most extensive use has been in downloading artistic files, such as music and video files. Because unauthorized duplication and use of such files violates copyright laws and deprives recording and film companies of revenue, these industries have sued some violators in court, and the U.S. Supreme Court ruled against organizations that provide file-sharing services. These actions and the proliferation of legitimate services that sell individual music tracks online for as little as 89 cents per track have reduced the use of file sharing for illegal copying, but have not eliminated it.

Web-Empowered Commerce

Increasingly fast digital communication enables millions of organizations to conduct business and individuals to research, market, educate, train, shop, purchase, and pay online. Entire industries, such as online exchanges and auctions, have been created thanks to the Web. Web-based commerce is covered in detail in Chapter 8, "The Web-Enabled Enterprise" and is illustrated with many examples throughout the book.

BANDWIDTH AND MEDIA

While people can enjoy technologies without understanding how they work, educated professionals often do need to understand some fundamental concepts to be able to participate in decision making when selecting networking equipment and services. This section introduces bandwidth and networking media.

Bandwidth

A communications *medium* is the physical means that transports the signal, such as a copper wire telephone line, a television cable, or radio waves. The **bandwidth** of the medium is the speed at which data is communicated, which is also called the **transmission rate** or simply the bit rate. It is measured as **bits per second (bps)**. Figure 6.1 shows common bit rate measurements. Bandwidth is a limited resource. Usually, the greater the bandwidth, the higher the cost of the communications service. Thus, determining the type of communications lines to install or subscribe to may be an important business decision.

FIGURE 6.1
Transmission speed measurement units

bps	=	Bits per second
Kbps	=	Thousand bps
Mbps	=	Million bps (mega bps)
Gbps	=	Billion bps (giga bps)
Tbps	=	Trillion bps (tera bps)

When a communications medium can carry only one transmission at a time, it is known as **baseband**. Dial-up connections through regular phone lines and Ethernet computer network connections are examples of baseband. When a line is capable of carrying multiple transmissions simultaneously, it is said to be **broadband**. Cable television, DSL (digital subscriber line), fiber-optic cables, and most wireless connections are broadband. In general, broadband offers greater bandwidth and faster throughput than baseband connections, and in common usage the term "broadband" is associated with a high-speed networking connection, which is required for fast transmission of large files and multimedia material. In contrast, the term *narrowband* refers to lower speeds, although the speed under which communication is considered narrowband has constantly increased.

Media

Communications media—the means through which bits are transmitted—come in several types. Media can be tangible, such as cables, or intangible, such as radio waves. The most available tangible media are twisted pair cable, coaxial cable, and optical fiber (see Figure 6.2). Intangible media include all microwave radio technologies, which support wireless communication. The electric power grid has also been added as a medium for communications. All can be used to link a business or household to the Internet. Later in the chapter we discuss the various Internet connection services and also refer to typical periodic cost of the services.

FIGURE 6.2
Networking media

Medium	Availability	Bandwidth	Vulnerability to Electromagnetic Interference
Twisted pair cable	High	Low to medium	High
Radio waves	High	Medium to high	Low (but vulnerable to radio frequency interference)
Microwave	Low	High	Low
Coaxial (TV) cable	High	High	Low
Optical fiber	Moderate but growing	Highest	Nonexistent
Electric power lines (BPL)	Very High	High	High

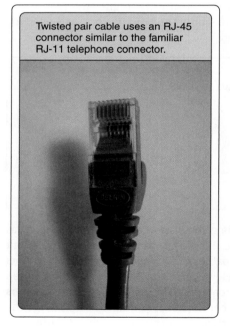

Twisted pair cable uses an RJ-45 connector similar to the familiar RJ-11 telephone connector.

Twisted Pair Cable

Twisted pair cable is a popular medium for connecting computers and networking devices because it is relatively flexible, reliable, and low cost. The most common types of twisted pair network cable today are Category 5 or Category 6 (Cat 5 or Cat 6), named for the cable standards they follow. Twisted pair cable connects to network devices with RJ-45 plug-in connectors, which resemble the RJ-11 connectors used on telephone wire, but are slightly larger.

Twisted pair cable is also used in telephone networks, but in the United States and many other countries, twisted copper wires are now used only between the telephone jack and the central office of the company providing the telephone service. The typical distance of this link is 1.5–6 kilometers (about 1–4 miles), and is often referred to as "the last mile." The central offices themselves are connected with fiber optic cables, but it is often the "last mile" media that determine the overall speed of the connection. In recent years many "last mile" connections have also been converted to optical cables. Most new buildings, including residential ones, are equipped with fiber optic cables rather than copper wires.

Coaxial Cable

Coaxial cable is sometimes called TV cable or simply "cable" because of its common use for cable television transmission. It is widely used for links to the Internet. Television companies use the same networks they employ to transmit television programming to link households and businesses to the Internet. Since telephone services can be offered on any broadband Internet link, cable companies also offer telephone service through this medium.

Optical Fiber

Fiber optic technology uses light instead of electricity to represent bits. Fiber optic lines are made of thin fiberglass filaments. A transmitter sends tiny bursts of light using a laser or a light-emitting diode (LED) device. The receiver detects the period of light and no-light to receive the data bits. Optical fiber systems operate in the infrared and visible light frequencies. Because light is not susceptible to **EMI (electromagnetic interference)** and **RFI (radio frequency interference)**, fiber optic communication is much less prone to error than twisted pair and radio transmission. Optical fibers can also carry signals over relatively longer distances than other media.

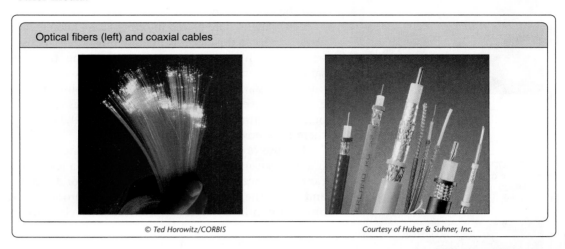

Optical fibers (left) and coaxial cables

© Ted Horowitz/CORBIS   Courtesy of Huber & Suhner, Inc.

The maximum speed attained with optical fibers has been 25.6 terabits per second (Tbps), enough to transmit the content of 600 DVDs in one second. Some optical carriers support bit rates of up to several Tbps. Such great bandwidth enables multiple streams of both Internet and television transmission. Some telecommunications companies, such as Verizon, have laid optical fiber lines to offer households both services, directly competing with TV cable companies such as Comcast. In other countries, such as Japan and South Korea, a greater percentage of households are offered broadband over optical fibers, and the bandwidth that subscribers can receive is significantly higher than that in the United States. However, an increasing number of U.S. communities are served with optical fibers, with speeds of several tens of megabits per second (Mbps). Such speeds permit the telecommunications company to offer television service on the same fiber that provides telephone and Internet service.

The potential of optical fibers is usually much greater than telecommunications companies actually provide. For example, Verizon's optical fibers—which are installed in some half a million U.S. households—can provide up to 644 Mbps, but the company does not offer more than 30 Mbps.

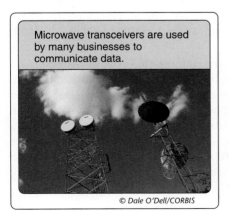

Microwave transceivers are used by many businesses to communicate data.

© Dale O'Dell/CORBIS

Radio and Satellite Transmission

Radio frequency (RF) technologies use radio waves to carry bits. Several wireless technologies can transmit through air or space. Some of the most popular for personal and business networking, such as Wi-Fi and Bluetooth, are discussed later in this chapter. **Microwaves** are high-frequency radio waves that can carry signals over long distances with high accuracy. You have probably noticed the parabolic antennas on the roofs of some buildings. They are so numerous on rooftops and high antenna towers because microwave communication is effective only if the line of sight between the transmitter and receiver is unobstructed. Clusters of microwave antennas are often installed on high buildings and the tops of mountains to obtain a clear line of sight. Terrestrial microwave communication—so-called because signals are sent from and received by stations on the earth—is good for long-distance telecommunications but can also be used in local networks in and among buildings. It is commonly used for voice and television communications. When radio communication is used outside buildings, it is vulnerable to weather conditions—thunderstorms, fog, and snow might degrade communication quality.

Signals can also be transmitted using microwaves via satellite links. The two major types of satellites are geostationary, also called GEO, and low earth orbit, also called LEO. Both types serve as radio relay stations in orbit above the earth that receive, amplify, and redirect signals. Microwave transceiver dishes are aimed at the satellite, which has antennas, amplifiers, and transmitters. The satellite receives a signal, amplifies it, and retransmits it to the destination.

GEO satellites are placed in orbit 35,784 kilometers (about 22,282 miles) above earth. At this distance the satellite is geosynchronized (synchronized with the earth); that is, once it starts orbiting, the satellite stays above the same point on earth at all times, without being propelled. Thus, a GEO satellite is stationary relative to earth. Because they orbit at such a great distance above the earth, three GEO satellites can provide service for every point on earth by relaying signals among themselves before transmitting them back down to their destinations.

LEO satellites blanket the earth to provide uninterrupted communications.

Courtesy of Teledesic LLC

Because of the distance from earth to satellites, the communication is fine for transmitting data because delays of a few seconds make no significant difference. However, a delay of even 2 or 3 seconds (due to the trip to and from the satellite and the time of processing the data) might be disturbing in interactive communication, such as when voice and pictures are communicated in real time. You might have noticed such delays when reporters use devices that communicate to a television station. When an anchorperson asks a question, the reporter on location receives the question with a noticeable delay.

LEO satellites minimize this shortcoming. These lower-cost satellites are placed about 800–1000 kilometers (500–600 miles) above earth. The signals' round-trip is short enough for mobile telephone and interactive computer applications. Unlike GEOs, LEO satellites revolve around the globe every few hours. Multiple LEOs are required to maintain continuous coverage for uninterrupted communication.

Large companies lease telecommunication satellite frequencies to transmit data across the globe.

Courtesy of NASA

Electrical Power Lines

One medium that had been available for years but has only recently been tapped for telecommunications is the electric power grid. The bits in an electric power grid are represented by electric impulses, but they must be distinct from the regular power that flows through the grid. Engineers have succeeded in overcoming this technical challenge. In some regions of the United States, broadband service is offered through power lines. The service is referred to as **Broadband over Power Lines (BPL)** or Power Line Communication (PLC). BPL is covered in more detail later in the chapter.

From the point of view of organizations, among the important factors in choosing a networking medium are availability, current and potential bandwidth, and vulnerability to electromagnetic interference (EMI) or radio frequency interference (RFI). Your business's current and future needs for data security, as well as compatibility with an already installed network, are also factors. Cost is another important consideration. For example, one of the benefits of optical fiber is that it is practically immune to EMI. However, it is more expensive than other options. Another point to consider is the availability of a specific service on an available medium. For instance, you might have a telephone line on a remote farm, but no company offers broadband service to it.

In the context of data communications, a **network** is a combination of devices or **nodes** (computers or communication devices) connected to each other through one of the communication media previously discussed. We will often use the word "computer" for a device that is networked, but this is only for convenience. Any compatible device that can transmit and receive on a network is part of it.

Types of Networks

Computer networks are classified according to their reach and complexity. The three basic types of networks are LANs (local area networks), which connect computers, printers, and other computer equipment for an office, several adjacent offices, an entire building or a campus; MANs (metropolitan area networks), which span a greater distance than LANs and usually have more complicated networking equipment for midrange communications; and WANs (wide area networks), which connect systems in an entire nation, continent, or worldwide. Some people also include a fourth category: PANs (personal area networks), which encompass connections between personal digital devices such as a computer and its keyboard or mouse, or a mobile phone and a hands-free headset.

LANs

A computer network within a building, or a campus of adjacent buildings, is called a **local area network**, or **LAN**. LANs are usually established by a single organization with offices within a radius of roughly 5–6 kilometers (3–4 miles). LANs are set up by organizations to enhance communications among employees and to share IT resources. Households might set up LANs to share a broadband link to the Internet and to transmit digital music, pictures, and video from one part of a home to another.

In office LANs, one computer is often used as a central repository of programs and files that all connected computers can use; this computer is called a **server**. Connected computers can store documents on their own disks or on the server, can share hardware such as printers, and can exchange e-mail. When a LAN has a server, the server usually has centralized control of communications among the connected computers and between the computers and the server itself. Another computer or special communications device can also exercise this control, or control can be distributed among several servers. A **peer-to-peer LAN** is one in which no central device controls communications.

In recent years the cost of wireless devices has decreased significantly, and many offices as well as households now network their computers wirelessly, or create networks in which some of the computers are wired and some are not. **Wireless LANs (WLANs)** offer significant benefits: installation is easy because there is no need to drill through walls to install wires, and equipment can be moved to wherever it is needed. Wireless LANs are less costly to maintain when the network spans two or more buildings. They are also more scalable. **Scalability** is the ease of expanding a system. It is easy to add more nodes, or clients, to a WLAN, because all that is needed is wireless circuitry in any device that comes within range of a wireless network.

However, wireless LANs have a significant drawback: they are not as secure as wired LANs unless some measures are taken. On a wired network, one needs to physically connect a device to access the network resources. On a wireless network, security measures must be taken to prevent connection by unauthorized wireless devices within range of the network. Some of these measures are covered later in the chapter.

MANs

A **metropolitan area network (MAN)** usually links multiple LANs within a large city or metropolitan region and typically spans a distance of up to 50 kilometers (about 30 miles). For example, the LAN in a chemistry lab might be linked to a research hospital's LAN and to a pharmaceutical company's LAN several miles away in the same city to form a MAN. The individual LANs that compose a MAN might belong to the same organization or to several different organizations. The high-speed links between LANs within a MAN typically use fiber optic or wireless broadband connections.

POINT OF INTEREST

From Your Scale to the Clinic

Tens of thousands of U.S. patients use remote monitoring devices, such as blood cuffs and weight scales, that can transmit readings of physical conditions from a patient's home to a healthcare facility. The wireless devices transmit data to a device connected to the telephone line. These devices are prescribed primarily for patients with chronic illnesses to ensure that their situation does not become worse. For example, a heart patient's sudden weight gain is often an indication that the heart is failing, because when the heart stops pumping blood normally, fluids accumulate in the lungs, abdomen, and lower limbs. When the cuff or scale detects a suspicious indication, it transmits the data via the telephone line to a computer monitor attended by a nurse. Research shows that remote monitoring of chronic heart failure reduces patient admission to hospitals and lowers mortality rates by almost 20 percent.

Source: Baker, M. L., "Bathroom Scales Aim to Save Lives (and Money)," *CIO Insight*, May 12, 2005; University of Alberta, April 20, 2007.

WANs

A **wide area network (WAN)** is a far-reaching system of networks. One WAN is composed of multiple LANs or MANs that are connected across a distance of more than approximately 48 kilometers (or 30 miles). Large WANs might have many constituent LANs and MANs on different continents. The simplest WAN is a dial-up connection to a network provider's services over basic telephone lines. A more complex WAN is a satellite linkup between LANs in two different countries. The most well-known WAN is the Internet.

WANs can be public or private. The telephone network and the Internet are examples of public WANs. A private WAN might use either dedicated lines or satellite connections. Many organizations cannot afford to maintain a private WAN. They pay to use existing networks, which are provided in two basic formats: common carriers or value-added networks.

A common carrier provides public telephone lines that anyone can access or dial up, and leased lines, which are dedicated to the leasing organization's exclusive use. The user pays for public lines based on time used and distance called. Verizon and AT&T are common carriers. Leased lines are dedicated to the leaseholder and have a lower error rate than dial-up lines, because they are not switched among many different subscribers.

Value-added networks (VANs) provide enhanced network services. VANs fulfill organizational needs for reliable data communications while relieving the organization of the burden of providing its own network management and maintenance. Many businesses use VANs for their electronic data interchange (EDI) with other businesses, suppliers, and buyers. However, due to cost considerations, an increasing number of organizations prefer to conduct commerce via the Internet rather than through VANs. VAN services cost much more than those offered by **Internet service providers (ISPs)**. (Many VAN providers also provide Internet links.) This issue is discussed in Chapter 8, "The Web-Enabled Enterprise."

PANs

A **personal area network (PAN)** is a wireless network designed for handheld and portable devices such as PDAs, cell phones, and tablet or laptop computers, and is intended for use by only one or two people. Transmission speed is slow to moderate, and the maximum distance between devices is generally 10 meters (33 feet). For example, Maria and Simon meet at a conference and exchange electronic business cards using their Bluetooth-enabled PDAs. When Maria gets back to her office, the PDA automatically synchronizes with her office notebook computer, updating the address book on the notebook with Simon's information. (Bluetooth and other wireless technologies are covered later in the chapter.)

LAN routers have become a common device in offices and households.

Courtesy of Linksys, a division of Cisco Systems

Networking Hardware

Networks use a variety of devices to connect computers and peripheral devices (such as printers) to each other, and to connect networks to each other. Each computer or device connected to a network must have a **network interface card (NIC)** or proper networking circuitry, which connects through a cable or a wireless antenna to a hub, switch, bridge, or router, which in turn connects to a LAN or WAN. A **hub** is a common device often used as a central location to connect computers or devices to a local network. A **switch** is like a hub, except that it is more "intelligent." Communications that go through a hub are broadcast to all devices attached to the hub; communications through a switch go only to designated devices on the network. A **bridge** is a device that connects two networks, such as a LAN, to the Internet. A **router** routes data packets to the next node on their way to the final destination. It can connect dissimilar networks and can be programmed to also act as a firewall to filter communications. Routers keep tables of network addresses, known as Internet Protocol (IP) addresses, which identify each computer on the network, along with the best routes to other network addresses. You are not likely to see a WAN router, but you might have seen a router used to support a LAN in a small office or in a household. A **repeater** amplifies or regenerates signals so that they do not become weak or distorted.

Another type of networking hardware that might be familiar to home computer users is the modem. A **modem**—a word contracted from *mod*ulator-*dem*odulator—in traditional usage is a device whose purpose is to translate communications signals from analog to digital, and vice versa. For many years the only way to link to the Internet was to dial up, meaning connecting over regular telephone lines. These lines were originally designed for analog—continuous—signals rather than for digital signals, which consist of discrete bursts. A modem turns the digital signal from your computer into an analog signal that can go out over the phone lines. A modem on the receiving computer transforms the analog signal back into a digital signal the computer can understand. The former transformation is called modulation and the latter is called demodulation.

A **dial-up connection** with a modem is very slow (usually no faster than 56 Kbps), so most users and small businesses have turned to faster connections that use digital signals throughout the connection, such as DSL and cable connections. Even though the medium transfers digital signals, the word "modem" is now used for the devices that connect computers to the Internet with these technologies. Thus, for example, if you use a cable company to link to the Internet, the device connecting your computer's network card to the cable is called a cable modem. If you use a DSL service, the device used is called a DSL modem, and if you use a power line, the device is called a BPL modem.

Virtual Private Networks

A LAN is a private network, because it only provides access to members of an organization. Though a firm does not own the lines it leases, the network of leased lines might be considered a private network, because only members authorized by the organization can use it. Many companies cannot afford or do not wish to pay for a private network. By implementing special software (and sometimes also hardware) they can create a **virtual private network (VPN)**. Although the Internet is discussed in Chapter 8, "The Web-Enabled Enterprise," VPNs are important in the context of the current discussion.

A virtual private network (VPN) can be thought of as a "tunnel" through the Internet or other public network that allows only authorized users to access company resources. The "virtual" in VPN refers to the illusion that the user is accessing a private network directly, rather than through a public network. VPNs enable the use of intranets and extranets. An intranet is a network that uses Web technologies to serve an organization's employees who are located in several sites that might be many miles apart; an extranet serves both the employees and other enterprises that do business with the organization. It is important to understand that once a LAN is linked to a public network, such as the Internet, technically anyone with access to the public network can obtain access to the LAN. Therefore, organizations that link their LANs to the Internet implement sophisticated security measures to control or totally deny public access to their resources.

Consider, for example, ITW Foilmark, a company located in Newburyport, Massachusetts. The company manufactures hot stamping foils for the design and packaging industries and serves customers such as Gillette (a Procter & Gamble subsidiary), AOL, and Hallmark. The company uses a VPN to provide corporate units in multiple sites access to its manufacturing system: they can enter orders, print work orders, and create reports. Once a month, all units send financial reports to the corporate offices. All this communication requires users to log in with a user name and password, and the communication itself is encrypted so that if it is intercepted it cannot be decoded easily by intruders.

Switching Techniques

Imagine that your telephone could connect to only one other telephone. Of course, this limitation would render the telephone impractical. The same is true of communications when using computers. You want to be able to link your computer to every other computer on a network. Or, imagine that you can link to any other computer, but you have to wait for a specific communications path to open to conduct a conversation; no other path is available to you. So you might wait a long time until no one is using any segment of that path to make your call. Obviously, this wait would be very inconvenient. To avoid such inconveniences, data communications must have mechanisms to allow your messages to be routed through any number of paths: if one is busy, then another can be used. These mechanisms, called switching techniques, facilitate the flow of communications and specify how the messages travel to their destination. The two major switching techniques are circuit switching and packet switching.

Circuit Switching

In **circuit switching**, a dedicated channel (a circuit) is established for the duration of the transmission. The sending node signals the receiving node that it is going to send a message. The receiver must acknowledge the signal. The receiving node then receives the entire message. Only then can the circuit be allocated for use of two other communicating parties. Traditional telephone communication is the most common type of circuit-switching communication. The advantages of circuit switching are that data and voice can use the same line and that no special training or protocols are needed to handle data traffic. One disadvantage is the requirement that the communications devices be compatible at both ends.

Packet Switching

In **packet switching**, a message is broken up into packets. A **packet** is a group of bits transmitted together. In addition to the data bits, each packet includes sender and destination information, as well as error detection bits (see Figure 6.3) and a packet number that indicates the packet's place in the file transmitted, that is, in the packets' sequence. Each of the message's packets is passed from the source computer to the destination computer, often through intermediate nodes. At each node, the entire packet is received, stored, and then passed on to the next node, until all packets, either kept together or reassembled, reach the destination.

FIGURE 6.3
A packet

Destination Address	Source Address	Packet Number	Data	Error Detection Bits

On their way to their final destination, the packets are transmitted independently to intermediate nodes. Different packets of the same message might be routed through different paths to minimize delay and are then reassembled at their destination. At the receiving device, the packet numbers are used to place each packet in its place so that the file transmitted is reconstructed accurately. This type of switching offers some advantages. Sending and receiving devices do not have to be speed-compatible, because buffers in the network might receive data at one rate and retransmit it at another. The lines are used on demand rather than being dedicated to a particular call. With packet switching, a host computer can have simultaneous exchanges with several nodes over a single line. The main disadvantage of packet switching is that it requires complex routing and control software. When the load is high, delays occur. When the network is used for voice communication, a conversation with long delays might sound unnatural. Therefore, voice communication in traditional telephone systems uses circuit switching.

Frame relay is a high-speed packet-switching method used in WANs. The frames are variable-sized packets. The service provider's software determines the route for each frame so it can arrive at the destination as quickly as possible. The variable size of packets allows more flexibility than with fixed-sized units; communication lines can be used more efficiently. One reason is that the higher ratio of data bits to nondata bits (such as destination and source addresses) in each packet is greater. Larger packets also enable lines to stay idle for less time.

Circuit switching is ideal for real-time communications, when the destination must receive the message without delay. Packet switching is more efficient, but it is suitable only if some delay in reception is acceptable, or if the transmission is so fast that these delays do not adversely affect the communication. The switching rules in a network are part of the communication protocol. These protocols, along with increasingly faster Internet connections, enable the growing use of the Internet for packet-switching telephony, known as VoIP, which we discuss later.

Multi-Protocol Label Switching (MPLS) is a relatively recent packet-switching technology that enhances services such as VoIP. Messages are broken up into packets, and packets are still transmitted independently, but all are routed through the same path on the network. This minimizes the time gaps between receptions of the packets. Therefore, content that must be communicated in real time—such as voice and video—is received at higher quality than if the packets are routed through different paths.

PROTOCOLS

A communications **protocol** is a set of rules that govern communication between computers or between computers and other computer-related devices that exchange data. When these rules govern a network of devices, the rule set is often referred to as a *network protocol*. If a device does not know what the network's agreed-upon protocol is, or cannot comply with it, the device cannot communicate on the network.

Some protocols are designed for WANs, others are designed for LANs, and some are designed specifically for wireless communications. This discussion addresses only some of these protocols. Protocols, often called "standards," do not necessarily compete with each other. They often work together or serve different purposes. The most important and pervasive set of protocols for telecommunications and networks today is called TCP/IP.

TCP/IP

Communication on the Internet follows mainly **TCP/IP (Transmission Control Protocol/ Internet Protocol)**, which is actually a set of related protocols. TCP ensures that the packets arrive accurately and in the proper order, while IP ensures delivery of packets from node to node in the most efficient manner.

A computer connected directly to the Internet **backbone**—the highest speed communication channels—is called a **host**. IP controls the delivery from one host to another until the message is received by the destination host. The host forwards messages to devices connected to it. Often, we call hosts servers. For example, your school has at least one e-mail server; it forwards to your computer e-mail messages addressed to you.

The current IP is IPv4 (Internet Protocol version 4). Under this version, every device on the Internet backbone is uniquely identified with a numerical label known as an Internet Protocol address, or **IP address**, a 32-bit numeric address, presented in four parts separated by periods, such as 146.186.87.220. Each of these parts can be a number between 0 and 255. If you know the IP address of a Web site, you can enter those numbers in the address box of a Web browser. However, it is easier to remember names and words, and therefore most organizations associate their IP addresses with names. The process of associating a character-based name such as *course.com* with an IP address is called domain name resolution, and the domain name resolution service is **DNS (Domain Name System)**. DNS servers are maintained by Internet service providers (ISPs) and other organizations. In large organizations, a server can be dedicated as a DNS server.

If a LAN is linked to the Internet through a router, the entire network has an IP address unique on the Internet. This number is stored in the router. To uniquely identify devices on the LAN, the router assigns local IP addresses to individual computers and devices. These IP addresses identify the computers only within the LAN. Only the router is identified uniquely on the Internet.

Servers and many other computers and devices are assigned permanent IP addresses, called a **static IP address**. A computer connected to the Internet intermittently might be assigned a temporary IP address for the duration of its connection only. Such a number is called a **dynamic IP address**. It is assigned by the host through which that computer is connecting to the Internet. Dynamic IP addresses give an organization flexibility with its limited number of assigned IP addresses: only devices seeking a connection to the Internet are assigned IP addresses. The number is disassociated from a device that logs off, and the server can then reassign the IP address to another device that has just logged on. Some broadband providers assign static IP addresses; others assign only dynamic IP addresses.

IPv4 poses several challenges that have been resolved in a new version, IPv6 (Internet Protocol version 6). One major problem was the limit that a 32-bit address poses on the number of unique addresses. It limits the number to 2^{32}, approximately 4 billion addresses. Under IPv6, IP addresses consist of 128 bits, allowing 2^{128}, which is approximately 3.4×10^{38} unique addresses. The new version also prescribes increased efficiencies in routing and transmitting messages on the Internet. The U.S. government ordered all federal agencies to deploy IPv6 by 2008 and purchase 247 billion IPv6 addresses. The People's Republic of China also started implementing IPv6, which it planned to showcase at the 2008 summer Olympics. Adoption of IPv6 would allow the huge country to have a much larger number of IP addresses than it could potentially secure under IPv4.

Ethernet

The Institute of Electrical and Electronics Engineers (IEEE) sets standards for communication protocols. IEEE 802.3, known as **Ethernet**, is the only LAN protocol of significance. Ethernet uses either coaxial cable or Cat 5 or 6 twisted pair cable. Different generations of Ethernet support speeds from 10 Mbps (10Base-T) to 100 Mbps (100Base-T or Fast Ethernet) to over 1 Gbps (**Gigabit Ethernet** and 10 Gigabit Ethernet). Ethernet is known as a contention-based protocol, because devices on the network "contend" with other devices on the network for transmission time. Each device constantly monitors the network to see if other devices are transmitting. A protocol called CSMA/CD (Carrier Sense Multiple Access with Collision Detection) ensures that if two devices want to transmit at the same time, they will detect the conflict and one will yield to the other.

Wireless Protocols

All wireless devices use radio transceivers (transmitter-receivers). The radio waves carry the digital signal, the bits. Depending on the protocol followed, the devices use different radio frequencies for their work.

IEEE 802.11 Wi-Fi

IEEE 802.11 is a family of wireless protocols, collectively known as **Wi-Fi** (for Wireless Fidelity). The term originally applied to the IEEE 802.11b standard that supports outdoor communication within about 100 meters (300 feet) of a wireless router at a maximum speed of 11 Mbps. The later 802.11g standard supports speeds of up to 54 Mbps for the same range. The 802.11a standard supports similar speeds to 802.11g, but in a different frequency range that is less susceptible to interference from cell phones and microwave devices. The 802.11n standard was expected to be approved in 2008, but products based on a draft version of the standard were already available years earlier. 802.11n supports maximum speeds of 248 Mbps and has about twice the range of 802.11b and g, about 70 meters (230 feet) indoors and 160 meters (525 feet) outdoors. The g standard is backward-compatible with the b standard, meaning that you can add b devices to a g network. The n standard is backward-compatible with the b, g, and a standards. However, in a mixed network, throughput will likely be at the speed of the lowest-speed device. The b and g standards use a radio frequency in the 2.4–2.5 GHz range, the 802.11a standard operates in the 5 GHz frequency, while the n standard can operate in either frequency. These radio frequency ranges do not require government licenses (referred to as "unlicensed"), and therefore are used for wireless communication. An additional standard, 802.11y, will operate on the licensed frequencies 3.65–3.7 GHz, and will increase the outdoor communication range to 5000 meters (3 miles), with a speed similar to that of the g standard.

A single Wi-Fi router can be connected to an **access point (AP)**, which in turn is connected to a wired network and usually to the Internet, allowing tens to hundreds of Wi-Fi-equipped devices to share the Internet link. A direct link to a wireless router or AP creates a **hotspot**. Hotspots allow Internet access to anyone within range who uses a wireless-equipped device, provided logging in is not limited by controlled access codes. Figure 6.4 illustrates a home wireless LAN (WLAN).

As mentioned earlier, security has been a concern for Wi-Fi networks. The earliest 802.11 standards had serious security flaws; 802.11g and 802.11a have improved security by offering the Wired Equivalent Privacy (WEP) protocol and the Wi-Fi Protected Access (WPA) and WPA2 security protocols. These protocols offer **encryption**, the ability to scramble and code messages through encryption keys that are shared only between the sender and receiver. Of course, to receive the protection of these protocols, they must be enabled on your wireless computer or device. Experienced "hackers" can break the codes of WEP and WPA within 10 minutes. WPA2 is a preferred measure.

Wi-Fi hotspots are appearing everywhere, from airports and restaurant chains to the local library and barbershop. Businesses also use wireless LANs for many types of operations. You will find a WLAN in almost every warehouse. Workers holding PDAs or specialized electronic units communicate with each other and receive information about the location of items by section, shelf, and bin. For example, General Motors equipped the forklifts in all its warehouses with Wi-Fi transceivers to help their operators locate parts. On sunny days retailers place merchandise and cash registers on sidewalks. The cash registers are linked to a central system through a WLAN. Conference centers and schools use WLANs to help guests, students, and staff to communicate as well as link to the Internet through a hotspot.

Many new airplanes for long flights are equipped with WLANs. Boeing started equipping its large airplanes with Wi-Fi in 2003. Lufthansa, British Airways, Japan Airlines, Scandinavian Airlines System, and other airlines have equipped their long-range jetliners with the technology to allow paying passengers to use a hotspot 12 kilometers (7.5 miles) above ground.

Utility companies have converted manually read electric, gas, and water meters to wireless meters. An employee need only pass by the client's building in a motor vehicle to record the reading. Newer meters use networks that relay the signal to the utility company's office and automatically update each customer's account in the company's computers. Wireless meters save millions of labor hours and overcome common problems, such as meters enclosed in locked places, inaccurate readings, and, occasionally, an aggressive dog.

FIGURE 6.4
A wireless home network

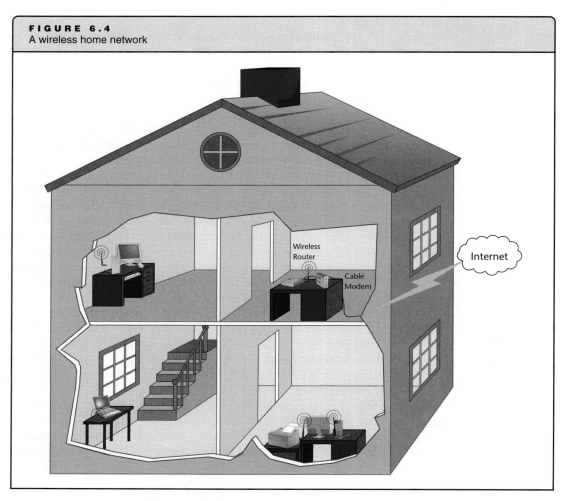

A growing number of electronic devices, such as cell phones, PDAs, digital cameras, and video game consoles, are equipped with wireless circuitry. This rids their owners of the need to physically connect a device to a computer or a router for communication. For example, with a wireless-enabled digital camera you can send digital pictures from your camera to your PC, or directly to a friend via a hotspot over the Internet.

Bluetooth supports a personal area network. The technology enables hands-free use of mobile phones.

Ed Hidden / istockphoto.com

IEEE 802.15 Bluetooth

Named after a Scandinavian king who unified many tribes, the **Bluetooth** standard was developed for devices that communicate with each other within a short range of up to 10 meters (33 feet) in the office, at home, and in motor vehicles. It transmits voice and data. Bluetooth was later adopted by IEEE as its 802.15 standard. Typical Bluetooth devices include wireless keyboards and mice, wireless microphones for cellular phones (especially for use in cars while driving), wireless headsets for hands-free mobile phone use, and increasingly, digital entertainment devices. For example, you can purchase a wrist-worn MP3 player that uses Bluetooth to transmit the music to earbuds or headphones, avoiding the wires that typically connect a portable player to headphones. Bluetooth is considered a personal area network (PAN) technology, because it typically supports a network used by only one person. Bluetooth uses the 2.4–2.5 GHz radio frequency to transmit bits at a rate of 1 Mbps.

IEEE 802.16 WiMAX

IEEE 802.16, Worldwide Interoperability for Microwave Access **(WiMAX)**, increases the range and speed of wireless communication. It might potentially reach up to 110 kilometers (about 70 miles) with a speed of 100 Mbps; however, it typically reaches 13–16 kilometers (8–10 miles). Experts say that with an investment of no more than $3 billion, WiMAX can cover 98 percent of American homes. This is a much lower investment than required for laying fiber optic cables. WiMAX uses licensed radio frequencies of 2–11 GHz. This standard can cover entire metropolitan areas and provide Internet access to hundreds of thousands of households that either cannot afford an Internet service or for some reason cannot obtain access. Many municipal governments wanted to establish such service for a fee or for free. However, this has created a threat to the business of ISPs, who count on subscriber fees for revenue, because an entire metropolitan area can become one huge hotspot, and the fees, if any, are collected by the local government rather than an ISP. Therefore, several states in the United States legislated against municipality-sponsored networks. However, some cities are using the technology, which enables households that cannot afford Internet connectivity to have access to this important resource. Philadelphia was the first American metropolis to do so. The city was exempt from a Pennsylvania law forbidding municipal networks.

WiMAX is a metropolitan area network (MAN) technology. Figure 6.5 shows how WiMAX works. A household, office, or public hotspot can use a router to link multiple devices either by linking directly to a WiMAX base antenna that is linked to the Internet, or by using a relay antenna that receives the signal and retransmits it to the Internet-linked antenna. If a mobile user's equipment included the proper WiMAX communication device, the user could communicate with the Internet moving at speeds of up to 150 Km/H (about 94 MPH), which enables convenient use of the Internet while sitting in a moving vehicle (though the driver should not be going that fast!). An extension of this standard, 802.16e, supports mobile Internet communication. The telecommunications company Horizon Wi-Com started the construction of 802.16e networks in Boston, New York, Philadelphia, Washington, D.C., Baltimore, Pittsburgh, Buffalo, Richmond, and Cincinnati. The installation was scheduled to be completed by the end of 2007. Similar efforts have taken place in other countries, notably Pakistan. However, a newer, special standard dedicated to mobile communications is 802.20.

IEEE 802.20 MBWA

Mobile Broadband Wireless Access (MBWA) functions similarly to cell phone communications, because it controls communication from stationary towers. The purpose of MBWA is to provide mobile communication that is compatible with IP services. This should enable worldwide deployment of affordable, always-on wireless access. The principle is simple: place wireless routers on towers so that mobile phones can use VoIP and access other Internet resources over wide areas, and, eventually, globally. MBWA is expected to work at speeds over 1 Mbps, using licensed radio frequencies below 3.5 GHz. If the standard is successfully implemented globally, it will reduce subscriber fees significantly and pose severe competition to providers of cell phone services.

The 802.20 standard is designed to be compatible with 802.11 (Wi-Fi) and 802.15 (Bluetooth). It can support Internet communication at a moving speed of up to 250 Km/H (156 MPH). MBWA promises to support practically everything that we now do with telephones and through the Internet: Web browsing, file transfer, e-mail, VoIP, video telephony and videoconferencing, audio streaming (such as listening to transmitted music), Web-based gaming, and file sharing. The technology includes security measures that meet the standards of the U.S. Department of Defense for protection of sensitive but unclassified information. To a large extent, this standard is still under development.

Figure 6.6 summarizes relevant features of the 802.xx wireless protocols discussed here.

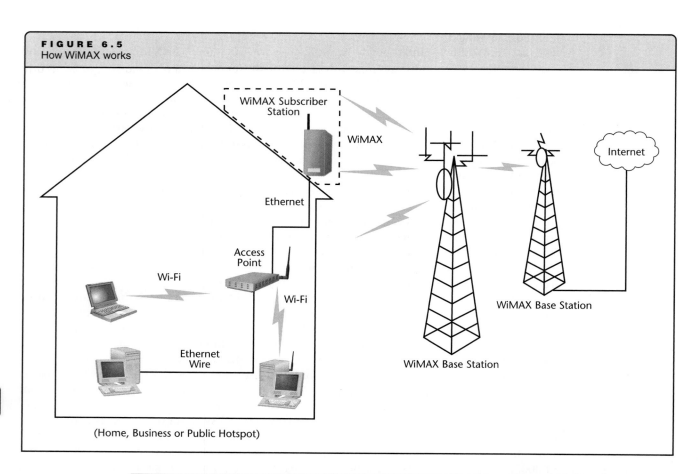

FIGURE 6.5
How WiMAX works

WiMAX Subscriber Station

WiMAX

Internet

Ethernet

Access Point

Wi-Fi

Wi-Fi

Ethernet Wire

WiMAX Base Station

WiMAX Base Station

(Home, Business or Public Hotspot)

FIGURE 6.6
Wireless networking protocols

Protocol	Max. Range	Max. Speed	Main Use
802.11a	75 meters (250 feet)	54 Mbps	LAN
802.11b	100 meters (330 feet)	11 Mbps	LAN
802.11g	100 meters (330 feet)	54 Mbps	LAN
802.11n	160 meters (530 feet)	248 Mbps	LAN
802.15 Bluetooth	10 meters (33 feet)	1 Mbps	PAN
802.16 WiMax	50 km (31 miles)	100 Mbps	MAN
802.20 MBWA	Global	4 Mbps	Mobile voice, data, and Internet communications

Generations in Mobile Communications

Networking professionals often refer to generations of mobile communication technologies. Each generation refers to a communication protocol or a combination of protocols. The differences among generations are mainly in capabilities (e.g., enabling a mobile phone to access additional resources) and transmission speed. The first generation, 1G, was analog and used circuit switching. Then 2G protocols became the first to provide digital voice encoding, and they worked at faster transmission rates. They include the GSM (Global System for Mobile) and CDMA (Code Division Multiple Access) protocols, the details of which are outside the scope of this discussion.

The 3G protocols support transmission rates of 1 Mbps. The protocols support video, videoconferencing, and full Internet access. 4G protocol devices operate only digitally and with packet switching, transmit at bandwidths of up to 100 Mbps, and include tighter security measures. In the U.S., Sprint Nextel Corp., along with Intel, Samsung, and Motorola, started development of 4G service relying on WiMAX networks. The high speed of the technology will enable the holder of a mobile phone handset to watch a DVD-quality video, listen to CD-quality music files, browse the Web, and make a telephone call at the same time.

POINT OF INTEREST

I Want to Talk, Not View

A GMI (Global Market Insight) survey of 15,000 consumers in 37 countries revealed that Americans are much less interested in receiving anything but good quality sound and connectivity through their mobile phones. Mobile phone users in less developed countries expressed a greater desire for "content": Web access, video, news, and other advanced features that come with modern mobile phones. The features were desired by 63.5 percent of respondents in South America, 56.4 percent in Asia, 53.9 percent in Eastern Europe, 30.4 percent in Western Europe, and only 22.6 percent in the United States.

Source: Burns, E., "Mobile Content Usage is Higher in Developing Countries," *ClickZ Stats* (www.clickz.com), March 2, 2007.

In a way, 3G and 4G cellular technologies compete with Wi-Fi, but it seems that eventually the technologies will complement each other: we will use 3G and 4G outdoors and Wi-Fi indoors. Wi-Fi is significantly less expensive to use than 3G and 4G, because mobile phone services involve a monthly fee for each phone, while using the Internet through a hotspot is generally free.

INTERNET NETWORKING SERVICES

Both organizations and individuals can choose from a variety of options when subscribing to networking services. Figure 6.7 summarizes the major services offered by telecommunications companies. Note that the bit rates shown are for **downstream**, which is the speed of receiving from the network; **upstream** speeds, the speeds of transmitting into the network, are usually much lower. Also be aware that these are typical speeds in the United States. They might be different in other countries. Monthly costs, too, are typical but vary from region to region. For some services, such as T1 and T3, companies also offer fractions of the speeds for lower fees.

For most individuals and businesses, a service that provides a much lower transmission rate (upstream speed) than reception rate (downstream speed) is suitable. This is because they rarely upload large files to Web sites or transmit large amounts of e-mail that must arrive at its destination in a fraction of a second. However, organizations such as online businesses and media companies that must upload large files quickly must also have high upstream speeds. Such organizations may opt for Internet communication lines that allow high speeds both downstream and upstream.

The proliferation of high-speed connection services, also called broadband services, is mainly the result of businesses' and individuals' rush to the Internet. Some of the services, such as cable, DSL, and satellite links, are offered both to businesses and residences. Others, such as T1 and T3 lines and the OC class, are offered only to businesses, largely because of their high cost. Note that some of the services are actually groups of services that differ in speeds. For example, some DSL services designed for businesses provide the same speed downstream and upstream, while options for households (see later discussion of ADSL) always provide a greater downstream speed than upstream speed.

FIGURE 6.7
Typical features of Internet services

Service	Downstream Speed	Availability	Monthly Fee
Dial-up	56 Kbps	Universal	$9–11
BPL	3 Mbps	Limited Availability	$30–40
Cable	0.5–3 Mbps	Widespread; available nearly everywhere TV cable service is offered	$30–50
DSL	0.5–8 Mbps	More limited than cable, but spreading faster; speed also depends on distance from telco office	$30–50
T1, T3	1.544 Mbps, 44.736 Mbps	Widespread	$300–1,000, $3,000–10,000
Satellite	1 Mbps	Widespread; practical only with view to the southern sky	$40–50
Fixed Wireless	100 Mbps	Limited, but spreading	$2,000
Fiber to the Premises	5–30 Mbps	Limited, but spreading	$30–180
OC-3	155.52 Mbps	Limited availability	$60,000
OC-12	622.08 Mbps	Limited availability	Several hundred thousand dollars
OC-48	2.488 Gbps	Limited availability	Several hundred thousand dollars

Cable

Cable Internet links are provided by television cable firms. The medium is the same as for television reception, but the firms connect the cable to an Internet server. At the subscriber's residence, the cable is split—one part is connected to the television set, and the other is connected to the computer via a bridge that is often called a cable modem. Both television transmission and data are transmitted through the same line. The cable link is always on, so the computer is constantly connected to the Internet. More than 90 percent of cable operators in the United States offer Internet access.

The major downside of cable is that cable nodes are shared by all the subscribers connected to the node. Therefore, at peak times, such as television prime time (7–11 p.m.), communication speed slows down. The speed also slows down as more subscribers join the service in a given territory.

Digital Subscriber Line (DSL)

With normal landline telephone service, the telephone company filters information that arrives in digital form and then transforms it to analog form; thus, it requires a modem to transform the signal back to digital form. This conversion constrains the capacity of the link between your telephone (or computer) and the telephone company's switching center to a low speed of 56 Kbps.

With **digital subscriber line (DSL)**, data remains digital throughout the entire transmission; it is never transformed into analog signals. So, the telephone company can transmit to subscribers' computers at significantly higher speeds of up to 8 Mbps (although speed rarely exceeds 1.5 Mbps). To provide DSL service, the telecommunications company connects your telephone line to a DSL bridge (often called a DSL modem). At the telephone company's regional central office, DSL traffic is aggregated and forwarded to the ISP or data network provider with which the subscriber has a contract. Often, the telephone company is also the ISP.

Detailing the various types of DSL is beyond the scope of this book, but they generally can be placed in one of two categories: symmetric and asymmetric. Asymmetric DSL (ADSL) allows reception at a much faster rate than transmission, that is, it is faster downstream than upstream. (Often, the respective terms "download" and "upload" are used.) The reason for the faster download is that home users and small businesses usually receive significantly more information (from the Web, for example) than they transmit. Symmetric DSL (SDSL) is designed for short-distance connections that require high speed in both directions. Some ADSL technologies let subscribers use the same telephone lines for both Internet connection and analog voice telephone service. Symmetric DSL lines cannot share lines with telephones.

The bit rates of DSL lines are closely related to the distance of the subscriber's computer from the regional central office of the telephone company. Telecommunications companies might offer the service to subscribers as far as 6,100 meters (20,000 feet) from the central office, but the speed then is usually no faster than 144 Kbps, unless the company has installed a DSL repeater on the line. Some companies do not offer the service if the subscriber's address is not within 4,500 meters (15,000 feet) of the central office. Most subscribers have ADSL, so the upstream speed is significantly lower than the downstream speed.

T1 and T3 Lines

T1 and T3 lines are point-to-point dedicated digital circuits provided by telephone companies. A T1 line is made up of 24 channels (groups of wires) of 64 Kbps each. T3 lines are made up of 672 channels of 64 Kbps. T1 and T3 lines are expensive. Therefore, only businesses that must rely on high speeds are willing to accept the high cost of subscribing to the service. Most universities, as well as large companies, use T1 or T3 lines for their backbone and Internet connections.

Satellite

Businesses and households in rural areas and other regions that do not have access to cable or DSL might be able to obtain satellite services, which use microwave radio transmission. In fact, satellite service providers target these households. The service provider installs a dish antenna that is tuned to a communications satellite. Satellite connections might reach a speed of 45 Mbps. The antenna for satellite communication can be fixed, as the ones you can see installed in the yards of private houses, or mobile, such as those installed on the roofs of large trucks. Most of the subscribers of fixed satellite dishes are households; most mobile dish users are shipping and trucking businesses. Subscribers to fixed satellite service must purchase the dish antenna, with a typical cost of $400, and pay a monthly fee of about $50. Trucking companies must have an antenna installed on each truck.

Many people use a free satellite service, the global positioning system (GPS). While a proper device is required to enable reception from the satellites (which were launched into orbit by the U.S. government), anyone can communicate free of charge. The satellite transmits back to any GPS device its location on earth by longitude and latitude.

Fixed Wireless

Another alternative for households and small businesses that cannot obtain cable or DSL connections to the Internet is fixed wireless. **Fixed wireless** is point-to-point transmission between two stationary devices, usually between two buildings, as opposed to mobile wireless, in which people carry a mobile device. Companies such as Sprint, AT&T, and many ISPs offer the service. ISPs that specialize in fixed wireless services are often referred to as WISPs, wireless ISPs. They install microwave transceivers on rooftops instead of laying physical wires and cables. Subscribers connect their computers to the rooftop transceiver. They can communicate at speeds up to 2 Mbps. Repeaters are installed close to each other to enhance the signal, which can deteriorate in the presence of buildings, trees, and foul weather. Transmission rates depend on the distance between the receiver and the base station. Up to 14 kilometers (9 miles) from the base station, the speed is 100 Mbps; speeds drop to about 2 Mbps at about 56 kilometers (35 miles) from the base.

Fixed wireless is highly modular—the telecommunications company can add as many transceivers as it needs to serve a growing number of subscribers. Unlike cable service, the company does not need franchise licenses. The technology is suitable for both urban and rural areas. For example, Daytona Beach, Florida, is served by a fixed wireless network that provides a broadband connection to anyone who is interested in the service. The local government of rural Owensboro, Kentucky, wanted to keep the town's businesses competitive. Since other options were not available, it built a fixed wireless network that provides broadband links to the Internet for $25 per month.

Fiber to the Premises

Fiber to the premises connects a building to the Internet via optical fiber. The service is widely available in the United States and other countries, but at varying speeds. In Hong Kong and South Korea, the maximum speed the providers of this service allow is 100 Mbps. In the United States, Verizon provides the service, which it calls FiOS (Fiber Optic Service), but limits the speed to 30 Mbps. While Verizon has deployed the service on a large scale, other companies such as AT&T provide similar service to some communities. When the optical fiber reaches the subscriber's living or work space, it is referred to as **Fiber to the Home (FTTH)**. Subscribers simply connect their computer, or LAN's router, to the optical fiber socket in the wall. In some communities, Verizon has also provided television programming on the same optical lines.

Optical Carrier

Companies willing to pay high fees can enjoy very high connection speeds. These services are denoted with **OC**, the acronym for **optical carrier**, because they are provided through optical fiber lines. The number next to OC refers to data speed in multiples of 51.84 Mbps, considered the base rate bandwidth. Thus, when available, the services are denoted as C-1, C-3, C-9, C-12, C-18, C-48, and so on through C-3072. For illustration, OC-768 (40 Gbps) enables you to transmit the content of seven CDs in 1 second. Typical businesses that purchase the services are ISPs, providers of search engines, and businesses that wish to support content-rich Web sites and high-volume traffic. However, media companies have also purchased such services because the high speeds support streaming video. Among companies that use OC-768, for instance, are Deutsche Telecom, NBC, Disney, the U.S. Department of Defense Advanced Research Projects Agency (the agency that developed the Internet), NASA, and Nippon TV.

Broadband Over Power Lines (BPL)

As mentioned in the discussion of communications media, electric power lines are capable of carrying digital signals. Subscribers simply plug their BPL modem into standard electrical wall outlets. Usually, utility companies partner with telecommunications companies to provide Broadband over Power Lines (BPL). For example, Cinergy, a Cincinnati-based utility company that serves 2 million customers in Ohio, Kentucky, and Indiana, partnered with Current Communications to provide broadband service. The service is offered to 50,000 households for a monthly fee of $30–40, based on transmission speed desired by the subscriber. Some experts estimate that the BPL market in the United States will reach $2.5 billion by 2010, while others expect only households that currently use dial-up to link to the Internet to adopt this type of service.

Interestingly, even if BPL service availability is to lag far behind cable and optical fiber in terms of subscribers and revenue, utility companies are likely to invest in the technology for their own use. They can use BPL to monitor power consumption down to the household, detect power failure in real time, track power outages by region, automate some customer services, and remotely control substations. Collecting and analyzing such business information might make the utility companies more efficient.

The speed and monthly service for BPL are similar to those of DSL, but the highest current speeds are lower than the highest speeds offered by DSL providers. The hope was that households in rural areas, where neither cable nor DSL service is available, could enjoy BPL. However, the density of households in rural areas is lower than the density of households where the other services are already offered. Utility companies have found that investing in the equipment required to provide BPL to a small number of households does not make business sense, and therefore it is unlikely that many rural areas will be offered BPL.

THE FUTURE OF NETWORKING TECHNOLOGIES

This section takes a look at networking technologies and trends that are likely to have a significant impact on businesses and the management of information in the near future: broadband telephony, radio frequency identification, and the convergence of digital technologies.

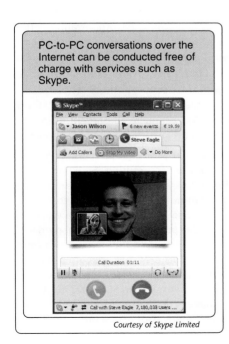

PC-to-PC conversations over the Internet can be conducted free of charge with services such as Skype.

Courtesy of Skype Limited

Broadband Telephony

While regular long-distance telephone companies charge according to the number of minutes a call lasts, Internet service providers (ISPs) charge customers a flat monthly fee for connection to the Internet. With the proper software and microphones attached to their computers, Internet users can conduct long-distance and international conversations via their Internet connection for a fraction of regular calling costs. The technology is called Internet telephony, IP telephony, or **VoIP (Voice over Internet Protocol)**. VoIP is a standard for software that digitizes and compresses voice signals and transmits the bits via the Internet link. Organizations can purchase the proper software or use the services of companies that specialize in providing IP telephony. Companies such as Vonage, Cablevision, Comcast, and many others offer inexpensive use of their VoIP telephone-to-telephone voice communication.

Computer-to-computer calls can be conducted free of charge by using the service of a company such as Skype or Jajah. Phone-to-phone service requires an additional modem, but it does not require a new phone or phone number, and it does not require routing calls through a home computer. Jajah also offers a free or low per-minute phone-to-phone service, if both the caller and recipient register at the company's Web site. The caller uses a computer to dial, the caller's phone rings, and when the caller picks up his phone, the service dials the recipient's number. Thus, no VoIP modem is required.

Telecommuting: Pros and Cons

When you are introduced to people, you usually mention your occupation, and then you might be asked, "Where do you work?" Many employed people now answer, "At home." They do not commute; they *telecommute*, or, as some prefer to call it, they *telework* . They have the shortest commute to work: from the bedroom to another room in the home that is equipped with a PC and a broadband Internet link. For an increasing number of workers, IT provides all that's needed to create the goods their employers sell: software, analysis reports, literature, tax returns, and many other types of output. If they need data from the office, they can connect to their office intranet using VPN software and retrieve the required information. If they need to talk to supervisors or coworkers, they use their computers to conduct videoconferencing. When they complete their product, they can simply e-mail it or place it on a remote server.

- **Telecommuting on the Rise.** Nearly a third of the U.S. workforce, more than 45 million individuals, work at home at least part time. Twelve million of them worked from home full time in 2006. Autodesk, Inc., a supplier of PC design software and digital content creation, established a pilot program in 1996 that allowed 20 workers to telecommute. Today, half the company's 3,000 workers telecommute, and every manager has the option to allow subordinates to telework. The program helps the company retain skilled workers and serves as an inducement when recruiting new employees. Telecommuting has increased productivity and reduced employee stress. Pitney Bowes, a business communications company with 32,000 employees, noticed increased productivity in employees on their telecommuting days. Managers there believe that telecommuting increases productivity because it accommodates both the "morning person" and "night owl" who can work at the time of day that best fits their preferences. Forty percent of IBM's 330,000 employees work from home, on the road, or at a client location on any given day. From a national economic perspective, telecommuting saves travel cost and time. It also decreases pollution.

- **Employment Opportunities.** Telecommuting enables people who could otherwise not work to join or rejoin the workforce. This includes not only people who live far away from the offices of companies that would like to hire them but also population groups that otherwise might not be able to join certain businesses. Disabled people and parents of small children can work from home. Older people who would rather retire than commute might stay in the workforce if allowed to work at home. Organizations hungry for labor can tap a larger supply of workers if they offer telecommuting.

- **Saving Time and Money.** Organizations offer telecommuting because it saves the cost of office space. Studies have shown that for each teleworker, the annual saving on office space is $5,000–10,000. When Nortel Networks allowed 4,000 of its 13,000 employees to telecommute, it saved $20 million per year on real estate. Studies also have shown that teleworkers are more productive by 15–50 percent than their office counterparts. Telecommuters like their arrangement because they save the time and money they would spend on commuting. Telecommuting reduces millions of tons of pollutants, saves billions of gallons of gasoline, and frees billions of personal hours for leisure time. AT&T, the telecommunications giant, reported in 2000 that its telecommuting program increased productivity by 45 percent and saved 50 percent on office space costs. Another report, by British Telecom and Gartner Group, said that telecommuting reduced office space and other costs equivalent to 17 percent of annual salary costs.

  State governments in the United States realize this and therefore offer tax incentives to companies that institute telecommuting. For example, Oregon allows tax deductions on expenditures for equipment and software required by companies that offer telecommuting options to employees.

  How well workers like their organizations often is associated with the option to telecommute. On the list of *Fortune* magazine's 100 Best Companies to Work for in 2007 you can find the following companies, with their proportions of regular telecommuters: Cisco Systems, 90 percent; Bain & Co., 76 percent; AstraZeneca, 75 percent; Bright Horizons, 60 percent; and Genentech, 57 percent.

- **The Downside.** However, not everyone is so enthusiastic about telecommuting. Sociologists have mixed opinions about the phenomenon. On one hand, telecommuting allows people to work who would otherwise remain outside the workforce, such as older professionals and many disabled people. On the other hand, it has been found that employers tend to pressure telecommuters to work harder than office workers. In the office an employee works a set number of hours, but the home worker has no defined workday; his or her workday is, the employer often assumes, 24 hours per day. In addition, telecommuters are more estranged from their fellow workers. For telecommuters, there is no office in which to foster new social ties and camaraderie.

The AT&T report said that teleworkers typically worked an hour more per day than their office-bound colleagues, which amounted to 250 hours per year. The British Telecom and Gartner Group report said that the average telecommuter works 11 percent more hours than his or her office-bound brethren. Perhaps this extra time is what companies observe as added productivity. Although this extra work time is good for corporations, it is not so good for workers: when you telecommute, you work more for no additional compensation.

Telecommuting might foster isolation. Teleworkers share fewer experiences with other people. In addition, leaving the workplace behind means leaving behind one more community that gives many people a sense of belonging, even if this belonging amounts only to having a sandwich together at lunchtime and complaining about the boss. At the same time, some managers might prefer to see their employees in the office and keep them in their "line of sight." The executive search firm Korn/Ferry International conducted a survey of 1,312 executives worldwide. Over 60 percent of them believed that telecommuters were less likely to advance in their careers than their peers who worked only in the office. Interestingly, 48 percent of the executives said they would consider a telecommuting job for themselves, and 78 percent opined that telecommuters were equally or more productive than noncommuters.

On a national level, telecommuting could severely affect some segments of the economy. Imagine the huge drop in revenue of New York City restaurants during lunchtime if only half of the 3 million or so commuters did not rush to grab lunch between 12 and 2 p.m. Some cities' dining industries could crumble if the telecommuting trend continues at the current pace. Many people live in cities mainly because of proximity to their offices, thus further movement to suburbs and remote residential areas would gut many other industries in central cities.

Many workers, given the option to work at home, have decided to return to the office. Interestingly, this also happens in the very industries that are so amenable to telecommuting, such as software development. These returning workers claim they missed social interaction with their peers, hallway chats, lunches with friends, and direct communication with fellow workers and supervisors. But telecommuting has grown, and will probably continue to grow, especially thanks to greater availability of broadband services and their declining monthly fees. Fewer than half of all Americans have broadband service at home. Among telecommuters, the proportion is greater than 90 percent. If the trend continues, offices occupied by organizations will be significantly smaller than they are now and will serve as the symbolic rather than physical centers of the organizations' activities.

Yet, the traditional 40-hour work week is already changing and will continue to change. A Gartner report published in June 2007 predicts that by 2015 most employers will offer 30-hour per week jobs, and that most workers will be "digital free agents." A Gartner researcher said: "As IT becomes woven into the fabric of people's lives and traditional work-home boundaries are rendered obsolete, digital free agency will emerge."

VoIP can save companies and households money. According to the research firm In-Stat, 20 percent of U.S. firms used VoIP in 2006, and two-thirds were expected to use VoIP by 2010.

Hamon Corp., a company that manufactures devices for control of air pollutants, noticed a significant increase in telephone costs as its staff grew from 130 to 500. The firm's CFO decided to subscribe the company to a VoIP service. The company's telephone cost decreased by at least $12,000 per month. The accounting firm Ernst & Young uses an Internet phone system it purchased from Cisco Systems to connect its 84,000 employees worldwide. Virgin Entertainment Group, which employs 1,500 people, saved $700,000 in the first year after switching to VoIP, and expects to save $1 million per year. Many households, especially in countries where telephone rates are high, use VoIP. Over 40 percent of all international calls from India use VoIP.

In 2006, the United States had about 7 million subscribers to VoIP services. The number is expected to triple by 2009.

In addition to the occasionally poorer sound quality, other differences exist between traditional and VoIP telephone services. Many VoIP services do not include the ability to call an emergency number such as 911. Also, when your link to the Internet is down, so is your VoIP service. Since the phone uses a modem that requires electric power, if power is out, the phone cannot be used. However, VoIP providers offer some advantages over traditional telephony. A subscriber receives a

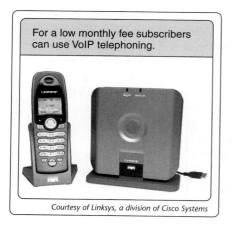

For a low monthly fee subscribers can use VoIP telephoning.

Courtesy of Linksys, a division of Cisco Systems

special converter into which the telephone number is programmed. The subscriber can take the converter anywhere there is a broadband link to the Internet and use it. This makes the VoIP telephone portable. Subscribers also have some options that they can control via the provider's Web site, such as routing calls to a different telephone number when on vacation or whenever VoIP service is not available.

Some experts see the future of telephony in a convergence of the cell phone and VoIP phone: you will use only one mobile phone. When outside the home or office, you will use the cell phone network; when back home or in the office, the phone will communicate through a VoIP service. This will reduce the higher cost of cell phone minutes.

While phone-to-phone VoIP is an attractive choice, so is computer-to-computer VoIP, especially because it is usually offered free of charge. Any business that needs only sound to sell its services can operate free of charge from anywhere in the world. Toniks Languages, a small company operated by a single owner, provides language tutoring via Skype. The owner recruited a dozen tutors who teach students foreign languages. Using traditional phones, he would have to pay $25 per hour for phone service. Using Skype, his communication cost is zero. The entrepreneur makes $40,000 per month. A piano teacher in Chicago uses Skype to give music lessons to students in places as far as Australia.

Radio Frequency Identification

In Chapter 3, "Business Functions and Supply Chains," you learned about the expanded efficiency and business intelligence that companies, especially in manufacturing and retail, can gain from one particular type of communications technology: radio frequency identification (RFID). This section explains in more detail how RFID works. RFID tags can be very tiny, about the size of a rice grain, or several square inches, depending on the amount of information they need to contain and the environment in which they are used. They are not always flat; they can be cylindrical. The tags need very little power. Passive tags use power from the reader that queries them; active tags have their own tiny batteries, which increase the range of the reading range. These tiny batteries last a long time.

POINT OF INTEREST

Predicting Traffic Jams

New York City has installed electronic signs that provide real-time traffic information to motorists. The system collects signals from *E-ZPass* tags and measures how long it takes drivers to travel between various points. The system uses the data to forecast travel time from point to point and displays the information on electronic message boards. The data collected from the *E-ZPass* tags enables drivers to know what to expect as they approach a bridge or another major location on their route.

Source: Transport Topics Magazine i-Tech Newsletter, March 6, 2006.

An RFID system works as follows: objects are equipped, often embedded, with a tag that contains a transponder. A transponder is a radio transceiver (transmitter-receiver) that is activated for transmission by a signal transmitted to it. The tag is equipped with digital memory that is given a unique code. If the tag is used to identify a product, it contains an **EPC (electronic product code)**. The interrogator, a combination of an antenna, a transceiver, and a decoder, emits a signal activating the RFID tag so the interrogator can read data from it and write data to it. Although the interrogator also writes to the tag, it is often called a reader. When an RFID tag enters the reader's electromagnetic zone, it detects the reader's activation signal. The reader decodes the data stored in the tag's memory, and the data is passed to a host computer for processing.

Wal-Mart, British Tesco, and German Metro AG, three of the world's largest retailers, embarked on a project that might radically change supply chains. They demanded that suppliers use RFID. Hundreds complied, among them Procter & Gamble, the world's largest supplier of consumer products. The companies use microchips that are embedded in products to replace the ubiquitous bar codes for tracking and checkout at store registers. Each microchip holds a product identification number. The microchips communicate with wireless computers, including hand-held and laptop computers, as they are moved in the production line, packed, picked, shipped, unloaded, shelved, and paid for by customers. As the item moves, the information about its location is communicated to a network of computers to which all businesses involved in the production and sale have access. This is often a Wi-Fi network. The benefits are a just-in-time (JIT) system that minimizes inventory throughout the supply chain to almost zero, and shelves that are always stocked. JIT, or a situation that is close to JIT, can be accomplished thanks to up-to-the-minute information about available inventory and when the next shipment from a supplier is needed. "Smart shelves," equipped with tiny wireless transceivers, alert employees whenever the shelf is running out of units, so they can put more units on the shelf immediately.

RFID is used for many other purposes as well, as Figure 6.8 shows. The investment in this technology yields efficiency rewards almost immediately to large companies, but is expensive for

FIGURE 6.8
RFID applications

Use		Example
Access Control		Cards used to replace door keys.
People Tracking	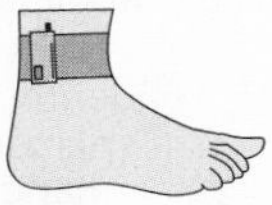	Keep children within school. Track prisoners on probation and prevent fleeing.
Animal Tracking		Track pets.
Livestock Management		Track life cycle of farm animals (e.g., feeding and immunization). Equip each cow with a unique ID to track diseases.
Antitheft Measures	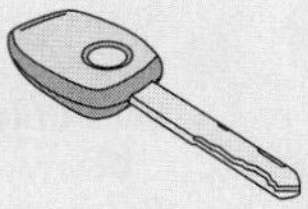	Transponders integrated into car keys. Only a legal key can start the engine.
Transportation		At airport, safety inspection of tagged luggage.
Retail		Tracking products in pallets and on shelves. Contactless payment.
Pharmaceuticals		Reduce drug counterfeiting.
Health Care		Tag people who enter and leave an epidemic zone.

Wal-Mart and other large retailers insist that suppliers use RFID electronic product code (EPC) tags like this.

Photograph courtesy of Intermec Technologies

small suppliers. The average price of an EPC tag of the standard used by Wal-Mart, Tesco, Metro, and the U.S. Department of Defense was 5 cents in 2007. It is expected that the price will continue to decrease to a cent or a fraction of a cent. When the price lowers sufficiently, you might begin to see many other uses of the technology, as listed in Figure 6.9.

Converging Technologies

Recall the discussion of converging hardware technologies in the previous chapter. Convergence occurs also in networking technologies. Cell phones used to be able to transmit and receive only through a dedicated network of analog or digital transceiver towers. Now many are constructed with dual technologies, so that they can serve both as a "traditional" cell phone and a wireless Web phone. When the circuitry detects that the phone is within the range of a hotspot, calling switches to VoIP to save cost. Eventually, we will be able to use the same phone as a landline phone, a VoIP phone, and a cell phone, depending on availability of service and the cost and quality we are willing to accept.

FIGURE 6.9
Future uses of RFID

Use in...		Activity
Shopping	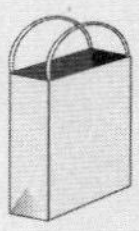	Identify dresses in your size, if you hold or wear a personal tag. Enhanced product information on your PDA/cell phone. Personalized customer service. Passive self-checkout. Dynamic pricing by demand. Return RFID tagged items without receipt.
Product Information		Scan an RFID tag of an item and download additional information about it from an Internet site to your cell phone. Use your cell phone to check the price of an item while in a competitor's store.
Manufacturer Serving Customers		Send recall message to customer cell phone or e-mail address. Send warranty and recall messages to customer.
Appliances		Washing machine automatically sets proper wash cycle based on information on tags attached to clothes. Refrigerator alerts you about expired or recalled foods, notifies you about items consumed, and prepares shopping lists. It can also log on to the Internet and search for recipes of dishes you can prepare with refrigerated items.
Agriculture		Tags attached to crops can transmit information about weather and soil conditions and trigger automatic irrigation.
Waste Management		Track hazardous materials to ensure proper disposal. Sort recyclable items.

New home television sets are being designed to connect to cable, satellites, and the Internet, not only alternately, but concurrently. Thus, we will be able to watch a sports game and chat online about it at the same time through the same device, using two different networking technologies. PDAs can already function as television sets and phones. Soon they will be able to do so simultaneously. For individuals, this means they can carry a single device that will connect them to any type of network, erasing the lines between radio, television, telephone, and Internet surfing. For businesses, this offers an opportunity to provide new information services and manage a more effective and efficient salesforce.

The Future Mobile Phone

Within two to five years, we may see a "cell phone" that has little resemblance to current mobile phones. New mobile phones will read biometric data, such as a thumbprint, to allow access to data stored on networks. They will allow access to multiple live television broadcasts. Small sensors will monitor heart rate, which the device will send to a doctor or fitness trainer. If the owner downloads a song, the device will maintain the license to prove that it was downloaded with permission, and may allow a limited time playing on another device to which the song will be transmitted. GPS circuitry will enable the owner to tag locations for friends. When the friends walk or drive by with their own devices, they know they are close to a "cool" spot. The device will also serve as a still and video camera, from which the photos or video can be transmitted to a flat-panel television set via Wi-Fi, so that family and friends can watch.

Source: Mehta, S.N., "Tomorrow's Cell Phone Will Entertain, Amaze – and Even Make Calls," *Fortune*, October 30, 2006, p. 148.

Wireless technologies can be combined in the same device to enhance functionality. For example, a portable digital music and video player can use Wi-Fi to communicate with your PC or another Wi-Fi device (possibly another music/video player) to download files. It can then use Bluetooth to transmit the music to your wireless earphones. When WiMAX is implemented, some local radio stations are likely to use MANs as additional broadcast channels. With proper software you can then select from the songs to which you have just listened and have them downloaded to your portable player or home computer.

In just a few years, you may be able use your phone to read the RFID electronic code of a product in a store and compare its price to the prices offered online by other retailers. Instead of asking for human help in finding an item in a supermarket, your phone may be able to guide you to the right aisle after it identifies the EPC of the product. And, as is already done in some countries, you will be able to pay for what you purchase by using your phone instead of a credit card.

- Telecommunications is communication over distance, primarily communication of bits representing many forms of data and information. In the past decade, telecommunications technology has driven the major developments in the dissemination and use of information.

- Telecommunications technology has changed the business environment. Businesspeople are increasingly more mobile; they can use cellular phones for greater availability to their employers and customers, using the phone for both voice and data communications. Videoconferencing brings together people who are thousands of miles apart. Peer-to-peer file sharing enables sharing of research, software code, and artistic works.

- Different media have different bandwidths, meaning that they are capable of carrying different numbers of bits per second (bps) without garbling messages. Wired media include twisted pair, coaxial cable, and optical fiber. Wireless media rely on radio waves, including terrestrial and satellite microwave.

- Networks are classified according to their reach and complexity. When computers are connected locally within an office, a campus, or a home, the arrangement is called a local area network (LAN). A metropolitan area network (MAN) connects LANs within a radius of about 50 kilometers (30 miles). When computers communicate over longer distances, the network is called a wide area network (WAN). Personal area networks (PANs) connect individual devices at short range.

- Although it uses the public Internet, a network can be turned into a virtual private network (VPN) by using advanced security measures.

- A communication line can be switched in two ways. In circuit switching, a message is communicated in its entirety from the transmitting device to the receiving device while the communication path is fully devoted to the exchange between the two nodes. In packet switching, data is divided into packets of bits and transmitted via several paths on the network. Internet protocols work with packet switching.

- Network protocols are sets of rules to which all devices on a network must adhere. Communication on the Internet adheres to a set of protocols called TCP/IP. Ethernet has long been a popular protocol for wired LANs. Wireless protocols offer many opportunities for more people to enjoy Internet links and for mobility while communicating. The most important are the IEEE 802.xx protocols, which include the popular Wi-Fi, Bluetooth, and WiMAX standards.

- Wireless technologies make it easy and affordable to create wireless LANs (WLANs) and hotspots. They allow workers mobility while retrieving information in warehouses and other work environments. They enable airline and retail customers to link to the Internet with portable computers, and make the reading of utility meters much less labor intensive and more accurate.

- Organizations and individuals have a variety of choices when subscribing to networking services. They can choose among digital subscriber line (DSL), cable, T1 and T3 lines, satellite links, fixed wireless service, optical fiber to the premises, optical carriers (OC), and Broadband over Power Lines (BPL).

- As Internet links become faster, Internet telephony, also known as Voice over Internet Protocol (VoIP), is gaining in popularity. Several companies offer the service, which is significantly less expensive than a landline service.

- Wireless technologies support the increasingly popular RFID technologies. RFID supports a variety of noncontact identification and payment mechanisms, from quick toll and gas payment to cattle tracking to sophisticated supply chain management, and many future uses are anticipated.

- Much like hardware, telecommunications technologies are merging. The same device can now use several different networks simultaneously, such as cellular telephone networks, the Internet, and television broadcasts.

- Increasing numbers of employees now telecommute. Telecommuting has advantages, but it does not serve some basic human needs, such as socializing during lunch break and the clear separation between work and family obligations.

QUICKBIZ MESSENGERS REVISITED

QuickBiz has upgraded its telecommunications systems—from cellular phones to GPS-capable cellular phones, from handheld computers to handheld computers with a wireless Internet connection, from a single local office to three networked offices, and from DSL Internet access to fixed wireless access. At each point, Andrew Langston has expanded his communications ability.

What Would You Do?

1. If QuickBiz could have connected its offices to the Internet via coaxial cable or a T1 line in time, should it have opted for either of those connections over fixed wireless access? If available in time, would optical fiber to the premises be preferred? Investigate the costs and capabilities of each and give your opinion.

2. Now that messengers carry GPS-capable cellular phones, QuickBiz can track their movements for every delivery and know where they are at all times. Consider QuickBiz's existing "family" culture. Should it follow its couriers' moves to check for efficiency? Why or why not?

3. A local marketing company has approached Andrew to see whether he would be willing to sell his customer list to them. Providing the information would be simple for QuickBiz. Do you think it should do so? Why or why not?

New Perspectives

1. During a conversation with a representative from the cellular phone company, the representative suggested to Leslie that QuickBiz purchase smart phones for their messengers. What would be the advantage of smart phones over the cellular phones that they are now carrying? Would this investment be financially worthwhile?

2. Andrew has seen customers at Starbucks wirelessly logging on to the Web through Wi-Fi connections. What factors should he consider to determine whether Wi-Fi devices would be a good option for his business?

3. A traffic-monitoring company has called on Andrew to offer its service. The representative says that his company can provide real-time traffic reports in the Seattle area. Dispatchers would be able to receive periodic or on-demand reports through their handheld Web-enabled devices and view traffic conditions as they happen. Should QuickBiz consider this service? What does it need to know to decide? Draw up a list of questions that Andrew should ask the representative.

KEY TERMS

access point (AP), 209
backbone, 207
bandwidth, 199
baseband, 199
bits per second (bps), 199
Bluetooth, 210
bridge, 205
broadband, 199

Broadband over Power Lines (BPL), 202
circuit switching, 206
coaxial cable, 200
dial-up connection, 205
digital subscriber line (DSL), 214

DNS (Domain Name System), 208
downstream, 213
dynamic IP address, 208
EMI (electromagnetic interference), 201
encryption, 209

REVIEW QUESTIONS

1. If all the paths of data communications were visible to the human eye, we might be overwhelmed. Why? Give some examples.

2. Data communications over long distances is carried out one bit after another. Why can't whole bytes be transmitted over a distance one byte per signal?

3. What makes one medium capable of greater data communication speed than another?

4. Which medium currently enables the fastest data communications?

5. Repeaters are used on many communication lines. What is their purpose? What does a repeater do?

6. Networking professionals speak of "the last mile." What is "the last mile," and what is its significance?

7. Would an astronomy observatory 20 miles away from a city or town likely be able to get DSL service? Why?

8. What risks to organizations does the growing use of networks pose?

9. What is a virtual private network? Why is it called "virtual"?

10. What is a network protocol?

11. What are the technical advantages of optical fibers over other communications media?

12. The same communication medium can transport three different services. This is true of two media. Which media? What are the three services?

13. What is the difference between circuit switching and packet switching?

14. Why does circuit switching accommodate voice communication more effectively than packet switching?

15. What is VoIP? Since VoIP uses packet switching, why is voice quality better now than several years ago?

16. What is ADSL? What does the A stand for, and what does it mean in terms of communicating with the Internet? Why do households receive only ADSL services and not other DSL services?

17. What is BPL? Why is the technology potentially available to almost every home?

18. Explain the notions of WAN, LAN, MAN, and PAN.

19. What are hotspots, and how can they help businesspeople?

20. What is the purpose of municipally provided WiMAX, and why is it in competition with subscriber broadband services?

21. Cellular phones are already wireless. Why should companies be interested in equipping employees with Wi-Fi-enabled mobile phones?

22. Wi-Fi is all around us. Is there any downside to its pervasiveness?

23. People express themselves differently when they speak (either face to face or via the telephone) versus when they send and receive e-mail. What are the differences? Which do you prefer when communicating with someone you don't know personally? Which do you prefer when you know the person?

24. Every home with access to the Internet can now inexpensively become a hotspot. How so? Are there any risks in turning a home into a hotspot?

25. What are the implications of telecommunications for group work?

26. As broadband services cover larger regions and become less expensive, the number of small businesses and home businesses grows. What is the relationship?

27. Some organizations stopped allocating offices to their sales representatives. Why, and is this a wise move?

28. List and explain the benefits of videoconferencing to an organization. List and explain the benefits to society.

29. Anything that does not take space can be traded solely via telecommunications networks. Do you agree? Explain your answer.

30. Do you see any undesirable effects of humans communicating more and more via computer networks rather than in person or over the telephone? What don't you like and why? What do you like about it?

31. List several jobs in which telecommuting would be infeasible. Explain why.

32. Wi-Fi circuitry is now embedded in consumer electronic devices such as digital cameras and cell phones. Give an example of what you could do with the Wi-Fi capability of a digital camera.

33. If you were given the opportunity to telecommute, would you? Why or why not?

34. Suppose that you are a middle manager. Would you allow the people who report to you to telecommute? Why or why not?

35. As a supervisor, would you be more inclined to promote your telecommuting or nontelecommuting subordinates, or would you be egalitarian? Why?

APPLYING CONCEPTS

36. Ima Jeenyes completed her book, *How to Become a Millionaire Upon Graduation*. She used a word processor to type the manuscript. She saved the book as a file of 5.7 MB. Ima lives in Philadelphia. The publisher asked that Ima transmit the book via the Internet to the publisher's office in Boston. Ima can transmit the file at a guaranteed speed of 400 Kbps. Because each packet of data transmitted must also contain some nondata bits, assume the total number of bits to transmit is equivalent to 6 MB.

    How long (in minutes) does it take to transmit the book? Ignore the distance between the cities. Remember how many bytes make up 1 MB. Show your calculations clearly using a spreadsheet. Use measurement units throughout your calculation. E-mail the spreadsheet file to your professor.

37. Justin Tyme uses a DSL modem to transmit a report from his office to headquarters. The DSL affords an average bit rate transmission (upload) of 250 Kbps. Since the transmission protocol adds additional bits to data bytes, assume that, on average, there is 1 additional bit for each transmitted byte. On average, a page contains 3,000 characters, including spaces. Justin is allotted only 3 minutes for the transmission. How many pages can he transmit?

38. Of the residential telecommunications services listed in Figure 6.7, find out which are available where you live and how much they cost. You might find several DSL and cable services, and perhaps also satellite and BPL services. Calculate the ratio of maximum bit rate per dollar (downstream) to monthly fee for each service. Which service provides the "biggest bang for the buck," that is, the greatest speed per dollar of monthly fee?

39. Broadband services provided in Japan, South Korea, and Canada are usually faster and less expensive than in the United States. Use the Web to research why this is so. Write a one-page report discussing the reasons.

40. Search the Web for a site that enables you to check your high-speed (broadband) link: DSL, cable, Fiber to the Premises, or (if you connect from school) T1 or T3 line. Follow the instructions. Usually, you simply have to click one button. Do so and wait for the response. Print out the response. Wait a minute, and repeat the process. The speeds are likely to be different. Why? Type up the answer, and submit with the two printouts analyzing the speed of your connection.

41. You are a telecommunications guru and love to help individuals and businesses. Assume that dial-up, cable, DSL, T3 line, and satellite links to the Internet are available everywhere unless the particular scenario indicates otherwise. Consider the following scenarios and suggest the best overall type of link (consider communication speed, cost, and any other factor you believe is relevant). Each scenario is independent of the others. For each scenario, explain why you selected the option.

    a. An author works at home writing articles for a magazine. Once per week she must transmit an article to her editor. She rarely uses the link for any other purpose.

    b. A large company maintains its own Web site for online catalogs and purchase transactions by its customers. Hundreds of customers visit the sites daily and make purchases.

    c. A small business uses the Internet for daily research. Owners have heard that some links are shared by other subscribers in the same area, which might slow down the connection or even pose security threats. Thus, they would like to avoid such a service. They do need a speed of at least 200 Kbps.

    d. A farm in New Mexico needs a link of at least 200 Kbps. People on the farm can receive television signals only through antennas. The closest telephone central office is 12 miles away.

    e. An Internet service provider specializes in hosting Web sites of small businesses.

    f. A cruise ship wants to provide Internet service to vacationers on the third deck. The ship cruises in the Caribbean. The link's speed must be at least 250 Kbps.

42. Team up with another student from your class. Select a bank branch close to your school. Interview the branch personnel about the telecommunications equipment used between the branch and (a) other branches, (b) headquarters, and (c) other institutions, such as credit information companies, if any. Use the discussion in this chapter to identify the various communications devices that the branch uses. List the devices and state their roles at the bank.

43. Team up with two other students from your class. Each of you should send an e-mail message to one other team member. One of you may use the school's facilities, but the other two should use a subscriber's address, such as an AOL or Comcast address. When you receive the messages, try to get the routing information. Which servers did the messages pass through on their way to you? How long did it take the messages to get to the server from which your own computer retrieves the messages? Print out the route your computer generated. Report your findings to your professor.

FROM IDEAS TO APPLICATION: REAL CASES

Let it Rain

The supermarket industry is known for its thin profit margin. The average profit is 1–2 percent. This means that every improvement in operation, even the smallest reduction in waste, can help keep a supermarket chain profitable. Perishable food items are especially vulnerable to real-time monitoring—which can be done only with a reliable communication network.

Hannaford Bros., headquartered in Scarborough, Maine, operates 159 supermarkets and combination food and drug stores in Maine, New Hampshire, Massachusetts, New York, and Vermont. It was established in 1883, and now employs more than 26,000 people. As a private company it does not disclose financial details, but its estimated revenue in 2006 was $3.4 billion.

Until 1996, inventory management and order data were decentralized. Each store ordered items separately. Electronic communication between stores and headquarters, let alone among stores, took place through several different technologies: TCP/IP, the standard for most commercial networks; Systems Network Architecture (SNA), IBM's proprietary protocol; X.25, a protocol for wide area networks that uses phone or ISDN lines with slow maximum speed of 128 Kbps; and the asynchronous transfer mode (ATM) protocol, which transmits data in packets at high speeds. Data was transmitted via three types of lines: satellite, telephone dial-up, and leased lines. This mix of technologies provided communication that was slow and unreliable. It did not support real-time monitoring of inventory.

Satellite communication was the least reliable because whenever it rained heavily, the link was not operable. In the northeastern region of the United States, where the chain operates, it rains where a Hannaford store is located almost daily. To ensure access to inventory and order data—which was stored on servers at headquarters—each store maintained four or five servers with its own inventory and order data. Data was often lost, and keeping it synchronized with headquarters was challenging. When conditions were good, data could be transmitted at a speed of no more than 19.2 Kbps.

In 1996, the company hired a CIO. The new executive put in place an IT strategy. In that year, Verizon installed a statewide network to connect government agencies and schools to the Internet at a speed of 10 Mbps. The CIO asked Verizon to connect headquarters to the network, but through a T1 line (which provides close to 1.5 Mbps). Verizon agreed. All 158 stores are connected to the Internet. Only one protocol exists: TCP/IP.

No other media or protocols are used. IBM installed a System p5 mainframe, on which all data of the chain is stored. Cisco Systems installed its Quality of Service (QoS) software, which monitors and maximizes network performance. The CIO was able to eliminate 1,000 servers at 100 locations. Only one or two servers remained in each store.

Point-of-sale (POS) terminals run on the Linux operating system. All are connected to the mainframe at headquarters. Verification of credit card information now takes four to five seconds less than before, and the CIO attributes 80 percent of this time reduction to the improved network. Customer lines move faster.

Over ten years the company grew by 50 percent, but the CIO could reduce the IT staff by 10 percent to 135. Remote management tools can diagnose IT problems and fix antennas, routers, switches, servers, and printers—reducing the need for a technician to visit a store every time an IT issue needs fixing.

Store employees can access the DB2 database that resides on the mainframe at headquarters and make queries through decision support software that was installed by the MicroStrategy company. Store managers walk around with handheld wireless devices, through which they access the mainframe to use applications. They use the devices to check inventory, adjust prices, and produce coupon stickers that are attached to discounted items and are scanned at the POS terminals. The adjusted price information is stored on the mainframe.

The system helps both store workers and headquarter managers. In a store, for instance, the butcher can see that a certain type of meat cut did not sell well yesterday and therefore cut less of it for today, preventing undesired discounting or sheer waste. At headquarters, category managers can plan better by examining data over the past few months or even years. If a certain item did not sell well last summer, they reduce the amount ordered for this summer.

In addition to these benefits, store managers now have a useful tool. Every POS terminal in any store channels all its transactions to a data warehouse that resides on the mainframe. The data is stored for further analysis. And the heavy rains? They no longer matter. The CIO says the network is up 100 percent of the time and that there are no network-related errors in data transmission any more.

Source: Bennett, E., "Digital Networking: Hannaford Brothers is a Cut Above," *Baseline*, October 2, 2006; (www.hannaford.com), April 2007.

Thinking About the Case

1. What were the disadvantages of the pre-1996 network?

2. What are the non-IT operational benefits of the newer networking system?

3. What are the IT operational benefits of the system?

Wireless Patient Care

A study by The Leapfrog Group, a Washington, D.C. voluntary organization of healthcare product purchasers, concluded that better information technology could prevent more than 50 percent of erroneous drug prescriptions. IT in general, and networking technologies in particular, could save lives and morbidity in hospitals. Most hospitals have caught up with new networking technologies only in recent years.

Children's Memorial Hospital in Chicago was no exception until recently. The hospital is part of Northwestern University's Feinberg School of Medicine. It is now ranked as the best children's hospital in Illinois and one of the best in the United States. In 2006, the 1,100 pediatric specialists treated over 100,000 patients and had more than 365,000 outpatient visits.

For many years the hospital had a hodgepodge of communications technologies: landline phones, a local network to support cell phones, a wireless surveillance system, pagers, a radio frequency system for tracking the electronic tags that doctors wear, and a variety of patient-monitoring systems. It did not have a way to ensure accurate drug administration, and this was one reason to reconsider the hospital's communication infrastructure.

Often, one signal interfered with another, creating several areas where cell phones and pagers could not function. Structural challenges also presented problems. Hospitals are built from steel floors and many concrete walls. Thick concrete and lead walls are built around radiation rooms. All of these materials weaken radio signals or block them out altogether.

The hospital's Director of IT started to look for a comprehensive solution. This would include not only better communications, but also improved technologies for the bedside staff and computerized drug prescription entry. He preferred a single system that would address their many challenges. After an extensive search, he selected a company called InnerWireless to deploy a broadband system. InnerWireless produces a system it calls Medical-grade Wireless Utility.

The system uses passive wireless, which means that the devices the staff uses activate the networking circuitry. In active wireless, electronic devices must provide electric power to convert radio signals. Passive systems do not need to be powered. Therefore, with Medical-grade Wireless Utility, fewer access points had to be installed and maintained. Other systems would require more access points and still would not totally eliminate dead spots. The InnerWireless system requires few access points but still provides uninterrupted communications throughout the building. The technology also includes a distributed antenna system. This allows the same wireless systems to support cellular phones, pagers, Wi-Fi (IEEE 802.11), two-way radio for facilities management, and first-responder radio for fire, police, and emergency medical teams.

InnerWireless specializes in in-building wireless communication, and has installed its system in several hospitals. It customizes the deployment for every hospital to ensure that communication is available throughout each building. Typically, a wireless router is installed in the basement. From the router a cable is run up through the building's "spine," and a distribution system is located on each floor.

Now, physicians enter drug prescriptions into a database for each patient. The hospital's pharmacy receives the transmitted prescription and prepares the drug, then attaches the proper bar-code to it. Nurses use carts equipped with a small networked computer that is also equipped with a bar-code scanner. On their rounds, before they administer drugs to patients, they scan the bar-code. The data is automatically communicated to the pharmacy database, and the nurse can see if the drug and dosage are the right ones for the patient. Nurses can also use e-mail through the same computers. The error rate of drug administration has decreased significantly.

The system cost the hospital $500,000, twice what it would pay for an active wireless system with multiple access points. The hospital hopes to recover the higher cost by saving money on the system's maintenance.

Source: Pettis, A., "Patient Care Goes Wireless," *eWeek*, April 10, 2006; (www.childrensmemorial.org), April 2007; (www.innerwireless.com), April 2007.

Thinking About the Case

1. The hospital already had bar-coding before the new networking system was installed. What can be done now that could not be done before to reduce drug administration errors?

2. As the case explains, passive wireless ensures full coverage of communications in an entire building. Why is this so important in hospitals?

3. The new communication network is more than just an Internet hotspot. It supports several modes of communication. What does this mean from a maintenance perspective?

SEVEN

Databases and Data Warehouses

LEARNING OBJECTIVES

As a professional, you will use databases and likely help design them. Understanding how to organize and use data is a way to gain responsibility and authority in a work environment. Data is usually collected in a way that does not make it immediately useful to professionals. Imagine building a model palace from a pile of building blocks. You have a good idea of what you want to build, but first you have to organize the blocks so it is easy for you to find and select only the blocks you need. Then you can combine them into substructures that eventually are integrated into your model. Similarly, data collected by organizations must be organized and stored so that useful information can be extracted from it in a flexible manner.

When you finish this chapter, you will be able to:

- Explain the difference between traditional file organization and the database approach to managing digital data.

- Explain how relational and object-oriented database management systems are used to construct databases, populate them with data, and manipulate the data to produce information.

- Enumerate the most important features and operations of a relational database, the most popular database model.

- Understand how data modeling and design creates a conceptual blueprint of a database.

- Discuss how databases are used on the Web.

- List the operations involved in transferring data from transactional databases to data warehouses.

QUICKBIZ MESSENGERS:
The Value and Uses of Databases

As QuickBiz grew, so did its reliance on databases. By the time the company had grown to 90 employees, Andrew was using databases to create weekly schedules for part-time and full-time employees, track customer orders, store and access employee and customer information, organize and report financial data, and provide crucial information for marketing strategies. As his database needs expanded, he transitioned from one database management system to another.

Moving Up: From Microsoft Access to Oracle

In the early days, Andrew had relied on Microsoft Access and Excel for his company's database needs. When he hired his first part-time messengers, he used an Excel spreadsheet to set up weekly schedules. He stored customer and order information in an Access database. As business grew, so did the size of the database. Kayla Brown, an IT consultant who worked in an office on the second floor of his building, told him that he should consider using a more powerful database management system (DBMS). When Leslie Chen suggested that QuickBiz create an intranet so that messengers could upload delivery information through their wireless connections, the need to switch to a more powerful DBMS became urgent. Microsoft Access wouldn't be able to handle the number of concurrent users that QuickBiz anticipated. Andrew decided to hire Kayla to help the office shift to Oracle. An Oracle database would be able to accommodate both the increased size of the database and the need for concurrent access.

Tapping the Power of Databases

Then Andrew turned his thoughts to using his data to improve his service—to maintain his existing customers and strengthen his relationships with them. He also wanted to find out who would be good potential customers. He hired Kayla to run SQL queries and create reports. Surely he could find valuable information by exploring customer information and buying patterns.

First, Andrew wanted to find out who his preferred customers were—those who used his service most often and provided the most revenue. The consultant used data-mining software to delve into the data and identified a profile. To his surprise, Andrew found that the legal and medical-supply industry clients were most profitable. He'd always thought the art gallery owners were his best clients because of the special handling their objects required. But lawyers and pharmacists needed faster delivery and special services, such as delivery confirmation, which commanded premium rates and generated additional revenue at no further cost per delivery to QuickBiz. Andrew designated those customers as VIPs and tagged their database files. VIPs would receive priority delivery on the routes from now on.

Also, Andrew was interested in the purchasing patterns of customers. He planned to target those opportunities with a promotion to gain new clients. Again, the consultant came back with interesting news: larger law firms with branches throughout the Puget Sound area used QuickBiz's service most often on weekdays between the hours of 10 a.m. and 1 p.m. Andrew decided to locate other similar firms and develop a direct-mail promotion to them—discounted deliveries for setting up an account and scheduling 30 orders in a month's time. Andrew also added additional messengers during that time frame to be sure to handle deliveries smoothly.

MANAGING DIGITAL DATA

You use your Web browser to go to your favorite online electronics store to search for high-definition flat screen television sets. You enter a price range and screen size. Within a few seconds, the screen is filled with details on available models complete with product photos and specifications. Where did this rich, well-organized information come from? It came from a database. A database management system responded almost instantly to your request.

Businesses collect and dissect data for a multitude of purposes. Digital data can be stored in a variety of ways on different types of media, as discussed in Chapter 4. They can be stored in what can be called the traditional file format, in which the different pieces of information are not labeled and categorized, but are stored as continuous strings of bytes. The chief advantage of this format is the efficient use of space, but the data is nonetheless difficult to locate and manipulate. By contrast, the database format, in which each piece of data is labeled or categorized, provides a much more powerful information management tool. Data in this format can be easily accessed and manipulated in almost any way desired to create useful information for decision making.

The impact of database technology on business cannot be overstated. Not only has it changed the way almost every industry conducts business, but it has also created an information industry with far-reaching effects on both our business and personal lives. Databases are behind the successful use of automatic teller machines, increased efficiency in retail stores, almost every marketing effort, and the numerous online search engines and Web-based businesses. Combined with interactive Web pages on the Internet, databases have made an immense contribution to commerce. Without them, there would be no online banking, consumer catalogs, search engines, stock brokerages, or chat rooms. Their impact on business has allowed fewer people to complete larger tasks, and their power has enabled organizations to learn more about us, as consumers, than we might realize. Imagine: every time you enter the address of a Web site, a special program performs a search in a huge database and matches your request with one of hundreds of millions of addresses. Every time you fill out an online form with details such as your address, phone number, Social Security number (SSN), or credit-card number, a program feeds the data into a database, where each item is recorded for further use.

POINT OF INTEREST

Your Lost Record, Our Lost Money

It is estimated that in 2006, U.S. corporations experienced $16 billion in lost productivity, additional paper work, and lost customers as a result of losing personal records. This may be a modest estimation, because only 31 states require corporations to report such incidents. On average, each lost record costs $182, of which the company loses $98 because of potential customers who shun it, $54 spent on incident response, and $30 due to lost productivity. Incident response includes costs such as customer notifications, free or discounted services to victims, and legal defense.

Source: DiJusto, P., "Your Secret Is Out: Data breaches cost companies billions each year," *Wired*, February 2007, p. 50.

In virtually every type of business today, you must understand the power of databases. The approaches to organizing and manipulating data presented in this chapter will help you gain this important knowledge.

Know About Data Management

You already use databases whenever you search the Web and on many other frequent occasions. You very likely will have to use, and perhaps participate in, building databases in your professional career. Search engines, customer loyalty programs, targeted marketing, customer services, and management of practically every corporate resource depend on databases.

Imagine a sales clerk who cannot immediately respond to a customer about the availability of an item, or an online shopper who cannot display the details of an item that is actually available for sale at the site. Customers experiencing this are not likely to patronize the business again. Imagine a treasurer who cannot figure out in real time how much cash the company has in the bank. The company might miss an important deal. Available and reliable information is the most important resource of any business, in any industry. Thus, professionals must understand at least the fundamentals of data organization and manipulation.

You will be a more productive professional if you know how databases and data warehouses are built and queried, and what types of information can be extracted from them. In any career you choose, you may be called to describe to database designers how data elements relate to each other, how you would like the data to be accessed, and what reports you may need. Knowledge of data management techniques and technologies will help you in your job.

The Traditional File Approach

Data can be maintained in one of two ways: the **traditional file approach**—which has no mechanism for tagging, retrieving, and manipulating data—and the **database approach**, which does have this mechanism. To appreciate the benefits of the database approach, you must keep in mind the inconvenience involved in accessing and manipulating data in the traditional file approach: program-data dependency, high data redundancy, and low data integrity.

Consider Figure 7.1, which shows an example of a human resource file in traditional file format. Suppose a programmer wants to retrieve and print out only the last name and department number of each employee from this file. The programmer must clearly instruct the computer to first retrieve the data between position 10 and position 20. Then he must instruct the computer to skip the positions up to position 35 and retrieve the data between positions 36 and 39. He cannot instruct the computer to retrieve a piece of data by its column name, because column names do not exist in this format. To create the reports, the programmer must know which position ranges maintain which type of data and insert the appropriate headings, "Last Name" and "Department," so that the reader can understand the information. If the programmer miscounts the positions, the printout might include output like "677Rapap" as a last name instead of "Rapaport." This illustrates the *interdependency of programs and data* of the traditional file approach. The programmer must know *how* data is stored to use it. Perhaps most importantly, the very fact that manipulation of the data requires a programmer is probably the greatest disadvantage of the file approach. Some business data is still processed this way. New data resources rarely are built this way, but the existing ones must be maintained with this challenge in mind.

Other challenges with traditional file storage are high data redundancy and low data integrity, because in older file systems files were built, and are still maintained, for the use of specific organizational units. If your last and first name, as well as address and other details, appear in the files of the department where you work as well as in the payroll file of the Human Resource department, data can be duplicated. This **data redundancy** wastes storage space (and, consequently, money) and is inefficient. When corrections or modifications need to be performed, every change has to be made as many times as the number of locations where the data

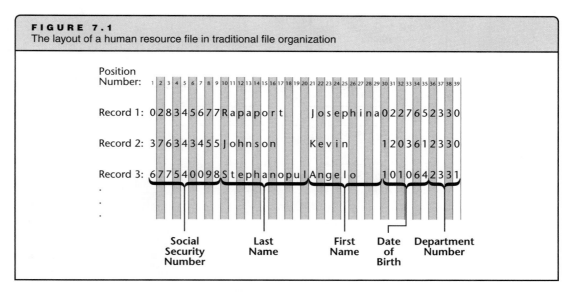

FIGURE 7.1
The layout of a human resource file in traditional file organization

appears, which takes time and might introduce errors. If the same data was entered correctly in one place but incorrectly in another, your record is not only inaccurate, but might appear to represent a different person in each place. Inaccuracies hurt **data integrity**—the characteristic that the data represents what it is supposed to represent and that it is complete and correct. Often, the traditional file approach to storing data leads to low data integrity. It is difficult to ensure that data is correct in all locations when there are myriads of places to insert data in files, as in the traditional approach.

The Database Approach

In the database approach, data pieces are organized about entities. An **entity** is any object about which an organization chooses to collect data. Entities can be types of people, such as employees, students, or members of fan clubs; events, such as sales transactions, sports events, or theatre shows; or inanimate objects, such as inventoried or for-sale products, buildings, or minerals. In the context of data management, "entity" refers to all the occurrences sharing the same types of data. Therefore, it does not matter if you maintain a record of one student or records of many students; the entity is "student." To understand how data is organized in a database, you must first understand the data hierarchy, described in Figure 7.2, which shows a compilation of information about students: their first names, last names, years of birth, SSNs, majors (department), and campus phone numbers. The smallest piece of data is a **character** (such as a letter in a first or last name, or a digit in a street address). Multiple characters make up a field. A **field** is one piece of information about an entity, such as the last name or first name of a student, or the student's street address. The fields related to the same entity make up a **record**. A collection of related records, such as all the records of a college's students, is called a **file**. Often, several related files must be kept together. A collection of such files is referred to as a database. However, the features of a database can be present even when a database consists of a single file.

Once the fields are assigned names, including Last Name, First Name, SSN, and the like, the data in each field carries a tag—a field name—and can be easily accessed by the field name, no matter where the data is physically stored. One of the greatest strengths of databases is their promotion of application-data independence. In other words, if an application is written to process data in a database, the application designer only needs to know the names of the fields, not their physical organization or their length.

Database fields are not limited to holding text and numbers. They can hold pictures, sounds, video clips, and even spreadsheets. Fields can hold any content that can be digitized. For example, when you shop online, you can search for a product by its product name or code, and then retrieve its picture or a video clip about the product. When you select a video at YouTube or MySpace Video, you retrieve a video clip from a database.

FIGURE 7.2
Data hierarchy

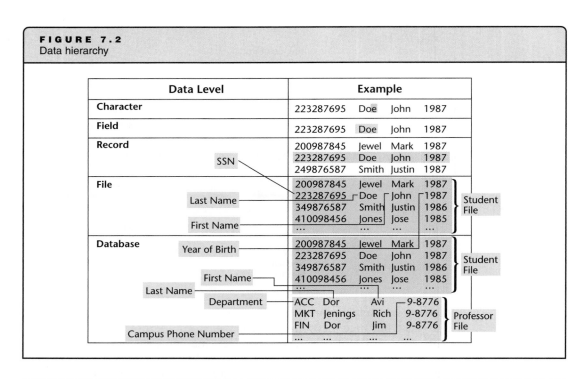

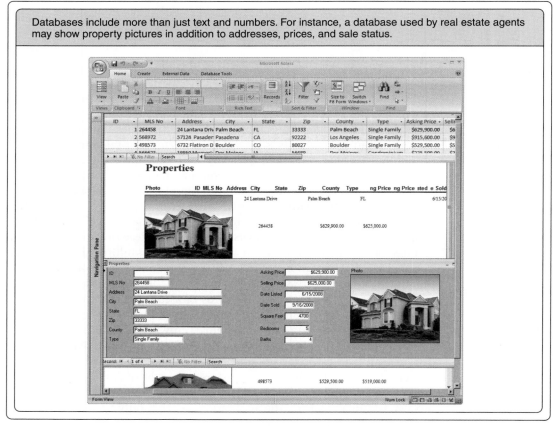

Databases include more than just text and numbers. For instance, a database used by real estate agents may show property pictures in addition to addresses, prices, and sale status.

While a database itself is a collection of several related files, the program used to build databases, populate them with data, and manipulate the data is called a **database management system (DBMS)**. The files themselves *are* the database, but DBMSs do all the work—structuring files, storing data, and linking records. As we described previously, if you wanted to access data from files that were stored in a traditional file approach, you would have to know exactly how many characters were designated for each type of data. A DBMS, however, does much of this work (and a lot of other work) for you.

If you are using a database, you want to be able to move rapidly from one record to another, sort by different criteria, select certain records or fields, create different types of reports, and analyze the data in different ways. Because of these demands, databases are stored on and processed from direct access storage devices, such as magnetic disks or DVDs. They can be backed up to sequential storage devices such as magnetic or optical tapes, but cannot be efficiently processed off such media because it would take too long to access the records. Note that storing databases on any device that is nonwritable, such as nonrewritable CDs or DVDs, may be suitable for a static database, such as a part list used by car repair shops, but is unsuitable for a database that must be updated.

Queries

Data is accessed in a database by sending messages called **queries**, which request data from specific records and/or fields and direct the computer to display the results. Queries are also entered to manipulate data. Usually, the same software that is used to construct and populate the database, that is, the DBMS, is also used to present queries. Modern DBMSs provide fairly user-friendly means of querying a database.

Security

The use of databases raises security and privacy issues. The fact that data is stored only once in a database for several different purposes does not mean that everyone with access to that database should have access to *all* the data in it. Restricting access is managed by customizing menus for different users and requiring users to enter codes that limit access to certain fields or records. As a result, users have different *views* of the database, as abstractly illustrated in Figure 7.3. The ability to limit users' views to only specific columns or records gives the **database administrator (DBA)** another advantage: the ability to implement security measures. The measures are implemented once for the database, rather than multiple times for different files. For instance, in the database shown in Figure 7.4, while a human resource manager has access to all fields of the employee file (represented by the top table), the payroll personnel have access only to four fields of the employee file (middle part of the figure), and a project manager has access only to the Name and Hours Worked fields. Views can be limited to certain fields in a database, or certain records, or a combination of both. We discuss security issues in detail in Chapter 14, "Risks, Security, and Disaster Recovery."

POINT OF INTEREST

Data (Mis)management

Unfortunately, database technology makes it easy for employees to lose, and for hackers to steal, millions of personal records. Personal information is compromised almost daily, but in some cases the numbers are astounding. In 2006, the U.S. Department of Veterans Affairs lost a laptop computer containing the records of 26 million veterans. In the previous year, CardSystems, a credit card-processing company, kept 40 million personal records in a database for research purposes. The records were stolen by intruders who used the information to defraud MasterCard International. This dubious record was broken by TJX, the parent company of T.J. Maxx and other retailers. Someone broke the security key to its database and stole 45 million credit and debit card records. The thieves later used the information to create dummy credit cards for purchasing Wal-Mart and Sam's Club gift cards. They used the gift cards to purchase $8 million of merchandise. How many personal records have been lost or stolen? The Privacy Rights Clearinghouse, a consumer advocacy organization, says that by early 2006 the number was over 93 million records. Adding just the TJX figure of 45 million, at least 138 million records have been compromised.

Source: Greenemeier, L., "T.J. Maxx Parent Company Data Theft is the Worst Ever," *InformationWeek*, March 29, 2007; Zeller, T., "93,754,333 Examples of Data Nonchalance," *The New York Times*, September 27, 2006.

DBMSs are usually bundled with a programming language module. Programmers can use this module to develop applications that facilitate queries and produce predesigned reports.

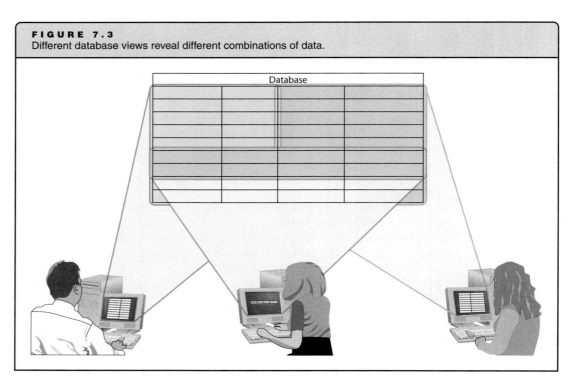

FIGURE 7.3
Different database views reveal different combinations of data.

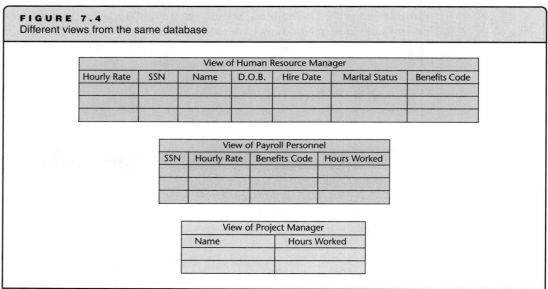

FIGURE 7.4
Different views from the same database

View of Human Resource Manager						
Hourly Rate	SSN	Name	D.O.B.	Hire Date	Marital Status	Benefits Code

View of Payroll Personnel			
SSN	Hourly Rate	Benefits Code	Hours Worked

View of Project Manager	
Name	Hours Worked

DATABASE MODELS

A *database model* is the general logical structure in which records are stored within a database and the method used to establish relationships among the records. The several database models differ in the manner in which records are linked to each other, which in turn dictates the manner in which a user can navigate the database, retrieve desired records, and create reports. The oldest models, the hierarchical and network models, are still used in some databases that were built in the 1970s and 1980s, but are no longer used in newly constructed databases. Virtually all new databases are designed following the relational and object-oriented models.

The Relational Model

The **relational model** consists of **tables**. Its roots are in relational algebra, but you do not have to know relational algebra to build and use relational databases. However, database experts still use relational algebra terminology: in a relational database, a record or row is called a *tuple*, a field—often referred to as a column—is called an *attribute*, and a table of records is called a *relation*. This text uses the simpler terms, as do the popular software packages: fields, records, and tables.

To design a relational database, you need a clear idea of the different entities and how they relate. For example, in a database for a DVD store, the entities might be Customer, DVD Rental, DVD, and Distributor. A single table is built for each entity (though each table can contain from only a few to potentially millions of records). DVD Rental is an associative entity; you can see in Figure 7.5 that the DVD Rental table associates data from the Customer and DVD tables.

Maintenance of a relational database is relatively easy because each table is independent of the others, although some tables are related to others. To add a customer record, the user accesses the Customer table. To delete a record of a DVD, the user accesses the DVD table. The advantages of this model make relational database management systems the most popular in the software market. Virtually all DBMSs currently on the market accommodate the relational model. This model is used in supply chain management (SCM) systems and many other enterprise applications as well as local, individual ISs.

To retrieve records from a relational database, or to sort them, you must use a *key*. A key is a field whose values identify records either for display or for processing. You can use any field as a key. For example, you could query the database for the record of John Smith from the Customer table by using the CustName field as a key. That is, you enter a query, a condition that instructs the DBMS to retrieve a record with the value of CustName as "John Smith." A key is *unique* if each value (content) in that field appears only in one record. Sometimes a key is composed of several fields, so that their combination provides a unique key.

As you can see, database design requires careful forethought. The designer must include fields for foreign keys from other tables so that join tables can be created in the future. A **join table** combines data from two or more tables. A table might include foreign keys from several tables, offering flexibility in creating reports with related data from several tables. The inclusion of foreign keys might cause considerable data redundancy. This complexity has not diminished the popularity of relational databases, however.

If a database has more than one record with "John Smith" (because several customers happen to have that same name) in the CustName field, you might not retrieve the single record you desire. Depending on the application you use for the query, you might receive the first one that meets the condition, that is, a list of all the records with that value in the field. The only way to be sure you are retrieving the desired record is to use a unique key, such as a Social Security number, an employee ID, or, in our example, a customer ID (CustID). A unique key can serve as a **primary key**. A primary key is the field by which records in a table are uniquely identified. If your query specified that you wanted the record whose CustID value is 36002, the system would retrieve the record of John Sosik. It will be the John Sosik you wanted, even if there are more records of people with exactly the same name. Because the purpose of a primary key is to uniquely identify a record, each record must have a unique value in that field.

Usually, a table in a relational database must have a primary key, and most relational DBMSs enforce this rule; if the designer does not designate a field as a key, the DBMS creates its own serial number field as the primary key field for the table. Once the designer of the table determines the primary key when constructing the records' format, the DBMS does not allow a user to enter two records with the same value in that column. Note that there might be situations in which more than one field can be used as a primary key. Such is the case with motor vehicles, because three different fields can uniquely identify the record of a particular vehicle: the vehicle identification number (VIN), its title number, and its state license plate number. Thus, a database designer might establish one of these fields as a primary key to retrieve records.

For some business needs you must use a **composite key**, a combination of two or more fields that together serve as a primary key, because it is impractical to use a single field as a primary key. For example, consider flight records of a commercial airline. Flights of a certain route are the same every week or every day they are offered, so the daily FlyOz Airlines' flight

FIGURE 7.5
A relational database

Customer Table

CustID	CustName	CustPhone	CustAddr
33091	Jill Bronson	322-4907	203 Oak Dr
35999	John Smith	322-5577	519 Devon St
36002	John Sosik	342-0071	554 Spring Dr
36024	Jane Fedorow	322-7299	101 Jefferson Ave

Primary key

Composite primary key

DVD Rental Table

CustID	CopyNum	Date Rented	Date Returned
35999	4452-1	5-1-08	5-3-08
36002	4780-3	5-3-08	
36024	5312-2	5-2-08	5-5-08

Copy Table

CopyNum	TitleNum
4452-1	4452
4452-2	4452
5312-1	5312
5312-2	5312
5312-3	5312
7662-1	7662
7662-2	7662
5583-1	5583

Primary key in *Title* and part of a
composite primary key in *Copy*

Title Table

TitleNum	Title	Category	DistribNum	RentPrice
4452	Enter the Dragon	Martial Arts	277	$4.00
5312	The Ring II	Thriller	305	$4.00
7662	Star Wars III	Sci-Fi	372	$5.00
5583	White Noise	Thriller	589	$2.50

Primary key in *Distributor* and foreign key in *Title*

Distributor Table

DistribNum	DistribName	Phone
277	HK Corp	1-877-555-0550
305	Columbia	1-888-222-3654
372	Lucas Films	1-247-233-6996
589	Booh Inc	1-866-222-9999

from Houston to Geneva—FO1602—for instance, cannot serve us well to retrieve a list of all the passengers who took this flight on May 3, 2008. However, we can use the combination of the flight number *and* date as a composite primary key. To check who sat in a particular seat, a composite key consisting of three fields is needed: flight number, date, and seat number.

To link records from one table with records of another table, the tables must have at least one field in common (i.e., one column in each table must contain the same type of data), and that field must be a primary key field for one of the tables. This repeated field is a primary key in one table, and a **foreign key** field in the other table. In the DVD store example, if you will ever want to create a report showing the name of every distributor and all the DVD titles from that distributor, the primary key of the Distributor table, DistribNum, must also be included as a foreign key in the Title table. The resultant table (Figure 7.6) is a join table. Note that although

DistribNum was used to create the join table, it does not have to be displayed in the join table, even though it could be.

Distributor	Telephone	Title
HK Corp	1-877-555-0550	Enter the Dragon
Columbia	1-888-222-3654	The Ring II
Lucas Films	1-247-233-6996	Star Wars III
Booh Inc	1-866-222-9999	White Noise

Since the relationships between tables are created as part of manipulating the table, the relational model supports both one-to-many and many-to-many relationships between records of different tables. For example, a **one-to-many relationship** is created when a group of employees belongs to only one department. All would have the same department number as a foreign key in their records, and none will have more than one department key. There is *one* department, linked to *many* employees. A **many-to-many relationship** can be maintained, for instance, for professors and students in a college database. A professor might have many students, and a student might have many professors. This can be accomplished by creating a composite key of professor ID and student ID. In our example of the DVD store, there is a many-to-many relationship between customers and the DVDs they have rented. The DVD Rental table enables the store manager to create a history report of customers and their rentals. It is clear that more than one customer has rented a certain DVD, and the same customer has rented many different DVDs.

The major vendors of relational DBMSs (RDBMSs) are IBM, Oracle, and Microsoft, with worldwide market share in licensing revenues of about one-third, one-third, and one-fifth, respectively. IBM licenses DB2, Oracle licenses DBMSs by the company name, and Microsoft licenses SQL Server and Access. MySQL, an open source DBMS, is also very popular. Evans Data Corporation, an IT market analysis firm, estimates that MySQL has a 44 percent share of the global installed relational DBMSs (leaving to the proprietary RDBMSs just a little over half the market). These DBMSs are an essential part of enterprise applications such as SCM and CRM systems.

The Object-Oriented Model

The **object-oriented database model** uses the object-oriented approach, described in Chapter 5, "Business Software," to maintaining records. In object-oriented technology, an object consists of both data and the procedures that manipulate the data. So, in addition to the attributes of an entity, an object also contains relationships with other entities and procedures to manipulate the data. The combined storage of both data and the procedures that manipulate them is referred to as **encapsulation**. Through encapsulation, an object can be "planted" in different data sets. The ability in object-oriented structures to create a new object automatically by replicating all or some of the characteristics of a previously developed object (called the parent object) is called **inheritance**. Figure 7.7 demonstrates how the same data maintained in a relational database at the DVD rental store would be stored and used in an object-oriented database. The relationships between data about entities are not managed by way of foreign keys, but through the relationships of one object with another. One advantage of this approach is the reduction of data redundancy.

Some data and information cannot be organized as fields, but they can be handled as objects, such as drawings, maps, and Web pages. All these capabilities make object-oriented DBMSs, also called object database management systems (ODBMSs) handy in computer-aided design (CAD), geographic information systems, and applications used to update thousands of Web pages daily, because they can handle a wide range of data—such as graphics, voice, and text—more easily than the relational model.

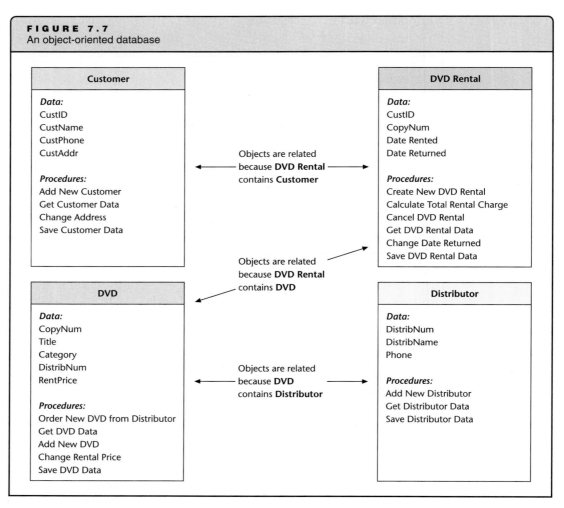

FIGURE 7.7
An object-oriented database

Customer

Data:
CustID
CustName
CustPhone
CustAddr

Procedures:
Add New Customer
Get Customer Data
Change Address
Save Customer Data

DVD Rental

Data:
CustID
CopyNum
Date Rented
Date Returned

Procedures:
Create New DVD Rental
Calculate Total Rental Charge
Cancel DVD Rental
Get DVD Rental Data
Change Date Returned
Save DVD Rental Data

Objects are related because **DVD Rental** contains **Customer**

Objects are related because **DVD Rental** contains **DVD**

DVD

Data:
CopyNum
Title
Category
DistribNum
RentPrice

Procedures:
Order New DVD from Distributor
Get DVD Data
Add New DVD
Change Rental Price
Save DVD Data

Distributor

Data:
DistribNum
DistribName
Phone

Procedures:
Add New Distributor
Get Distributor Data
Save Distributor Data

Objects are related because **DVD** contains **Distributor**

Similar to relational DBMSs, ODBMSs provide a graphical user interface (GUI) to manage the DBMS. The user can choose objects from "classes," which are groups of objects that share similar characteristics. Elements of ODBMSs are often incorporated into relational databases, and such databases are sometimes known as *object-relational databases*.

Object-oriented databases (ODBs) do not store records, but data objects, which is an advantage for quick updates of data sets and the relationships among them. For instance, in the example of the DVD store, in the ODB the relationship between a DVD and its distributor is not established through a foreign key; it exists because the DVD class contains the Distributor class. However, object-oriented databases also have some disadvantages, compared with relational databases. For example, there is dependence between applications and data; they are simply "wrapped" together. Changing the structures of tables in a relational database does not require changes in applications that use the data in those tables, while it would require changes in applications in an object-oriented database. This dependence also limits the ability to enter *ad hoc* queries in an ODB, that is, to enter queries at will. While not as popular or as well understood as relational databases, ODBs are gaining adopters.

Several software companies have developed popular ODBMSs. Among them are Objectivity/DB (Objectivity, Inc.), ObjectStore (Progress Software, Inc.), and Versant (Versant Corporation).

RELATIONAL OPERATIONS

As mentioned before, the most popular DBMSs are those that support the relational model. Therefore, you would benefit from becoming familiar with a widely used relational database, such as Access, Oracle, or SQL Server. To use the database, you should know how relational operations work. A **relational operation** creates a temporary table that is a subset of the original table or tables. It allows you to create a report containing records that satisfy a condition, create a list with only some fields about an entity, or produce a report from a join table, which combines relevant data from two or more tables. If so desired, the user can save the newly created table. Often, the temporary table is needed only for *ad hoc* reporting and is immediately discarded.

The three most important relational operations are *select, project,* and *join. Select* is the selection of records that meet certain conditions. For example, a human resources manager might need a report showing the entire record of every employee whose salary exceeds $60,000. *Project* is the selection of certain columns from a table, such as the salaries of all the employees. A query might specify a combination of selection and projection. In the preceding example, the manager might require only the ID number, last name (project), and salary of employees whose salaries are greater than $60,000 (select).

One of the most useful manipulations of a relational database is the creation of a new table from two or more other tables. As you might recall from our discussion of the relational model, the joining of data from multiple tables is called a *join.* We have already used a simple example from the DVD store database (Figure 7.6). However, join queries can be much more complex. For example, a relational business database might have four tables: SalesRep, Catalog, Order, and Customer. A sales manager might wish to create a report showing, for each sales rep, a list of all customers who purchased anything last month, the items each customer purchased, and the total amount spent by each customer. The new table is created from a relational operation that draws data from all four tables.

The join operation is a powerful manipulation that can create very useful reports for decision making. A join table is created "on the fly" as a result of a query and exists only for the duration the user wishes to view it or to create a paper report from it. Design features allow the user to change the field headings (although the field names are kept the same in the internal table), place the output in different layouts on the screen or paper, and add graphics and text to the report. The new table might be saved as an additional table in the database.

POINT OF INTEREST

Terrifying Terabytes

ChoicePoint, Inc., based in Alpharetta, Georgia, is in the business of collecting, maintaining, and selling data on the American population: names, addresses, Social Security numbers, listed and unlisted phone numbers, employment history, criminal history, driving records, DNA records, and much more. Its database includes more than 250 terabytes of personal data on 220 million people.

Source: "They're Watching You ," *BusinessWeek online,* January 24, 2005; *Business Intelligence Lowdown* (www.businessintelligencelowdown.com), February 2007.

Structured Query Language

Structured Query Language (SQL) has become the query language of choice for many developers of relational DBMSs. SQL is an international standard and is provided with most relational database management programs. Its strength is in its easy-to-remember intuitive commands. For example, assume the name of the entire database is DVD_Store. To create a list of all titles of thriller DVDs whose rental price is less than $5.00, the query would be:

    SELECT TITLE, CATEGORY FROM DVD_STORE
          WHERE CATEGORY = 'Thriller' and RENTPRICE < 5

Statements like this can be used for *ad hoc* queries or integrated in a program that is saved for repeated use. Commands for updating the database are also easy to remember: INSERT, DELETE, and UPDATE.

Integrating SQL in a DBMS offers several advantages:

- With a standard language, users do not have to learn different sets of commands to create and manipulate databases in different DBMSs.

- SQL statements can be embedded in widely used third-generation languages such as COBOL or C and object-oriented languages such as C++ or Java, in which case these languages are called the "host language." The combination of highly tailored and efficient 3GL or object-oriented statements with SQL statements increases the efficiency and effectiveness of applications accessing relational databases.

- Because SQL statements are portable from one operating system to another, the programmer is not forced to rewrite statements.

Some relational DBMSs, such as Microsoft Access, provide GUIs to create SQL queries; SQL queries can be placed by clicking icons and selecting menu items, which are internally converted into SQL queries and executed. This capability allows relatively inexperienced database designers to use SQL.

The Schema and Metadata

When building a new database, users must first build a schema (from the Greek word for "plan"). The schema describes the structure of the database being designed: the names and types of fields in each record type and the general relationships among different sets of records or files. It includes a description of the database's structure, the names and sizes of fields, and details such as which field is a primary key. The number of records is never specified because it might change, and the maximum number of records is determined by the capacity of the storage media.

Fields can hold different types of data: numeric, alphanumeric, graphic, or time-related. Numeric fields hold numbers that can be manipulated by addition, multiplication, averaging, and the like. Alphanumeric fields hold textual values: words, numerals, and special symbols, which make up names, addresses, and identification numbers. Numerals entered in alphanumeric fields, such as Social Security numbers or zip codes, cannot be manipulated mathematically. The builder of a new database must also indicate which fields are to be used as primary keys. Many DBMSs also allow a builder to positively indicate when a field is not unique, meaning that the value in that field might be the same for more than one record.

Figure 7.8 presents the schema of a database table created with the Microsoft Access DBMS. The user is prompted to enter the names and types of fields. Access lets the user name the fields and determine the data types. The Description section allows the designer to describe the nature and function of the fields for people who maintain the database. In the lower part of the window the user is offered many options for each field, such as field size, format, and so on. In Access the primary key field is indicated by a little key icon to its left.

FIGURE 7.8
Schema of the Employee table in an Access 2007 database. The Field Properties list on the bottom shows the property of the attribute (field) Salary.

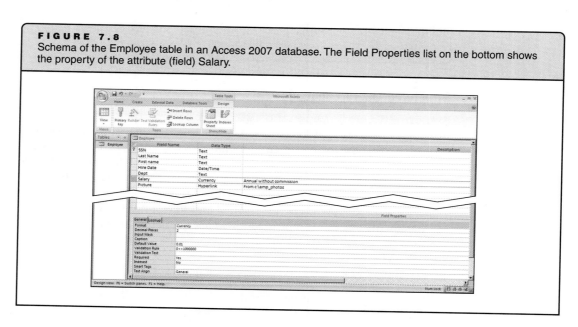

The description of each table structure and types of fields become part of a **data dictionary**, which is a repository of information about the data and their organization. Designers usually add more information about each field, such as where the data comes from (such as another system or entered manually); who owns the original data; who is allowed to add, delete, or update data in the field; and other details that help DBAs maintain the database and understand the meaning of the fields and their relationships. (Some people prefer to call this **metadata**, meaning "data about the data.") Metadata includes:

- The source of the data, including contact information.

- Tables that are related to the data.

- Field and index information, such as the size and type of the field (e.g., whether it is text or numeric), and the ways the data is sorted.

- Programs and processes that use the data.

- Population rules: what is inserted, or updated, and how often.

DATA MODELING

Databases must be carefully planned and designed to meet business goals. How they are designed enables or limits flexibility in use. Analyzing an organization's data and identifying the relationships among the data is called **data modeling**. Data modeling should first be done to decide which data should be collected and how it should be organized. Thus, data modeling should be proactive. Creating data models periodically is a good practice; it provides decision makers a clear picture of what data is available for reports, and what data the organization might need to start collecting for improved decision making. Managers can then ask experts to change the relationships and design new reports or applications that generate desired reports with a few keystrokes.

Many business databases consist of multiple tables with relationships among them. For example, a hospital might use a database that has a table holding the records of all its physicians, another one with all its nurses, another with all the current patients, and so on. The administrative staff must be able to create reports that link data from multiple tables. For example, one report might be about a doctor and all her patients during a certain period. Another might revolve around a patient, such as details of the patient, a list of all caregivers who were involved in his rehabilitation, and a list of medications. Thus, the database must be carefully planned to allow useful data manipulation and report generation.

Effective data modeling and design of each database involves the creation of a conceptual blueprint of the database. Such a blueprint is called an **entity relationship diagram (ERD)**. An ERD is a graphical representation of all entity relationships, an example of which is shown in Figure 7.9, and they are often consulted to determine a problem with a query or to implement changes. ERDs are a main tool for communication not only among professional DB designers, but also among users and between users and designers. Therefore it is important that professionals in all of these fields know how to create and read them.

In an ERD, boxes are used to identify entities. Lines are used to indicate a relationship between entities. When lines shaped like crow's-feet are pointing to an object, there might be many instances of that object. When a link with a crow's-foot also includes a crossbar, then all instances of the object on the side of the crow's-foot are linked with a single instance of the object on the side of the crossbar. A second crossbar would denote "mandatory," which means that the relationship must occur, such as between a book title and author: a book title must have an author with which it is associated. A circle close to the box denotes "optional."

- In Figure 7.9, the crow's-foot on the Department end of the Department/College relationship indicates that there are several departments in one college, indicating a one-to-many relationship between College and Department. In addition, the crossbar at the College end of the College/Department link indicates that a department belongs to only one college.

- A department has many professors, but a professor might belong to more than one department; thus, the relationship between Professor and Department is many-to-many, represented by the crow's-feet at both ends of the link.

- A course is offered by a single department, indicated by the crossbar at the Department end of the Department/Course link.

- A professor might teach more than one student, and a student might have more than one professor, thus the crow's-feet at both the Professor and Student ends of the many-to-many relationship between Professor and Student.

- However, the ring at the Student end indicates that a professor does not have to have students at all. The ring means "optional," and is there for cases in which professors do not teach.

A diagram such as Figure 7.9 provides an initial ERD. The designers must also detail the fields of each object, which determines the fields for each record of that object. The attributes are listed in each object box, and the primary key attribute is underlined. Usually, the primary key field appears at the top of the field list in the box. Figure 7.10 is an example of possible attributes of a Professor entity. Database designers can use different notations; therefore, before you review an ER diagram, be sure you understand what each symbol means.

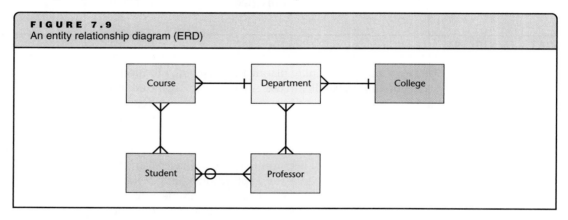

FIGURE 7.9
An entity relationship diagram (ERD)

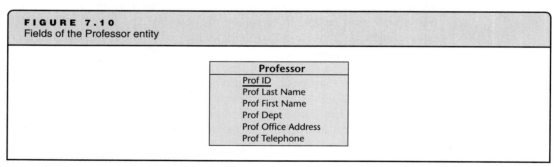

FIGURE 7.10
Fields of the Professor entity

The examples given here are fairly simple. In reality, the reports that managers need to generate can be quite complex in terms of relationships among different data elements and the number of different tables from which they are assembled. Imagine the relationships among data maintained in libraries: a patron might borrow several titles; the library maintains several copies of each title; a title might be a book, a videotape, a CD, or a DVD; several authors might have published different books with the same title; librarians must be able to see availability and borrowed items by title, by author, and by patron; they should also be able to produce a history report of all the borrowing of each patron for a certain period of time; and so on. All of these relationships and the various needs for reports must be taken into account when designing the database.

DATABASES ON THE WEB

The Internet and its user-friendly Web would be practically useless if people could not access databases online. The premise of the Web is that people can not only browse appealing Web pages but also search for and find information. Most often, that information is stored in databases. When a shopper accesses an online store, he or she can look for information about any of thousands, or hundreds of thousands, of items offered for sale. For example, when you access the site of Buy.com, or Target, or Overstock.com, you can receive online information (such as an image of an electronics item, price, shipping time, and consumer evaluations) for thousands of items offered for sale. Entering a keyword at YouTube results in a list of all video clips whose title or descriptive text contains the keyword. Wholesalers make their catalogs available online. Applications at auction sites receive inquiries by category, price range, country of origin, color, date, and other attributes, and identify records of matching items, which often include pictures and detailed descriptions. Behind each of these sites is a database. The only way for organizations to conduct these Web-based businesses is to give people outside the organizations access to their databases. In other words, the organizations must link their databases to the Internet.

From a technical point of view, online databases that are used with Web browsers are no different from other databases. However, an interface must be designed to work with the Web. The user must see a form in which to enter queries or keywords to obtain information from the site's database. The interface designers must provide a mechanism to figure out data that users insert in the online forms so that they can be placed in the proper fields in the database. The system also needs a mechanism to pass queries and keywords from the user to the database. The interfaces can be programmed in one of several Web programming languages, including Java servlets, active server pages (ASP), ASP.NET (the newer version of ASP processes within the .NET framework) and PHP (Hypertext Preprocessor), as well as by using Web APIs (application program interfaces). The technical aspects of these applications are beyond the scope of this book. The process is diagrammed in Figure 7.11.

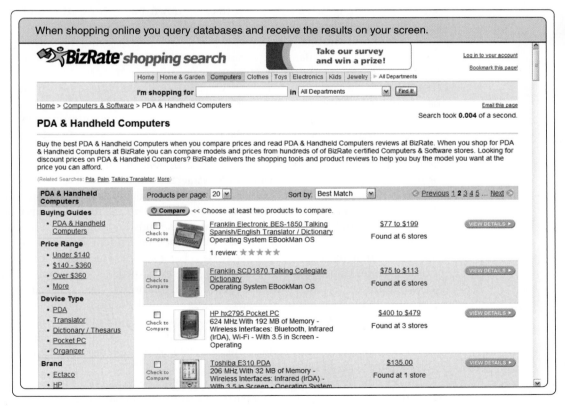

To ensure that their production databases are not vulnerable to attack via the Internet, organizations avoid linking their transaction databases to the Internet unless the databases are dedicated to online transactions, in which case the organization must apply proper security software. They must also be careful when linking a data warehouse (discussed next) to the Internet.

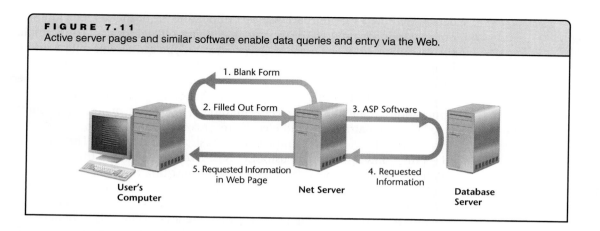

1. Blank Form

2. Filled Out Form

3. ASP Software

5. Requested Information in Web Page

4. Requested Information

User's Computer

Net Server

Database Server

DATA WAREHOUSING

The great majority of data collections in business are used for daily transactions and operations: records of customers and their purchases and information on employees, patients, and other parties for monitoring, collection, payment, and other business or legal purposes. The transactions do not stay in these databases long; usually only a few days or weeks. However, many organizations have found that if they accumulate transaction data, they can use it for important management decisions, such as researching market trends or tracking down fraud. Organizing and storing data for such purposes is called data warehousing.

A **data warehouse** is a large, typically relational, database that supports management decision making. The data warehouse is large because it contains data, or summaries of data, from millions of transactions over many years and/or from national or global transactions rather than from a short period or a single region. It might maintain records of individual transactions or summaries of transactions for predetermined periods, such as hourly, daily, or weekly. The purpose of data warehouses is to let managers produce reports or analyze large amounts of archival data and make decisions. Data-warehousing experts must be familiar with the types of business analyses that will be done with the data. They also have to design the data warehouse tables to be flexible enough for modifications in years to come, when business activities change or when different information must be extracted.

Data warehouses do not replace transactional databases, which are updated with daily transactions such as sales, billing, cash receipts, and returns. Instead, transactional data is copied into the data warehouse, which is a separate data repository. This large archive contains valuable information for the organization that might not be evident in the smaller amounts of data typically stored in transactional databases. For example, an insurance company might keep monthly tables of policy sales; it can then see trends in the types of policies customers prefer in general or by age group. Such trends are meaningful only if they are gleaned from data collected over several years. Data from transactional databases are added to the data warehouse at the end of each business day, week, or month, or it might be added automatically as soon as a transaction is recorded in a transactional database. While a transactional database contains current data, which is disposed of after some time, the data in data warehouses is accumulated and might reflect many years of business activities.

Organizations often set up their data warehouse as a collection of **data marts**, smaller collections of data that focus on a particular subject or department. If data marts need to be used as one large data warehouse, special software tools can unify data marts and make them appear as one large data warehouse.

Every Move You Make

The widespread use of database management systems coupled with Web technologies allows organizations to collect, maintain, and sell vast amounts of private personal data fast and cheaply. Millions of credit-card transactions take place in the world, each carrying private information. Millions of personal data items are routed daily to corporate databases through sales calls and credit checks. Millions of consumer records are collected and updated daily on the Web. For businesses, such data is an important resource. But for individuals, such large data pools and the ways they are used threaten a fundamental human right: privacy.

- **Out of Hand—Out of Control.** You have just received a letter from John Doe Investments. In the letter, the president tells you that at your age, with a nice income like yours, the company could provide you with innovative investment services. How did the company know about your existence? About your annual income? Could it be that some time ago you applied for a credit card? The company receiving the information sold part of it, or all of it, to John Doe Investments. You now enjoy your credit card, but you paid a hidden cost for it.

- **The Web: A Source of Data Collection.** In the preceding example, you were at least aware that you gave somebody information. But many consumers provide information routinely without being aware of it. A huge amount of personal data is collected through the Web. You might wonder why the home pages of so many Web sites ask you to register with them. When registering, you often provide your name, address, and other details. The site asks you to create a user ID and password. If the pages you are accessing contain private data such as your investment portfolio, a user ID and password protect you, but if you are accessing news or other nonpersonal pages, a user ID and password actually serve the site operator. From the moment you log on to the site, the server can collect data about every move you make: which pages you are visiting and for how long, which icons you click and in which order, and which advertising banners you click. In many cases, the organization that collects the data doesn't even own the site. The site owner hires a business such as DoubleClick, FastClick, and Avenue A to collect data. When you click an advertisement, that information is channeled into one of these organization's huge databases. What does the firm do with the database? It sells parts of it to other companies, or it slices and dices the information to help other companies target potential buyers belonging to certain demographic groups. And,

no, it does not bother to tell you. While the software of such companies as DoubleClick can only identify the computer or IP number from which you logged on to a site and not you, personally, the information can be matched with you, personally, if you also use your personal ID and password.

In addition to Web cookies, companies also use Web bugs to track our Web movements. A Web bug, also known as a "Web beacon" or "clear GIF," is a graphic image on a Web site used to monitor a surfer's activity. The image is usually undetectable because it usually consists of a single pixel. The bug links the Web page to the Web server of a third party, such as DoubleClick. Much as other ads appear on a page you view from a server different from the site you accessed, a Web bug comes from a different server, the server of a third party. This happens because the original site's page contains code that calls the bug (the same way as some ads) from the other server. The same technique is used in e-mail. The third party's server obtains the URL (Web address) of the user as well as the URL of the site from which the user views the page. As long as the bug is "displayed" by the user's computer, the third-party server can request session information from the user's Web browser. Session information includes clickstream and other activities performed by the user while visiting the site.

- **Our Finances Exposed.** Everyone is sensitive when it comes to finances. In the United States, the Gramm-Leach-Bliley law, which went into effect on July 1, 2001, was supposed to protect consumer privacy. The law entitles consumers to opt out of having their private information shared with "nonaffiliated" third parties. It requires companies to tell consumers what information they collect and how they might use it, and to establish safeguards against fraudulent access to confidential information.

  Yet, critics claim that the law does not provide the most important protection it was supposed to provide: not allowing companies to share private financial information with other organizations. Whether you opt out or not, the law allows companies that reside under the same corporate umbrella to share your information. "Companies under the same umbrella" include a bank and its subsidiaries or sister companies such as an insurance company and a bank owned by the same parent company.

Also, companies are allowed to share information with unaffiliated companies if they have service or marketing agreements with those unaffiliated companies.

Consider this sentence from the privacy policy of one bank: "We recognize that an important benefit for our customers is the opportunity to receive offers for products and services from other companies that may work with us." Consumer advocates read the sentence this way: "Whether you like it or not, we will share your information with other companies, and they can do with it whatever they wish, including bombarding you with unsolicited mail and e-mail."

- **Our Health Online.** Allowing medical staff and pharmacists to share patient medical information might help them help us. Imagine being injured on a trip thousands of miles from your home. If the doctor treating you can immediately receive information about your allergies to certain medications, it might save your life. However, any electronic record residing on a database that is connected to a public network is potentially exposed to unauthorized access by people who do not have a legitimate need to know.

The Health Insurance Portability and Accountability Act of 1996 (HIPAA) is the U.S. federal law that was enacted to—among other purposes—mandate how health-care providers and insurance firms are to maintain records and disclose information so that patient privacy is not violated. The law restricts who accesses your medical records. Yet, even this law recognizes the inability of organizations to ensure patient privacy. For example, you can ask your doctor not to share your medical record with other doctors or nurses in the clinic, but they do not have to agree to do what you ask.

- **The Upside.** In spite of the downside of collection of personal data, there is also a positive side. Database technology enables companies to provide us with better and faster services. It also makes the market more competitive. Small firms often cannot afford the great expense of data collection. For much less money, they can purchase sorted data—the same data that is available to the industry leader. So, the wide availability of data contributes to a more egalitarian and democratic business environment. The beneficiaries are not only vendors but also consumers, who can purchase new and cheaper products.

And while many of us complain that these huge databases add to the glut of junk mail and spam, better information in the hands of marketers might actually save consumers from such annoyances. After all, those annoying communications are for products and services you don't need. With more specific information, marketers can target only those individuals that might be interested in their offerings. While you shop, special tracking software can tell the online business, at least indirectly, what you do not like about the site. This enables businesses to improve their services. For example, many online retailers discovered that a hefty proportion of shoppers abandoned their virtual shopping carts just before the final purchase. Analysis of collected information discovered that some people wanted to know the handling and shipping charges before they charged their credit cards. Now, most online retailers provide clear shipping information and charges up front.

From Database to Data Warehouse

Unlike data warehouses, transactional databases are usually not suitable for business analysis because they contain only current, not historical, data. Often, data in transactional databases are also scattered in different systems throughout an organization. The same data can be stored differently and under other names. For example, customer names might be recorded in a column called Name in one table and in two columns—First Name and Last Name—in another table. These discrepancies commonly occur when an organization uses both its own data and data it purchases from other organizations, or if it has developed more than one database that contains the same data under a different label. When management decides to build a data warehouse, the IT staff must carefully consider the hardware, software, and data involved in the effort.

The larger the data warehouse, the larger the storage capacity, the greater the memory, and the greater the processing power of the computers that are needed. Because of capacity needs, organizations often choose mainframe computers with multiple CPUs to store and manage data warehouses. The computer memory must be large enough to allow processing of huge amounts of data at once. The amount of storage space and the access speed of disks are also important. Processing millions of records might take a long time, and variations in disk speed might mean the difference between hours or minutes in processing time. And since a data warehouse is considered a highly valuable asset, all data must be automatically backed up. Keep in mind that data warehouses grow continually, because their very purpose is to accumulate historical records. Retail chains such as Wal-Mart and Costco record millions of sales transactions daily, all of which are channeled into data warehouses. Some have data warehouses that hold tens or hundreds of terabytes of data. In addition to retailers, banks, credit-card issuers, health-care organizations, and other industries have augmented their hardware for large data warehouses. Many organizations accumulate not only sales transactions but also purchasing records, so they can produce information from which to make better purchasing decisions, such as which suppliers tend to offer lower prices for certain items at certain times of the year.

POINT OF INTEREST

The World's Largest Data Pool

The world's largest data bank is the World Data Center for Climate (WDCC) in Hamburg, Germany, operated by the Max Planck Institute for meteorology and the German Climate Computing Centre. In all, the center holds six petabytes (6 PB = 6 quadrillion bytes) of data, stored on magnetic tapes. In addition, 220 terabytes (trillion bytes or TB) of climate research and anticipated climatic trends is accessible through the Web. The center also maintains 110 TB of climate simulation data. Six PB of data is about three times the contents of all the U.S. academic research libraries.

Source: *Business Intelligence Lowdown* (www.businessintelligencelowdown.com), February 2007.

The data from which data warehouses are built usually comes from within an organization, mainly from transactions, but it can also come from outside an organization. The latter might include national or regional demographic data, data from financial markets, and weather data. Similar to metadata in any database, data-warehouse designers create metadata for their large data pools. To uncover the valuable information contained in their data, organizations must use software that can effectively "mine" data warehouses. Data mining is covered in Chapter 11, "Business Intelligence and Knowledge Management."

Designers must keep in mind scalability: the ability of the data warehouse to grow as the amount of the data and the processing needs grow. Future growth needs require thoughtful planning in terms of both hardware and software.

Phases in Data Warehousing

Three phases are involved in transferring data from a transactional database to a data warehouse: extraction, transforming, and loading (ETL). Figure 7.12 describes the process.

In the *extraction* phase, the builders create the files from transactional databases and save them on the server that holds the data warehouse. In the *transformation* phase, specialists "cleanse" the data and modify it into a form that allows insertion into the data warehouse. For example, they ascertain whether the data contains any spelling errors and fix them. They make sure that all data is consistent. For instance, Pennsylvania might be denoted as Pa., PA, Penna, or Pennsylvania. Only one form would be used in a data warehouse. The builders ensure that all addresses follow the same form, using uppercase or lowercase letters consistently and defining fields uniformly (such as one field for the entire street address and a separate field for zip codes). All the data that expresses the same type of quantities is "cleansed" to use the same measurement units.

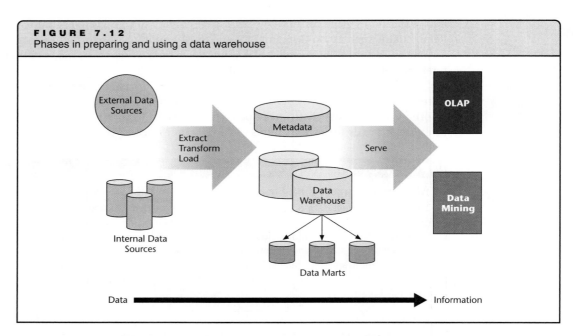

FIGURE 7.12
Phases in preparing and using a data warehouse

In the *loading* phase, the specialists transfer the transformed files to the data warehouse. They then compare the data in the data warehouses with the original data to confirm completeness. As with any database, metadata helps the users know what they can find and analyze in the data warehouse.

A properly built data warehouse is a single source for all the data required for analysis. It is accessible to more users than the transactional databases (whose access is limited only to those who record transactions and some managers) and provides a "one-stop shopping" place for data. In fact, it is not unusual for a data warehouse to have large tables with fifty or more fields (attributes).

Much of the ETL activity can be automated. Depending on the needs of its users, the structure and content of the data warehouse might be changed occasionally. Techniques such as data mining and online analytical processing (OLAP) can be used to exploit it. Managers can then extract business intelligence for better decision making. Data mining, OLAP, and business intelligence are discussed in Chapter 11, "Business Intelligence and Knowledge Management."

- In their daily operations, organizations can collect vast amounts of data. This data is raw material for highly valuable information, but data is useless without tools to organize it, store it in an easily accessible manner, and manipulate it to produce that information. These functions are the strength of databases: collections of interrelated data that, within an organization and sometimes between organizations, are shared by many units and contribute to productivity and efficiency.

- The database approach has several advantages over the more traditional file approach: less data redundancy, application-data independence, and greater probability of data integrity.

- The smallest piece of data collected about an entity is a character. Multiple characters make up a field. Several fields make up a record. A collection of related records is a file, or in the relational model, a table. Databases usually contain several files, but the database approach can be applied to a single file.

- A database management system (DBMS) is a software tool that enables us to construct databases, populate them with data, and manipulate the data. Most DBMSs come with programming languages that can be used to develop applications that facilitate queries and produce reports. DBMSs are also a major part of enterprise applications.

- A database model is the general logical structure of records in a database. The various database models are: hierarchical, network, relational, and object-oriented. The most popular model is the relational model, which is used to build most new databases, although object-oriented databases are gaining popularity. Some vendors offer DBMSs that accommodate a combination of relational and object-oriented models, called object-relational.

- The links among entities in a relational database are maintained by the use of key fields. Primary keys are unique identifiers. Composite keys are combinations of two or more fields that are used as a primary key. Foreign keys link one table to another within the database.

- In an object-oriented database, data sets, along with the procedures that process them, are objects. The relationship between one set of data and another is established by one object containing the other, rather than by foreign keys.

- SQL has been adopted as an international standard language for querying relational databases. SQL statements can also be embedded in code that is produced using many programming languages.

- To construct a database, a designer first constructs a schema and prepares metadata, which is information about the data to be kept in the database.

- To plan databases, designers conduct data modeling. Before they design a database, they create entity relationship diagrams, which show the tables required for each data entity and the attributes (fields) it should hold, as well as the relationships between tables. Then they can move on to constructing a schema, which is the structure of all record structures of the entities, and the relationships among them.

- Many databases are linked to the Web for remote use. This arrangement requires Web server software, such as active server pages and Java servlets, which allow users to enter queries or update databases over the Internet.

- Data warehouses are huge collections of historical transactions copied from transactional databases, often along with other data from outside sources. Managers use software tools to glean useful information from data warehouses to support their decision making. Some data warehouses are made up of several data marts, each focusing on an organizational unit or a subject.

- In each addition of data from a transactional database to a data warehouse, the data is extracted, transformed, and loaded, a process known by its acronym, ETL.

- The low price of efficient and effective database software exacerbates a societal problem of the Information Age: invasion of privacy. Because every transaction of an individual can be easily recorded and later combined with other personal data, it is inexpensive to produce large dossiers on individual consumers. This poses a threat to privacy. However, commercial organizations insist that they need personal information to improve their products and services and to target their marketing only to interested consumers.

QUICKBIZ MESSENGERS REVISITED

QuickBiz gathers and maintains many types of data in its database. The company has tried to ensure the data is secure and safely backed up while still being accessible to customers and employees. Let's explore some of the issues QuickBiz faces in managing its database.

What Would You Do?

1. QuickBiz's database is vital to its operations. The case at the beginning of the chapter didn't mention its supplier data. QuickBiz has suppliers for its fleet of cars and trucks and for its office supplies. What sorts of data would QuickBiz likely keep about its suppliers? What controls and limits should it put on its supplier data? Make a recommendation to Andrew Langston on who should be able to review and change this data and where the data should be maintained.

2. Andrew had run into IT consultant Kayla Brown many times and began talking to her about his IT concerns. When he realized that he was going to need a database management system, he decided to take her advice and purchase Oracle. What sort of research should Andrew have done to make sure that Oracle was the best solution? What advantages and disadvantages should he have considered when purchasing a new DBMS?

New Perspectives

1. QuickBiz has used SQL queries and reports to identify VIP customers and discover its most profitable clients and services. QuickBiz also has a Web site. How could it use Web site tracking data to enhance its services? What departments would be interested in this information? Discuss and list as many as you can.

2. Andrew has heard that databases can also store digital images. Are there any parts of QuickBiz's data operations that might use digital images? If so, what are they?

KEY TERMS

1. It is easier to organize data and retrieve it when there is little or no dependence between programs and data. Why is there more such dependence in a file approach and less in the database approach?

2. Spreadsheets have become quite powerful for data management. What can be done with database management systems that cannot be done with spreadsheet applications? Give several examples.

3. What is the difference between a database and a database management system?

4. DBMSs are usually bundled with powerful programming language modules. Why?

5. DBMSs are a component of every enterprise application, such as a supply chain management system. Why?

6. What are the advantages and disadvantages of object-oriented databases?

7. What is the relationship between a Web site's local search engines and online databases?

8. When constructing a database, the designer must know what types of relationships exist between records in different data sets, such as one-to-many or many-to-many. Give three examples for each of these relationships.

9. Give an example of a one-to-one relationship in a relational database.

10. What is SQL? In which database model does it operate? Why is it so popular?

11. What is a data warehouse? How is it different from a transactional database?

12. Why is it not advisable to query data from transactional databases for executive decision making the same way you do data warehouses?

13. What are the phases of adding data to a data warehouse?

14. What does it mean to cleanse data before it is stored in a data warehouse?

15. What are data marts? How do they differ from data warehouses?

16. Retail chains want to ensure that every time a customer returns to purchase something, the record of that purchase can be matched with previous data of that customer. What objects that consumers often use help the retailers in that regard?

17. Increasingly, corporate databases are updated by the corporations' customers rather than their employees. How so?

18. Can you think of an industry that would not benefit from the promise of a data warehouse? Explain.

19. Shouldn't those who build data warehouses trim the data before they load it to data warehouses? Why do they usually not cut any data from transactions?

20. The combination of RFID and database technology will soon enable retailers to record data about consumers even when they have not purchased anything at the store. Can you think of an example and how the data could be used?

21. A retailer of household products maintains a data warehouse. In addition to data from sales transactions, the retailer also purchases and maintains in the warehouse daily weather data. What might be the reason?

22. Many organizations have posted privacy policies at their Web sites. Why do you think this is so? How is this related to databases and data warehouses?

23. Consider the following opinion shared by some people: database management systems and data-warehousing techniques are the greatest threat to individual privacy in modern times. What is your opinion?

24. The proliferation of organizational databases poses a threat to privacy. After reading the following passage, what would you say to someone in response to these statements: "I'm a law-abiding citizen and pay my taxes promptly. I don't care if anyone reviews my college grades or my income statements, because I have nothing to hide. I have no reason to worry about violation of my privacy. All these complaints about violation of privacy are not valid. Only individuals who have something to hide need to worry."

25. Privacy rights advocates demand that organizations ask individuals for permission to sell personal information about them. Some also demand that the subjects of the information be paid for their consent. Organizations have argued that they cannot practically comply with these demands and that the demands interfere with the free flow of information. What is your opinion?

26. Organizations whose Web sites offer visitors some control of how their personal information is collected and used offer one of two options: "opt out" or "opt in." Explain each term.

27. Some people say that as long as the concept of "informed consent" is applied, individuals should not complain about invasion of their privacy. What is "informed consent"? Do you agree with the argument?

28. Some people say that the affordability of sophisticated DBMSs and data warehouses makes the business world more "democratic" and puts all businesses almost on an equal footing. Assume they are right, and explain what they mean.

29. Businesses in the United States and many other countries rarely allow customers to scrutinize and correct records that the organizations keep about them. Technologically, does the Web make it less expensive for organizations to allow that?

APPLYING CONCEPTS

30. Direct your Web browser to *www.zillow.com.* Enter a real address. What is displayed comes from at least one database. Prepare a short report answering these questions: What information elements must Zillow pull from databases to display what you see? (street address, town, etc.) Are all the data elements textual? Explain.

31. Acxiom is a data services firm. Browse this company's site and research its activities at its own site and at other Web sites. Write a two-page summary of the company's activity: What does the company sell? How does it obtain what it sells? Who are its customers, and how do they use what they purchase from Acxiom?

32. Research the business of DoubleClick, Inc. What type of data does the company collect and sell? How does it collect the data? Who are the company's customers, and how do they use the services or data they buy from DoubleClick? At the company's Web site, you may notice that the company presents a privacy policy *at this Web site*. Explain why only *at this Web site* and not a general privacy policy.

33. Research Web resources to write a two- to four-page research paper titled "Object-Oriented Databases," in which you explain the differences and similarities between relational databases and object-oriented databases as well as their comparative advantages and disadvantages.

34. Mid-County Hospital holds data on doctors and patients in two tables in its database (see the following tables): DOCTOR and PATIENT.

DOCTOR					
ID#	LIC#	Last Name	First Name	Ward	Salary
102	8234	Hogg	Yura	INT	187,000
104	4666	Tyme	Justin	INT	91,300
221	2908	Jones	Jane	OBG	189,650
243	7876	Anderson	Ralph	ONC	101,800
256	5676	Jones	Ernest	ORT	123,400
376	1909	Washington	Jaleel	INT	87,000
410	4531	Carrera	Carlos	ORT	97,000

PATIENT					
SSN	Last N	First N	Admission Date	Insurance	Doc ID
055675432	Hopkins	Jonathan	4/1/08	BlueCross	221
101234566	Bernstein	Miriam	4/28/09	HAP	243
111654456	McCole	John	3/31/08	Kemper	221
200987898	Meanny	Marc	2/27/09	HAP	221
367887654	Mornay	Rebecca	4/3/08	HAP	410
378626254	Blanchard	George	3/30/09	BlueCross	243
366511122	Rubin	David	4/1/08	Brook	243

Use your DBMS to build the appropriate schema, enter the records, and create the reports described.

a. A report showing the following details for each doctor in this order: Last Name, First Name, and Ward. Arrange the report by ascending alphabetical order of the last names.

b. A report showing the entire record with the original order of columns of all the doctors whose salary is greater than $100,000 who work for one of the following wards: Internal (INT), Obstetric-Gynecological (OBG), Oncology (ONC).

c. A report showing the following details for all of Dr. Anderson's patients: Dr. Anderson's first name, last name, and Doctor's ID, and ward (from the DOCTOR table) should appear once at the top of the report. Each record on the list should show the Patient's Last Name, First Name, and Date of Admission (from the PATIENT table).

35. Mr. Lawrence Husick is an inventor who, with other inventors, obtained several U.S. patents. Find the site of the U.S. Patent and Trademark Office. Conduct a patent search at the site's online patent database. Find all the patents that mention Lawrence Husick as an inventor. Type up the patent numbers along with their corresponding patent titles (what the invention is). E-mail the list to your professor. Find and print out the image of patent No. 6469. Who was the inventor and what was the invention?

36. Your team is to design a relational database for an online pizza service. Customers log on to the site and provide their first and last names, address, telephone number, and e-mail address. They order pizza from a menu. Assume that each item on the menu has a unique number, a description, and a price. Assume there is one person per shift who receives orders and handles them, from giving the order to the kitchen to dispatching a delivery person. The system automatically records the time at which the server picked up the order. The business wants to maintain the details of customers, including their orders of the past six months. The following are reports that management might require: (1) a list of all the orders handled by a server over a period of time; (2) summaries of total sales, by item, for a period; and (3) a report showing all of the past week's deliveries by server, showing each individual order— customer last name and address, items ordered, time of order pickup, and last name of delivery person. (You can assume the last names of delivery people are unique, because if there is more than one with the same last name, a number is added to the name.)

    a. Chart the table for each entity, including all its fields and the primary key.

    b. Draw the entity relationship diagram.

37. Your team should contact a large organization, such as a bank, an insurance company, or a hospital. Interview the database administrator about the database he or she maintains on customers (or patients). What are the measures that the DBA has taken to protect the privacy of the subjects whose records are kept in the databases? Consider accuracy, timeliness, and appropriate access to personal records. Write a report on your findings. If you found loopholes in the procedures, list them and explain why they are loopholes and how they can be remedied. Alternatively, log on to the site of a company that posted a detailed privacy policy and answer the same questions.

FROM IDEAS TO APPLICATION: REAL CASES

Unearthing the 36-Hour Day Billing

Law firms charge their clients by the hour. The greater number of billable hours, the greater the revenue. However, even the most talented attorney does not work more than 24 hours per day. Yet, clients sometimes find that they are billed for more hours than there are in a day.

Stuart Maue is a firm that specializes in helping the clients of legal firms ensure that they are billed only for the work done for them and only for work that the legal firm was asked to perform. To this end, Stuart Maue maintains a large database containing details of thousands of legal relationships. For example, it might discover that a deposition that could be taken in an hour was billed for four hours.

The firm was established in 1985 to provide legal auditing and litigation consulting services. Over the years it has adopted increasingly sophisticated hardware and software to offer its clients—usually corporations—a range of cost-management consulting services. Thus, it should come as no surprise that the Careers section of its Web site lists at least as many IT specialist openings as lawyers and accountants. Between 2000 and 2006, the firm spent over $10 billion on IT.

In the 1980s, the work was mostly manual. Accountants and lawyers pored over bills and searched for inconsistencies, double billings, and noncompliant charges. In addition to common sense, the analysts also used the rules that corporations set for the law firms representing them. For example, a guideline may be that lawyers do not fly first class, or that no more than two lawyers take a deposition. The manual work to discover noncompliance was effective, but labor-intensive.

In 1988, Maue purchased its first Oracle database management system and hired software developers. The database served to store the details of thousands of bills. The software was designed to search and analyze legal bills, fees, and expenses.

To be able to analyze how its clients are billed by their attorneys, Stuart Maue fed all billing details of the client into a data warehouse. Its staff used an optical character recognition (OCR) system to read and feed the data warehouse. It then used statistical and other proprietary software applications to find irregular and inappropriate billing.

One client was a golf course developer in Texas. The innovative course had some holes that mimicked famous holes of world-renowned courses, such as Pebble Beach, Pinehurst, and Augusta National. An attorney involved in the case thought that to better understand the case it would be a good idea to test

these professional courses. Combing the legal bills details stored in the data warehouse, Stuart Maue analysts found that the law firm billed the developer for the expensive games at those golf courses.

Stuart Maue is the oldest business in the legal audit industry, but its success attracted competitors. In 2004, the company had another wave of technology overhaul, partly because of mounting competition and partly because clients wanted to access reports through the Internet. Clients used to call the technical staff and ask for reports such as a list of all the legal firms serving the client ranked by billable hours or overall dollars charged. The staff produced the reports, but this typically took at least a day. Several clients threatened to switch to competitors if Maue did not provide self-service reporting. They also wanted to perform some analyses of their own.

Maue upgraded its DBMS to the latest Oracle system that offers a business intelligence tool (which we discuss in detail in Chapter 11) called Discover and Oracle Portal, which links a database to the Web. The system cost $2 million. The data warehouse is installed on a Dell 6800 server with a storage capacity of 500 GB. Online data entry and retrieval takes place on a Hewlett-Packard Itanium 2 server, and the firm's proprietary software runs on a variety of Dell servers. The Web portal is managed by the open source Apache application, which runs on a Red Hat Linux operating system.

Maue hoped that the technological overhaul would allow the firm to continue to serve customers with the same number of 17 IT staff members. It wanted to accommodate a growth rate of at least 20 percent per year in audited billing. It wanted to enable customers to access their own data through a self-service Web site and produce reports by themselves. In addition to satisfying client demands, Maue also hoped that adding self-service would reduce the amount of IT staff labor by 80 to 90 percent.

Shortly after the system upgrade, Maue was approached by Steadfast Insurance, a unit of the Swiss company Zurich Insurance. The company insured Purdue Pharma, a pharmaceutical corporation that was sued over its OxyContin, a pain killer. Purdue Pharma claimed it incurred over $400 million in legal fees to defend against nearly 1,400 lawsuits over injuries attributed to use of the pain killer. It demanded that Steadfast, Purdue's insurer, reimburse the company. Steadfast refused to reimburse some of the money because it suspected the billing was exaggerated. Purdue Pharma sued. Steadfast hired Maue to audit the legal bills.

The task was huge. The legal defense for Purdue involved 70 law firms in 32 states, 322 partners, 849 associates, and 1,032 paralegal workers. Steadfast was served with invoices for 1.2 million billed hours and associated expenses. Maue passed this test with flying colors. Using the OCR system, its staff took only six weeks to feed the data from 200 boxes of paper documents into the new database. The task would take many more months if it were performed manually. Maue provided Steadfast with reports that were used to successfully challenge some of the bills. Purdue Pharma and Steadfast reached a confidential settlement. The company and three of its current and former executives pled guilty to criminal charges of misleading doctors and patients by claiming OxyContin was less likely to be abused than traditional narcotics, and both the company and its executives had to pay hefty fines.

Experts say that the use of business intelligence tools is spreading from retailers to other industries. Maue's success can be attributed in part to the fact that the business intelligence software it uses is preintegrated with the DBMS. This eliminates the need to fit analysis tools to a database. The same is true of the Web site that serves clients through a standard Web browser. Clients are happy that they can see their legal expenses in different perspectives by sorting them in various ways and at different levels of detail.

Maue, a privately held company, hoped to grow the business to analyzing $700 million of legal billing by 2006. It actually handled $2.2 billion in that year, and enjoyed revenue of $20 million. The clerical staff now spends a fraction of the time originally spent on the same tasks before the implementation of the data warehouse and its upgrade. The use of OCR technology alone reduced labor by 30 percent. Since the upgrade, the turnaround time from invoice submission to audit and report has been reduced from 10 days to 5 days. The IT staff has remained at its 2003 size: 17 people.

Source: Duvall, M., "No Lawyer Joke," *Baseline*, December 18, 2006; Maue, B., "Stuart Maue Wins Big with Oracle Business Intelligence Solution," *DMReview*, January 2007; (www. stuartmaue.com), May 2007; Meier, B., "Narcotic Maker Guilty of Deceit Over Marketing," *The New York Times*, May 11, 2007.

Thinking About the Case

1. Consider the type of data entered into Maue's data warehouse. In what sense is it different from data entered in retail enterprises?

2. One benefit of the self-service capability that the system now affords the clients was to satisfy client demand. What was the other benefit?

3. What technologies (hardware, software, networking) save labor for Stuart Maue when compared to the situation in the 1980s?

4. Modern DBMSs are usually bundled with other applications. Identify those applications in this case, and the purpose they serve.

United They Stand

Southside Electric Cooperative (SEC) is an electric power distributor in south-central Virginia. As a cooperative, it is a nonprofit, member-owned organization. Customers, all of whom are also members, can choose among competing producers of electricity, but the actual distribution, regardless of producer, is performed by SEC. The cooperative was incorporated in 1937 and is dedicated to "Helping rural families live better electrically." In addition to power, SEC also provides wiring and electrical consulting services to members, as well as energy audits, heat loss/heat gain estimates, safety education programs, electronic bill payment, third-party notification bill payment services, budget billing, and security lighting.

Good service attracted a growing number of customers. Between 2002 and 2005, the number of customers grew from about 30,000 to 52,500. SEC wanted to continue its good service, from inquiry about opening new accounts to sending a crew to resolve an outage. However, the near doubling of the customer base complicated a situation that was already challenging: SEC was using six disparate databases. Each database served a different purpose: outages, dispatches, electricity usage, geographic mapping, billing, and accounts receivable. The databases were not connected to each other.

A typical business cycle of service was as follows: A clerk received the call, filled out and printed a service order form, and placed it in a supervisor's tray. The supervisor sorted and prioritized orders, and placed them in a technician's tray. The technician picked up the service order, typically the next day, and drove out to the field to perform the repair or other service, such as connecting a new customer to the grid. Upon completion, the technician brought the service order back to the office, where a clerk documented the completion in the task and entered the proper information in the billing and accounts receivable databases.

The disconnect between the databases caused inaccuracies and inefficiencies. For example, a technician sent to fix a problem with a line could use Qualcomm OmniTRACS—a device with two-way satellite communication link—to exchange information with the dispatch database. The technician could receive the customer's address and the status of a repair for the customer. However, when the repair was complete, no information, such as the repair details and the charge for it, could be entered into the accounts receivable database. A clerk had to receive the information from the technician and manually enter it into the accounts receivable database.

In 2005, SEC decided to build a real-time integrated system accessible by employees of all departments. The cooperative turned to what is popularly called a service-oriented architecture (SOA), in which systems are integrated to better serve customers. The IT staff integrated a customer information system, geographic information system (GIS), automated meter reading, financial management, materials management, and mobile data. The GIS, a database of maps and other data, runs on an Oracle DBMS. The six databases were integrated using IBM's WebSphere.

WebSphere consists of three software components: messages, adapters, and broker. Messages are data. Adapters are used to retrieve relevant data from a database and send it to and from the broker. The broker is connected to all the databases and is programmed to know which data needs to be sent to which database. Users see only a single interface. One of the reasons SEC decided to select the IBM software was IBM's willingness to help analyze the system requirements before the purchase.

SEC was helped by a consultant who accompanied the project. Three IT staff members did the coding. Design and planning took more than a year. Coding and implementation took three months. Testing took another two months.

With the new system, when a service order arrives, it is displayed on the TRACS mobile unit in the technician's vehicle. The technician performs the required work, completes the service order form electronically in the vehicle, and sends the data to the integrated system. The system automatically posts the data in the various databases. All of the details are linked. For example, a clerk can retrieve a customer record and see a list of all the repairs ordered and completed for the customer over a specified period of time. The accounts receivable database is updated automatically, and linked to the customer record. The GIS can show where a customer resides and which tasks where performed for that location.

Information gathered by the technician in the field is updated across systems in real time. There is no need for paper forms to wait for a clerk from which to enter data. The data entry, which typically took a half hour, was totally eliminated. Since no manual data entry is

involved except the data entry from the technician, data is less prone to errors. The results: all users, including service people, have the most accurate information. This, in turn, reduced the typical power outage for a customer from as long as a week to only one or two days.

Since all customers are also owners of the cooperative, surpluses are returned to the customer owners. Perhaps the benefits of the new system contributed to SEC's ability to pay its members back $1.2 million for 2006.

Source: Violino, R., "How One Electric Company Stepped Into the Light," *CIO Insight*, February 22, 2007; "IBM Turns On Southside Electric Cooperative to Software Recycling; Coop Delivers $1.2 million Capital Credits Refund to Member-Owners," *ArriveNet* (press.arrive.net), March 06, 2007; (www.sec.coop), May 2007.

Thinking About the Case

1. What were the faults with the old system?

2. Were the original databases changed in any way? Explain.

3. What are the benefits of the re-architectured system?

4. Why is it important that users are not aware of the disparate databases?

Rescued by Data

Not knowing enough about yourself might be dangerous. One company learned this lesson in time to come out of bankruptcy with the help of IT. Leiner Health Products, Inc., the world's largest manufacturer of private label vitamins, minerals, and nutritional supplements, is also the second largest manufacturer of private label over-the-counter (OTC) pharmaceuticals in the United States. Private label products are the same products sold by leading brand-name manufacturers, but under another name and for a lower price. The company markets more than 480 vitamins and stocks more than 6,000 items. It holds a 50 percent share of the private label vitamin market (more than twice the market share of its next largest competitor), and a 25 percent share of all mass-market vitamin product sales in the United States. However, despite its market position, inefficiencies and lack of access to critical information almost brought the company to its knees, and it recovered only thanks to implementation of new information technologies.

Management knew the situation was bad. Leiner finished the previous year with revenues of $662 million, 60 percent of which came from large retailers such as

Wal-Mart, Sam's Club, and Costco. Maintaining a profit on vitamins and food supplements is not easy, because the profit margins are low. Thus, constantly pursuing efficiency is critical. Yet, Leiner's operations were far from efficient.

Ostensibly, Leiner had every reason to be in good shape. It had 150 customers to whom it sold 4,000 different products manufactured in five plants. However, customer service was unsatisfactory. Thirty percent of deliveries were either not on time or incomplete. Its inventory of finished goods turned over only 2.5 times per year, which is half the industry's typical turnover in profitable years.

Managers did not have the information they needed. They could not figure out who their best and worst customers were. They did not have the information needed to schedule deliveries based on customers' needs. The financial situation was not good. In 1999, an international cartel of 27 vitamin companies was found guilty of price fixing, an event that depressed prices just when Leiner was holding $150 million of inventory. It had to cut the prices on that inventory to well below cost, and ended the year 2000 with a loss of $2 million before interest and taxes. The firm was left with only $8 million in cash and was about to default on its bank loans of $280 million and its own bonds of $85 million. To top off its woes, Wal-Mart threatened to stop purchasing from Leiner, because the company was often late in restocking Wal-Mart's shelves.

The executive team called in a consulting firm that generated the proper reports from Leiner's databases. The main report was a list of customer accounts and the profit margin derived from each of them. Executives discovered that many customers were costing Leiner more than its revenue from them. The firm asked those customers to choose between paying more and being dropped from its clientele. It was left with only half of the customers, all now profitable. Similarly, it produced reports on profit by product, and decided to drop 40 percent of the products it made. Now that Leiner produced much less, it shut down three of its five plants, saving $40 million annually.

Although Leiner had an MRP II (manufacturing resource planning) system, its MRP (materials requirement planning) component was not in sync with the production process. A new order triggered lists of materials to be purchased, but without regard to manufacturing capacity and future orders. To save money, purchasing officers decided which of the system's recommended materials to purchase and which to hold off. Consequently, the plants could not produce some of the ordered lots, and some customers could not receive completed shipments. It was clear that data on manufacturing capacity was missing from the decision-making process.

Another consulting firm was hired, which put in place a new database. Over a period of six months, the database collected data from point-of-sale systems of Leiner's most important customers as well as from its own manufacturing facilities. More than 17,000 pieces of data were collected, which the consultants fed into the MRP system. The MRP system was modified to receive up-to-the-minute data on customer orders and delivery timetables. Now, the amounts of raw materials ordered were not too high or too low. The combined costs of overstocking raw materials and warehousing finished products decreased by $50 million.

From the data collected, executives discovered that they had based pricing on the fastest machines Leiner had in its two plants. Slower machines meant greater cost, and therefore offering products for higher prices so that profit is not eroded. The new data helped produce models for pricing of the various products at different quantities and timetables. Managers could use the models to price profitable contracts when existing ones expired.

Timely collection of money from customers is extremely important. Leiner had too much money tied up in disputes with customers who often required details on billing. Because Leiner accountants and salespeople did not have easy access to such data, collection often took up to three months. To solve the problem, management hired a third consulting firm. The consultants established a database and applications that replaced the manual process. Instead of handing a typed or written contract to the accounting people, salespeople now had to enter contract data into the database.

The database and applications enabled both sales and billing people to keep track of payments from invoicing to collection. There were no more paper orders. Every change in pricing or quantities ordered could be made only after the change was made to the cash management system. Whenever a customer asked to verify a bill, the software could immediately determine who originated the order and where, and then e-mail the salesperson the details, which the salesperson could forward to the customer. The customer then had all the necessary information to pay immediately. Within six weeks of installation, the software reduced the number of backlogged payments by 75 percent. The improved inventory and accounts receivable systems increased cash in the firm's coffers from $8 million to $20 million.

The IT makeover helped the company escape from bankruptcy. By mid-2002, output per employee increased 63 percent. Ninety-five percent of shipments were accurate and on time. Shipping costs decreased 15 percent. Accounts receivable were collected in fewer days than the industry's average. Inventory turnover is up from 2.5 to 4 times per year. After losses in 2000 and 2001, Leiner had a profit of $40 million in 2002 and $70 million in 2003. In 2006 and 2007, it continued to be profitable, at approximately $40 million per year despite mounting competition.

How did the experience impact executives' own behavior? Leiner's CEO now has a monitor on his desk that shows continuously updated key financial information: working capital, accounts receivable, accounts payable, cash flow, and inventory. As one observer said, Leiner might see difficult times again, but at least management will know what is going wrong and what should be fixed.

Source: Rothfether, J., "How Leiner Health Cured Its IT Woes," *CIO Insight* (www.cioinsight.com), March 1, 2003; (www.leiner.com), May 2003; "Leiner Sustains Healthy Market Share with High-Volume Warehouse Management Solutions from Apriso," (www.apriso.com), May 2003; (www.leiner.com), May 2007.

Thinking About the Case

1. One of Leiner's executives likened the firm's situation in 2001 to an injured person, saying it was bleeding but didn't know from where. Explain this observation in business terms.

2. Was all the data required for better operations and decision making available within the company? Which data was not?

3. What information is required for fast collection of accounts receivable, and what data can it be derived from?

4. The title of this case is "Rescued by Data." Was the collection and organization of proper data alone enough to save the company? Explain.

5. How could the company use a data warehouse to improve operations?

© Bob Torrez/Getty Images

PART THREE

Web-Enabled Commerce

CASE III: IT FITS OUTFITS

Shari Steiner, the chief executive of It Fits Outfits, was calling to order the quarterly strategic planning meeting of her top managers. Today was a big day for her company. After four years, they would begin phase two of Shari's original mission plan: to expand operations to the Internet.

A New Concept in Teen Clothing

Shari was a young executive who'd studied fashion design and merchandising at the Fashion Institute of Technology in New York City. She started in the business by designing her own clothing line and selling it on commission to other chains. But she wasn't satisfied with that small niche and needed to accomplish more. No stores in which her fashions were sold had the image that she wanted her clothes to convey—they were part of a culture that she thought "preached or beseeched." Some stores used their brand names to establish fashions that were emerging from Paris, New York, and Milan. Other stores tried to court teens by playing loud music and catering to the latest fads.

Shari wanted to sidestep conformity and connect directly to teens, involving them in the design

process. So, she took the next step and opened her own store. She actively recruited teens from nearby high schools to work for her. She chose a location that was next door to a coffee shop that was popular with teens. She set up secured computer stations around the store. In the front, she put up a bulletin board where people could post messages. Touch screen monitors invited teens to vote on different issues each week. At first, weekly votes focused on the coolest band or the lamest movie, but soon Shari and her staff came up with more creative ideas. They would ask weekly questions such as, "If you could rename Everest High School, what name would you give it?" Then they would post the top 10 answers.

Every two months, she and her staff would organize an after-hours fashion show. Using the computer stations or a suggestion box, teens would submit any ideas or phrases that came to mind and Shari would design outfits that reflected their thoughts. The staff and other volunteers would organize the show. When someone dropped the expression "totally tubular," which had become popular at one of the local high schools, into the suggestion jar, Shari and her staff had a hoot coming up with ideas. Other suggestions, like "casual lace" and "understated," helped Shari create lines that sold well nationally. She called the design line "Teen Voices."

The Next Step

It Fits Outfits quickly became popular and received local and regional media attention. Within four years, she'd built a small chain of 24 stores concentrated in major urban centers in the East. Now it was time to launch phase two.

Shari looked up at Martin Tate. It would not be easy for him to step aside and let Adina Silverman take over some of his IT responsibilities. Martin had set up the computer system the chain currently used for its marketing, sales, and financial needs. He had contracted with a local graphic designer to set up the company's first Web site. He had trained local store managers. But Martin didn't have the expertise Shari needed for this second phase of her plan. Adina had helped manage the Web site of the most technologically innovative clothing chain in the country for the past three years. Adina would hire the staff, set up the standards, and oversee the creation of a Web site that would allow Shari to reach beyond local groups of teens to college campuses. Most of her original customers were now in college. She could reach out to them by establishing an online storefront based on the same principles of It Fits Outfits.

"As you all know, Adina is going to help us move into the college market by establishing an It Fits Outfits storefront that is entirely Internet-based. Let me start this meeting off by telling you what we are *not* going to do. We are not simply going to take our merchandise, policies, supply, and manufacturing structure and dump it on to a Web site. What we are going to do is take our original idea—customer participation in design—and create a new operation that will meet the needs of our loyal clientele as they leave home and head off to school. These customers aren't going to have as much time to shop, and most of them won't have as much

expendable income. Many are going from big cities into small college towns. Their needs are changing. How can we meet them? We're going to have to work together to figure this out. We'll start by setting up work groups."

Warehousing, Supply, and IT

"Martin, you're going to have to work with Jun and Adina to figure out how to avoid a warehousing nightmare. You're also going to need to focus on how we can use IT to tighten up our supply chain for our current operations."

Most of It Fits Outfits' manufacturers were located in China. Shari had met Jun Kaui at her favorite coffee shop when she was still in college. Before their first conversation, Shari had thought that he was shy and studious. One day when the coffee shop was crowded, he asked if she would share her table. From there, she found out that his father had owned and managed a number of textile mills in Asia before retiring. She also discovered that Jun was both smart and assertive. They remained close friends even after he had completed his business degree. When she shared her dream of opening her own store with him, Jun had made it possible by taking charge of manufacturing. He visited textile mills, set up contracts, and dealt with shipping and customs information.

Using the Web to Build Networks for the Future

"Suzanne, James, and Tony, you need to work with Adina to figure out the needs of our college-aged customers. I have a list of our former employees who've gone off to school. I want you to contact them and involve them in this process. Use the Internet—maybe start with an online survey—to gather their feedback.

Shari had planned to expand into college campuses from the start. Rather than losing touch with former employees as they left for college, she had deliberately kept in touch. It Fits Outfits' current Web site hosted a bulletin board where former employees could contact each other, and an area where they could request references from their former bosses. About twice a year, the company organized an It Fits Outfits online reunion that Shari herself attended. While only store managers might know staff from other offices, the reunion offered college students a chance to visit with old friends and make connections that might help them in the future. Now, it was time to utilize this network she had built up.

Shari had set a deadline of six months to get the online storefront up and operational. It was now February. She wanted it running smoothly by the first semester of the following school year.

"We're going to meet monthly until the launch, so make sure your schedules are updated on the intranet. During lunch, you can break up into your work groups and start talking details. And now Jun is going to discuss the next item on our agenda."

BUSINESS CHALLENGES

It Fits Outfits is facing some potential opportunities and problems. Most of its design and retail functions are directly tied to its information systems. So, the success of the company's ISs are central to its continued survival. Some of these issues are explored in the following chapters:

- In Chapter 8, *"The Web-Enabled Enterprise,"* you learn how businesses use the Internet to achieve strategic advantage and how It Fits Outfits can use the Internet to extend its reach and develop a college-aged clientele.

- In Chapter 9, *"Challenges of Global Information Systems,"* you learn how sharing electronic information and operations among companies and across international boundaries can bring tremendous efficiencies—and challenges—to operations such as It Fits Outfits.

© Bob Torrez/Getty Images

8

EIGHT

The Web-Enabled Enterprise

LEARNING OBJECTIVES

The Web continues to be the most exciting development in the field of information systems and telecommunications. The combination of advanced telecommunications technology and innovative software is revolutionizing the way people communicate, shop, make contracts and payments, educate, learn, and conduct business. Numerous companies throughout the world have been established thanks to the enabling power of the Web, and existing businesses have used the Web to extend their operations. Firms conduct business electronically with each other and directly with consumers, using a variety of business models. This chapter focuses on Web technologies and businesses on the Web.

When you finish this chapter, you will be able to:

- Describe how the Web and high-speed Internet connections are changing business operations.

- Explain the functionality of various Web technologies.

- Compare and contrast options for Web servers.

- Explain basic business-to-business and business-to-consumer practices on the Web.

- Explain the relationship between Web technologies and supply chain management.

- Give examples of features and services that successful business Web sites offer.

- Learn about online annoyances such as spam and adware, and how to protect against online identity theft.

IT FITS OUTFITS:
Setting Up Operations on the Internet

It Fits Outfits was holding their second quarterly strategic planning meeting of the year. It was one of the company's most important meetings to date. Martin Tate, the company's CIO, had hired an outside contractor to create the original Web site for It Fits Outfits. The Web site was little more than an advertisement, but things were changing. It Fits Outfits was reaching out to the college market, and the company needed to establish a site that was a fully functional online storefront.

Connecting to College Students

From the beginning, Shari Steiner, CEO of It Fits Outfits, had planned to launch an online storefront for college students. First, she had established a teen market through her concept of involving teens in the design process. Then she had carefully established a network of former employees who had gone off to college. Finally, she hired Adina Silverman away from the most innovative online clothing retailer. Shari knew that college students rarely had the time or opportunity to shop for clothes. With Adina on board, Shari felt confident that she could meet the needs of her former clientele as they went off to school.

Adina took charge of Web operations for the new It Fits Outfits online storefront. She had spent the past two months working with the directors of sales, marketing, and design to discover the needs of their college-bound clientele and figure out ways to meet those needs. They had e-mailed a link to an online survey to all former employees in college asking them for their ideas, advice, and experiences. They had conducted online weekly chats with former employees, former customers, and college students who were interested sharing their ideas. Now, Adina was certain she had a plan for an online store that would be as successful as the retail chain—or even more successful.

The Virtual Fitting Room

Adina looked around at the other managers. She was the newbie, but she knew that by the time the meeting was over and she had explained the basic functionality of the online store, she would have won the confidence of every member of the senior team.

"First, the survey told us what we already knew—that college students don't have time to shop—and often, if they are in a small college town, they don't have the opportunity. Second, we confirmed our suspicion that freshmen often need to make adjustments to their wardrobe. They often gain weight within the first month or two and go up a size. College dress is different from high school dress and, as we know, each crowd dresses differently. Many of our customers often mix with a different crowd when they reach college.

"To meet this need, we've established a prototype called the Virtual Fitting Room. Customers enter their height, weight, coloring, shape, and other measurements, and the program saves this information. Then the customer can go shopping, trying on different combinations of items and viewing the model from the front, sides, and back. They select the items that they want to purchase and enter them into a shopping cart. When they are ready, they check out. The Virtual Fitting Room comes complete with a virtual salesperson, who makes suggestions. For example, if a customer is trying out a skirt, the salesperson might suggest a series of shirts or shoes that would go nicely with the skirt. She might say, 'Need accessories? I've got something that will look fabulous on you!' We're trying to take this concept further than other online clothing retailers by virtually re-creating the buying experience—without the hassles of actually going to the store in person.

"We also have a number of former employees on campus who are willing to rejoin us and serve as customer sales representatives who customers can chat with in real time. So if customers run into any serious snags, they can talk to a real, live person."

Travel Blog and Fashion Chats

The Virtual Fitting Room would not be enough to provide It Fits Outfits with the advantage it needed over well-established clothing e-tailers. The operation would have to take the ethos of the retail business that targeted teens and somehow re-create it within the college environment. The idea behind It Fits Outfits was to channel teen voices and spirit into the store and the product. Adina's workgroup had asked former employees how they could do this. What was the best way for college students to participate in the design process?

Adina explained the answers her group had come up with. "The first thing that we discovered that kind of surprised us was that the college students were very interested in Shari's professional life. They wanted to know what shows and conferences she'd been to, how she got her start, and what were her future plans. A lot of our former employees spend part of their downtime looking over adult fashion magazines, and they're beginning to wonder about things like how the world works—including the fashion world. They're considering how they should adjust to professional attire and lifestyle as they move toward the careers of their choice. They want to know how they can take the part of them that is unique—the part that Teen Scene gave a voice to—and integrate it into the adult world.

"That's *our* role," Shari interrupted. "That's what we have to help them do!"

"Exactly," Adina continued. "So, we've established an area for you, Shari, to record a travel blog. Next week, you're going to Paris. You'll describe Paris, discuss the shows, and post photos.

You'll talk about what people are wearing on the streets, in the shops, and in the offices. You'll describe the personalities and the nuts and bolts of the industry."

Shari looked flabbergasted. "And since," Adina went on, "we all know that you have no time to do this and that you can't write to save your life, Hector from marketing will do the actual writing." The group laughed. "Hopefully, though, you'll agree to participate in prearranged chat sessions to discuss major events in the industry."

Campus Clothing E-Zine

"We'll also have an e-zine on the site. We're going to have students report on dressing fashions at their universities. We'll do a story a week and archive them. At first, we'll have our former employees write the stories, but we'll open it up right away to volunteers. Students can then respond—telling Shari what they liked about the styles at the university. At the end of the semester, we'll list the top 10 or 20 designs—and of course, allow students to purchase them.

"Since college is the time of life when people stay up all night talking philosophy, we'll also directly address issues related to conformity, modesty, and symbolism in dress. We'll ask students to submit personal stories and then set up discussion groups to talk about the personal story and the philosophical issue behind it. We're hoping to design products connected to the stories or ideas that are shared in this forum.

"We've been brainstorming other ideas. Each week, we might hold a competition where students vote for the best dressed physicist or the best dressed park ranger. Not everything is squared away yet, and we're open to all ideas and feedback. So why don't you all jump in with questions and comments."

Adina looked around the table. She could feel the excitement of the senior team. They knew the company was on the verge of another great enterprise.

WEB BUSINESS: GROWING AND CHANGING

DLA Piper LLP is the second largest law firm in the world. It employs 3,200 lawyers located in 24 countries and 63 offices throughout Asia, Europe, Africa, Australia, New Zealand, the Middle East, and the United States. Some of the firm's cases involve up to 300 lawyers from 40 different law firms. The attorneys needed a way to collaborate and share documents among themselves, with clients, and with attorneys from other law firms. DLA Piper decided to adopt eRoom, a Web-based collaboration and document-sharing tool offered by EMC Corp. Attorneys and clients can use the tool from their browsers anywhere in the world where they have access to the Internet. They can initiate projects, track project status, have direct access to data, and know which project team members are available at any time. They can also notify all parties involved of project and document updates, use project management tools, and drag and drop files from and to their own local computers. All databases, contracts, and other documentation are stored centrally. More than 3,000 DLA Piper attorneys, other attorneys, and client individuals use 1,500 eRooms to manage cases, contracts, and projects. Using eRoom, the firm has saved 15,000 labor hours annually, used and mailed significantly fewer paper documents, and reduced redundant work.

The Web has been a great enabler for conducting business within organizations, between organizations, and between organizations and consumers. Vanguard, one of the world's largest mutual fund management companies, receives over 80 percent of its new clients through the Web. Social networking on sites like Facebook and MySpace has exploded over the past few years, and content delivery of video clips and feature-length movies has boomed. The spread of broadband links, new ideas of Web use for commerce, and continued development of Web technologies help business on the Web to grow and change all the time.

WEB TECHNOLOGIES: A REVIEW

Several standards and technologies enable the Web to deliver rich information. The following is a review of some nonproprietary standards and technologies.

HTTP

In Chapter 6, "Business Networks and Telecommunications," you learned about protocols. The protocol used to transfer and download Web information is **Hypertext Transfer Protocol**, or **HTTP**. A secure version of the protocol for confidential transactions is **HTTPS (HTTP Secure)**. Under these protocols, each Web server is designated a **Uniform Resource Locator (URL)**, which is a unique address for a Web site. The address is the IP address assigned to the site, but in most cases the site also has a **domain name** made up from letters. The term "URL" also refers to the domain name. Domain names are used for convenience, because it is easier to remember domain names than IP addresses. Each Web page has its own URL, which contains the IP address or domain name of the site. Because the domain name must be unique, when an owner of a Web site reserves a domain name to be associated with an IP address, no other site can be associated with that domain name. Note that domain names often start with—but do not have to include—*www*.

The last part of a URL, such as the ".com" in www.pinzale.com, is the top-level domain (TLD). In addition to .com, .org, and .edu, many other TLDs can be requested for a domain name, some of which are reserved for certain types of organizations and professions, and some that are not. Country codes such as .ca for Canada or .uk for the United Kingdom can also serve as TLDs. The only organization that is authorized to approve new TLDs is the Internet Corporation for Assigned Names and Numbers (ICANN), a not-for-profit organization established specifically for this purpose. Usually, a Web site with any TLD can be viewed in the same way regardless of technology. However, in 2007 ICANN approved .mobi as a TLD and standard for mobile devices. Currently, it is the only TLD that requires the use of special software to access the domains.

While domain names consisting of catchy and meaningful words were considered prized assets, companies such as Amazon.com and Google have demonstrated that the name itself is worthless unless the service provided is excellent. Few people know what these site names mean (abundant like the Amazon rainforest; and googol, an impossibly large number), but everybody knows of these sites and the purpose of their business. New Internet companies do not spend as much energy seeking an attractive domain name as they did in the past.

POINT OF INTEREST

The Importance of a Domain Name

Suppose you mean to direct your browser to that famous video Web site and see this: "Universal Tube & Tollform Equipment Corporation specializes in buying and selling Used Tube Mills, …" Huh? You were looking for YouTube and got Utube. In August 2006 Utube sued YouTube for brand degradation and technology costs, because the company received 68 million visits from online video fans. The unintentional high traffic crashed Utube's site, making it unavailable to potential customers. In the lawsuit, the small company said: "Due to confusion in the minds of consumers, the spillover of nuisance traffic to Plaintiff's neighboring Web site at utube.com has destroyed the value of Plaintiff's trademark and Internet property, repeatedly caused the shutdown of Plaintiff's Web site, increased Plaintiff's Internet costs by thousands of dollars a month, and damaged the Plaintiff's good reputation. Plaintiff seeks preliminary and permanent Injunction, the transfer of the youtube.com domain to Plaintiff, damages, costs and attorneys' fees…" However, the company of 17 employees and $12 million in annual revenue is trying to make money through advertising. The top of its home pages lists links to sites that sell anything from dating and broadband services to casino games. Utube.com now receives about 120,000 unique visits per day.

HTML and XML

Hypertext Markup Language (HTML) is the most common programming language for creating Web pages and other information viewable in a Web browser. It determines the look and location of text, pictures, animations, and other elements on a Web page. Extensible Markup Language **(XML)** enables the creation of various types of data. It is most often used not for determining the *appearance* of visual elements on a Web page but to convey the *meaning* or content of the data. The World Wide Web Consortium (W3C), the organization responsible for Web standards, has combined the two markup languages HTML and XML into a standard called Extensible Hypertext Markup Language **(XHTML)**.

Every file displayed on the Web is coded with a markup language such as HTML or XML. Simply put, markup languages provide a system of standardized "tags" that format elements of a document, including text, graphics, and sound. Formatting includes opening and closing tags preceding and following a part of the document, such as <b> at the start of bold text, and </b> at the end of bold text. Some tags are marked to link to another page either at the same site or another site, and others create links to e-mail addresses. Browsers interpret HTML and XML tags and display the text in the fashion defined by the tags, or allow other software to pick up data from the page and process it or copy it into the proper place in a database.

As in HTML, tags are used in XML to mark data elements. However, XML tags define "what it is," as opposed to "how it looks." Figure 8.1 illustrates the difference between HTML and XML tags. XML tags can be used in the same page with HTML tags to indicate both what the data means (which is not visible to the user) and how each element should be displayed.

FIGURE 8.1
HTML and XML code: XML provides a method for describing or classifying data in a Web page.

Visible Web Page Content	HTML code	XML code
Reebok® Classic Ace Tennis Shoe was $56.00; Now $38.99 Soft leather tennis shoe. Lightweight EVA molded midsole. Rubber outsole. China.	<strong><font face="Verdana, Arial, Helvetica, sans-serif">Reebok® Classic Ace Tennis Shoe </font></strong> <strong>Was $56.00; Now $38.99 </strong><table width="100%" border="1"><tr><td>Soft leather tennis shoe. Lightweight EVA molded midsole. Rubber outsole. China.</td></tr></table>	<product type="shoes"> <name> Reebok® Classics Ace Tennis Shoe </name> <price>$38.99</price> <description> Soft leather tennis shoe. Lightweight EVA molded midsole. Rubber outsole. China. </description> </product?

Source: Succeeding with Technology, 2nd edition (Course Technology, 2007)

File Transfer

File Transfer Protocol (FTP) is a common way of transmitting files from one computer to another. Every time you download a file from a Web site or attach files to e-mail, you are using an FTP application. The file transmitted can be of any type: text, graphics, animation, or sound. FTP is embedded in browsers and therefore is "transparent" to the users. You can also use a separate FTP utility, with many available as shareware, to manage transmitting files.

Businesses use FTP to place files on a server for sharing among professionals. FTP is also useful for placing files on a server that hosts a Web site. It's also convenient for retrieving large files that might exceed an e-mail system's size limits. For example, authors can place large chapter and figure files in a folder on a server maintained by their publisher. Manufacturers often place full assembly and maintenance manuals or videos at their Web site so customers can download them any time.

FTP has already changed the way in which many software firms sell their products. Instead of spending millions of dollars copying new software on storage media, packaging it, and shipping it, developers simply post their software products on their Web sites and let buyers download them for a fee. Music lovers can use FTP to download music files.

RSS

Really Simple Syndication (or in a newer version, Rich Site Summary) (**RSS**) is a family of XML file formats that allows automatic downloads of content such as news, sports, or other information of particular interest to users. An RSS file is installed at Web sites to help users check updates to the site. When users subscribe to the RSS service of a site, the software communicates to their computers short descriptions of Web content along with the link to the site. Users can instruct the software to automatically transmit new or updated information to their own computers. This software is especially useful for news Web sites and sites that host blogs and podcasts (see the next sections). Subscribers to mass media sites such as newspapers and news services such as Reuters can receive the latest news without actively going to the site or receiving e-mail messages. At some Web sites you might see a button with the letters RSS or XML. If you click them, you can arrange for the site to automatically send your computer updates of designated information by topic.

Blogs

A **blog** (a contraction of "Web log") is a Web page that invites surfers to post opinions and artistic work as well as links to sites of interest. Blog sites focus on a topic or a set of related topics, and provide an easy way to post Web pages or update existing ones. Most blogs contain commentaries and humorous content. Users can simply click a button to open a window in which they type text, and click another button to post it. The text is added to the Web page either automatically or after a review by the blog's operators. Some blog sites simply let "bloggers" add comments on a topic, with the most recent comment appearing at the top, similar to the way online newsgroups work. Many companies have established blogs, and invite employees to use them for self-expression. The policy might encourage new ideas from which the company can benefit. Some, however, shun the idea, because management believes blogs are too informal and uncontrolled.

One interesting feature of some blogs is *trackback*. Trackback software notifies bloggers when their posts have been mentioned elsewhere on the Web, so they and their readers can extend the discussion beyond the original blog. Below each post there is a TrackBack button or similar option. When it is clicked, a new window pops up listing the sites mentioning the post.

The commercial potential of blogs has not escaped businesspeople. As traffic grows at some popular blogs, entrepreneurs have started selling advertising space at the sites. The old rule on the Web is still much in force: the greater the number of eyeballs, the greater the commercial potential of the site.

The importance of blogs to commercial organizations is primarily to find out what blog participants think and say about the organizations. Many organizations use special software that combs blogs for postings that mention the organizations' names. PR people then read the content and relay feedback to others in the organization as needed. For example, an anonymous blogger boasted that he could break Kryptonite bicycle locks with a pen. Within a week the posted item was mentioned in *The New York Times*, and Kryptonite recalled the locks. Some companies offer blog mining applications, which is software that combs blogs, identifies company names, and automatically tracks discussions. Such tools can turn blog data into useful market research information.

Wikis

Many Web sites invite visitors or subscribers not only to read, view, and listen to their content, but also participate in the site building and editing process. In the past, to do so would require access codes and at least some knowledge of Web editing software. Now, wikis make the process easy and fast.

A **wiki** (from Hawaiian: quick) is a Web application that enables users to add to and edit the content of Web pages. The term also refers to software that enables collaborative software used to create and revise Web sites. All the software required to edit the pages is embedded in the pages. Visitors do not need any software of their own, and do not need to upload saved pages. The additions and revisions are performed on the page, using tool icons that are provided at the site. The popular online encyclopedia Wikipedia demonstrates the concept well. For example, if you enter the term "Internet," you will notice that at the end of each section you can click on "[edit]." When you click an edit link, a new window opens, displaying both the text and a set of tool icons to help you edit the text. Except for some protected entries, anyone with Web access can participate in improving Wikipedia.

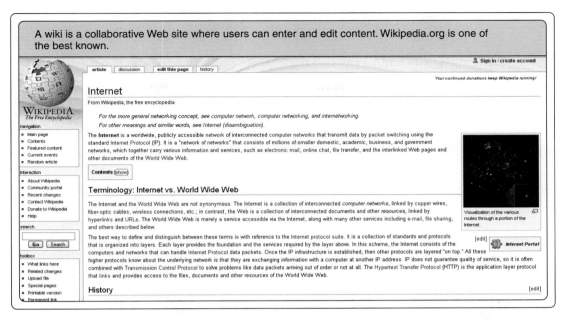

A wiki is a collaborative Web site where users can enter and edit content. Wikipedia.org is one of the best known.

Popular wiki applications include the free MediaWiki, Corendal Wiki, Clearspace, Yawwiki, XWiki, and VQ Wiki. The WikiMedia Foundation (*www.wikimedia.org*) provides information and links to sites that teach how to install and use wikis. The features of wiki technology make it a popular ingredient in groupware, software that helps groups collaborate on projects.

Podcasting

While blogging is publishing text and other visual material, podcasting is publishing sound and video. To **podcast** is to make a digital audio recording, usually of voice, and post the file on the Web so that people can download it and listen to it. RSS software called an *aggregator* or *feed reader* automatically checks for new content and downloads files from a designated site in the same way as is done for text files from online newspapers. Similarly to subscribing to such newspapers, users can subscribe to a podcast site to receive the latest audio files. The files are usually in MP3 format, which can be played on any portable player, including Apple Computer's iPod, from which the word "podcast" was born. However, one does not need this specific MP3 player to enjoy podcasts.

Podcasting has several potential uses. It already serves as "time-shifted" broadcast of radio stations that post their programs for later listening. It is used by some museums for audio tours. Some schools have experimented with the concept to deliver lessons to remote students or to post recordings of lessons for students to review. Whatever the use, people can listen to their favorite content wherever they can obtain a link to the Internet, without paying radio license fees.

Podcasting opens business opportunities. For example, *garageband.com* is a Web site that invites aspiring musicians to post their music tracks free of charge so they can be podcast. This exposes to the world talented people who could not otherwise afford to broadcast their work. Podcasting does more than post MP3 files for downloading. By allowing computers to automatically check for new music tracks, the method helps create a following for an artist, which might result in a future fan base willing to pay for a CD, concert, or downloaded music files.

Instant Messaging

Instant messaging (IM) offers users real-time online interactivity. It might be thought of as "real-time e-mail," because, unlike e-mail, it is synchronous. IM allows a user to detect whether another person who uses the service is currently online, and the user can then exchange information with an entire group (referred to as a "chat room"), or with only one other "chatter" in privacy. Some IM applications include two-way video, which turns the chat into a video-conference, and most also include FTP to allow sending and receiving files.

Many organizations have added instant messaging to the contact options listed at their Web sites.

REALTOR.org

ories | Need Help?

Account | NRDS

Tools

» Instant Help Online
(online chat, for registered members)

Chat with us Online
click here

» Schedule a time for us to call you
(for registered members)

» E-mail:
InfoCentral@realtors.org

» Fax us at
312/329-5960

» Voice-enabled online chat via Net Meeting
(Voice over the Internet, for registered members)

Free IM applications are operated through a server, or a group of connected servers, which provides a directory and functions as the hub for all callers. Some IM setups, such as AOL Instant Messenger (AIM), Yahoo! Messenger, MSN Messenger, and ICQ, have become the electronic meeting places for millions of people, making them an attractive target for online advertisers. To overcome the need to use multiple IM applications, some software developers produced universal IM applications that allow, for example, an AIM user to chat with an MSN Messenger user. Trillian and Pidgin are two of these applications. Meebo, at *www.meebo.com*, enables people to use its own and the four most popular IM applications from the Web site. Like Web-based e-mail, this is Web-based IM.

While IM serves social purposes, it also can serve an important business purpose. Many online retailers post a special button on their Web pages that lets shoppers establish real-time communication with a sales representative. This instant access fosters more personal service and saves telephone costs. For example, Venus Swimwear, a company that specializes in direct mail junior bathing suits, uses InstantService, a chat application that enables employees to answer customer questions in real time online. Venus's director of e-commerce marketing added this option to the three sites operated by the company because customers often abandoned the site when they could not get answers while shopping. Using the telephone was a bad option for those who used their telephone line for a dial-up Internet connection, and e-mail was inefficient and time-consuming. IM enables sales agents to handle up to five inquiring customers at a time. The application also enables the company to "push" answers from a library of answers such as a sizing chart, instead of typing them. The live chat reduced the amount of e-mail employees have to handle and decreased the customer abandonment rate by 15 percent despite the increasing traffic at the sites. Many businesses have added "Chat Online" options to the mail, telephone, and e-mail contact information listed on their Web sites.

Cookies

If you have ever surfed the Web, your computer probably contains cookies. A cookie is a small file that a Web site places on a visitor's hard disk so that the Web site can remember something about the surfer later. Typically, a cookie records the surfer's ID or some other unique identifier. Combined with data collected from previous visits, the site can figure out the visitor's preferences. The user can opt to allow cookies; the option is exercised by checking a box in the browser's configuration window. On the user's hard disk, the cookie subdirectory (folder) contains a cookie file for each cookie-using Web site that the surfer has visited. Cookies might hold server URLs. When you instruct the browser to reach a URL from which you have a cookie, the browser transmits the information from the cookie to the server.

Cookies have an important function in Web-based commerce, especially between businesses and consumers. They provide convenience to consumers; if the cookie contains your username and password for accessing a certain resource at the site (e.g., your bank account), you do not have to reenter the information. Cookies often help ensure that a user does not receive the same unsolicited information multiple times. For example, cookies are commonly used to rotate banner ads that a site sends so that a surfer receives different advertisements in a series of requested pages. They also help sites to customize other elements for customers. For example, when a retailer's site identifies a returning customer, it can build a page showing a list of items and information in which the customer might be interested based on previous purchases.

Some cookies are temporary; they are installed only for one session, and are removed when the user leaves the site. Others are persistent and stay on the hard disk unless the user deletes them. Many cookies are installed to serve only first parties, which are the businesses with which the user interacts directly. Others serve third parties, which are organizations that collect information about the user whenever the user visits a site that subscribes to the service of these organizations. These organizations include DoubleClick, ValueClick, and Avenue A.

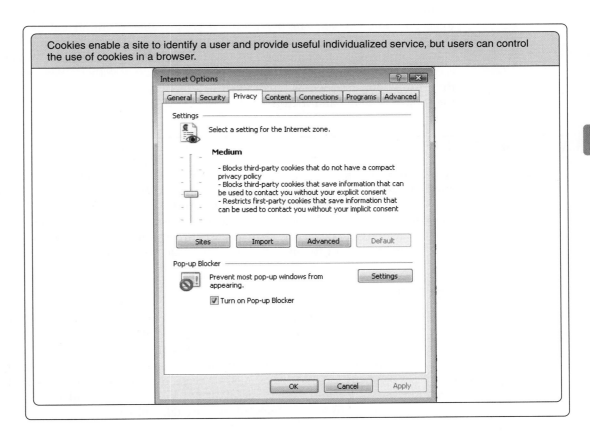

Cookies enable a site to identify a user and provide useful individualized service, but users can control the use of cookies in a browser.

While cookies can make online shopping, investing, and reading more convenient, they also open the door for intrusion into a person's privacy. Remember that every piece of information you provide while your browser is configured to permit cookies can be recorded and kept for further use—use over which you have no control. Choices you make when selecting menu items, clicking buttons, and moving from one page to another are also recorded. Such activities are called **clickstream tracking**. Although some organizations post privacy policies at their Web sites and tell you what they will or will not do with the information they gather, you cannot see what information they have compiled by using cookies and how they use it. Especially worrisome are third-party cookies, which collect your browsing and shopping habits across many Web sites. This is akin to a spy who follows you from one store to another. Software designed to trace and report your online behavior without your knowledge is called **spyware**. It includes cookies and other, more sophisticated applications that are installed on your computer unbeknownst to you and transmit information about you while you are online.

It seems that the public is gradually becoming more aware of cookies and taking steps to control them. In March 2007, Jupiter Research disclosed the results of a study showing that 58 percent of users delete their cookies regularly, and 40 percent delete them monthly. However, the great majority of cookies deleted are those installed by third parties. Only 1 percent of Web visitors delete first-party cookies.

Proprietary Technologies

In addition to these and other widely used and usually free Web technologies, many companies offer proprietary technologies. A proprietary technology is the intellectual property of its developer and is not free for all to use. These software packages include local search engines for finding information about specific items; shopping cart applications for purchase, including selection of items to place in a virtual cart and credit-card charging; wish lists, which allow shoppers to create lists of items they would like others to purchase for them; video streaming tools; and a host of software packages that are invisible to visitors but help the site owner to analyze and predict visitor behavior, especially shopper behavior. The latter technologies might not be considered Web technologies per se, but they analyze data that is collected from visitors accessing Web sites. For example, Amazon.com uses software that follows the estimated age of those for whom a shopper purchases items, and offers new items that fit the progressing age of the shopper's family and friends.

WEB-ENABLED BUSINESS

Web-enabled business is often classified by the parties involved in the interaction: business-to-business (**B2B**) and business-to-consumer (**B2C**). Some people also add government-to-consumer and government-to-business. Auction sites are sometimes referred to as consumer-to-consumer (C2C) sites, but we consider them B2C, because the business does intervene in several parts of the transaction and also charges the parties commissions. The following sections describe the business models of the most pervasive types of Web-based business.

B2B Trading

Business-to-business (B2B) trading takes place only between businesses. Consumers of the final goods and services are not involved. In general, the volume of e-commerce between businesses is about ten times as great as that of business-to-consumer e-commerce. And although not all electronic B2B transactions take place on the Internet, most do. Estimates vary, but by 2007, businesses conducted transactions valued at several trillion dollars annually. In the United States, it is estimated that about 40 percent of all B2B commerce is conducted online. This section discusses the major forms in which this activity takes place.

Advertising

Online advertising is done mainly in two ways: through search engines and through banners. Although advertising on the Web is not just aimed at consumers, most of it is directed to them. However, selling and buying Web ads occurs between businesses: the Web site operators sell advertising "real estate" to another business. Regardless of media, advertisers are interested in reaching as many people who might buy their goods or services as possible. On the Web, advertisers are interested in what they call "traffic volume," that is, the number of people who come across their messages. As the number of people who log on to the Web increases, so does advertiser interest in this medium. Internet World Stats (*www.internetworldstats.com*) estimated that of the 6.6 billion world population, 1.1 billion—17 percent—used the Internet in 2006. With this traffic volume, advertisers are willing to spend a lot of money on Web advertising. About $19 billion was spent on Web advertising in 2006, and $20 billion was expected to be spent in 2007. The research firm Piper Jaffray & Co. predicts that by 2011 the spending will grow to $81 billion, a sum greater than the $80 billion spent annually on television advertising.

Search advertising, which is any form of advertising through an online search site, is regarded by businesses as highly effective. Shoppers have discovered that the fastest way to find a business that can sell them the product or service they need is by looking up the product or service on the Web, and the most effective searches are through the best-known services and those that identify the largest number of Web pages: Google, Yahoo!, MSN, and AOL (see Figure 8.2). All of these sites have the same advertising patterns. Whenever you search for an item, the top and right-side links are "sponsored," that is, paid for by advertisers.

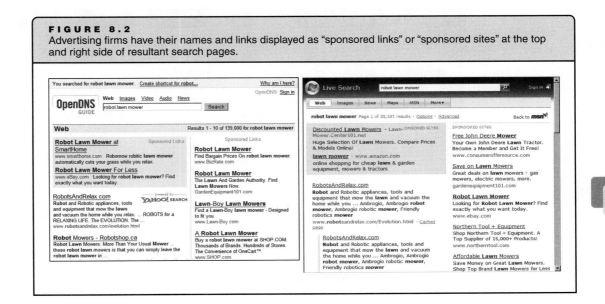

FIGURE 8.2
Advertising firms have their names and links displayed as "sponsored links" or "sponsored sites" at the top and right side of resultant search pages.

Banners are images placed on a Web site that link to the site of the company selling the product or service. In the early days of the Web, the ads were shaped like a banner at the top of the page. However, any image placed for advertising is now referred to as a banner.

How does a potential advertiser know how much traffic a site attracts? The most basic metric that can be measured at a site is the number of impressions. An **impression** occurs whenever a browser downloads the page containing the banner. More useful metrics are provided by several companies that rate Web site visits similar to rating television viewing. For instance, comScore, an online rating firm, maintains a panel of more than 2 million English-speaking online consumers whom the firm polls periodically. The companies produce several metrics for subscribing companies. Usually, subscribers of online rating firms are high-traffic sites that generate large revenues from advertising.

In addition to impressions, rating companies measure other metrics. One is *unique visitors per month*. If the same person visited a site several times during a given month, the person is counted only once. The reason? Advertisers are interested in reaching many people, not the same people with many visits. Another metric is *unique visitor pages*, which is the number of different pages at the site that a single visitor accessed. The reason for this metric is that the same visitor is exposed to different ads at the site. **Reach percentage** is the percentage of Web users who have visited the site in the past month, namely, the ratio of visitors to total Web population.

As in print advertising, the site owner charges the advertiser by the ad's size (so many pixels by so many pixels) and by the amount of traffic, usually measured in number of impressions. Using the IP addresss of computers accessing the pages with the banners, the advertiser can easily count impressions. The charge is per month. Placing a banner at a heavily trafficked site may cost hundreds of thousands of dollars.

To find out which sites engage visitors more than others, advertisers count on the ranking of two main competing measurement firms: Nielsen//NetRanking and comScore. Prior to July 2007, both companies ranked sites by the number of pages viewed by visitors. In July 2007, Nielsen//NetRanking decided to rank sites by the amount of time visitors spend at sites.

It is said that word of mouth might be the most effective advertising, and it comes at no cost. Free advertising on the Web often takes the form of multiple mentions on blogs. Keen, the seller of athletic sandals, went from idea to market in two months. The four men who started the business spent almost no money on advertising. However, within months of the product launch, some of the most trafficked Web sites and outdoor-gear blogs praised the shoes. The free advertising helped. In 2004, the first year in which the sandals were on sale, Keen sold 700,000 pairs for $30 million.

POINT OF INTEREST

No Sales Tax in Cyberland

If an online seller does not have a physical presence in your state, it does not have a legal obligation to collect sales tax. If you live in a state that collects sales tax, you ought to pay a use tax of the same rate in lieu of the sales tax that was not collected by the seller. However, many Americans purchase online to avoid paying sales tax. State and local governments estimated that they would lose $33.7 billion in sales tax revenue in 2008 as result of online shopping. Even conservative estimates peg the loss at $5 billion.

Source: Thompson, C., "Should I be paying taxes on my online purchases?" *Wired*, March 2007, p. 46.

Exchanges and Auctions

In the old days, a meeting place of buyers and sellers had to be tangible: a marketplace, an annual fair, or a store building. Finding a buyer for scrap metal, used scientific equipment, or any other commodity might have taken a long time. Also, the buyer and seller had to pay high finder's fees to individuals and firms that specialized in such intermediary trade. On the Web, the marketplace can include as many sellers and as many buyers as wish to participate, as long as they have access to the Internet.

An **intranet** is a network used only by the employees of an organization. An **extranet** limits site access to the employees of particular organizations, usually business partners. An extranet might be viewed as connecting intranets of business partners.

An exchange is an extranet for organizations that offer for sale and bid on products and services of a particular type. Unlike a public auction site, such as eBay or uBid, access is usually limited to subscribers who often pay a periodic fee to the site's operator. Auction sites whose purpose is to serve as a meeting place of buyers and sellers in a particular industry are sometimes operated by an industrial association. Others, like askart.com, are established by entrepreneurs for the sole purpose of making profit. When the purpose is only to provide a place where sellers compete for the business of a single buyer, the buyer operates the site.

When the site is established by a private business as a meeting place for multiple buyers and sellers, the operator is impartial and profits from transaction fees paid either by one party or both—the seller and buyer—whenever a sales transaction is signed. One of the largest of these exchange sites is ChemConnect, where sellers can auction off chemicals, plastics, oil, natural gas, and renewable energy. ChemConnect has 9,000 corporate members in more than 150 countries. The president of a petroleum company who makes 15 percent of his natural gas purchases and sales at the site summarized the advantage of such exchanges to businesspeople. In the past he had to spend a day and use the phone to find buyers or sellers for a single purchase or sale. Now, he posts the information online and 150 interested people see it.

Auction sites sell a great variety of items, including live ones. PEFA.com, in Zeebrugge, Belgium, is a private auction site for fresh fish. More than 500 buyers, usually large companies from all over Europe, use PEFA.com to purchase 60 different species of fish—fresh, farmed, processed, and frozen—worth several hundred million dollars from sellers in 18 ports. The site accommodates buyers and sellers in seven languages. It also allows subscribers to use PEFA's database, so they can receive rich information on market conditions and statistics. The database is available in 11 languages.

Some electronic markets are established by a single buyer or by an organization that represents many buyers. For example, ChoiceBuys.com is a site operated by Choice Hotels International, a company that franchises the hotels Comfort Inn, Comfort Suites, Quality, Clarion, Sleep Inn, Econo Lodge, Rodeway Inn, Cambria Suites, Suburban Extended Stay Hotel, and MainStay Suites. The company franchises over 5,200 establishments in the United States and 48 other countries. In 1999, the company established the Web site so it could concentrate all purchases for the hotels through a single channel. The site invites sellers to offer their products. For the sellers, this is an opportunity to obtain big contracts. For the hotels, this is a way to enjoy substantial discounts, which ChoiceBuys.com obtains through its buying power. For Choice Hotels, it is a way to generate revenue from the transaction fees that the bidders pay, and indirectly a way to attract more franchisees that, of course, pay the company franchise fees. Independent hotels are allowed to make their own purchases, and they do so for a total of $1.5 billion annually, but an increasing number of purchases are made through the site. In 2004, ChoiceBuys.com processed 5,236 orders for toilet paper, towels, and other supplies for more than 1,000 hotels in its franchise. Processing orders through the site not only saved money for the hotel operators but also enriched Choice Hotels. Since all transactions are electronic, they are automatically recorded and provide valuable data from which useful information can be gleaned. As of June 2006, ChoiceBuys.com still served as a hub for buyers and sellers, but stopped processing orders.

Many exchanges require businesses to register as members and to pay an annual fee. Many guarantee sellers that they will receive payment even if the buyer defaults, an important and attractive consideration for sellers. Electronic marketplaces bring markets closer to what economists call perfect markets. In a perfect market, no single buyer or seller can affect the price of a product. On the Internet, all buyers and sellers have access to the same information at the same time. Thus, no single buyer or seller has an information advantage over competitors.

Online Business Alliances

Companies in the same industry—competitors—often collaborate in establishing a Web site for one or several purposes. One major purpose might be to create buying power by consolidating purchases. Another might be to create a single place for customers, assuming that expanded choice will benefit the group. The concept is not new: real estate agents have collaborated in the Multiple Listing Service (MLS), whereby multiple agencies have access to real estate that is registered for sale with one of them. This system, which was in place many years before the Web, is now managed on the Web.

Interestingly, real estate agents are now under competitive pressure from operators of Web sites. Sellers who wish to reduce the traditional 6 percent agent commission increasingly turn to online real estate markets. Many sellers are willing to do much of the work done by agents, and list their property with sites such as Redfin.com and ziprealty.com instead of with an agent who is a member of the National Association of Realtors® (NAR), the organization that operates MLS. Many buyers prefer to search for houses on their own. They can use a site such as Zillow.com to estimate house prices in an area of interest. They may save thousands of dollars when purchasing from a seller who pays a low flat fee instead of a high percentage commission to a real estate agent. In other words, the Web is removing the expensive middleperson in this industry, too. It is no wonder that NAR tried to limit the access of such sites to MLS.

In some cases, the purpose of an alliance site is the same as an auction site operated by a single company, but the operator is a business that works for the allied companies. The purpose of such a site is to set the prices of purchased products and services. Big players in an industry, such as airlines or automakers, establish a shared company that operates the site. Suppliers are invited to sell through the site and compete among themselves. The competition drives prices down. The allies may enjoy lower costs and greater profit margins. A grand attempt by the auto industry to use this method through an alliance called Covisint failed because suppliers refused to compete against each other online. The Covisint software was sold to Compuware, a software development company in Detroit. However, Star Alliance, an alliance of 16 airlines, has done well with its online joint purchasing.

Star Alliance is one of several airline alliances, such as OneWorld, SkyTeam, and Orbitz. Star Alliance established an extranet for two purposes: to concentrate purchases from parts and service providers, and to represent the group to its clients, airline passengers. The alliance includes Air Canada, Air New Zealand, Austrian Airlines, Lufthansa, Scandinavian Airlines, United, Varig, and other companies. On the consumer side, the airlines collaborate in frequent flier programs: you can fly with any of them and accrue miles with the entire alliance rather than with a single airline. The Star Alliance site provides several useful services for travelers of all member airlines. On the B2B side, the alliance solicits bids from suppliers of aircraft parts and maintenance services, food, ground equipment, office supplies, and other products and services. The allies use the extranet to share information about inventory levels, facilitate joint planning and forecasting for material requirements, and facilitate communication and business transactions between the airlines and suppliers. The hub for joint purchasing has saved the allies millions of dollars annually.

Orbitz was established by United Airlines, Delta Air Lines, Continental Airlines, Northwest Airlines, and American Airlines to serve customers through a single Web site. The allies wanted to establish a site that "would provide comprehensive and unbiased travel information" as well as one that would make planning and buying travel on the Internet easy and hassle free instead of the usual "scavenger hunt." The site provides a comprehensive search engine, a list of flights by lowest price and number of stops without favoring any airline, and special Internet-only fares with over 455 airlines. By using their own Web site, airlines can stop paying the $5 to $10 commission they usually pay to online brokers such as Expedia, Travelocity, and Priceline.com. They can also eliminate fees to online database companies that link travel agencies, airlines, and other travel companies. In addition to the airlines, the site serves 22 rental car companies, cruise lines, and tens of thousands of lodging establishments.

Similar alliance sites have been established by firms in the general retail industry, the food industry, and the hospitality industry (hotels).

B2C Trading

Although business-to-business trading on the Internet is much larger in volume, online business-to-consumer (B2C) trading is more visible to the general public. Online consumer shopping and buying has become a daily activity for many consumers, much like mall shopping and buying. As Figure 8.3 indicates, in 2006, in the United States alone consumers spent $219.9 billion online, a growth of 25 percent from the previous year. European Union consumers spent $174 billion.

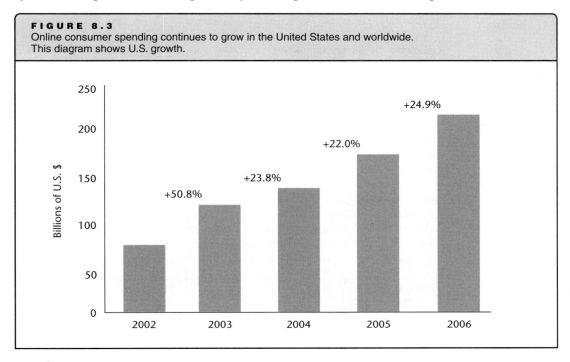

FIGURE 8.3
Online consumer spending continues to grow in the United States and worldwide. This diagram shows U.S. growth.

E-Tailing

You can shop the Web for virtually any item you want, from collectibles to automobiles, order these items online, and receive them at your door. In 2005, the scope of online retail was $88 billion. By some forecasts it is predicted to grow to $144 billion by 2010. At various retail sites, shoppers can use sophisticated search applications to find information on desired items, read other shoppers' reviews of items such as books and music CDs, post their own opinions, drop an item in a virtual shopping cart, change their minds and remove the item, or decide to buy the item and pay by providing a credit-card account number or through a debit service such as PayPal. Online retailing—*e-tailing*—continues to grow throughout the world for several reasons: greater availability of faster communication lines to households, growing confidence in online purchases, and the increasing ability to find the item one searches for and rich information about it.

Amazon.com, Buy.com, and the e-tailing arms of brick-and-mortar retail chains such as Wal-Mart and Target are among thousands of e-tailers. Their sites combine user-friendly and enticing Web pages, database management systems, and transaction software to provide shopping and buying convenience. Companies that sell online have smaller expenses than those selling from stores because they do not have to buy or rent store buildings and use labor to operate cash registers. They do, however, maintain large warehouses and pay for picking, packing, and shipping, three activities known as **fulfillment**. In some cases, however, they do not even need to maintain their entire inventory because they can simply route orders directly to manufacturers, who ship the items directly to the customer.

An important element of online retailing is selection. Compare the variety of products that can be offered at a Web storefront with the selection offered by a brick-and-mortar store: a typical CD store offers about 25,000 different titles, while Amazon offers 500,000 titles, which is the total number of CD titles offered for sale in the United States. E-tailers also experience fewer returns because shoppers have more information at hand before they make their purchases. So, shoppers are more satisfied with their buying decisions, and Web businesses save the costs of dealing with returns. Another explanation for fewer returns is the creation of what marketers call one-to-one relationships. Using cookies, the Web site software can track the clickstreams of

POINT OF INTEREST

Enabling the Disabled

Much of the information appearing on Web pages is visual. Visually impaired users do not just miss the beauty of pictures and colors. When they visit a commercial Web site, their ability to get information may be severely reduced because access to information and activation of transactions are often done by clicking pictures and icons. One way to help the visually impaired is to attach text to every image when composing Web pages. The user can then depress the Alt key to invoke text-to-voice software that can read the information aloud. In 2006, a landmark lawsuit by the National Federation of the Blind (NFB) against Target allowed disabled people to sue companies that do not provide the tools required to enable disabled people to use their Web sites as effectively as other people.

shoppers who browse the site and track their interests. When the same shopper logs on again, the software offers new items that fall within his or her interest area. And with every visit, the software learns more about visitors and their preferences.

As Figure 8.4 summarizes, consumers find several advantages to shopping online: convenience, time-savings, wide selection, search mechanisms, comparative shopping, and product reviews. E-tailers face mounting challenges, however. The competition is just a click away, so it is critical to offer a wide selection and excellent service, in addition to low price. E-tailers provide easy-to-use online tools to track shipments, and many make returns easy. Some brick-and-mortar retail chains allow purchases made on their Web sites to be returned to physical stores.

FIGURE 8.4
Web-shopping benefits

Benefit	Because shoppers can...
Convenience	Shop from anywhere at any time of the day.
Time-saver	Visit numerous online stores in a few minutes; it would take hours to do so at shopping malls.
Wide selection	Choose from a selection of products that no shopping mall can offer.
Search mechanism	Find who sells a specific item within seconds by using search engines.
Comparative shopping	Quickly compare quality and price across multiple sellers.
Product reviews	Read product reviews by independent experts and other shoppers, often on the same Web page that describes the item.

The greatest challenge for e-tailers is turning shoppers into buyers, and then turning the buyers into repeat buyers. To this end, many e-tailers have linked their customer relationship management (CRM) applications, discussed in Chapter 3, "Business Functions and Supply Chains," to their Web sites. Using cookies and other software, they not only collect large amounts of information about individual shoppers but also constantly update their profiles. The purpose of **consumer profiling** is to know the consumers better so the business can serve them better, while also streamlining its marketing and sales operations.

While consumer profiling might sound benign, many privacy advocates claim that it violates privacy rights. Imagine that every time you log on to a site, this fact is recorded. Then, when you click on an icon, it is recorded, too. The software at the server side also records the time you spent at each specific page, assuming the longer you spent, the more interested you are. Chances are the site will send you promotional e-mail about the items displayed on that carefully viewed page. Also, the next time you log on to the site, you might find that this particular page appears on your computer monitor faster than before. These subtle changes result from intelligent analysis of the information you provided knowingly or unknowingly to the site—and perhaps other sites that forwarded the information to this retailer. Many retailers also sell the information they collect to data brokers such as ChoicePoint and Acxiom, who combine personal data and sell the records to other companies. Privacy advocates object to such observation and sales of data without the user's consent.

Affiliate Programs Many online businesses offer affiliate programs to Web site owners. The affiliate, the Web site owner, places a link, usually a banner, to the e-tailer at the site. Affiliates are compensated in one of several ways: *pay per sale*, in which only if a visitor ended up purchasing something is the affiliate paid a fee; *pay per click*, in which the affiliate is paid a small fee (usually a few cents) whenever a visitor clicks the banner; or *pay per lead*, whereby a lead means that the visitor clicked through to the advertiser's site and filled out a registration form to receive periodic information. Retailers usually use the pay per sale model.

Some e-tailers have hundreds or even thousands of affiliates. Amazon.com and other large e-tailers probably have tens of thousands of affiliates. These programs provide huge, effective advertising for online businesses.

Some companies make money by being affiliate aggregators. LinkShare (www.linkshare.com) and Commission Junction (www.cj.com) let you choose from hundreds of affiliate advertisers, some offering commissions of up to 40 percent. You can pick the ones you want to mention at your Web site.

Coopetition Amazon.com has taken a step beyond affiliate programs to cooperate with competitors, a model we may call "coopetition": Amazon includes its competitors on its own site. When you use the search engine at the company's site for a certain item, it brings up the product description and price from Amazon.com's database and also the same type of information from other companies' databases. Although these companies are direct competitors, Amazon benefits from this cooperation in two ways: it attracts more shoppers to visit its site first, because they know there is a high probability that they will find the item they want at the site, even if they end up buying from another company; and it receives a fee from these affiliated companies whenever they sell through Amazon's site. Other e-tailers also give Web presence to competitors.

Auctions and Reverse Auctions

Similar to auctions among companies, some Web sites serve as auction hubs for individuals. The most prominent of the sites is eBay, but there are others, such as uBid and AAANDS. The business model is simple: sellers list information about the items or services they offer for sale, and if a sale is executed, the site owner collects a fee. Because the sites provide only a platform for a transaction that eventually takes place between two consumers, some people like to call online auctions consumer-to-consumer e-business. To participate in auctions, one needs to register as a member. To help bidders know to what extent they can count on the sellers' integrity, eBay publishes the number of feedback comments it received on a member, and the number and percentage of positive feedbacks. (However, eBay's feedback may be abused. For example, a seller may earn positive feedback on many small sales, then use the high rating to cheat buyers on large sales.)

The ability of Web sites to serve as prompt exchanges of information has supported another popular business model, the **reverse auction** or **name-your-own-price auction**. Consumers at Priceline.com are invited to post requests for services and name the prices they are willing to pay. Although they also deal in home mortgages, the services are mostly for travel, such as flights, cruises, lodging, and car rentals. Customers can post the destination, day, and time of a flight as well as the maximum price they are willing to pay. Then, airlines are invited to consider the requests. The first airline to accept the terms wins the offer. Shoppers are required to transmit a credit-card account number. The account is charged as soon as an airline accepts the deal. Priceline's revenue comes from the fees that airlines and other businesses pay to use the service.

Content Providers

On the Web, content means information, such as news, research results, statistics, and other useful information as well as artistic works such as music, pictures, and video clips. Some put in this category classified ads, including job postings and online dating services. Over the years, individuals and organizations have spent increasing amounts of money on content. Although most news can be obtained free of charge, many articles cannot. Some audiences welcome for-fee content, especially if it is highly specialized. Given a choice, many people prefer to read the same information online rather than on paper because they can use search operations to quickly find specific articles. This might be one reason why many prefer to subscribe to the electronic version of a newspaper. Content revenues also have grown since companies such as Apple and Wal-Mart started selling individual song files online.

In recent years, video has become a popular type of content. Several sites enable anyone to upload video clips. Some of the sites, such as YouTube, have attracted millions of people daily. The high volume of traffic translates into dollars through advertisements, as the site owners provide the content free of charge.

Bill Presentment and Payment

Because it is so easy to transfer funds online from one bank account to another, and it is so easy to send information, including bills, by e-mail, many utility companies try to convince customers to accept electronic bills and pay them online. Some customers accept the option of electronic bill presentment but refuse to sign an agreement that would enable the company to automatically charge their bank account. Obviously, banks are always a participant in electronic payment if the charge is to a bank account (which is how most utility and mortgage companies want to be paid), but some banks, for their own reasons, refuse to join such trilateral initiatives.

FIGURE 8.5
Phishing plagues Web commerce. An e-mail arrives (below) that prompts the recipient to update personal information at a fictitious but legitimate-looking Web site (left image on the next page). Note how similar the features in the fictitious (left) and legitimate Web site (right) are.

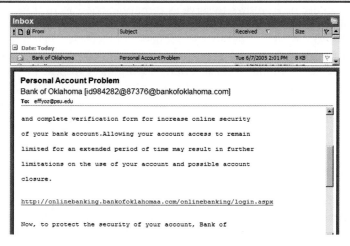

Electronic bill presentment and payment (EBPP) saves utility companies and financial institutions that bill customers regularly—mainly for loan payments—millions of dollars. The bills are presented automatically, directly from the companies' information systems to payers' e-mail addresses, and therefore save labor, paper, and postage. Direct charge to a bank account saves the labor involved in receiving and depositing checks. Yet, EBPP is spreading slowly. Most people still prefer to pay their bills by check and through the mail, partly because fraud on the Internet has increased in recent years, especially through a practice called **phishing**, discussed in the Ethical & Societal Issues feature. Figure 8.5 illustrates the practice. Note that the fictitious site is not secure. The legitimate one is secure; note the "s" in https://. Also note the bank warnings to customers. This particular e-mail was sent to a person who does not even have an account with the bank.

Despite the public's growing awareness, phishing sites grew from fewer than 5,000 in November 2005 to over 37,000 in November 2006. The schemes are increasingly sophisticated.

Extra-Organizational Workforce

The Web enables companies to purchase labor from many more people than their own employees. For example, companies can augment their intellectual pool by using the Web to employ talent beyond their own employees. They can enjoy more labor for less money by offering cash for research and development (R&D) solutions provided by researchers outside their organizations.

InnoCentive, Inc. is a subsidiary of the pharmaceutical company Eli Lilly and Company. It operates InnoCentive.com, a Web site connecting companies ("seekers") with scientists ("solvers"). Companies whose R&D staffs cannot find a solution to a biological or chemical problem can post the challenge at the site and offer a cash reward for a practical solution.

Notice that the real Web site on the right uses https:// while the faked site on the left does not.

Online Annoyances and Worse

The Web provides excellent opportunities but its wide availability combined with ingenuity have created some practices that range from mildly annoying to criminally dangerous.

- **Spam.** Spam is the term for any unsolicited commercial e-mail. The reason for spam is simple: it is the most cost-effective marketing method. Even if a fraction of a percent of the recipients end up purchasing the product or service touted, the spammer profits. Both individuals and organizations dislike spam. Individuals have to wade through a long list of unsolicited e-mail to get to useful e-mail. Organizations face an increasingly costly burden. Consider that if spam makes up half of the e-mail the organization receives, the organization must employ twice the bandwidth it really needs for communications and twice the space on e-mail servers. Obviously, it will pay twice as much as necessary to operate an e-mail system. Spam-filter software has helped to some extent, but spam is still on the rise and still wastes resources. By some estimates, spam constitutes about 75–80 percent of all e-mail on the Internet. About 50 percent of spam originates in North America, 30 percent in Europe, and the rest in Asia and other regions.

    The Direct Marketing Association (DMA) defends the right of businesses to send unsolicited commercial e-mail as a legitimate and cost-effective way of doing business. Indeed, the method gives small, entrepreneurial businesses a chance to compete. The DMA sees no difference between junk "snail" mail, which most of us reluctantly tolerate, and spam.

    On the other hand, the Coalition Against Unsolicited Commercial E-mail (CAUCE) calculated the amount of spam we might soon receive unless new laws stop the phenomenon. There are 24 million small businesses in the United States. If just 1 percent of those businesses sent you just one e-mail message per year, you would receive 657 messages in your inbox each day. And this number does not include e-mail from large and non-U.S. companies. Opting out is not a realistic option: who has the time to respond to 657 organizations with a request to stop? CAUCE does not believe that opting in—avoiding sending e-mail unless you specifically ask for it—would solve the problem either. Antispam legislation in many countries has not reduced spam. In the United States, the CAN-SPAM law has not reduced the practice. According to this law, spam is a crime only if the sending party hides its true identity or does not provide an opt-out option.

Is unsolicited commercial e-mail a legitimate marketing tool, or is it a nuisance that should be eradicated by strictly enforced laws? Would it be fair to outlaw an efficient way for businesses, especially entrepreneurial businesses, to approach potential customers?

- **Pop-Up Windows.** You browse the Web, stop to read an interesting article, and a few seconds later a window pops up, partially covering the text you were reading. The pop-up contains an advertisement. You look for the icon to close the window. It is not in its normal location. You finally manage to close the window, but as soon as you do, another one pops up. And so on, and so forth, more and more windows. When you finally close the main site's window, you discover that several other windows popped up *behind* the window. The site owner is paid by advertisers to run these pesky windows, which is legitimate. However, many people are quite annoyed by the practice. Some employ special applications or turn on a browser option that prevents pop-up windows. Is this a good solution? Not always. Many sites have links that open a little window to provide further information, such as a window with help or explanation of a term. If you block all pop-up windows, such useful windows do not open. If you use a selective pop-up "killer," you have to program it to allow pop-up windows for individual sites. Thus, even with a solution, pop-up windows waste surfers' time. Web surfers might not like pop-up windows, but advertisers love them, because they are an effective marketing tool.

- **Adware.** A growing number of organizations use adware, software that delivers ad banners or pop-up advertising windows on the Web. Often, the banners hide large parts of the information on the page. Adware is often tailored to users, based on their profiles, such as previous interests. Some companies use adware that pops up deliberately to cover banners of competing companies that paid to advertise at the site a user visits. The visitor might not even know that the ad is not originating from the Web site or its legitimate advertising clients, but from another one.

- **Spyware.** A more disturbing "ware" is spyware. As discussed in this chapter, spyware is software that uses the Internet connection of a computer to transmit information about the user without the user's knowledge or permission. Usually, the software transmits information about users' activities with their computers, including their every move on the Internet. It sits on the computer's hard disk, secretly collects information, and transmits it to the computer of a company, usually for marketing purposes, but also for industrial espionage. Some surreptitious software is also designed to pop up windows. Some countries have criminalized adware and spyware, but in much of the world the software does not violate any law. A bill against spyware and adware has been introduced in the U.S. Senate, but so far the practice is legal in most of the United States.

- **Phishing.** A growing number of Web users receive a special kind of spam that intends not to sway them to buy something but to defraud them. The practice is called phishing, a play on "fishing." Criminals send thousands of messages that look as if they were sent from a bank, a credit-card company, or any other financial institution or an organization where the recipient has authority to withdraw funds. The e-mail provides a Web link where the recipient is urged to go and supply personal information, including codes that are used to withdraw or transfer funds. One of many "reasons" is "explained" in the message: your account must be renewed, the bank lost your details, you should verify your personal information or the

account will be revoked, and many others. Thousands of people have fallen prey to the con artists, who used the information to withdraw funds. The most obvious sign that an e-mail message tries to phish is a message from an institution with which you have never transacted, such as a bank where you do not have an account. A more subtle sign is the URL that appears when you move your mouse to the link provided: the domain name is not the one of the legitimate organization. Suspect every e-mail message that asks you to update your personal information online. Call the organization using the legitimate number you have on file and ask if the message is genuine. Banks and other institutions rarely use e-mail to ask for "account information update." Phishing continues to grow. According to the Anti-Phishing Working Group—an industry association that works to eliminate phishing and identity theft—well over 20,000 phishing campaigns take place every month, involving millions of potential victims and numerous financial institutions. In March 2007, 24,853 phishing campaigns were conducted. The number of phish sites in that month was 20,827.

POINT OF INTEREST

Phighting Phishing

When you receive an e-mail message that sends you to your bank's Web site to reenter your personal details, you should ensure that the site is indeed your bank's. SpoofStick is a free add-on to your browser, which you can download from *www.corestreet.com/spoofstick* and install on your computer. Whenever you direct your browser to a site, a line underneath the URL box will tell you the correct site, such as: "You're on usbank.com" or "You're on 218.97. 169.53." The former is the legitimate site. The latter is the site for the link the phishers sent you. The designer slogan is a paraphrase of Teddy Roosevelt's famous statement: "Browse freely but carry a SpoofStick." If the notice you receive is not suspicious, it is still a good policy to log on to your bank's site independently—not from the link provided—or to telephone the bank.

Scientists and researchers from around the world can register with the site and work on solutions. The site is operated in seven languages to accommodate scientists and organizations from all over the world. So far, Eli Lilly and 30 other companies, including Boeing, Dow Chemical Co., and the giant consumer product company Procter & Gamble, have awarded sums of $4,000 to $100,000. The site has more than 120,000 registered scientists located in 125 countries. At any given time the site posts about 50 scientific challenges. InnoVentive.com is not alone. Yet2.com and NineSigma.com offer similar arrangements.

When a company employs a staff of researchers it must pay them regardless of how fruitful their efforts are. When offering cash for solutions, many more scientists might work for the company, but the company pays only the scientist who solves the problem. This enables the company to tap many more creative minds and to reduce the high risk involved in R&D. The solution to a scientific or technological problem may arrive sooner, and the cost savings can be huge.

M-Commerce

In Chapter 6, "Business Networks and Telecommunications," you learned about the many wireless technologies that enable people to access the Web while away from the office or home. Wireless technologies enable what some people call mobile commerce, or **m-commerce**. Mobile devices already let users log on to the Internet, but they can also provide an additional benefit to businesses: a device can be located with an accuracy of several feet, much like locating a cellular phone. As soon as you come within a few blocks of a store, your handheld computer or phone could beep and display a promotional message on its monitor.

M-commerce allows people to use their mobile devices to experience an event and react immediately. For example, they might view a horse race and place bets from their seat. Or, they can see a demonstration of a product at a public place and order it online. Impulse shopping will no longer be limited to shopping malls. Recall our discussion in Chapter 6 of future uses of RFID. Mobile devices might be equipped with RFID readers so their owners can use a product's electronic product code (EPC) to download information about it from the Web.

Smart mobile devices might be helpful in salesforce automation. Traveling salespeople are able to access data through the mobile device almost anywhere. They are able to access corporate databases through their company's intranet. Both traveling salespeople and consumers already practice m-commerce whenever they transact while using a hotspot or a Web-capable cell phone.

Experts believe that the most attractive mobile application might not be online buying, but the delivery of highly relevant information, custom-tailored to the user's current location and activity. Location services include downloading coupons at the store in which the consumer has just entered, finding out about nearby restaurants, or reading product reviews while shopping at an appliance store.

In the United States, cell phones must, by law, include global positioning system (GPS) capability, so that people can be located in case of emergency. As telephoning and other technologies are merged into a single device that can also link to the Web, the potential for marketing and pushing information might be too tempting for businesses.

So far, however, predictions about the growth of m-commerce in North America and Europe have not materialized. The only countries where m-commerce has become popular are Japan and South Korea. In Japan, subscribers to the DoCoMo's i-mode service use their smart cell phones to purchase soda cans from vending machines, buy food at fast-food restaurants, and shop at Web sites of online retailers. Purchases are charged to the cell phone service provider, NTT. Analysts expect that the United States will catch up to Japan and South Korea. In Coral Gables, Florida; Vancouver, British Columbia; and several other cities in North America and Europe, drivers can already pay for parking by keying into their cell phones the number of the parking meter. When time is up, the meter calls the cell phone, and the driver can add funds without rushing to the meter, by dialing a number.

Privacy proponents have already voiced concerns about m-commerce. Apparently, not many people are happy to find out that commercial organizations can track them down anytime when their mobile device is on. These devices not only allow consumer profiling, as already practiced by many online retailers, but can also tell retailers and other organizations your exact location at any given time. The result might be "we know who you are, what you have done on the Web, and where you are now."

POINT OF INTEREST

Thou Shalt Not Steal... My Identity!

Gartner, Inc. reported that in the 12 months between mid-2005 and mid-2006, about 15 million Americans fell prey to crimes as result of identity theft. This number represents a 50 percent increase since 2003, when the Federal Trade Commission estimates that 9.9 million Americans fell victim to identity theft. Most identity theft is done through the Internet, by phishing and similar schemes. A Gartner survey of 5000 online U.S. adults in August 2006 revealed that the average loss was $3,257, an increase from the $1,408 average of 2005.

Source: Gartner, Inc., March 6, 2007.

To encourage m-commerce and other mobile Web activities, the Internet Corporation for Assigned Names and Numbers (ICANN) approved .mobi as a top-level domain (TLD) specifically for mobile access to the Web. Companies that create Web sites with .mobi domain names are expected to use special technologies that optimize content for easy viewing on mobile devices. Bango, GoDaddy, 1and1, and mobisitegalore.com are some of the companies that offer software for building mobile-optimized sites.

However, a Web site can accommodate mobile devices with any domain name, using the proper software. World Wrestling Entertainment (WWE) is one company that caters to its mobile clientele. In June 2006, WWE created its "made for mobile" content, which provides television clips and merchandise promotions to mobile phones. Consumers from over 40 countries can use the service of any of 110 mobile phone vendors to purchase television clips and merchandise, and pay for them through the phones. The service is available to more than 100 million mobile phone users.

Virtual Worlds

You walk around, look at a map, like a piece of land, and decide to buy and develop it. You agree to pay the price, $2,000. You develop it and later sell it for a profit of $4,800. Where is that? Nowhere on planet Earth. It's all in a virtual world accessed through Web browsers and additional proprietary software that can be downloaded free of charge. Transactions like this take place daily (or hourly) at SecondLife, a virtual world on the Internet. And, no, the dollars we mentioned are not virtual. They are real dollars that you and the buyer paid using a real credit card. A **virtual world** is a combination of images, video, sound, and avatars that resemble the real world created by software and accessible for interaction by subscribers. An **avatar** is a three-dimensional graphical character that subscribers use to represent them in the virtual environment.

Virtual worlds enable individuals and corporations to purchase and develop real estate, and use virtual structures and avatars for many activities.

While Second Life is one of the best known virtual worlds, it is not unique. Other virtual worlds include There.com and Virtual Laguna Beach (VLB, operated by MTV). Many organizations, including IBM, Cisco, and several universities, have purchased "islands" in virtual worlds. They "developed" the "real estate" to build meeting facilities where employees can meet using their avatars. Universities use the environment for innovative ways to deliver classes. Apple, Dell, and other companies opened virtual stores in which they display and sell their products.

The hosts of virtual worlds make money by selling "real estate" and services for developing the purchased "real estate" as well as additional features. Owners are free to sell their "real estate" and "facilities" to other owners. They use credit cards—real money—to purchase "local" money to pay for the purchases. For example, at Second Life you can purchase Linden Dollars (L\$). Virtual worlds seem to be popular. At least 1.3 million people have tried Second Life. Some have made money by designing virtual clothes, jewelry, and other products, selling them to other visitors for virtual dollars, and converting the proceeds to real money.

How people behave in virtual environments is being studied. It seems some differences exist that are advantageous to employers. For example, IBM found that new employees were more likely to ask questions about the IBM culture when they interacted with their new mentors on IBM's island at Second Life than they did in real life. The company uses its virtual island to expedite orientation.

Employee training costs in the United States alone totaled \$55 billion in 2006. Only a small fraction of that training was executed in virtual worlds, but it is expected that such training will grow in the coming years.

SUPPLY CHAINS ON THE WEB

Supply chains extend from commercial organizations to both suppliers and buyers. Organizations connect their supply chain management (SCM) systems to their suppliers at one end, and to their buyers at the other end. Thus, an organization might be a participant among other buyers in an extranet managed by one of its suppliers, and a participant among several sellers in an extranet of a buyer. Large retailers manage extranets through which their suppliers' SCM systems can provide useful information to their own, so they can track orders and shipments as well as collect useful information for decision making on which supplier to select for which order. In this regard, a large retailer's extranet becomes a marketplace for many sellers and a single buyer.

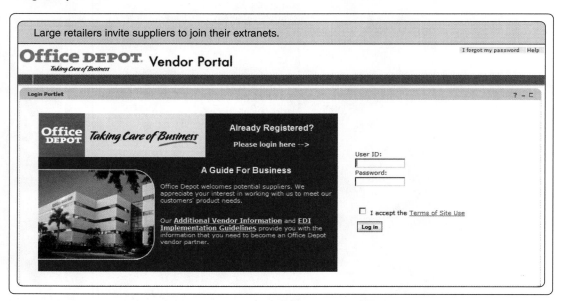

In the years before the Internet opened to commercial activities, many companies invested in Electronic Data Interchange (EDI) systems to exchange documents electronically with business partners. EDI consists of certain standards for formatting documents such as orders and invoices, software that translates the data properly, and the networks through which the information flows between subscribing organizations. The networks are owned and managed by value-added network (VAN) companies, telecommunications companies that manage the traffic of EDI between the business partners. Subscribers pay for this service. However, EDI can also be executed on the Internet. Although EDI provides some advantages, such as a high degree of data security and nonrepudiation (inability to deny sent messages), companies that want to connect to establish similar data exchange with business partners can use the Web technologies on the Internet. XML, in particular, enables business partners to set standards for data formats in Web pages. Dynamic page technologies, the software that links Web pages with databases, automate much of the business activity with business partners. Orders can automatically trigger notices to warehouse personnel on their stationary or handheld computers to pick and pack items for shipping. The information automatically flows into the accounting ISs as well as SCM systems of both the buyer and seller. Figure 8.6 illustrates how information flows between organizations.

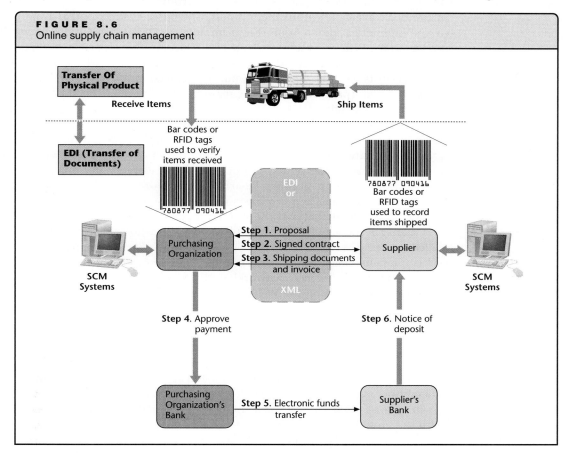

FIGURE 8.6
Online supply chain management

Companies encourage their suppliers to join their extranets. For example, the bookstore chain Barnes & Noble uses an extranet to do business with thousands of its 30,000 suppliers. So does Office Depot, one of the world's largest retailers of office supplies. It uses an extranet to order 7–8 billion items annually. The extranet saves the company much paper and administrative labor. Wal-Mart, the world's largest retailer, uses an extranet with Procter & Gamble and hundreds of other manufacturers. In addition to saving both labor and paper, the results are smaller inventory and greater in-stock availability of products.

XML is used extensively in Web technologies so that the SCM systems of two organizations can "speak to one another." This ensures that the meaning of data exchanged between the organizations can not only be displayed for employee eyes through Web browsers, but that the received data can be interpreted correctly by systems that automatically capture and store it in a database for further processing.

OPTIONS IN ESTABLISHING A WEB SITE

A Web site is, practically speaking, the Web pages that make up the information and links to Web technologies that the site provides. To establish a Web business, an organization must have access to an Internet server and the ability to control its content. Recall that an Internet server is a computer that is connected to the Internet backbone. Businesses have two options when establishing a Web site: installing and maintaining their own servers, or contracting with a Web hosting service.

Owning and Maintaining a Server

Installing and maintaining a server at the business's own facility is a costly option, but it gives the business the greatest degree of control. Setting up a server requires expertise, which may or may not be available within the business. The business must obtain a high-speed physical link to the Internet backbone. It must also employ specialists to maintain the server or many servers on which the Web site resides. In large organizations, these specialists might be employees of the company; in smaller ones, they might be contract personnel whose services the company hires. The specialists purchase a server (or multiple servers) for the company, connect it to the Internet through a high-speed dedicated line, register a domain name for the site, and install the proper software for managing the server and creating Web pages. The specialists "scale up" the server system when the business grows and handle issues such as load balancing to ensure quick response and to minimize the probability of site crashing. A site crashes when too many people try to log on and the software stops responding to anyone. **Load balancing** transfers visitor inquiries from a busy server to a less busy server for identical information and services. Thus, the specialists often must connect **mirror servers**—servers on which the same content and applications are duplicated—to speed up and back up the process.

A large company that uses the Web for much or all of its business usually has its own servers and manages them fully. This may be a company whose entire business is done online, often called a **pure-play** Web company, or a **brick-and-mortar** company that owns stores but also offers the same, many of the same, or additional items for sale online. These companies employ crews that manage Internet networking, the hardware and software of the site, and the people responsible for updating the Web pages.

Using a Hosting Service

A majority of organizations that have a commercial presence online either do not own servers or own servers but let someone else manage at least some aspect of the site. These organizations use **Web hosting** services. Web hosting companies specialize in one or several types of Web hosting: shared hosting, virtual private server hosting, dedicated hosting, or co-location.

In **shared hosting**, the client's Web site is stored on the host's same physical server along with the sites of other clients. The hosting company owns the server and the server management software. It offers space on the servers for hosting Web sites. This is a relatively inexpensive option. The client can use templates provided by the hosts for building pages, or, for an extra fee, have the host's designer design the Web site. However, many clients prefer to design and upload their own Web pages. The service includes transaction and payment software for use by the subscribing businesses' clients. If the server is shared, the host might not be able to allow a client to maintain its own domain name, such as *www.myownco.com*, but only a subdomain that contains the host's domain name, such as *myownco.myhost.com*. However, special software employed by many hosts allows clients to use their own domain names, and although the server has only one IP address, the software directs traffic to the proper site on the server. If an independent domain name is important, this is a factor that a business must consider before selecting a hosting service.

Small businesses with a limited number of products to sell can select a host such as Bigstep for shared hosting. The company invites you to "create an online business" for as little as $29.95 per month. When your business grows and has more products to sell, the company promises to "grow with you" by providing more disk space. Large search engine and portal companies, such as Yahoo!, offer similar services. Yahoo!, for example, offers to build and host a fully functioning online store for $39.95 per month. Clients have access to easy-to-use Web design software tools to create the pages of their new site. This type of option is often a "turnkey" solution for a small business that wishes to go online almost overnight. In addition to disk space and help with Web site design, the hosting company typically also provides a number of e-mail addresses and a control "dashboard," a mechanism for the client business to have remote control over content and other aspects of the site. Some hosts also offer to list the new site on frequently used search sites, such as Yahoo! and Google. Many of the hosts also help with domain name registration.

In shared hosting, hundreds of businesses might share the same server and storage space. Therefore, the host often limits the storage space allotted to each client, the number of transactions performed per month, or simply the amount of data, in megabytes, that the site transmits per month. Also, a technical problem in one site could affect the functionality of the other sites residing on the server.

Many companies offer Web hosting to support online businesses.

The purpose of a **virtual private server** is to create the impression that the client maintains its own server. Virtual private server technology enables one server to be virtually split into many addressable servers, each for a different client and with its own domain name. This option is usually less expensive than renting a dedicated server, while enjoying the same benefits, including full control of the content of the virtual server.

Some companies might want to use entire physical servers all for themselves, and therefore opt for dedicated hosting. In **dedicated hosting**, the host dedicates a server to the client, and the client can fully control the content on the server's disks. The host is responsible for networking management. For example, iValueHost, Verio, and aplus.net offer such a service and allow the client to select from several servers. The greater the server's power, the higher the monthly fee. This service is more expensive than shared hosting, but it comes with several advantages. Dedicated hosting has fewer restrictions on storage space and transactions, and since only one site resides on the disks, no other site can affect its functionality. Renters of dedicated servers usually have *root access*, which means they act as unrestricted administrators of that computer. The greater control of the dedicated server option comes with a price: this option is more expensive than shared hosting or a virtual private server.

In some cases, a company might want to fully manage its own Web servers but prefers the expertise of a hosting company in managing networking and security. Some hosting services accommodate such demand by offering a **co-location** service. The client owns the servers and manages their content, but the servers are co-located with the servers of other clients as well as those of the host's in a secure (physical) site. This approach has been taken by some online retailers, such as Overstock.com, because it affords some advantages: the client does not have to employ hardware and network specialists, spend money on building a special secure location for the server, or ensure power supply. All these concerns are transferred to the hosting company. Co-location is usually the most expensive of hosting options. The client must purchase and run the servers, as well as pay for the co-location.

Selecting the proper host may determine the difference between a faltering site and a thriving one. ClawfootCollection.com was operated by an online retailer Vintage Tub & Bath. It was spun off from VintageTubs.com. The site offered reproduction antique bathtubs and bathroom fixtures. The site sold to no more than a dozen buyers in its first year online. Vintage decided to hire the services of a new hosting company, Demandware, Inc. Demandware not only hosted the site, but also redesigned it. It maintains the site's "back end"—the connections to software such as billing, accounting, and other business functions invisible to shoppers—and ensures the shopping cart works. Within a few months the site was listed as number one for the keywords 'acrylic clawfoot tub' and sales increased. Interestingly, Demandware charges its clients by the number of site visitors.

Considerations in Selecting a Web Host

A majority of businesses do not maintain their own Web servers or co-locate them; they use host services. When a decision is made to use such a service, managers must consider several factors. Figure 8.7 lists the major factors. Hosts can be compared using points, for example, on a scale of 1 for the best and 5 for the worst. A simple evaluation method is for managers to compare each factor for the prospective host, compare the total scores, and then make a decision. The evaluators might wish to assign different weights to the various items based on how important each item is to the business.

The business should be able to use a database management system (DBMS) for cataloging its products and enable online shoppers to perform searches. Thus, the DBMS offered is important. It also might need to use **dynamic Web pages**, pages that enable communication between the shopper's browser and the database. Such pages can be built with several programming tools: CGI, Java servlets, PHP, and ASP (Active Server Pages). Since the functionality of databases and dynamic pages, as well as some features on the pages, depend on the operating system that the hosts use, all this software must be considered. For example, if the client elects to build and maintain the Web pages and prefers to use ASP, they should be aware that such software will run only on a server running Microsoft Windows. Similar restrictions apply to some page features that can be developed with the Web developing tool FrontPage. Most hosting companies offer the use of a combination of software popularly called LAMP, which is an acronym: Linux for operating system, Apache for server management software, MySQL for DBMS, and PHP or Python or Perl for developing dynamic Web pages. All of these resources are open source software, and therefore do not require license fees for the host, who can thus make the service more affordable.

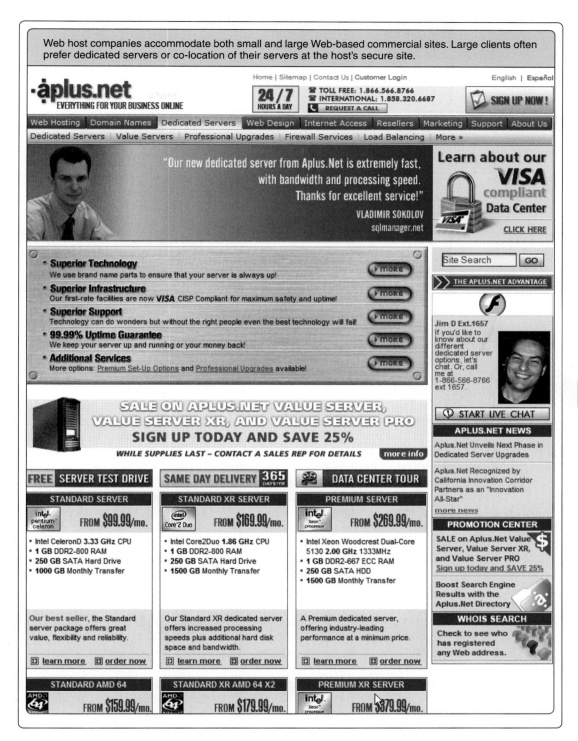

Web host companies accommodate both small and large Web-based commercial sites. Large clients often prefer dedicated servers or co-location of their servers at the host's secure site.

However, many hosts also offer other software, including Windows, for higher fees. In addition to these issues, the client should ensure satisfactory shopping cart, credit-card processing, and other applications at the site. If the client business needs integration of the Web software to its back end systems, it must ensure that the hosting company can execute and maintain the integration.

Storage space limitations might become a serious inhibitor, especially if the business expects to offer a growing number of products and augment the information provided through the site. The client should enquire about options to increase storage space on demand and its cost.

Most hosts provide technical support 24 hours per day, 7 days per week throughout the year, known as 24/7/365. The client should ensure such service is provided and know exactly what support services are included in the contract.

Technical support involves the quality of the equipment that the hosting company provides, security measures it maintains, the sophistication of server and load management, and the technical skills of its personnel. Companies should inquire about past downtimes and recovery time frames for the hosting company because they are an important part of technical support. If the client needs help in developing and updating Web pages, the evaluators should explore the appearance and functionality of current clients of the hosting company.

Some hosting companies charge extra fees for shared hosting if the site experiences activity above a predetermined amount of data that is transferred (downloaded from the site or uploaded to it) or number of visits from Web surfers, known as *hits*. In such arrangements, additional fees are charged for over-the-limit data transfer or hits. Web hosting companies price their services this way because the greater the number of hits, the more bandwidth they must allocate. The size of every file that is downloaded or uploaded is recorded, and if the limit—say 200 GB—has been exceeded, additional fees are charged. If the client's business grows, the cost might end up much greater than planned.

All hosting firms provide subscribers with several e-mail addresses. Some also provide forwarding to other e-mail addresses. Clients should examine these factors as well as the size of e-mail boxes, autoresponding (automating e-mail reply), and access to mailing lists.

Scalability is the ability of an organization to modify the capabilities of IT to accommodate growing needs. In this context, it is the ability for a Web site to grow—an important factor for most businesses. It is best to select a hosting company that has the hardware, software, and expertise to accommodate varying traffic levels and that can demonstrate its ability to develop a site from a simple, static one (one that does not require interactivity) to a heavily trafficked, interactive one. This applies to disk space, growing sophistication of software used, faster backup mechanisms, and other resources.

Smaller businesses often need help with the design of their Web pages. They need to discover whether the hosting company maintains experienced and available Web design personnel.

The host's physical site must be well secured against physical entry as well as intrusion through the Internet. Clients should ask for information about security measures. Some hosts are so careful as to not even advertise where they keep servers. Yahoo! is one such host.

Businesses want their Web sites available to users all the time. Downtime denies them business and damages their reputation. Hosting companies usually advertise their uptime as a percentage. For example, they might say they guarantee 99 percent uptime. This means that the client should expect the site to be down 1 percent (87.6 hours) of every year. Companies that need a higher uptime, such as 99.9 or 99.99 percent, should take notice of this and ensure that the host has the resources to claim such a high number of "nines." Such resources often include subscriptions to the services of two electric power companies or the availability of gas generators or other type of power backup. Redundancy, placing the site on two or more servers, is also a measure to ensure uptime.

Setup and monthly fees are self-explanatory. Monthly fees can range from several tens of dollars to several hundred dollars. Some hosting companies offer large discounts to clients that sign multiyear contracts.

More than Meets the Eye

However a business chooses to run its Web site, several elements must be present to conduct business, which are illustrated in Figure 8.8. While the shopper (a consumer or corporate purchasing officer) sees only Web pages, several applications and databases actually support online shopping and purchasing: an application that provides an inquiry interface for the shopper, which is connected to a catalog that is actually a database consisting of visual and text product descriptions; an application that takes the order, which is connected to an inventory application that is also connected to the product catalog database; a credit-card application that verifies authenticity of credit-card details and balance; and, in many cases, an order-fulfillment system that displays on monitors located in warehouses which items are to be picked off shelves and where they should be shipped. The latter system might include an automated conveyor system that picks the items with little human labor.

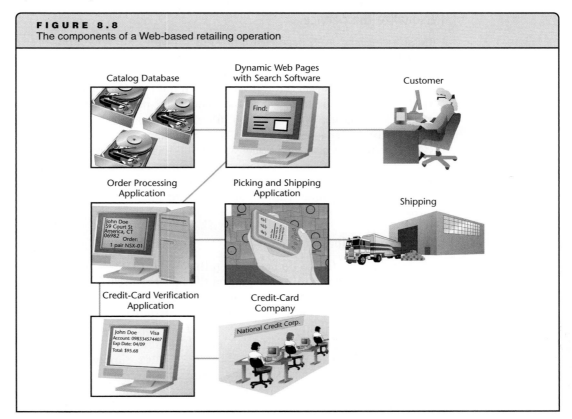

FIGURE 8.8
The components of a Web-based retailing operation

RULES FOR SUCCESSFUL WEB-BASED BUSINESS

Most organizations that operate a Web site do so to sell either products or services. Web software and the ability to connect Web servers to organizational information systems open numerous opportunities. Often, whether online business succeeds depends not only on availability of the proper software but how it is used. Several elements must be considered, especially if the site is to support B2C commerce.

Target the Right Customers

Targeting the people or organizations that are most likely to need the products and services you offer has always been the most important effort in marketing, with or without the Web. On the Web, targeting includes identifying the sites your audience frequently visits. For instance, a business that sells sporting goods should create clickable links at sites that cover sporting events and provide sports statistics. Banks that offer mortgage loans should create links at realtors' sites. And any business that targets its products to young people should do so at popular music sites. This principle should also apply to blogs and popular podcasts. Podcasting can include visual advertisements displayed by the player software.

Capture the Customer's Total Experience

By using cookies and recording shoppers' movements, CRM software can create electronic consumer profiles for each shopper and buyer. The shopper's experience with the site then becomes an asset of the business. Such marketing research fine-tunes the portfolio of products that the business offers and tailors Web pages for individual customers. It also can be used to "market to one" by e-mailing the shopper about special deals on items in which he or she has shown an interest.

Personalize the Service

CRM software and Web page customization software can be combined to enable customers to personalize the pages and the service they receive when they log on to the site. Letting shoppers and readers select the type of e-mail content they want is welcome, but sites should respect privacy by letting customers opt in rather than opt out. Opting in means that the customer can actively check options to receive e-mail and other promotions, while opting out requires the customer to select *not* to receive such information—an annoyance to many customers.

The Web also enables companies to let consumers tailor products. Land's End's Web site invites men to dress a virtual model with a build like theirs and order pants online. Although a pair costs significantly more than one a customer would purchase in a store, Land's End has been very successful with the concept. It has acquired many loyal customers because there is little reason to return a pair of pants that is made to order, although the company's policy allows returns.

Shorten the Business Cycle

One reason people like to do business on the Web is that it saves them time. Businesses should keep looking for opportunities to shorten the business cycle for their customers, from shopping to paying to receiving the items they ordered. Fulfillment, the activities taking place after customers place orders online, is one of the greatest challenges for online businesses.

Those who can ship the ordered products fastest are likely to sustain or increase their market shares. Some have decided to outsource the entire fulfillment task to organizations that specialize in fulfillment, such as UPS's e-Logistics and FedEx's Supply Chain Services. E-Logistics, for example, offers to receive and store the business's merchandise in its warehouses, receive orders online, and then pick, pack, and ship them to the online business's customers. It also offers a product return service. A shorter business cycle is not only important for customer satisfaction but also enables the company to collect payments faster because credit cards are usually charged upon shipping.

Let Customers Help Themselves

Customers often need information from a Web-enabled organization. Such information includes the status of an order, the status of a shipped item, and after-sale information such as installation of add-on components and troubleshooting. Placing useful information and downloadable software at the site not only encourages customer loyalty but also saves labor.

Practically every online business now sends e-mail messages with the status of the order, a tracking number, and a link to the shipping company for checking the shipping status. Hardware companies can post online assembly instructions for their "assembly required" products. In addition to including Frequently Asked Questions (FAQs) information, some companies have used knowledge management software (discussed in Chapter 11, "Business Intelligence and Knowledge Management") that can answer open-ended questions.

Be Proactive and De-Commoditize

Expecting customers to visit your Web site every time they need your service might not be enough in today's competitive marketplace. Customers now demand not only prompt e-mail replies to their queries but also proactive alerts. For example, the travel Web sites Orbitz and Travelocity e-mail airline customers gate and time information if a customer's flight is delayed or if gates change. Some manufacturers e-mail customers about product recalls or to schedule periodic service appointments. Online drugstores invite customers who regularly take a certain medication to register for automatic replenishment of their drugs. The company's software calculates when the next lot is to be shipped and ensures that it arrives in time.

All these initiatives, as well as many others, are efforts to *de-commoditize* what companies sell. A commodity is any product that is sold for about the same price by a multitude of vendors in a highly competitive market, usually with a thin margin of profit. By adding a special service or additional information, the company keeps the products it sells from becoming a commodity. Adding an original service or information to the product differentiates the "package" that online shoppers purchase from the "package" sold by competitors.

E-Commerce Is Every Commerce

You might have noticed that the title of this chapter does not contain the term "e-commerce." You might have also noticed that this is not the only chapter in which Web-enabled business activities are discussed. In fact, every chapter in this book gives examples of what is popularly referred to as e-commerce. Web technologies have been integrated into the business world to a degree that makes it difficult at times to realize which activities take place inside the organization and which involve information flowing from other places through the Internet. We have become so accustomed to the integration of the Web into our daily activities, especially the commercial ones, that the lines between commerce and e-commerce have been blurred. We will eventually stop using the term "e-commerce" and simply consider the Web another means of supporting business, much the way we consider technologies like the telephone and fax.

- Some industries have changed dramatically and continue to change thanks to Web technologies. This includes activities between and within organizations.

- HTTP is an Internet standard that enables addressing of Web servers with domain names. HTTPS is a secure version of the protocol and is used for confidential transactions. HTML is a markup language for presentation of Web pages. XML is a markup language for delivery of information about data communicated through Web pages. XHTML combines features of HTML and XML. FTP is a protocol for uploading and downloading files. RSS is software that uses XML to automatically update text and audio from the Web site that posts it to subscriber online devices. It is the main enabler of podcasting. Blogs enable people to conveniently create discussion Web pages by posting comments and responding to them. Instant messaging online chat services enable people to correspond in real time and help businesses serve online customers. Cookies help Web sites to personalize the experience of visitors. Along with other software that spies on unwitting Web surfers, they might provide detailed information about Web users.

- In addition to a large number of nonproprietary Web technologies, many more are developed and licensed to organizations by software vendors.

- An organization has two options when deciding to do commerce online: own and maintain its own Web servers at its own facilities, or contract with a Web hosting company. When contracting with a Web host, there are several degrees of service: shared hosting, virtual private servers, dedicated hosting, and co-location.

- When selecting a Web hosting company, organizations should consider several factors: type and quality of application provided, storage space, quality of technical support, traffic limits, availability of e-mail accounts and services, scalability, support of page design, security, uptime ratio, setup fee, and monthly fee.

- Web-enabled commerce can generally be classified as business-to-business (B2B) or business-to-consumer (B2C). In the former, businesses use networks to trade with other businesses, possibly through an extranet. In the latter, businesses advertise and sell goods and services to consumers via the Web. The higher volume of e-commerce is conducted between businesses.

- Business-to-business trading often relies on electronic data interchange (EDI), which is conducted over value-added networks. XML facilitates inter-organizational online trading similar to EDI. When linked to internal ISs, Web technologies enhance supply chain management. Online interorganizational commerce often takes place through an extranet.

- With the proliferation of wireless handheld computers and smart mobile phones, the next wave in B2C might be mobile commerce, popularly called m-commerce. It is already popular in Japan, but to a much smaller degree in the United States and Europe.

- To be successful, an online business must target the right customers, capture the customer's total experience, personalize the service, shorten the business cycle, let customers help themselves, and be proactive.

- Virtual worlds have become a popular way to meet and conduct social and business activities on the Web.

- Spam, and to a lesser degree spyware, adware, and pop-up windows, have become online annoyances. Society is trying to strike a balance between allowing these phenomena to continue as a form of commercial promotion and free speech, and curbing them to reduce the public's waste of resources. Phishing has become a pervasive crime, defrauding people and stealing their identities.

IT FITS OUTFITS REVISITED

It Fits Outfits is expanding onto college campuses by establishing an online storefront. Let's explore some of the issues it faces in managing its Web site.

people consider consumer profiling to be a violation of individual rights. Should It Fits Outfits profile its customers?

What Would You Do?

1. The online store has arranged for customer service representatives to answer questions in real-time chat rooms. What can the company do to make sure that these representatives are not wasting time answering the same questions over and over again?

2. Consumer profiling would help the online store target its customers individually and offer products that they would be most interested in. Some

New Perspectives

1. Shari Steiner opened the very first It Fits Outfits store next door to a coffee shop that was popular with local high school students. She also hired students from local high schools to run the store. What steps should It Fits Outfits take to attract college students? How can It Fits Outfits encourage students to drop by their storefront on their way to and from their favorite online hangouts?

KEY TERMS

avatar, 291
B2B, 278
B2C, 278
banner, 279
blog, 274
brick-and-mortar, 294
clickstream tracking, 277
co-location, 296
consumer profiling, 284
cookie, 276
dedicated hosting, 296
domain name, 271
dynamic Web page, 296
extranet, 280
File Transfer Protocol
   (FTP), 273

fulfillment, 283
Hypertext Markup Language
   (HTML), 272
Hypertext Transfer Protocol
   (HTTP), 271
Hypertext Transfer Protocol
   Secure (HTTPS), 271
impression, 279
instant messaging (IM), 275
intranet, 280
load balancing, 294
m-commerce, 290
mirror servers, 294
phishing, 287
podcast, 275
pure-play, 294

reach percentage, 279
reverse auction (name-your-
   own-price auction), 285
RSS, 273
search advertising, 279
shared hosting, 294
spyware, 277
Uniform Resource Locator
   (URL), 271
virtual private server, 295
virtual world, 291
Web hosting, 294
wiki, 274
XHTML, 272
XML, 272

1. The Web has been an enabler of new business methods. What does "enabler" mean in this context? Give an example of a business activity that is enabled by the Web and that would not be possible without the Web.

2. What is HTML, and why is it needed to use the Web?

3. What is XML? How is it different from HTML, and what purpose does it fulfill in Web commerce?

4. What is the relationship between a domain name and an IP address?

5. When you visit a Web site and click a Download button, you activate software that adheres to a certain protocol. What protocol is that?

6. What is instant messaging (IM), and how can it support business operations? Which technology does IM replace in online retailing?

7. What is RSS, and for which industry is it especially useful?

8. What is blogging, and what potential does it have for businesses?

9. What is podcasting, and how is it different from radio broadcasting?

10. In the context of the Web, what is a cookie? What is the benefit of cookies for online shoppers? What is the risk?

11. What is the difference between first-party cookies and third-party cookies? Which is usually disliked more by consumers, and why?

12. What is an intranet, and what purposes does it serve?

13. What is an extranet, and what purpose does it serve?

14. When contracting with a Web hosting company, what is the difference between a shared server and a dedicated server?

15. What is co-location? What are its benefits?

16. When selecting a Web hosting company, one of the important factors to consider is uptime ratio. What is it, and why is it important?

17. What does "unique monthly visitors" mean in online lingo? Who uses this metric and for what purpose?

18. What is a reverse auction? Would it be practical without the Web? Why or why not?

19. What is phishing? How do people get "phished"?

20. What business purposes do virtual worlds fulfill?

21. What is an avatar, and what function does it fulfill?

22. Recall our discussion in Chapter 6, "Business Networks and Telecommunications." What is the single most important factor enabling streaming video on the Web?

23. Sun Microsystems Corp. coined the slogan "The network *is* the computer." What does this mean?

24. Some top-level domains (TLDs) are reserved for certain organizations. Why is this important? Would you prefer that anyone could register a TLD of his or her choice?

25. Some people criticized ICANN's decision to create the .mobi TLD, arguing that no TLD should not be associated with special software for optimizing it. Explain this argument. What is your opinion?

26. Podcasting is said to allow subscribers to "time-shift." What does this mean, and does this give listeners a benefit they do not have with radio programs?

27. E-tailers can use their software to charge different shoppers different prices. This is called price discrimination, and it is legal. Some observers say that shoppers discriminate based on price when they decide from whom to buy, and therefore it is ethical for e-tailers to price-discriminate. Do you agree?

28. Do you see blogging and podcasting as a threat to the written and broadcast media? Explain.

29. One of the most frustrating types of events to an e-tailer is shopping cart abandonment. From your own online shopping experience, what are the things that would cause you to abandon an online shopping cart?

30. M-commerce will give organizations the opportunity to send location-related advertising, that is, they will send to our handheld devices advertising based on where we are. What is your feeling about this?

31. Some states (e.g., Washington) have legislated voting through the Internet. Some observers believe we will move toward a wider "teledemocracy." In addition to electing their representatives and other officials, citizens will be able to vote via the Internet on other issues, such as international alliances, tax cuts or new taxes, federal and state budgets, and other issues that are now dealt with only by their representatives. Do you favor this teledemocracy? Why or why not?

32. Gambling on the Web is growing fast. Do you see a danger in this phenomenon more than in gambling in traditional ways? Does Web gambling have more or less social impact than casino gambling?

33. You have a new home business. You sell a consumer product for which you have a patent. You believe there will be much demand for it. To promote it, you decide to purchase a list of 2,000,000 e-mail addresses of people who fall in the demographic groups that are likely to purchase the product. The seller told you that these were only addresses of people who did not opt out from receiving messages from businesses. After you e-mailed the promotional message, you received hundreds of angry e-mail messages, including one from the Coalition Against Unsolicited Commercial E-mail (CAUCE). Was there anything wrong in what you did? Why or why not?

34. The owners of a small business tell you that they would not be able to reach enough customers to survive if they couldn't use mass, unsolicited e-mail. You strongly object to spamming. How do you respond to them?

35. Scott McNealy, former CEO of Sun Microsystems, said: "You already have zero privacy. Get over it!" Some observers say that expecting privacy when using the Internet is ridiculous; the Internet is a public network, and no one should expect privacy in a public network. Do you agree? Why or why not?

36. A student established a Web site that serves as an exchange of term papers. Students are invited to contribute their graded work and to search for term papers that other students contributed. When criticized, the student claims that this, too, is a way to do research. He argues that the moral responsibility rests with those who access his site, not with him. Do you agree? Why or why not?

37. There have been international efforts to harmonize laws addressing free speech on the Web. Do you think such efforts can succeed? Why or why not?

38. What do you think makes social networking Web sites so popular?

39. If you have not tried a virtual world site, try one. What attracts you in it? If you are not attracted by its features, explain why not.

40. Web advertising is directed mostly at consumers, yet the topic is discussed in this chapter as a business-to-business commercial interaction. What could be the reason for this?

APPLYING CONCEPTS

41. Find three commercial sites that operate in three different markets and offer affiliate programs. Write up a summary: What do they sell? What do their affiliate programs promise, and in return for what? Classify each program as pay per sale, pay per click, pay per lead, or another type, and explain why you classified the way you did.

42. Choose a topic in which you are interested. Select three different search engines (e.g., Google, Yahoo!, and MSN) and use them to look for information about the subject. Rank the performance of each site. A long list of sites that provide too broad a range of information is bad; a shorter list of sites that provide more narrowly defined information is good. Explain your ranking.

43. You have been hired by a pizza delivery service to design a Web site. The site should be attractive to families and young professionals and should allow them to order home delivery. Use a Web page development application to build the home page of the business. Submit your page to your instructor.

44. A hosting company offers the following features for $9.99 per month. (This is a partial list of what they offer for this fee.) Explain each item offered. If you are not familiar with an item, research it on the Web.

Included Domains	3
Web Space	250 GB
Monthly Transfer Volume	2,500 GB
E-mail Accounts	2,500
Mailbox Size	2 GB
Search Engine Submission	Yes
Website Builder	18 Pages
Chat Channels	Yes
E-mail Newsletter Tool	Yes
Form Builder	Yes
Support	24/7 Toll-free, E-mail

HANDS-ON ACTIVITIES

45. Prepare your résumé as an HTML document. If you wish, include your scanned photograph. Submit your work by e-mail, or post it to your Web site and e-mail the Web site link to your professor.

46. Consider the following options for a business that wishes to use a Web site: (1) maintaining their own server at their facility, (2) using a host for a shared server, (3) using a host for a virtual private server, (4) using a host for a dedicated server. You are a consultant. Consider each of the following scenarios independently of the others, and recommend the best hosting option to the business. Consider all the relevant factors, such as purpose and cost.

a. *A family-owned store at a shopping mall.* The owners want to make the public aware of what the store offers. They want to pay as little as possible for this Web presence.

b. *A large retail company.* Management wants to be able to execute purchases from suppliers through the new Web site, and to allow their own customers to shop and buy through the site. It is willing to employ its own team and facility for the servers.

c. *A small business.* The owners insist on having their own domain name. They wish visitors to have every sense that the site is run and controlled by the owners.

d. *A small business.* Management does not want to register and pay for its own domain name.

e. *A large pure-play (Internet-only) e-tailer.* It needs to change the list of products daily. Its Web design team might want to change the DBMS, shopping cart application, and other applications when the need arises. It already owns the servers but no longer wants to manage networking, backup, redundancy, and security.

f. *A brick-and-mortar retailer that wants to extend its sales operations to the Web.* It has a Web design crew that is capable of changing content and is expert at using and modifying Web applications such as dynamic pages and shopping cart applications. However, management does not want to purchase servers or manage their networking and security.

47. Devote one day as "Low Technology Day" and express your experience. For one full day (24 hours) do not use the Internet, do not use a mobile phone or any other communication device. Write a 1–2 page report on how this affected your mood, social experiences, time management, and any other aspect of your life during that 24-hour period.

48. You and three other students should create avatars at Second Life (www.secondlife.com). Choose your appearances. Decide to convene at a certain public place, and have a discussion for at least 10 minutes. Summarize your experience in a two-page report. How easy or difficult was the process up to completing your avatars? How easy or difficult did you find navigation and movement (walking, running, flying)? How easy or difficult was it to exchange ideas?

49. Team up with another student to analyze the privacy policies of three companies that specialize in collecting consumer information on the Web. All must be companies that install third-party cookies. There are at least 10 such companies. List the common factors of the three companies. Then list the factors in which they differ. For each of the differing elements, which company treats consumers better, in terms of less invasion of privacy and more disclosure of its activities? Among other factors, see if the companies offer opt-in or opt-out options.

50. With two other team members, prepare a rationale for an original business idea that could generate revenue on the Web. Prepare the rationale in a way that would convince a venture capitalist to invest money in this new business.

FROM IDEAS TO APPLICATION: REAL CASES

A Blooming Web Business

Flowers are a highly perishable product. Most pass their useful commercial life within hours. Therefore, ensuring timely arrangement, packaging, and delivery is of utmost importance for the survival of any florist, let alone one that operates on the Web.

1800Flowers.com, Inc. is the oldest online florist. Jim McCann opened a single flower shop on Long Island in 1976. Within a few years he augmented it into a chain of 14 flower shops in the New York City metropolitan area. In 1986, he acquired the telephone number 1-800-356-9377, which allowed him to advertise the company as 1-800-FLOWERS to receive orders by phone. He registered it as a trademark. In 1995, he acquired the domain name 1-800-FLOWERS.COM, registered it, too, as a trademark, and opened a Web site. The domain name has since become one of the most recognized brands in gift retailing.

From its flowers-only beginning, the company expanded into plants, gourmet foods, candies, gift baskets, and other unique gifts, all offered at the Web site. Non-flower products are shipped to customers worldwide. To accommodate flower customers anywhere in the United States, the company created BloomNet®, a network of about 9,000 florists. This network supplements the company's own 100 stores from coast to coast in the United States.

To cast a wider net, the company also operates an affiliate program, enabling any online business that joins it to earn up to 12 percent of sales that originate at that business' site. The company offers a "10-Day Return Cookie." If a customer comes to 1-800-FLOWERS.COM and decides not to make a purchase right away, the affiliate can still earn the commission if that customer purchases from 1-800-FLOWERS.COM within 10 days.

Visitors who enter 1-800-flowers.com into their browsers may think they are taken to a single site, but they may actually be routed to one of several servers. If they look carefully at the address field, they may see "ww11.1800flowers.com" or "ww32.1800flowers.com" or another one of six "wwxx.1800flowers.com" addresses. All these addresses reside on three physical servers hosted by AT&T at locations that the company would not disclose. Thus, the "site" is actually three sites.

Web transactions constitute 75 percent of the company's transactions. The Web site is built to accommodate the typical surges in the weeks of Valentine's Day and Mother's Day. On Valentine's Day and the 48 hours before the day, the number of transactions increases to 10 times the normal number. Each of the three physical locations maintained by AT&T hosts several servers, and therefore the different numbers succeeding the "ww" after the domain name is entered. Customers may not even notice this, and it should not be important to them, but this architecture is extremely important for 1-800-FLOWERS.COM.

This arrangement helps balance the load when transactions increase, especially on those two high-demand days. They also provide backup should a location go down. Before every busy week, the IT staff goes over a checklist of 37 items—from network connections to database accessibility and response speed—to ensure smooth and uninterrupted operations.

Starting in January, the IT staff takes each hosting facility offline for a few days and tests it. With help from Hewlett-Packard, the staff simulates a transaction load similar to what it expects on its busiest day, Valentine's Day, and fixes any problems that emerge. During this time all transactions are handled by the other two sites. The IT staff stops completely any changes to the systems several weeks before Valentine's Day. The company carries out most of its applications development and maintenance over the summer, a time when transactions do not peak and long enough to prepare for Christmas and New Years Day.

This attention to detail and corporate discipline may be the reason for the company's success. From a single store in 1976, it has grown to a combination of brick-and-mortar and Web-based business with revenue of close to $800 million. Chris McCann, the company's president, says that the Web section of the business is the company's most profitable not only because it eliminates the use of telephones, but also because it is easier to market to Web customers, and these customers are typically easy to turn into repeat customers.

Source: Hertzberg, R., "Surviving Valentine's Day at 1-800-Flowers.com," (www.cioinsight.com), February 14, 2007; (www.1800flowers.com), May 2007.

Thinking About the Case

1. 1-800-FLOWERS.COM promises its Web affiliates that it will pay them a commission even if a customer made a purchase at 1-800-FLOWERS.COM several days after the visit at the original site. How can 1-800-FLOWERS know that a customer visited the original business and when?

2. What is the purpose of maintaining three sites at three different locations?

3. What is the purpose of freezing all changes to the sites some time before Valentine's Day?

4. Business that comes through the Web is the most profitable for the company. Why?

5. How difficult is it at 1-800-FLOWERS.COM to plan upgrades, testing, and changes to the Web sites?

Banking on Web Technologies

Sovereign Bank, the 18th largest bank in the United States, is based in Reading, Pennsylvania and has financial assets of $82 billion. It employs more than 10,000 people, operates more than 800 branches throughout the northeastern region of the United States, and maintains over 2,000 ATMs. Sovereign has experienced a dramatic increase in the number of customers doing business through its Web site. Management believes that a key to increasing its customer base is to offer a large financial institution's services in a personal manner typical of small community banks. To that end, the bank is relying on consumer-centric multichannel marketing. In marketing, a channel is a manner in which the marketing effort is carried out: mail, e-mail, Web advertising, newspaper advertising, and so forth.

Executives know that most customer interactions occur online, so the bank has decided to restructure its contact procedure with a focus on its Web site. The bank's Director of Online Business and Affinity, Marianne Doran-Collins, believes that online banking will improve customers' interaction and increase their business relationship with Sovereign. Results from a recent study by comScore, an online rating and research firm, confirm Doran-Collins's view. Surveying more than 1,500 people who regularly bank online, the researchers found that online banking customers tend to be more satisfied with their bank and in turn refer their friends to their bank's Web site. The study also found that users of online banking demonstrate greater bank loyalty and higher cross-sell rates than offline customers. Cross-selling is the selling of a service not related to the service a customer already pays for.

In January 2004, the bank launched its redesigned Web site, developed jointly for Sovereign by Agency. com, Ltd., and Tallán, Inc. Bringing together a decade-old interactive marketing and technology company and an application and systems development firm with 20 years of experience to create a new, easier-to-use site earned Sovereign high marks from industry reviewers.

But Sovereign sought to reap deeper rewards from its site in the long-term, as well. So the bank turned to highly specific customer analysis and personas.

A persona is a model archetypical customer representing groups whose members have common needs and goals. Thus, a persona includes particular values and attributes that guide individuals' behavior in the marketplace. Identifying these values and attributes through a combination of behavior and demographic analyses enables vendors not just to better frame their products and services but to determine which ones to offer. By deciphering what drives its customers' behaviors, Sovereign can influence those behaviors.

The bank hired the services of Claritas, a firm that specializes in gleaning knowledge from customer data for marketing purposes. Claritas used the bank's customer database to analyze geographic, demographic, and psychographic information and help the bank build personas. In 2003, shortly before its revamped Web site went live, Sovereign completed its development of four personas. Each persona is built on the occupational, residential, and purchasing attributes of its members.

The personas enable Sovereign to predict its customers' motivations, financial needs, and ways in which the bank can further increase their banking experience. With this new information, the bank can appropriately mold its online presence. Doran-Collins acknowledges that personas, rather than products, dictate the bank's online appearance and conduct. Words and images specifically cater to personas. When a young mother logs on she might see a picture of a young couple with two children on the front lawn of a new house, because young people are likely to want to finance a new home. A couple that plans to retire early might receive a Web page that features a couple in their 50's golfing or relaxing by the pool.

Building on personas to interact with customers, creating customer profiles can reap benefits across and between Sovereign's marketing channels. As soon as a customer logs on to the Web site, the site "knows" which services customers already have and what might interest them. This helps to avoid redundancy in delivering information and improve efforts to cross-sell. The customers enjoy a better Web site experience, and the bank sells more services.

To implement personas, management had to overcome tradition. Traditionally, the marketing people for each product used similar promotions for all customers, and they rarely shared information with the marketers of other products. Marketing with personas in mind forces marketers to share information.

Managers of the various product lines meet regularly with the Web program manager to ensure that all communication and data collection strategies work together rather than as internal competitors. Since adopting the use of personas, Sovereign customers have launched more online applications for additional services and referred more people to Sovereign than before. To enhance its service, the bank provides a chat option on the Web site, so customers can receive help immediately whenever they bank online.

As the personas have proved their worth online, Sovereign plans to use them increasingly to improve mail and e-mail campaigns. The bank sends close to six million pieces of direct mail each year, an effort that costs much money. Reducing mail by better knowing customer needs would save costs. Although e-mailing is not as expensive as mail, it, too, could benefit the bank if optimized along personas' lines. Eventually, management hopes to personalize each letter and e-mail sent to customers.

The bank contracted with Click Tactics, a firm that specializes in using information collected through Web sites for multichannel marketing. The Web site helps collect rich information about customers and their preferences. Combining this information with the four personas, Click Tactics develops marketing campaigns for Sovereign.

Although the information is gleaned mostly from customer interaction with the Web site, the marketing campaigns are not limited to the Web. They are also conducted through other channels, based on customer preferences as expressed at their visits. For example, customers are asked to indicate how they prefer to receive information about short-term investments. If the answer is "by e-mail," this is how information about short-term investment products will be communicated to the customer.

Click Tactics' automated integration of customer information streaming through the Web site also helps Sovereign create different campaign versions to fit each of the four personas. The bank can communicate 50,000 versions of a marketing effort for the same cost of producing 5,000 different versions in its pre-persona days. Instead of focusing so much on production, Sovereign can concentrate more on content, timing, and ideal channel selection. With the use of personas and improved multichannel marketing, response rates to direct mail campaigns have significantly increased. More customers feel that the bank seeks to serve them better rather than simply trying to get as much business from them as possible.

Collecting so much data from the Web site has helped the bank in other ways, too. Because the bank has gained more information about its customers, it

has become better prepared to answer their questions and propose solutions to needs previously unacknowledged.

In light of its demonstrable success, Sovereign has plans to add more technologies to the Web site. One is a search engine that supports free-form inquiries. The purpose of such search engines is to help customers find information with their own words rather than go to a FAQ page or engage bank employees with their inquiries. Software will capture the questions and analyze them to give the bank better business intelligence about customer interests. Another effort is to use analysis software to localize content presentation at the Web site for customers in different regions.

What happens when the bank has little or no information about visitors to its Web site? Prospective customers, about whom data is limited, and existing customers who do not fit into any of the four personas receive less personalized Web site information. In other words, if this is your first visit to the site, you might receive a "plain vanilla" Web presentation.

Sources: Mummert, H., "Channeling the Customer," *Catalog Success*, 2005; (www.sovereignbank.com), August 2007; (www.claritas.com), 2005.

Thinking About the Case

1. What is a marketing channel? How does the bank use the Web as a marketing channel?

2. Most of the information that the bank collects about customers is collected at the Web site. Why?

3. How does the Web site help marketing in other channels?

4. Explain what personalized Web presentation is and how the bank uses it for its purposes.

5. The bank intends to install a free-form search engine at the Web site. Apart from helping customers with their searches, how can this help the bank's marketing effort?

6. A growing number of people have added their telephone numbers to "do not call" lists. Do you think this has anything to do with the bank's increased Web efforts?

Using the Web for a Sea Change

The world's largest aquarium opened in Atlanta, Georgia, on November 23, 2005. It occupies over 505,000 square feet and holds 8 million gallons of fresh and marine water in its tanks. The aquarium is home to more than 120,000 animals, representing 500 species. The aquarium includes a 16,400 square-foot Oceans Ballroom that accommodates 1,100 guests for a sit-down dinner or 1,600 people for a reception. The entire building can be reserved by groups to host 10,000 people. Georgia Aquarium was built for $200 million, money that was donated by Bernie Marcus, a co-founder of The Home Depot and his wife, Billi. The aquarium is operated by 220 employees and over 1,000 volunteers.

Like all organizations, Georgia Aquarium needed well-functioning information systems. Because the facility was new, the small five-person IT staff could implement ISs from scratch. Management hired Accenture as consultant, both for its experience with nonprofit organizations and because Accenture agreed to provide some of its service free of charge. Accenture approached Unisys, the IT service company, to propose hardware, software, and networking solutions that would satisfy two requirements: be flexible enough to grow as demand grows, and be manageable by a small staff.

The hardware installed includes six servers. The software supports ticketing, financial activities, membership management, content management, a Web site, and e-mail. The Web sites enables anyone to purchase tickets online. The ticketing software is similar to that of a sports organization. It has a control to allow only the maximum of 4,800 guests at any time. It keeps track of the time each ticket was sold, and the location of the visitors in the building. It is interfaced with the membership database to ensure that members receive a reduced rate when purchasing tickets.

The Web-based ticketing system is so convenient that 125,000 tickets were purchased online before the aquarium was opened. Being a new attraction, management expected the day of opening to be extremely busy. It was, but all went smoothly. Atypical of such organizations, 70–80 percent of tickets are purchased online. In fact, the aquarium's site encourages visitors to purchase their tickets online and warns that the number of walk-up tickets is limited.

Patrons may purchase tickets online and have them placed for pickup at entrance boxes, but they are encouraged to use their e-mail receipts as tickets. When you purchase tickets online and provide your e-mail address, you receive a bar-coded message. You print it out, and this is your ticket. Lines at the gates are short, because wireless scanners scan the bar-coded tickets. Only on weekend middays do lines of ticketed visitors extend to more than a few minutes. Although tickets are sold out almost daily, the only waiting time is for security checking.

If patrons lose their online-ordered tickets, they can go to "my account" at the Web site and reprint it. If they want to adjust their visit date, they can do so via the Web site. If the call center is too busy, they can e-mail the aquarium to receive answers to their questions. Visitors are invited to use the Web also to order a "behind the scenes" guided tour, as well as sign up for birthday parties and closed group events.

Since the aquarium relies so much on IT, Accenture and Unisys made critical components of the system redundant. If a server goes down, the system is still operable, and most employees and volunteers will not even know of the mishap. All data is automatically duplicated.

The Web site provides practically any information in which a visitor or a planner of a special event may be interested. This saves much staff time. Apparently, this and the ability to purchase tickets via the Web site is one reason for the aquarium's popularity. Many people in other countries learned about the new aquarium through the Web and decided to visit it on their next trip to the United States. In the year after its opening in April 2005, more than 1.5 million people visited the facility. By early 2007, it received more than 4 million visitors. On average, 5,000–6,000 people visit the aquarium daily on weekdays, and about 10,000 on weekends.

Source: Waxer, C., "Aquarium Profits from a Whale of a System," *eWeek*, June 19, 2006; Williams, E.E., "No fish story: Aquarium draws million in 3 months," CNN, March 1, 2006; (www.georgiaaquarium.org), May 2007.

Thinking About the Case

1. What major functions does the Georgia Aquarium's Web site fulfill?

2. How does the site help maintain high revenues for the aquarium?

3. What types of costs does the Web site save?

9

NINE

Challenges of Global Information Systems

LEARNING OBJECTIVES

A growing number of organizations operate globally or, at least, in more than one geographic market. These organizations face some challenges that have a considerable impact on their information systems. The organizations have to meet the demands of global operations by providing international ISs to accommodate the free flow of information both within a single company's divisions and between multinational corporations. These issues are important because so many companies operate a Web site, and Web sites are accessible globally. For professionals, this means a growing need to understand other cultures, standards, and legal systems when applying and using information systems.

When you finish this chapter, you will be able to:

- Explain why multinational corporations must use global information systems.

- Provide elementary advice for designing Web sites for an international audience.

- Cite the cultural, legal, and other challenges to implementing international information systems.

IT FITS OUTFITS:
Expanding Globally

Jun Kaui, director of manufacturing for It Fits Outfits, turned away from his computer and put his head in his hands. A moment later, his cell phone rang. It was Shari Steiner, his old friend and CEO. "I've got news," Jun said. An hour later, they were sitting at a table in Cuppa Joe, the coffee shop where they had first met.

Global Connections

Shari Steiner had studied design at the Fashion Institute of Technology in New York City and later designed clothing for many different companies. When she decided to open her own retail chain, she turned to Jun Kaui. As a designer, she had interfaced very little with the manufacturing end of the textile business. Jun had been a business major, and his father had managed textile mills in Asia. He had the experience necessary to contact, evaluate, and contract with textile factories. Jun also spoke fluent Mandarin, which made it possible for It Fits Outfits to use Chinese textile factories to manufacture outfits based on Shari's designs. Manufacturing clothing was cheaper in China, but a far more complicated process. With Jun's language skills, business experience, and personal contacts, they were able to establish relationships with reliable manufacturers and jump through the bureaucratic hoops involved in exporting goods from China to the United States.

The News

"Remember in January," Jun said, "when the World Trade Organization's export quotas were lifted and the textile factories were free to increase exports? Well, the textile industries in the European Union and the United States became very unhappy about that. They didn't want to lose business to China—understandably. However, in response to U.S. pressure, the Chinese government just increased export taxes by 400 percent."

Shari leaned forward. "So, what can we expect?"

"If their profit margins plummet, some of our manufacturers will close—and maybe quite suddenly. Others might try to pass the cost on to us."

"Depending on how much it is, passing the cost to us might be doable. Unexpected closings would be a disaster." Shari leaned back. "What do we do?"

"Order lunch," Jun smiled.

Analyzing the Impact of the Chinese Tariff Hike

Over the next two days, Jun and his staff read and analyzed the reports coming out of China. Seventy-four products were to be affected by the hike in tariffs. They needed to find out which of their manufacturers would be subject to the rate increase. For each manufacturer, would the hike squeeze out enough of their profit so that they would be forced to close shop? If the manufacturers passed this expense onto foreign retailers, would It Fits Outfits be able to afford to remain with the manufacturer?

After a careful analysis, Jun determined that the hardest hit would be manufacturers of cotton shirts. The tariff was to go into effect June 1. By August and September, many manufacturers would be forced to close. It Fits Outfits would have to consider looking for cotton shirt manufacturers outside of China.

Considering Alternatives

"We need to brace ourselves," Jun told his team. Fortunately, Jun was in a much better position to look elsewhere than he would have been four years earlier, when Shari had first persuaded him to come aboard. At that point, he had contacts only in China and Asia. Today, he knew people in the textile industry from all over the world. The global textile market had turned to the Internet to facilitate trade. Business-to-business Web portals reported the latest textile news on the national and international scene. A company in the states or the European Union could bid on a shipment of jeans produced in Bahrain or Indonesia. Retailers could review textile manufacturer directories from all over the world. They could then make contact and visit the facilities. National consulting businesses could help retailers deal with the legal issues of exporting the goods.

Still, a number of obstacles would need to be overcome if It Fits Outfits moved out of China. Jun and his team would have to bridge language and cultural differences, study tariff laws and export legislation, and keep an eye on political and economic factors that could influence the industry.

Back at Cuppa Joe a week after the tariff hike had been announced, Jun explained to Shari, "In the short term, I'm going to check out a couple of domestic manufacturers. It would be expensive, but it'll cover us if our shirt manufacturers close down at the end of the summer. In the long term, we're looking into Bangladesh. They've decided not to continue cash incentives to the industry, so the market should be stable—at least for a while."

Shari smiled. "Until the next crisis...." Then she looked at Jun seriously. "You realize your proactive approach to tackling potential manufacturing problems keeps our company stable in the short term and the long term."

"You're welcome." Jun smiled, "Do you think you're grateful enough to split that chocolate éclair?"

MULTINATIONAL ORGANIZATIONS

A software development firm in India sends an electronic greeting card to employees of business partners worldwide. The card includes a swastika, a symbol sacred in Indian culture. In protest, 14 international teams stop working for 11 days. An international company discovers that its invoices are electronically stamped with a date that is one day before a shipment is made from Singapore to the United States. An international team using Web tools to collaborate discovers near the completion of the project that some of the measures in electronic drawings are metric and some are English. A Web-based business learns that the law in some countries forbids the sale of an item it offers at the site. International corporations regularly encounter these types of problems, and realize that they need to overcome cultural, legal, and other challenges.

An increasing number of the world's corporations have branched into countries all over the globe, becoming true multinationals. While they might have headquarters in a single country, they operate divisions and subsidiaries in different countries to take advantage of local benefits. For instance, a company might establish engineering facilities in countries that offer large pools of qualified engineers, build production lines in countries that can supply inexpensive labor, and open sales offices in countries that are strategically situated for effective marketing.

Because of these dispersed operations, a company's nationality is not always obvious. For example, consider IBM and Philips. While IBM is known as an "American" corporation because its headquarters and most of its research activities are in the United States, the company has numerous subsidiaries in other countries. These subsidiaries are registered and operate under the laws of the respective countries, and they employ local workers. Likewise, not many Americans realize that Philips' headquarters is in the Netherlands and that it owns one of the largest U.S. sellers of electric razors, Norelco. Similarly, Intel, an American company, has major research and development facilities in Israel, where some of its latest microprocessors have been developed. Accenture, Tyco International, Ingersoll-Rand, and Cooper Industries—perceived to be American companies—are all headquartered in Bermuda. Lenovo, the world's third largest maker of PCs, is legally a Chinese corporation. However, the company rotates its headquarters between Beijing, Singapore, Paris, and Raleigh, North Carolina.

One hundred of the 500 largest Canadian companies have majority U.S. ownership, and 90 percent of U.S. multinational companies have Canadian offices. Japanese companies own U.S. subsidiaries in every imaginable industry. British companies have the largest foreign investment in the United States. Thanks to the North American Free Trade Agreement (NAFTA) and agreements between the United States and the European Union, we might witness the internationalization of many more American, Canadian, Mexican, and European corporations.

Multinational corporations must use **global information systems**, which are systems that serve organizations and individuals in multiple countries. These companies might have unified policies throughout their organizations, but they still have to abide by the laws of the countries in which each unit operates, and be sensitive to other local aspects of their interaction with businesses as well as consumers. Therefore, unlike organizations that operate in a single country, multinational companies have the burden of ensuring that their information systems and the information flowing through the systems conform to laws, cultures, standards, and other elements that are specific to countries or regions.

THE WEB AND INTERNATIONAL COMMERCE

The emergence of the Web as a global medium for information exchange has made it an important vehicle for both business-to-business (B2B) and business-to-consumer (B2C) commerce. In 2007, more than 888 million people regularly logged on to the Internet across the globe. Over 70 percent of them come from non-English-speaking countries, as Figure 9.1 shows, and more than half of all e-commerce revenues come from these countries. The ratio of non-English speakers to English speakers has steadily grown over the years. As Figure 9.2 indicates, a growing number of Web users come from regions other than North America.

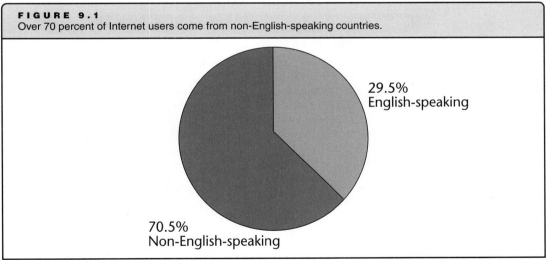

FIGURE 9.1
Over 70 percent of Internet users come from non-English-speaking countries.

29.5% English-speaking

70.5% Non-English-speaking

Source: Internet World Stats (www.internetworldstats.com)

The spread of Internet use opens enormous opportunities for businesses the world over. Some of the countries with current low participation rates have the greatest potential for expanding accessibility to the Internet, such as China. About 137 million citizens of the People's Republic of China logged on in 2007, but more than a billion of them might do so in the future, and the Chinese market is expected one day to be the world's largest in terms of consumer spending.

The Web offers opportunities not only to increase revenue but also to save on costs. Consider, for example, how much money is saved when instead of printing product and service manuals on paper and shipping them to customers, companies publish them on the Web, ready to be downloaded at a user's convenience. Furthermore, imagine the convenience if the manuals were prepared not only using hypertext and graphics but also animation for easier and more informative use. Some companies place video clips to instruct buyers how to assemble the products they purchased. Many companies have stopped enclosing manuals with their retail

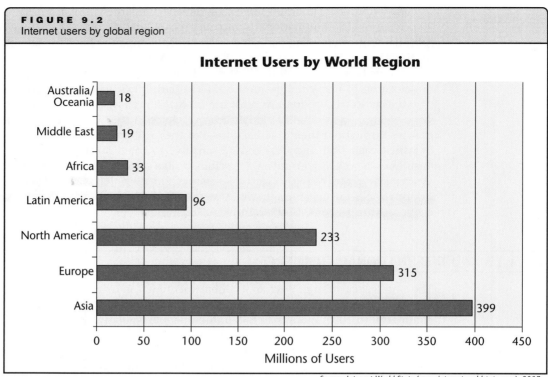

FIGURE 9.2
Internet users by global region

Internet Users by World Region

Region	Millions of Users
Australia/Oceania	18
Middle East	19
Africa	33
Latin America	96
North America	233
Europe	315
Asia	399

Millions of Users

Source: Internet World Stats (www.internetworldstats.com), 2007

POINT OF INTEREST

Low Penetration

In 2007, the world population was estimated at 6.7 billion people. About 16.9 percent of the population used the Internet on a regular basis. Africa is second only to Asia in population size, with 933.5 million people. Yet, only 3.6 percent of Africa's inhabitants used the Internet regularly. In contrast, 69.7 percent of the inhabitants of North America and 38.9 percent of the inhabitants of Europe used the Internet regularly. The country with the largest penetration of the Internet is Iceland. Over 86 percent of this island's population regularly uses the Internet. The country with the lowest penetration is Iraq. Only 0.1 percent of the 27 million Iraqis use the Internet.

Source: (www.internetworldstats.com), June 2007.

---

products. They invite you to log on to their Web site and peruse the product's manual in your own language. This saves not only paper and printing but also much of the labor involved in customer service. By placing maintenance manuals in multiple languages on their Web sites, some companies cut as much as 50 percent of their customer service costs.

Organizations that wish to do business globally through their Web sites must be sensitive to local audiences. Thus Web sites should be tailored to the audiences they are meant to reach. A majority of Web users prefer to access the Web using a language other than English, so organizations must provide their online information and services in other languages, as well as English. As Figure 9.3 shows, organizations must plan and carefully design their global sites so that they also cater to local needs and preferences, a process sometimes called **glocalization**.

FIGURE 9.3
Imperatives to heed when designing Web sites for an international audience

Plan	Plan the site before you develop it. A site for an international audience requires more planning than a national one.
Learn the Preferences	Learn the cultural preferences, convention differences, and legal issues, or use experts who know these preferences. Tailor each local site (or the local section of your site) to the way in which the local people prefer to shop, buy, and pay.
Translate Properly	Use local interpreters to translate content for local audiences. Do not use software or other automated methods, unless humans review the translated material. Experienced translators are attentive to contemporary nuances and connotations.
Be Egalitarian	Do not let any audience feel as if it is less important than other audiences. Keep all local sections of your site updated and with the same level of information and services.
Avoid Cultural Imperialism	If the local language or culture has a word or picture for communicating an idea, use it; do not use those of your own country. Give the local audience a homey experience.

Glocalization is a combination of universal business models and management philosophy with some adaptations for local audiences. One example of an organization that glocalizes is McDonald's. While the restaurant chain's logo and many other features are the same throughout the world, it makes some menu changes to appeal to local palates. Sometimes, other elements are changed. For example, in France, the restaurant chain replaced its familiar Ronald McDonald mascot with Asterix the Gaul, a popular French comic book character. Much like the presence of a global restaurant chain, Web sites are present everywhere someone can link to the Web. Therefore, Web site designers must keep glocalization in mind.

Think Globally, Act Locally

Marketing experts often advise companies that operate internationally to "think globally, act locally." Acting locally means being sensitive to regional customs and language nuances. When interest in the company's business increases, especially from consumers, it is advisable to open a local office and let a local team handle both the Web site and fulfillment operations. Recall that fulfillment in online business includes picking, packing, and shipping. When most of the business comes from one country or region, the business, its Web site, and its information systems are managed centrally, but when a growing proportion of transactions takes place in other regions, businesses find that they must decentralize control.

Thinking globally and acting locally might sound like contradictory ideas, but they are not. Recall our discussion of strategies in Chapter 2, "Strategic Uses of Information Systems." Thinking globally has to do with the company's strategic planning. It involves decisions such as product lines and business alliances. However, the same strategy can be followed with a local flavor. For example, the same product, in whose design and production the company holds a competitive advantage, can be packaged and advertised with local motifs. The local branch of the company might still recruit engineers with the same excellent qualifications as those of their peers in other countries, but apply different interview tactics and social benefits suited to the customs and holidays of that country.

Learn About Challenges of Global ISs

The growing globalization of business means that chances are high you will be employed by a company that operates outside your country. Even if your employer does not have offices in another country, you may be involved in global business. Being aware of the challenges involved in global business and the information systems supporting the business may determine your professional success. One does not have to work for a multinational corporation to need to be aware of the challenges discussed in this chapter. One only has to work for any organization that has a multilingual workforce or which operates in multicultural markets to have to care about these issues. And there is a high probability that you will work for such an organization.

By default, every business that establishes a Web site in some way uses a global information system. Many organizations use additional types of global ISs. All face challenges. Neglecting to pay attention to such issues as different cultures, language nuances, conflicting national laws, and different standards can hurt the business's reputation and cause loss of revenue. As a professional who is knowledgeable about these issues, you can be a valuable asset to your organization.

CHALLENGES OF GLOBAL INFORMATION SYSTEMS

While the Web offers tremendous opportunities for establishing international ISs, global ISs are not without their challenges, both for B2B and B2C commerce. Some of the challenges that businesses must address are technological barriers, regulations and tariffs, electronic payment mechanisms, different languages and cultures, economic and political considerations, different measurement and notation standards, legal barriers, and different time zones. These challenges are discussed in the following sections.

Note that we discuss differences among world regions and among countries, but much of the discussion applies to regions within the same country. For example, there are legal and cultural differences among states in the United States, Germany, India, and Brazil, as well as legal, cultural, and linguistic differences among provinces in Canada and cantons in Switzerland.

Technological Challenges

Not all countries have adequate information technology infrastructure to allow resident companies to build an international information system. International ISs, especially those using the Web, often incorporate graphics to convey technical or business information, and those applications, as well as interactive software, require increasingly fast (broadband) communication lines. The bandwidth available in some countries is too narrow for high-volume transmission of graphically and animation-rich Web pages. Thus, companies might have to offer two versions of their sites, one for wide bandwidth and another for narrow bandwidth. Often, companies use one site but provide the same content in both graphically rich and text-only pages, or the same video for download at different speeds.

Language issues present another technological challenge. You might recall the earlier discussion of how characters are represented by bytes in computers. This setup is fine for languages with up to 256 (2^8) characters, such as English and other languages whose alphabetic root is Latin, and for other languages whose characters represent individual phonemes rather than words, such as Cyrillic, Hebrew, and Arabic. But eight-bit bytes are not sufficient for languages with larger numbers of characters, such as Chinese, in which characters represent whole words. The solution for this obstacle is to ensure that computers can use Unicode, with double-byte characters—allowing for up to 65,536 (2^{16}) characters. However, if only the servers are programmed to accommodate Unicode, while the other systems (such as databases and applications on computers interacting with the servers) work with single-byte characters, then these back-end systems will record and display gibberish. Thus, entire systems must be reprogrammed or use special conversion software. As computers convert to operating systems that support Unicode, displaying different character sets should be less of a problem.

Businesses that cater to international audiences must "glocalize" their Web sites.

Other points that might sound trivial can also wreak havoc in international ISs or prevent individuals and companies in some world regions from transacting with companies that did not make their Web sites and applications flexible. For example, fields such as telephone numbers should be set for variable length, because the number of digits in telephone numbers varies by country. Many sites still offer forms that limit telephone numbers to 10 digits and do not accept shorter or longer numbers even when they are meant for audiences outside the United States and Canada. Similarly, postal codes are organized differently in different countries and are not called zip codes, and yet some U.S. sites are still designed with only a 5-digit (or extended 9-digit) postal code field.

In some cases, no elegant solution can be found even if every effort is made to localize information systems. AES Corp. generates and distributes electric power to 27 countries on five continents. It uses SAP's ERP system on servers that connect all of its sites. SAP does not support the Ukrainian language, so the company decided to use Russian in its Ukrainian office.

Regulations and Tariffs

Countries have different regulations on what may or may not be imported and which tariff applies to which imported product. While many executives know they might be missing out on lucrative deals with overseas businesses, they are afraid that exploring international opportunities would entail too many hassles. They are also afraid that even with the proper research, employees might not know how to comply with the laws of destination countries, let alone calculate how much the organization would have to pay in taxes, tariffs, custom duties, and other levies on exported or imported goods.

Companies such as NextLinx help exporters and importers who use the Web for commerce. The NextLinx software is integrated with a company's ERP systems and Web site. When a business from another country places an order, the information—such as type of item and destination country—is captured by the software, and an export manager can see how much the company will have to pay in tariffs, receive an estimate of how long the goods will stay in the seaport or airport before they are released from customs, and, if the manager wishes, also receive information on regulations, license required, shipping companies in the destination country, and other useful information. Since the software is linked to the Web, it is continuously updated and provides useful information immediately. The software also calculates, on the fly, the total cost of delivering the goods to the buyer's door. It also provides more than 100 forms that exporters can fill out and save electronically. The logistics component of the application offers shipping options with land, sea, and air carriers; books shipping space; and tracks shipping status. Several studies have shown that U.S. companies have turned away about 80 percent of online orders that come from other countries because they are not familiar with export regulations. This service can expedite the process.

Differences in Payment Mechanisms

One of the greatest expectations of e-commerce is easy payment for what we buy online. Credit cards are very common in North America and are the way businesses prefer to be paid online. However, this practice is not widespread in other regions of the world. The high rate of stolen credit cards, especially in Eastern Europe, attaches risk to such payments and deters potential online customers. Also, most Europeans prefer to use debit cards rather than credit cards. (The holder of a debit card must maintain a bank account from which the purchase is immediately deducted; the holder of a credit card receives a grace period of up to a month and pays the credit-card issuer in any way he or she prefers.) Americans are more willing to give credit-card details via the Web than people from other nations. Until citizens of other countries become willing to do so, payment through the Web, and therefore B2C trade, will not reach its full potential.

Americans pay with credit cards in 20 percent of all transactions and in almost all of online transactions. In Japan, on the other hand, only 8 percent of transactions involve credit cards, and most Japanese are reluctant to use credit cards for online purchases. This calls for a different mechanism of payment. In Japan, many people who order merchandise online prefer to pick it up at convenience stores called "konbini," and pay there for what they purchase. Since shipping companies are reluctant to leave parcels unattended when the recipient is not home, the alliance of e-tailers and konbini affords not only payment confidence but also convenience. E-tailers from other countries who want to operate in Japan must be aware of these preferences.

Language Differences

To communicate internationally, parties must agree on a common language, and that can create problems. For instance, data might not be transmittable internationally in real time because the information must first be translated (usually by human beings). Although some computer applications can translate "on the fly," they are far from perfect. Another hurdle is that national laws usually forbid businesses to run accounting and other systems in a foreign language, leading to an awkward and expensive solution: running these systems in two languages, the local one and English, which is the *de facto* international language.

Companies that are in the forefront of Web-based e-commerce have translated their original Web sites into local languages. They localize their sites by creating a dedicated site for each national audience. But translation can be tricky. For instance, the Taiwanese use the traditional set of Chinese characters, but people in the People's Republic of China prefer the simplified character set. Spanish terms in Spain might be different from those in Latin America, and even within Latin America. In some parts of South America people do not even call the language Español (Spanish) but Castellano (Castilian). In addition, mere linguistic translation might not capture cultural sensitivities. Therefore, some companies prefer to leave Web design and translation to their local overseas offices.

Several companies, such as TRADOS, Inc., offer translation software and services to companies involved in global commerce. TRADOS' software package by the same name translates Web pages into many languages, including those requiring special characters such as Hindi, Chinese, Greek, and Hebrew, but also ensures consistency of terms and sentence structure in different languages. When Web pages are translated, the software ensures that the XML tags and statements are retained from the original languages, so that the company maintaining the Web site can continue to use the same XML code for online transactions with companies and shoppers in its new markets. Other tools translate MS-Word documents to multiple languages. One such tool is Wordfast.

Cultural Differences

XL Capital is a global insurance firm operating 77 offices in 28 countries and proclaiming to be "one company without borders." At one point, the company had seven different e-mail addressing standards at local offices. When the company's CIO decided to adopt a single universal naming format, he faced resistance. In South America, for instance, a person might use five names: his first and middle names, and his parent's middle and last names. That caused some people to have long e-mail addresses. The CIO's suggestion to use employee ID numbers as their e-mail addresses (with the company's suffix) was received with resentment in South America and Europe because it was impersonal. To mitigate these unexpected cultural differences, the CIO established a system that greets each employee by name in a personal manner as soon as the employee logs on to a computer.

Cultural differences refer in general to the many ways in which people from different countries vary in their tastes, gestures, preferred colors, treatment of people of certain gender or age, attitudes about work, opinions about different ethical issues, and the like. ISs might challenge cultural traditions by imposing the culture of one nation upon another (cultural imperialism). Conservative groups in some countries have complained about the "Americanization" of their young generations. Governments might be inclined to forbid the reception of some information for reasons of undesirable cultural influence. An example of such fear is the French directive against use of foreign words in government-supported mass media and official communications. A similar example is the ban by the Canadian province of Quebec on the use of non-French words in business signs. These fears have intensified with the growth of the Internet and use of the Web. Because the Internet was invented and first developed in the United States and is still used by a greater percentage of Americans compared with any other single nation, its predominant culture is American.

As mentioned previously, companies that use the Web for business must learn cultural differences and design their sites accordingly. Web designers need to be sensitive to cultural differences. People might be offended by the use of certain images, colors, and words. For example, black has sinister connotations in Europe, Asia, and Latin America; the index-finger-to-thumb sign of approval is a rude gesture meaning "jackass" in Brazil; the thumbs-up sign is a rude gesture in Latin America, as is the waving hand in Arab countries; and pictures of women with exposed arms or legs are offensive in many Muslim countries.

Conflicting Economic, Scientific, and Security Interests

The goal of corporate management is to seize a large market share and maximize its organization's profits. The goal of a national government is to protect the economic, scientific, and security interests of its people. Scientific information is both an important national resource and a great source of income for foreign corporations, so occasionally those interests conflict.

For instance, companies that design and manufacture weapons have technical drawings and specifications that are financially valuable to the company but also valuable to the security of their country. Hence, many governments, including the U.S. federal government, do not allow the exchange of weapon designs. Transfer of military information to another country, even if the receiving party is part of an American business, is prohibited. Often, products whose purpose has

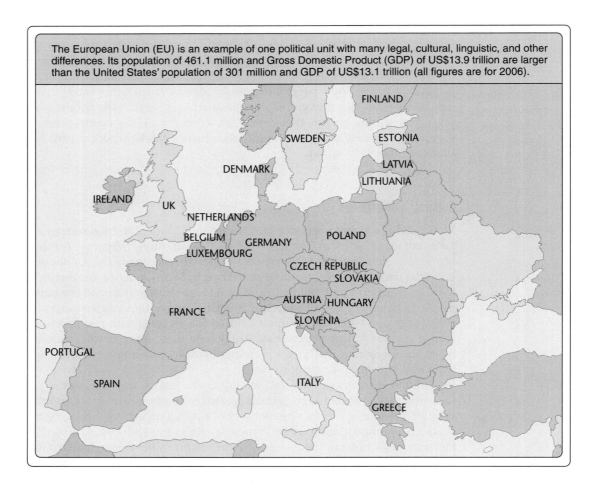

The European Union (EU) is an example of one political unit with many legal, cultural, linguistic, and other differences. Its population of 461.1 million and Gross Domestic Product (GDP) of US$13.9 trillion are larger than the United States' population of 301 million and GDP of US$13.1 trillion (all figures are for 2006).

nothing to do with the military are included in the list of prohibited trade items, because of the fear that they could be converted for use against the country of origin. In recent years, the list has included some software packages. The result is that, although American divisions of a company can use such software, their sister divisions in other countries cannot.

Consider some of the encryption applications offered by software developers. When Phil Zimmermann, developer of PGP (Pretty Good Privacy), offered this encryption application for free downloading, he was faced with federal criminal charges and severe penalties. His purpose was to allow individuals and companies to scramble their communications via computer networks. Companies use such software to protect corporate information. However, strong encryption methods are on the U.S. federal government's list of restricted exports because, like weapons, they could compromise America's national security. Under public pressure, the government dropped the charges. In 2001, when it was found that the 9/11 terrorists used the software to encrypt their communications, Zimmermann expressed regret.

Another problem that arises with international information interchange is that countries treat trade secrets, patents, and copyrights differently. Sometimes business partners are reluctant to transfer documents when one partner is in a country that restricts intellectual property rights, while another is in a country that has laws to protect intellectual property. On the other hand, the employees of a division of a multinational corporation might be able to divulge information locally with impunity. Intellectual property is tightly protected in the United States and Western Europe, and American trade negotiators and diplomats have pressured some countries to pass and enforce similar laws. Reportedly, the legislatures of several Asian nations have passed such laws or have revised existing laws in response to U.S. pressure.

Some nations are afraid that cross-border information flow promotes cultural imperialism.

© Michael Lassman/Bloomberg News/Landov

Political Challenges

Information is power. Some countries fear that a policy of free access to information could threaten their sovereignty. For instance, a nation's government might believe that access to certain data, such as the location and quantity of natural resources, might give other nations an opportunity to control an indigenous resource, thereby gaining a business advantage that would adversely affect the resource-rich country's political interests.

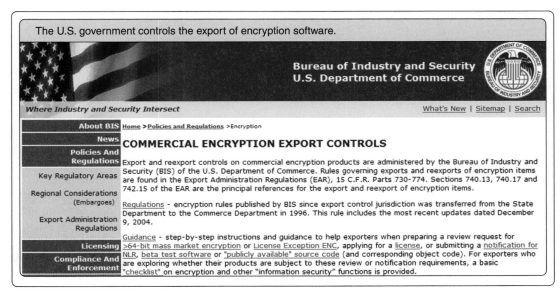

The U.S. government controls the export of encryption software.

As mentioned in Chapter 5, "Business Software," however, the recent trend in less rich countries is to adopt free open source software to avoid high costs. National governments in South America as well as local governments in Asia and Europe have adopted policies of using only open source software whenever it is available. Global corporations must ensure compatibility with the software adopted by governments and corporations in such locales.

Companies must also be aware of limits that some governments impose on Internet use. China, Singapore, and many Arab countries impose restrictions on what their citizens can download, view, and read. Free speech is not a universal principle. In practical terms, this means that executives might want to rephrase or cut out some content from their Web sites or risk their sites being blocked by some governments. This is an especially sensitive issue if a company enables employees or customers to use blogs at its Web site, in which they express their personal opinions.

Some corporations have found themselves in uneasy positions in countries that have limited civil rights. Microsoft, Yahoo, and Google were warned that they could not do business in China unless they collaborated with the government. This collaboration started by blocking certain search terms (e.g., "Taiwan" and "Falun Gong"), but were extended to providing the government the identities of people who searched for certain information, which the government took as a sign that they were dissidents. In some cases, individuals were imprisoned and tortured. The ethical dilemma for these companies is how to balance the business interest of their shareholders with moral principles of privacy, serving all Web users equitably, and not helping dictatorships to violate civil rights.

POINT OF INTEREST

Collaborating with the Censors

Companies that wish to do business in a large market where political interests limit freedom of speech often yield to government pressure. Microsoft Corp. cooperates with China's government in censoring the company's Chinese-language Web portal. The company's policy of cooperation affects blogs. It works with the authorities to omit certain forbidden language. Bloggers are not allowed to post words such as "democracy," "human rights," and "Taiwan independence." Attempts to enter such words generate a message notifying the blogger that such language is forbidden. Yahoo has been accused of releasing e-mail of Chinese dissidents who were later imprisoned and tortured. Several groups have tried to pressure executives at Microsoft, Yahoo!, Google, Cisco Systems, and other companies to urge the Chinese government for reforms on free expression, but free speech is no match for the economic interest of these companies. China's estimated online population is 137 million, second only to the United States. Some of the criticized companies justify their compliance by saying that their presence in such countries gradually encourages the free exchange of ideas.

Different Standards

Differences in standards must be considered when integrating ISs internationally, even within the same company. Because nations use different standards and rules in their daily business operations, sometimes records within one company are incompatible. For instance, the bookkeeping records of one division of a multinational company might be incompatible with the records of other divisions and headquarters. As another example, the United States still uses the English system of length and weight measures (inches, feet, miles, quarts, pounds, and so on), while the rest of the world (including England) officially uses the metric system (centimeters, meters, liters, kilograms, and the like). There are also different standards for communicating dates, times, temperatures, and addresses. The United States uses the format of month, day, year, while the rest of the world records dates in the format of day, month, year—so a date recorded as 10/12/08 might be misinterpreted. The United States uses a 12-hour time notation with the addition of a.m. or p.m., while other parts of the world use a 24-hour notation (called "military time" in the United States because the U.S. military uses this notation). The United States uses Fahrenheit temperatures, while other countries use Celsius temperatures. Americans communicate addresses in the format of street number, street name, and city name. Citizens of some other countries communicate addresses in the format of street name, street number, and city name.

Not resolving different standards can be extremely costly. In 1999, NASA lost track of a spacecraft that it sent to Mars. Reportedly, an investigation found that an error in a transfer of information between the Mars Climate Orbiter team in Colorado and the mission navigation

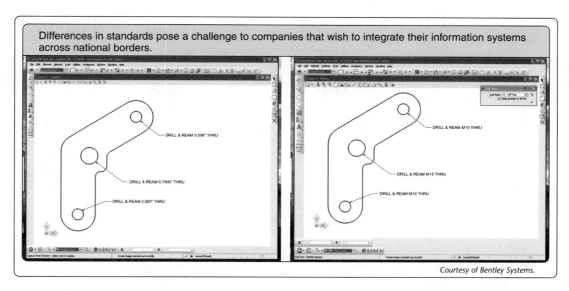

Differences in standards pose a challenge to companies that wish to integrate their information systems across national borders.

Courtesy of Bentley Systems.

team in California led to the spacecraft's loss. Apparently, one team used English units and the other used metric units for a key spacecraft operation. The information was critical to the maneuvers required to place the spacecraft in the proper Mars orbit. The cost to U.S. taxpayers was $125 million.

Companies that want to operate globally must adapt their ISs to changing formal or *de facto* standards. In recent years the growing number of countries joining the European Union (EU) imparted significant power to this bloc. Corporations in non-EU countries have grown accustomed to adapting their systems to those of the EU. For example, in 1976, Europeans adopted the 13-digit **European Article Number (EAN)**, while American companies used the 12-digit **Universal Product Code (UPC)**. The additional bar in the EAN bar code identifies the product's country of origin. For seven years, the American **Uniform Code Council (UCC)** promoted the use of the European standard. In 2004, the organization officially adopted it. Retailers embarked on a hectic effort to modify information systems to recognize, record, and process UPCs of 13 bars instead of 12 bars so they could meet the January 2005 deadline. Most bar-code readers could already read the extra bar, but the software in back-office systems—such as sales, shipping, receiving, and accounting systems—had to be modified. Best Buy, the large electronics and appliance retailer, spent 25,000 hours of staff and consultant time to ensure that cash registers, software applications, and databases could process and store the extra digit.

The UCC is trying to expand product codes to 14-digit **Global Trade Item Numbers (GTINs)**. This code is large enough to identify more than 100 times the number of products and manufacturers that the 12-digit UPCs could. GTINs are designed to support global supply chains. Eventually, manufacturers and retailers might have to use either GTINs or another standard of larger codes embedded in RFID tags. A major push for using RFID tags is taking place in the United States, and American standards could expand to Europe and the rest of the world.

Legal Barriers

The fact that countries have different laws has a significant impact on global business in general, and on e-commerce in particular. The differing laws can pose serious challenges to international transfer of data, free speech, and the location of legal proceedings when disputes arise between buyer and seller.

Privacy Laws

Although many of the challenges involved in cross-border data transfer have been resolved through international agreements, one remains unresolved: respect for individual privacy in the conduct of international business. Interestingly, despite the importance attached to privacy, its value is not even mentioned in the constitutions of the United States and many other countries. Nonetheless, a majority of the democratic nations try to protect individual privacy.

Legal Jurisdictions in Cyberspace

Imagine you are surfing the Web and come across a virulent site that preaches hatred and violence. You file a complaint in court, but the court cannot do anything because the site is maintained in another country that does not uphold your country's law. Or, you shop on the Web and purchase an item from a site that is physically maintained on a server in another country. When you receive the item, you discover that it is of a lower quality than promised. When you contact the site, the owners are rude and unresponsive. You decide to sue, but under which country's laws? These problems are two examples of the legal challenges in today's electronic global markets.

- **Global Free Speech.** In the spring of 2000, the International League against Racism and Anti-Semitism (LICRA), the Movement against Racism (MRAP), and the Union of French Jewish Students (UEFJ) filed a lawsuit against the American Internet company Yahoo! in a French court. The organizations complained that Yahoo!'s auctioning of more than 1,200 Nazi-related items amounted to "banalization of Nazism," which violates French law. The Nazi items offered for sale on the site included everything from Nazi flags and uniforms to belt buckles and medals. In November 2000, Judge Gomez ruled that a French court had jurisdiction over Yahoo! for violations that occur within France. He ordered Yahoo! to block French citizens' access to auctions of Nazi items within three months or face a fine of 15,000 Euros ($13,000 at the time) per day.

    Like many global e-commerce sites, Yahoo! did not require its Webmasters to maintain a dedicated site for each language. French users merely saw a customized overlay using the main Yahoo! pages that are viewed by all visitors. This technique has enabled Yahoo! to offer country-specific and language-specific versions of its site at relatively low cost. Yahoo! screened out the items from its French site, but this did not satisfy the court because French citizens could still view the items via the general site.

    To countries that have been subjected to a ruthless occupation, free speech is less important than preventing offenses such as "banalization of Nazism." In the United States, however, free speech is legally protected even when we dislike someone's opinion or trade of objects that we find offensive. Yahoo! decided to remove such items from all its servers but also received a California court decision that nullifies the decisions of non-U.S. courts regarding U.S. firms.

Consider this situation: a ruling against a company registered in India by a court in Germany, and an appeal in a court in India of a court decision made under the laws of Germany. This legal tangle is the result of doing business globally. More than one law might govern a business practice or communication of ideas. The legal environment was once confined to national boundaries, and jurisdiction referred to a territory. This is no longer so. Now the "territory" is cyberspace. It is difficult to define jurisdictions in cyberspace. The lingering question is, "Whose law applies?"

- **Consumer Protection by Whom?** Where can consumers sue for e-commerce transactions gone wrong? Suppose you purchased an item from a site located in another country, and the item has a defect or arrived after the time promised. Because your request for compensation or another remedy has not been answered satisfactorily, you decide to sue the e-tailer. Where do you file the lawsuit? Your own country? The e-tailer's country? The venue of e-commerce lawsuits is still undecided in many parts of the world.

    In November 2000, the European Union (EU) passed a law that lets consumers file lawsuits against an online business in any of the member countries composing the EU. Before the amendment to the 1968 Brussels Convention (which regulates commercial-legal issues in the EU), consumers could sue an online business only in courts in the country of the online business. If a Web site has directed its business at consumers in a certain country, the consumers can sue the Web site's owner in their own national courts. Businesses vehemently opposed the move, but consumer advocates said people would be more confident about online shopping if they knew they could get redress in their local courts.

- **Two Approaches to Jurisdiction.** As you have seen, the issue of e-commerce jurisdiction is broad. The U.S. Federal Trade Commission and European government organizations have examined the issue in an attempt to reach an international agreement such as the one reached within the EU.

    There are two approaches to such agreement. One approach is the country-of-origin principle, whereby all legal matters are confined to the country from which the site operates. Under this principle, the laws of that country apply to the operations and conduct of the site and whoever interacts with the site, regardless of their own location. Therefore, a lawsuit could be brought only in the country of the Web site and would be adjudicated according to that country's laws. Under this principle it is likely many firms would opt to establish Web sites in countries with lax consumer protection laws.

The other approach is the country-of-destination principle, whereby dealings with the site, regardless of the site's country of operation, are guided by the laws of the country to which the site caters. The EU adopted this approach within its territory, however achieving broad international agreement on e-commerce jurisdiction might take several years.

POINT OF INTEREST

International Hall of Shame

Privacy International (PI) is an international organization that monitors governments and commercial organizations around the world for violations of privacy. Each year, the members and affiliated organizations of PI present the "Big Brother Award" to the most invasive company, worst public official, most intrusive government, and most appalling project or technology for threatening privacy. Between 1998 and 2007, more than 74 ceremonies had been held in numerous countries and hundreds of awards had been given out. A "lifetime menace" award is also presented. The award is a golden statue boot stamping upon a human head. The organization rated Russia and the U.K. as "endemic surveillance societies," the worst ranking. The United States was ranked among "extensive surveillance societies," the second worst ranking. Germany and Canada were ranked as countries with "significant protections and safeguards" of privacy. To counterbalance the shameful awards, PI also gives awards to individuals and organizations that have made an outstanding contribution to the protection of privacy. You can read more about it at www.privacyinternational.org/bba.

Countries differ in their approaches to the issue of privacy, as reflected in their laws. Some are willing to forgo some privacy for the sake of a more free flow of information and better marketing. Others restrict any collection of personal data without the consent of the individual.

The European Union enforces a privacy law called the Directive on Data Privacy. Member countries have crafted their laws according to the Directive. Usually the law is titled "Data Protection Law." The EU defines personal data as "any information relating to an identified or identifiable natural person; an identifiable person is one who can be identified, directly or indirectly, in particular by reference to an identification number or to one or more factors specific to his physical, physiological, mental, economic, cultural, or social identity." Some of the principles of the directive are in stark contrast to the practices of U.S. businesses and therefore limit the free flow of personal data between the United States and the EU. For example, consider the following provisions and how they conflict with U.S. practices:

- Personal data can be collected only for specified, explicit, and legitimate purposes and not further processed in a way incompatible with those purposes. However, in the United States, businesses often collect data from people without having to tell them how the data will be used. Many U.S. corporations use personal data for purposes other than the original one, and many organizations purchase personal data from other organizations, so subjects do not even know that the data is used, let alone for what purpose. Obviously, these activities would not be allowed under the EU directive.

- Personal data can be processed only if the subject has given unambiguous consent or under other specific circumstances that the directive provides. Such circumstances are not required by American laws. In the United States, private organizations are allowed to process personal data without the subject's consent, and for practically any purpose.

- Individuals or organizations that receive personal data (the directive calls them "controllers") not directly from the subject must identify themselves to the subject. In the United States, many organizations purchase personal data from third parties and never notify the subject.

- People have the right to obtain from controllers "without constraint at reasonable intervals and without excessive delay or expense" confirmation that data about them is processed, to whom the data is disclosed, and the source that provided the data. They are also entitled to receive information on the "logic involved in any automatic processing of data concerning" them, at least in the case of automated decision making. Decision making, practically speaking, means using decision-support systems and expert systems to make decisions on hiring, credit extension, admittance to educational institutions, and so forth. None of these rights is mandated by any U.S. law.

- People have the right to object, "on request and free of charge," to the processing of personal data for the purpose of direct marketing, or to be informed before personal data is disclosed for the first time to third parties or used for direct marketing. Furthermore, controllers must expressly offer the right to object free of charge to disclosure of personal data to others. American companies use personal data *especially* for direct marketing, never tell subjects that they obtain data about them from third parties, and rarely offer subjects the right to object to disclosure of such data to other parties.

American companies are very busy collecting, buying, and selling personal data for decision-making and marketing purposes. The American view is that such practices are essential to efficient business operations, especially in marketing and extension of credit. Thus, this huge discrepancy between the European and American approaches does not allow unrestricted flow of information.

The EU directive is only a framework within which member states may maintain their own, more restrictive, laws. Consider, for example, the French law, which states, "An individual shall not be subject to an administrative or private decision involving an assessment of conduct which has, as its sole basis, the automatic processing of personal data defining his profile or personality." This provision limits the use of a computer as a sole decision aid in certain circumstances. For instance, this law forbids automatic decisions for credit applications or admittance to a college. While the latter decision is often accompanied by human intervention, the former is often fully automated in the United States and other countries.

The EU directive recognizes that countries outside the European Union use personal data that are transferred from the EU. It therefore provides that when a "third country does not ensure an adequate level of protection within the meaning of [the directive], member states shall take the measures necessary to prevent any transfer of data of the same type to the third country." This provision has created an interesting situation: agents of the European Data Protection Authorities (DPAs) arrive at least monthly in the United States to monitor American companies that process personal data of European citizens to ensure that the EU Directive on Data Protection is obeyed regarding these citizens. These representatives monitor the ISs of companies such as Visa, MasterCard, American Express, and other credit-card issuers. Companies that want to do business in EU member states must accept the restrictions of the directive on their practices. Business leaders on both continents hope that a way can be found to bridge the gap between the two approaches to data privacy, but it seems that a legal solution will not come before a change in culture.

In the meantime, a practical solution has been sponsored by the U.S. federal government. The EU agreed that the U.S. Department of Commerce could establish a **Safe Harbor**, an arrangement for U.S. companies that have agreed to comply with the EU directive regarding EU citizens so that European companies can trade with these U.S. companies without fear of violating the directive. By June 2007, 1184 U.S. companies had joined the list. You can view information about the Safe Harbor arrangement and the list of companies that have joined at www.export.gov/safeharbor.

Since privacy laws regarding employees—not just consumers—are also different in the United States and European Union, American companies employing European citizens must comply with EU laws regarding transfer of employment information. They must comply with the DPAs. Under the Safe Harbor arrangement, claims of European citizens against U.S. companies regarding privacy are heard (with some exceptions) in the United States. To be sure they are not breaking the law, European companies that wish to transfer personal data to U.S. companies can simply check at the Web site previously noted to see if those U.S. companies are listed.

Applicable Law

As discussed in the Ethical & Societal Issues box in this chapter, countries have differing laws regarding free speech, which can significantly impact what a company may or may not display from its servers. Other laws affecting online business include those that address gambling, auctioning, sales of alcoholic beverages and drugs, and other areas. After establishing online business in another country, some companies discovered that their practice was not in compliance with a local law. For example, eBay discovered that Dutch and Italian laws required that a certified auctioneer be present at any auction. This made its online auctions illegal in these countries. Some countries have changed their laws to accommodate online business, but others have not. Such legal discrepancy among jurisdictions should not come as a surprise to executives; they must research the legal environment in every jurisdiction where they intend to do business. The lessons of Yahoo!, eBay, and other online pioneers prompted many companies to employ legal research experts before they start business in a new jurisdiction. Often, this effort is part of a larger effort to research the local culture and practices. Some companies have hired local experts to help them in assessing local considerations, and in some cases executives decided to avoid doing online business in certain countries altogether.

As mentioned before, legal barriers to online business often exist within a country. For example, states in the United States have different laws regarding purchase and delivery of alcoholic beverages. A company selling wine online to individual consumers must ensure that the buyer's state allows home delivery of wine. Shipping to a state forbidding such a transaction is criminal.

Different time zones must be considered by all organizations that do business in multiple countries.

© Eddie Gerald/Alamy

Different Time Zones

Companies that operate in many global regions, especially multinational corporations, must craft policies for the work of both their employees and information systems. Teleconferencing systems must be available much of the day, and in many cases 24 hours per day, so that employees many time zones apart can communicate to discuss problems that need immediate resolution. Teams in support centers might have to work in shifts to accommodate clients worldwide. When scheduling teleconferencing sessions, managers in North America should remember, for instance, that scheduling a session for Friday afternoon with their Australian counterparts will force the Australians to come to the office on Saturday morning.

In their global supply chain management systems, managers must be aware of what might seem to be incorrect time stamping in shipments and payment records. For example, consider interaction between a corporation's Pennsylvania manufacturing plant and its South Korean assembly plant. Because South Korea is 14 hours ahead of Pennsylvania, shipping records could show that subassemblies were shipped from Pennsylvania a day before they were ordered in South Korea. To eliminate confusion, the systems at both locations can be designed to record the local times of both locations, or only that of a single location, such as the company headquarters' time.

SUMMARY

- As more companies use the Web for both B2C and B2B business, they realize that they must accommodate non-English-speaking audiences and tailor their sites to local preferences. They also must be carefully attuned to the cultural differences and payment preferences of different world regions as well as be aware of legal and tariff issues.

- Organizations that engage in international trade, especially through the Web, must also be aware of the linguistic, cultural, economic, and political challenges involved in such trade.

- One important unresolved issue is the discrepancy between the laws governing the collection and manipulation of personal data in two economic powers, the United States and the European Union, which have incompatible data privacy laws. This difference restricts the flow of personal data between the United States and the EU. The Safe Harbor arrangement enables EU companies to do business with U.S. businesses that comply with EU policies on handling personal data of its citizens.

- Several cases have demonstrated that the old legal approach of territorial jurisdiction is inadequate when so much information is communicated and so much business is conducted on the Internet. Issues such as free speech and consumer litigation of e-tailers have brought to light the need for an international legal reform for cyberspace.

IT FITS OUTFITS REVISITED

It Fits Outfits is a global enterprise. The company's clothing is manufactured in China and exported to the United States. They would also like to extend their e-tail and retail operations to other countries. Let's explore some of the issues it faces in operating globally.

2. It Fit Outfits recently created an online storefront for the college market. They would like to be able to sell these products to college students in Europe and Asia as well. How should they deal with payment and currency issues?

What Would You Do?

1. Contracting with manufacturers in another country involves communicating with the company, understanding the laws of the country, and uncovering political, economic, and technological issues that might affect trade. Describe the obstacles you think It Fits Outfits might face as it expands to other countries such as Bangladesh.

New Perspectives

1. In the future, Shari Steiner hopes to open branches of It Fits Outfits in large cities around the world. How do you think the company's unique business principle, customer participation in design, will be received in different countries? What adjustments should the company make to accommodate other cultures?

European Article Number
    (EAN), 325
global information
    system, 315

Global Trade Item Number
    (GTIN), 325
glocalization, 316
Safe Harbor, 328

Uniform Code Council
    (UCC), 325
Universal Product Code
    (UPC), 325

REVIEW QUESTIONS

1. What is meant by the term "global information systems"?

2. Executives of multinational corporations are advised to think globally and act locally. What does this mean?

3. Manufacturers and retailers have used product bar codes for many years. What information does the 13-digit European Article Number (EAN) contain that the 12-digit Universal Product Code (UPC) did not, and why is this information important?

4. Is every Web site a form of global IS? Why or why not?

5. Using software for automatic translation of Web pages into other languages for local audiences saves much labor cost and time. If you were an executive for a company that maintains a multilingual Web site, would you settle for software-based translation only? Why or why not?

6. Many organizations, especially multinational corporations, must consolidate reports to ensure smooth operations. These reports include currency, measurements, and dates. How would you help them receive reports "on the fly" that are in the desired currency and format?

7. Many European countries have stricter privacy laws than the United States. What is the impact of this discrepancy on multinational corporations with offices on both continents? In terms of business functions, which activities, in particular, are affected?

8. Give three examples of cultural imperialism. Why do you think your examples reflect cultural imperialism?

9. American companies whose main business is Web search have encountered political challenges in some countries. Give two examples of such challenges.

10. What are the implications of different time zones for global supply chain management systems?

11. Countries can adopt either a country-of-origin law or a country-of-destination law. What is the difference between the two approaches? Which is more helpful to consumers and which is more helpful to e-tailers? Explain.

DISCUSSION QUESTIONS

12. Ask yourself: what are the "nationalities" of the following corporations? Consider nationality to be the country where the corporation is registered: SAP (software), Bull (computers), BP (gasoline), CheckPoint (security software), LG (electronics), Corona (beer), Heineken (beer), Goodyear (tires), JVC (electronics), Braun (small appliances), Siemens (electronics), Nokia (mobile telephones), Business Objects (software). In your opinion, if a company has its headquarters in Bermuda, is it a "Bermudan" company? If so, in what respect? If not, explain why not.

13. Several technologies have been practically given away by the United States to the world. Name at least two such technologies. Do you think that this was "charity" or that the United States reaps some benefits from having made the technologies widely available? Explain.

14. The U.S. Department of Commerce has relaxed restrictions on the export of encryption (scrambling) software for communications, but it still bans the export of many such applications. Do you agree with such bans? Why or why not?

15. Almost all of the European Union countries use the Euro as their common currency. Does this help or hinder international ISs? Explain.

16. Allegedly, Yahoo handed over e-mail correspondence of suspected dissidents to the Chinese government, and those dissidents were arrested and tortured. Yahoo, Microsoft, and Google have complied with the Chinese government (restricting search engines, delivering private records, etc.), claiming that they comply with local laws in order to do business there. Do you agree with this approach? If you do, why? If you don't, what would you recommend to these companies?

17. Consider sensitivity to privacy in the United States. Are Americans more sensitive to *government* handling of private information or to *business* handling of private information? Now answer the question regarding Western European countries.

18. Apparently, the European Union has stricter privacy laws than the United States, and not many U.S. companies are willing to comply with the EU Directive on Data Protection. The Safe Harbor arrangement is one way to resolve the issue, but only several hundred companies have subscribed. How would *you* resolve the conflict?

19. If a non-English-speaking country had established the Internet, do you think that country would impose its own "cultural imperialism" on the Web? Why or why not?

20. Which legal approach do you prefer for e-commerce: country of origin or country of destination? Answer the question as a businessperson, then answer it as a consumer.

21. If some countries clearly adopt the country-of-origin approach for legal issues of e-commerce, online retailers might relocate to operate from those countries. Why?

22. An American company employs engineers in California and in several Asian and European countries. The engineers exchange e-mail and communicate via VoIP, teleconferencing, and collaborative project management tools. The Americans often use phrases such as "Let's touch base in a week," "Right off the bat...," and "...all the way to the end zone." An executive instructs them to avoid such phrases in communication with colleagues from other countries, and perhaps even with any colleague. Why?

APPLYING CONCEPTS

23. You are an executive for Bidway.com, an auction site that has successfully competed with eBay and Yahoo! in the United States. Management decided to open use of the site to residents of all countries. You were given an important assignment: collect intelligence that will help ensure a smooth transition from a national business to an international business. If you envision that there might be too many difficulties in certain countries, management will accept your recommendation to block bidding by residents of those countries, but you must be careful not to miss potentially profitable markets. Prepare an outline of all the aspects about which you will collect intelligence for each country, and explain why this item is important.

24. Why does the United States still use the English system? When was the last attempt to officially move to the metric system? Does the use of English measurement units put U.S. companies at a disadvantage when competing on international contracts? How has software solved the challenge? Give examples of engineering software that resolves this challenge. Research on the Web and summarize your findings in two pages.

25. You are the international sales manager for Aladdin Rugs, Inc., a multinational company headquartered in the United States. At the end of every month, you receive reports from the national sales managers on your company's operations in England, Germany, and Japan. The products are sold by area. The managers report the units sold and income from sales in their national currencies: pounds sterling (£), euros (€), yens (¥), and U.S. dollars (US$). Use your spreadsheet program to consolidate the sales reports you received, as follows.

    a. Under "Totals," enter formulas to convert square yards to square meters and enter another formula to total the area in square meters for all four countries.

    b. In a financial newspaper such as the *Wall Street Journal* or on the Web, find the rates of exchange for the three currencies against the US$ on the last business day of last month. Enter a formula that will convert all non-U.S. currencies to US$. (Extra challenge: program a macro to do the calculations.)

    c. Test all formulas with actual numbers.

26. Google and other sites offer Web-based translation services. Test the quality of such tools. Write a message of 50 words in English. Use the tool to translate it to German or another language with which you might be familiar. Copy the translated text, paste it to be translated, and use the tool to translate it back to English. Compare the original and translated English messages. Write a short report and e-mail it to your professor. How good is the translation tool? About how much of the text in the translated version came out identical to the original? Was the *spirit* of the message accurate (even if in different words)? Did you find anything funny in the back-translated text?

27. Team up with three other students. Decide on three keywords with which the team will conduct a Web search. All of you should use the same search engine. One team member should record the number of sites found in the United States, another in Germany, another in France, and another in the Netherlands. Also, record the sites the team found whose domain name is non-U.S. but that used English rather than, or in addition to, the local language. (Note that many non-U.S. companies use the .com and .org top-level domains. Before you start this assignment, research the Web for ways to determine the location of Web sites by their IP address rather than the TLD.)

    Prepare a brief report detailing what you recorded. Write your own conclusion. How dominant is English on the Web? Do you think the Web is "Americanized"? Do you consider what you found to be cultural imperialism?

28. "Electronic immigrants" are residents of one country who are employed by a company in another country. They are the result of what some people call "offshoring" of jobs. They deliver the results of their work through the Internet or private communications networks. Your team should conduct research with two companies in four different industries, one of which is in software development. The title of your research is "The Electronic Immigrant: Economic and Political Implications." Contact the human resource managers of the two companies, present the issue, and ask for the managers' opinions on the following issues. Can the company use "electronic immigrants"? Can it be hurt if competitors use them? Do the HR managers think the national economy can gain or lose from the phenomenon? Do they foresee any political ramifications? Your team should prepare a report starting with half a page of background on each company.

FROM IDEAS TO APPLICATION: REAL CASES

Export with Confidence

The products of Fairchild Semiconductor are installed in a large array of items, from satellites and motor vehicles to cell phones, medical devices, and home appliances. The company is a world leader in design and manufacturing of microprocessors that control power. Fairchild calls itself The Power Franchise. The company was established in 1957, and is the world's largest supplier of power semiconductors. The company has manufacturing facilities in the United States, South Korea, and Singapore, and has assembly and testing facilities in China, Malaysia, and the Philippines. It maintains 36 offices in 17 countries. It sees the entire world as its market. Fairchild ships more than 17 billion units of products annually. Some products are shipped to as few as 6 countries, while others are shipped to as many as 45 countries.

The microchips are designed in the United States and South Korea. Manufacturing starts in the United States. Then, the chips are sent to plants in Asia for assembly and testing. The products are then shipped to customers around the globe. Customers can order products online at the company's Web site. Over the past decade, Fairchild established additional design and manufacturing facilities in Korea and China. With facilities as well as customers in a growing number of countries, complying with both U.S. and other countries' export and import laws became complex. Work-in-progress is often shipped from one country to another, and then to another or back to the original country for further processing. Logistics managers decided to use the services of a company that specializes in software that helps manage such complex operations. Fairchild approached Management Dynamics, Inc. (formerly Nextlinx, Inc.), a company with expertise in software that supports online logistics.

Together, the companies configured and implemented Management Dynamics software called Trade Export Solution, which automates Fairchild's global logistics. The software provides information on laws and regulations of each country where Fairchild transacts, as well as customs duties. For each shipment it determines the fastest and least costly carriers as well as the minimum duties to be paid. The application provides digital forms that enable employees to enter details on content, value, and destination of a shipment. The various costs are calculated automatically for the country and particular seaport or airport. The system ensures full compliance with the country's regulations. In recent years, many new U.S. regulations have focused on homeland security issues. They may forbid the export of certain types of microchips to some countries, or the export of certain items from some countries. All this information is closely monitored by Management Dynamics and added to the documentation and automated forms.

The software has been implemented in every facility of the company in the world. It is integrated into Fairchild's enterprise resource planning (ERP) systems as well as the shipment application of its major shipping carrier. Shipping clerks can easily retrieve trade documentation and be sure that all shipping complies with the destination country's regulations. This saves the typical labor-intensive search for trade compliance documentation, helping the company to clear 90 percent of its shipments with the proper authorities before the shipment reaches the destination country. The system also shortened shipping durations and reduced shipping delays. As a result, Fairchild could reduce the amount of raw materials inventory it carries.

The use of the new system reduced the number of employees involved in shipping processes, but it also had another positive effect—it standardized shipping procedures and records worldwide. The same shipping procedures and records are maintained at every company facility, anywhere in the world. This enables Fairchild to easily implement the procedures and documentation in new facilities it might establish in the future.

Sources: (www.nextlinx.com/news/casestudies/case_fairchildsemi. shtml), 2005; (www.nextlinx.com/html/case/mdi_case_fairchild. shtml), August 2007; (www.fairchildsemi.com), August 2007.

Thinking About the Case

1. The software Fairchild integrated into its ERP system reduces risk. What risk?

2. List the cost savings provided by the software.

3. Why is it important to integrate an application such as Management Dynamics' to the ERP system of the adopting global company?

4. Why is standardization of processes so important to a company such as Fairchild?

If You Want to Do Business Here...

Sometimes standards at a foreign market force a business to improve its products or leave that market. The higher standard could require better information systems, and the results could have long-lasting positive effects on the products worldwide. Such seems to be the case with Kia Motors, the Korean car manufacturer.

In 1995, when Kia Motors first started selling cars in North America, it found itself the butt of many jokes. Kia Motors America, based in Irvine, California, is the Korean company's U.S. subsidiary. The company offered its cars for low prices. Prices were so low that Kia managed to increase sales from 12,000 cars in 1995 to 270,000 in 2004. Yet, the cars were notoriously low quality, which translated into costs involved in fixing repeated defects. Until 2002, Kia was ranked at the bottom of J.D. Power and Associates' annual quality survey. J.D. Power's survey reports on car quality are considered the most trusted in the United States and many other countries. The reports rely on responses of vehicle owners after 90 days of ownership of a new car. In 1997, a car manufactured in North America had 1.1 defects on average, whereas Kia had 2.75 defects. In 2002, the auto industry average was 1.33 defects per vehicle, whereas Kia cars had 2.12. An expert noted that such improvement over just five years is impressive in the auto industry, but Kia's ratio was still significantly worse than the average. In addition, the expert said, it takes a long time to change consumer perception of quality.

The CEO of Kia Motors America, Peter Butterfield, was determined to change the defect ratio and remove Kia's stigma as a manufacturer of low-quality vehicles. He set a goal: by 2007, Kia's defect average would reach the auto industry average, and by 2010, Kia's quality would be equal to that of the top Japanese manufacturers.

The CEO announcement came after a U.S. federal mandate was declared. In 2000, the U.S. Congress passed the Transportation Recall Enhancement, Accountability, and Documentation (TREAD) Act. Under the new law, all manufacturers of motor vehicles sold in the United States were ordered by the National Highway Traffic Safety Administration (NHTSA) to operate systems that can report all defects, accidents, and injuries involving their vehicles. The deadline for implementing the system was December 1, 2003. Under the Act, if senior executives failed to include any of the required details in the quarterly reports, they could be prosecuted and sentenced to time in prison. The key to improving quality was monitoring defects, and monitoring defects required better information systems. Many

of the elements of such a system were required by the new law anyway. But Kia decided to do more, for its own sake.

Management faced daunting challenges. The data from which the company could glean the information required by the new law resided in seven different databases on different computer systems, mainly because each system was owned and operated by a different department. The warranty department, the parts department, and the legal and consumer affairs department managed their data on different computers that were neither networked nor integrated. For example, if a consumer complained about failed brakes, the complaint was recorded by the legal and consumer relations department in its database, but the other departments did not have access to the data. Other customers might have had the same problem, but had it fixed without complaining to that department, so the department had no record of those occurrences.

To create the quarterly reports for NHTSA it was possible to retrieve the disparate information from the databases and combine the pieces manually. However, this would not only entail much labor but also cause Kia to miss a great opportunity. Management's purpose was not only to comply with the law but also to ensure that the CEO's commitment to quality was fulfilled. One expert noted that the problem of maintaining information in separate places is that managers can never receive a complete picture. A car owner might call the consumer relations department and file a complaint only after she had the air conditioner in her car fixed three times. This is the first time that department knows about the problem. By that time the owner might have decided never to purchase a Kia again. Other owners might not file a complaint with the department at all but reach the same decision. However, if the customer relations department learns about the problem at first occurrence, managers can see an overall picture created by these puzzle pieces, and draw the engineers' attention much sooner.

In late 2002, Kia engaged Infogain, a software consulting firm. Infogain professionals implemented a central application that links to each of the databases. It can reach out to each database and break down and categorize all the data around individual car components, such as power train, steering assemblies, or headlights. The aggregated information is stored on a Microsoft SQL Server database. The application retrieves data from the disparate databases on a daily basis, combining data from areas such as warranty claims, parts sales, vehicle identification number master storage files, and vehicle inventories. It is also connected to a customer relationship management (CRM) system that monitors consumer complaints. The system automatically creates reports of repeating problems. For example, if a dealership tried three times to fix a steering system and still cannot find out why the failure recurs, Kia would send out one of its engineers to the dealership to investigate. The engineer files the report in the CRM system.

Analyzing parts used in repairs is easy, because this is structured data. Customer relations are much more difficult to analyze, because they are not structured. Customers often telephone Kia or send e-mail messages to complain. Under the TREAD Act, such communication must be reviewed to see if it contains details that must be reported to NHTSA. Evaluating recorded communication manually could double the agents' amount of labor.

Infogain implemented an application that uses keywords to search recorded text reports. Keywords include "fire," "burn," "spark," "combustion," and "smolder." If a keyword is found, the application prompts the customer relations agent to conduct further investigation and see if the incident needs to be included in the NHTSA reports. If a pattern emerges, engineers might be called to investigate and propose changes to designs. However, the agent might find out that a customer was just angry and suggested they "fire the salesperson." This "fire" might trigger an investigation of the dealership but is not reported to NHTSA.

Kia employs 50 agents at its main customer call center in Irvine, California. With the help of the new system it did not have to add a single employee to comply with the new law. Managers now have a rich database in SQL Server. They use a business intelligence application called Crystal Analysis, which is sold by Business Objects SA, a company that specializes in online analytical processing (OLAP) and data-mining software. (These techniques are discussed in detail in Chapter 11, "Business Intelligence and Knowledge Management.") Managers can get information about a world region or drill down to focus on a recurring problem with a specific part across all dealerships. They can retrieve information across departments by daily, weekly, or quarterly reporting periods, and by car models, model years, and components.

Managers have found that part sales are the first indicator of a defect and that warranty claims are the second. For example, if Kia receives a monthly order that is 15 percent or larger than the historical average for a particular part, an investigation is triggered because this is a good indication that something must be wrong with that item. Managers might notice that brake pads are inordinately ordered only for the four-wheel drive version of a model but not for the two-wheel drive models, and therefore the problem might have to do with vibrations of the vehicle, not the pads. The decision will then be made to look into the structural design of the vehicle. With 60,000 different parts that go into assembling a car, decisions such as these can save the company much money and help improve the design of new cars.

The new law and pressure for higher quality in the United States might have helped this Korean company. J.D. Power reports show that Kia had 1.53 defects per vehicle for its 2004 model year, and that the ratio decreased to 1.40 for the 2005 model year. Kia did particularly well in the compact car category. The Kia Spectra was ranked second after the Toyota Prius and ahead of the Honda Civic and Toyota Corolla. By 2007, the company's vehicles were no longer the subject of jokes. The company's popular minivan, Sedona, as well as another three cars, earned five stars (the highest rank) for safety from the U.S. NHTSA. As Kia's Web site says, the company now has so much confidence in the quality of its cars that it includes in the car sale a warranty for 10 years or 100,000 miles.

Sources: Duvall, M., " Kia Motors America: Lemon Aid," *Baseline* (www.baselinemag.com), June 10, 2005; (www.kia.com), 2007; (www.businessobjects.com), 2007.

Thinking About the Case

1. Do you think market forces would push Kia Motors to invest in IT the way it did if the U.S. government had not passed the TREAD Act? Why or why not?

2. What was the role of IT in raising the quality of Kia's cars? Is IT alone the reason for improvement?

PART FOUR

Decision Support and Business Intelligence

CASE IV: DEBOER FARMS

The DeBoer family's roots run deep in their South Dakota farm. Carl DeBoer's great-grandfather Johann began farming in the Dakota Territory just before it became a state in 1889. Through the years, each generation worked hard, saved, built improvements, and added to the farm, until it grew from its original several hundred acres to its current size of 12,000 acres. His great-grandfather would hardly recognize the place these days, Carl thought. DeBoer Farms' acreage is planted in corn, soybeans, wheat, oats, and alfalfa. Carl inherited the farm last month when his father passed, although he had been managing the farm on his own for the past ten years. Big changes lay ahead for both Carl and DeBoer Farms.

Relying on Scientific and Technological Advances

The farm industry had always relied on the benefits of scientific discoveries and inventions. Advances in machinery, fertilizers, insecticides, land use, and plant breeding had fostered a healthy agricultural industry since—well, even before Johann DeBoer's time. Carl scanned his wheat field. In the late 1830s,

when John Deere started manufacturing the first steel plows, it would have taken him about 300 hours to produce 100 bushels of wheat. Today, Carl could do it in just under three hours.

Like other businesses caught in the industrial revolution, farms had to adopt new innovations or they would become less productive than their competitors. Johann DeBoer had planted the first hybridized corn and spread the first mixed chemical fertilizers on the farm. Carl's grandmother Elizabeth had bought the first tractor during the Second World War. His father had introduced new pesticides and no-till methods to curb erosion. Now, Carl had brought the farm into the computer age.

The IT Revolution

The computer revolution came to farming in the early 1990s. Several years ago, only a little more than half of all farms used personal computers. The DeBoer farm couldn't have afforded to wait that long if they were to keep up with other mid-sized and large farms. Back in the late 1980s, Carl had converted the farm's manual financial and accounting systems to computer applications. They began to use spreadsheets and then databases to track information such as crop yield and soil testing results. Then the DeBoer family discovered the Internet. Carl could get the latest weather information and investigate pricing issues as well as order farm equipment, parts, and repairs online.

The real revolution came in the mid-1990s. Carl remembered taking his 15-year-old daughter Allie to an exposition that featured a global positioning system (GPS). The GPS was part of precision farming, a new approach to farming that manages a variation of factors, such as soil nutrients, within a field. Up until that point, farmers had practiced whole-field management. Now they could program their equipment to spread more fertilizer in one area and less in another. Allie thought the system was cool. Carl moaned about the headache of converting databases and installing new software. They both agreed that precision farming would decrease nitrogen runoff and be better for the environment.

After the expo, Carl had taken a "wait-and-see" approach for a few years, but when other farmers began to report increased profits with precision farming, Carl decided that it was time to try it out for himself. Since Allie was now helping him manage the farm, she could help him sort out technological issues.

Growing Information in the Farming Industry

In the past, Carl often drove over to the South Dakota Cooperative Extension Service to discuss soil and other crop-related issues with a staff member, usually Steve Janssen. The Cooperative Extension Service provided invaluable information to the farming community: what counties were being affected by the corn borer moth; whether heavy rains were depleting the soil of nitrogen; how to record restricted use of pesticides in accordance with federal law. The service provided worksheets to determine whether farming practices might have affected groundwater and drinking water. It also provided information on drinking water treatment systems. The service also helped farm owners with strategic planning, commodity marketing, and production and risk.

Today most of this information was available on the service's Web site. In a couple of clicks, Carl could download a PDF publication on weed control in soybean fields. He could read an announcement from the Environmental Protection Agency (EPA) that an exemption had been granted to South Dakota farmers, allowing them to use tebuconazole on wheat to combat Fusarium head blight. He could access maps of soil temperature. He could even research the history of soybean aphids. Nowadays, he and Steve exchanged e-mails regularly. If the extension service didn't also hold expositions, Carl would hardly ever see his old friend.

As a matter of fact, Carl was heading to the regional manure-handling exposition this afternoon. He should probably start getting ready now. Carl looked at the watch his grandfather had given him as a high school graduation gift. It seemed like every time he looked up from this watch something was gone or replaced—farmhands were implementing some new technique, planting a new type of grain, or using a new piece of machinery. Carl sighed. It was all part of life.

Carl's watch said 10:45. He shook his wrist. The watch had stopped. He bent over and picked up his cell phone. It was 11:38—exactly.

BUSINESS CHALLENGES

If you were Carl DeBoer, how would you decide which technologies to use to improve your farm production? Technology is costly, but it also provides substantial benefits, as the DeBoers have seen on their farm. In the upcoming chapters, you learn how to recognize and evaluate decision support systems, expert systems, and knowledge management systems.

- In Chapter 10, *"Decision Support and Expert Systems,"* you learn how to determine the characteristics of businesses and decisions that can benefit from decision support systems and what is involved in creating and using them.

- In Chapter 11, *"Business Intelligence and Knowledge Management,"* you learn how knowledge management systems support the business process and how data mining can be used to establish strategic advantage.

© USDA Photo by: Tim McCabe

10 TEN

Decision Support and Expert Systems

LEARNING OBJECTIVES

Decision making plays a key role in managerial work. Managers often have to consider large amounts of data, extract and synthesize only relevant information, and make decisions that will benefit the organization. As the amount of available data grows, so does the need for computer-based aids to assist managers in their decision-making process.

When you finish this chapter, you will be able to:

■ List and explain the phases in decision making.

■ Articulate the difference between structured and unstructured decision making.

■ Describe the typical software components that decision support systems and expert systems comprise.

■ Give examples of how decision support systems and expert systems are used in various domains.

■ Describe the typical elements and uses of geographic information systems.

DEBOER FARMS:
Farming Technology for Information

Carl DeBoer was finishing some paperwork for the day in his farm office when his computer beeped. Steve Janssen from the South Dakota Cooperative Extension Service had sent him an instant message. Carl knew many of the service staff, but Steve was an old friend. Carl had consulted with Steve for the past 20 years.

Running Risk Simulations

"Saw you were logged in. Have you checked out the risk management calculator? They just uploaded the latest version," Steve's message read in the pop-up box.

Carl shifted his swivel chair in front of the keyboard and positioned his index fingers to begin typing, "No, but I was…"

The computer beeped again and Carl read Steve's new message. "It's very neat. You can run simulations. You can adopt different risk strategies such as crop insurance or government price protection programs and see the potential outcomes. That's the Strategies Grid. There's also a Revenue Grid. The calculator can help you decide what risk strategies to take!"

Carl erased his unsent message and began to type, "Is it a spreadsheet that…"

The computer beeped again. Steve wrote, "There's an instruction sheet in Word available online and the calculator is still in Excel, so you should have no problem downloading it. If you have any questions, just give me a buzz."

Carl again deleted his message and began to type, "Maybe I'll call…" when the computer beeped again. "Oh, for Heaven's sake," Carl said out loud.

"Gotta go, but I'll call you tomorrow afternoon," Steve wrote.

Carl deleted his message hurriedly and typed three letters, "Bye," and hit Enter. He smiled and wiped the sweat from his brow.

Obtaining Expert Advice from a Machine

Carl logged on to the state university's extension and outreach Web site. He planned to check out the calculator, but first he wanted an update on a corn rootworm infestation in his fields. Through his participation in the university's research study, Carl had access to a new expert system that the university was building. The program was called PestPRO and had been helping the DeBoers and other participants manage their fields to track, diagnose, and control pests.

Dr. Neil Wildes, an entomologist at the university, had been conducting studies of common pest infestations in corn crops since the mid-1980s. He was particularly interested in corn rootworm because of its widespread and devastating effect on crop yields. Dr. Wildes had constructed the expert system and was responsible for maintaining it. He had made the system available to the extension and outreach service to continue his research.

Eliminating Pests and Increasing Yield

Carl DeBoer logged on to the PestPRO site and pulled up information on his fields. He had always tracked rootworm populations by visual inspection, but last year the PestPRO system had recommended he place a dozen sticky traps in his fields during the insect's egg-laying period to get a more accurate count. The system told him that often farmers overestimate the number of pests in fields. PestPRO also told him that if he trapped more than six rootworm beetles per day per trap, then he should rotate that field to soybeans or apply soil insecticide during planting the next year. The site also recommended a new hybrid corn seed, which was bred to resist corn rootworm. Unluckily, Carl did trap more than the required number of beetles. He followed the system's advice, rotating crops in some fields and applying more insecticide in others.

Soon the DeBoers would be harvesting their corn crop, and they were particularly interested in the fields treated with additional insecticide. Would the soil insecticide be effective this year in reducing corn rootworm? Did the resistant hybrid seed help reduce the pests? They hadn't experienced a drought this year, which could hamper the insecticide's effectiveness in the soil. From their casual inspections and collected data so far, the rootworm population did seem to be lower. After the harvest, they would feed the yield information into the system. It would factor in the weather and general soil condition readings and formulate additional recommendations for the next year. It had taken a long time to develop the system—it was refined based on multiple readings over the years, and the results were continually compared with the judgments of Dr. Wildes and other expert entomologists. But the system seemed to be working. As Carl entered the data, counting the number of beetles in the sticky traps he had collected and replaced, the system reported that he had trapped fewer beetles than last year at this time.

Carl immediately e-mailed Steve. "Nice chatting with you this afternoon. Entered the data from the beetle traps and PestPRO is working. You, my friend, are outdated." Carl chuckled to himself as he clicked Send.

DECISION SUPPORT

The success of an organization largely depends on the quality of the decisions that its employees make. When decision making involves large amounts of information and a lot of processing, computer-based systems can make the process efficient and effective. This chapter discusses two types of decision support aids: decision support systems (DSSs) and expert systems (ESs). In recent years applications have been developed to combine several features and methods of these aids. Also, decision support modules are often part of larger enterprise applications. For example, ERP (enterprise resource planning) systems support decision making in such areas as production capacity planning, logistics, and inventory replenishment.

Furthermore, many vendors of computer-based decision support tools, such as Pilot Software and Cognos, Inc., no longer call their applications decision support systems. (Pilot Software was acquired by SAP, the world's largest supplier of ERP systems. The acquisition may be an indication of how decision support software is becoming an integral part of enterprise software.) They prefer to call them business analysis tools, business intelligence applications, or other names. In a way, almost any system that produces useful information is a decision aid. Decision support systems and expert systems are especially designed to streamline the decision-making process by providing either a single optimal solution to a question or problem, or a narrow set of solutions from which decision makers can select.

The emergence of data warehouses and online processing (OLAP) technologies has enhanced the abilities of employees at all levels to effectively use data for decision making. We discuss OLAP in Chapter 11, "Business Intelligence and Knowledge Management." The pervasive use of the Web prompted software developers to make practically all decision support applications accessible through Web browsers.

THE DECISION-MAKING PROCESS

When do you have to make a decision? When you drive your car to a certain destination and there is only one road, you do not have to make a decision. The road will take you there. But if you come to a fork, you have to decide which way to go. In fact, whenever more than one possible action is available, a decision must be made. If you have to decide based only on distance, making a decision is easy. If you have to choose between a short but heavily trafficked road and a longer road with lighter traffic, the decision is a bit more difficult.

A decision is easy to make when one option will clearly bring about a better outcome than any other. Decisions become more difficult when more than one alternative seems reasonable and when the number of alternatives is great. In business, there can be dozens, hundreds, or even millions of different courses of action available to achieve a desired result. The problem is deciding on the best alternative. (You can see why problem solving and decision making are so closely related.)

Herbert Simon, a researcher of management and decision making, described decision making as a three-phase process (see Figure 10.1). First, in the *Intelligence* phase, decision makers collect facts, beliefs, and ideas. In business, the facts might be millions of pieces of data. Second, in the *Design* phase, the method for considering the data is designed. The methods are sequences of steps, formulas, models, and other tools that systematically reduce the alternatives to a manageable number. Third, in the *Choice* phase, when there is a reduced number of alternatives, decision makers make a choice; that is, they select the most promising alternative.

FIGURE 10.1
The three phases of decision making

Intelligence	• Collect data from inside the organization. • Collect data from outside the organization. • Collect information on possible ways to solve the problem.
Design	• Organize the data; select a model to process the data. • Produce reasonable, potential courses of action.
Choice	• Select a course of action.

Businesses collect data internally (from within the organization) and externally (from outside sources). They use models to analyze data. Generally speaking, a **model** is a representation of reality. For instance, in architecture, a tabletop representation of a building or a city block is a model of the full-sized structure. A map is a small-scale representation—a model—of a particular geographic area that can include topographic information and political boundaries. And in business, mathematical equations that represent the relationships among variables can be models for how businesses respond to changes, such as: what happens to profits when sales and expenses go up or down? Decision makers either use universal models, such as certain statistical models, or design their own models to analyze data. Then they select what they perceive as the best course of action.

STRUCTURED AND UNSTRUCTURED PROBLEMS

A **structured problem** is one in which an optimal solution can be reached through a single set of steps. Since the one set of steps is known, and since the steps must be followed in a known sequence, solving a structured problem with the same data always yields the same solution. Mathematicians call a sequence of steps an **algorithm** and the categories of data that are considered when following those steps **parameters**. For instance, when considering the problem of the shortest route for picking up and delivering shipments, the parameters are shipment size, the time when shipments are ready for pickup, the time when shipments are needed at their destinations, the distance of existing vehicles from the various destinations, the mandatory rest times of the drivers, the capacities of the trucks, and so on.

Most mathematical and physical problems are structured. Finding the roots of a quadratic equation is a structured problem: there is a formula (an algorithm) you can use to solve the problem. For the same equation the roots are always the same. Predicting how hot a liquid will get in a particular setting is a structured problem: if you know the properties of the liquid, the size of its container, the properties of the energy source heating the liquid, and the exact length of time the energy will be applied, you can figure out what temperature the liquid will reach. Unfortunately, most problems in the business world cannot be solved so easily.

An **unstructured problem** is one for which there is no algorithm to follow to reach an optimal solution—either because there is not enough information about the factors that might affect the solution or because there are so many potential factors that no algorithm can be formulated to guarantee a unique optimal solution. Unstructuredness is closely related to uncertainty. You cannot be sure what the weather will be tomorrow, let alone two months from now; nobody can guarantee what an investment in a certain portfolio of stocks will yield by year's end; and two physicians might diagnose the same symptoms differently. These are all areas where unstructured problems predominate.

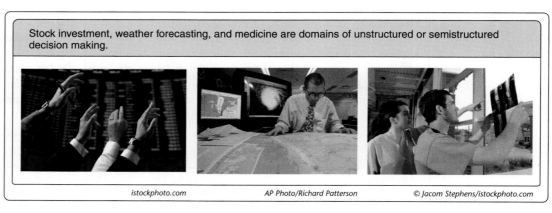

Stock investment, weather forecasting, and medicine are domains of unstructured or semistructured decision making.

| istockphoto.com | AP Photo/Richard Patterson | © Jacom Stephens/istockphoto.com |

Some management scientists refer to semistructured problems. A **semistructured problem** is one that is neither fully structured nor totally unstructured. The problem "Should I invest for two years $100,000 in municipal bonds that pay 3 percent per annum tax free or should I invest in CDs (certificates of deposit) with 4 percent taxable interest?" is structured. To find the solution you have to follow a simple algorithm that takes as parameters your $100,000, the two years, and the 3 percent interest rate. Unless the city that issued the bonds goes bankrupt, your calculated income is guaranteed. Similarly, it is easy to calculate the after-tax yield of the same amount invested in the CDs. However, the problem "Should I invest $100,000 in the stock of XYZ, Inc. and sell the stock after two years?" is semistructured. Too many factors must be taken into account for it to be considered structured: the demand for the company's products, entrance of competitors into its market, the market of its products in this country and overseas, and so on. So many factors affecting the price of the stocks might change over the next two years that the problem is semistructured at best and totally unstructured at worst.

Professionals encounter semistructured problems almost daily in many different industries and in many different business functions (see Figure 10.2).

A manager solving a typical semistructured problem faces multiple courses of action. The task is to choose the one alternative that will bring about the best outcome. For example:

- In manufacturing, managers must provide solutions to semistructured problems such as: (1) Which supplier should we use to receive the best price for purchased raw materials while guaranteeing on-time delivery? (2) Assembly line B has a stoppage; should we transfer workers to another assembly line or wait for B to be fixed? (3) Demand for product X has decreased; should we dismantle one of the production lines, or should we continue to manufacture at the current rate, stock the finished products, and wait for an upswing in demand?

- Managers of investment portfolios must face semistructured decision making when they decide which securities to sell and which to buy so they can maximize the overall return on investment. The purpose of research in stock investing is to minimize uncertainties by trying to find patterns of behavior of stocks, among other trends. Managers of mutual funds spend much of their time in semistructured decision making.

FIGURE 10.2
Examples of structured and semistructured problems

Structured Problems	Semistructured Problems
How many workers are needed to fully staff production line A?	What are the benefits of merging with XYZ, Inc.?
What is our optimal order quantity for raw material Z, based on our production?	Where should we deploy the next five stores of our retail chain?
How many turbines are needed to supply power to Hickstown?	How will the consumer react if we lower the price of our product by 10 percent?
Which of our regions yields the highest revenue per salesperson?	What is the best advertisement campaign to launch our new financial service?
Which money market fund currently yields the highest return?	What are the benefits of opening an office in Paris, France?
How much would the implementation of pollution-preventing devices cost us?	Which stock will yield the highest return by the end of the year?

- Human resource managers are faced with semistructured problems when they have to decide whom to recommend for a new position, considering a person's qualifications and his or her ability to learn and assume new responsibilities.

- Marketing professionals face semistructured problems constantly: should they spend money on print, television, Web, e-mail, or direct-mail advertisements? Which sector of the population should they target?

Because of the complexities of the problems they face, managers in many functional areas often rely on decision support applications to select the best course of action.

Why You Should Be Familiar with Decision Aids

The terms "decision support systems" and "expert systems" are mentioned less frequently these days. However, the concepts of modeling decision-making processes and automating them and the transformation of human expertise into software are alive and thriving. While many situations exist in which only an experienced professional can make good decisions, much of the decision-making process can be automated through use of computer-based decision aids. The raw materials for many decisions are already in corporate databases and data warehouses, and they can be accessed through ISs such as supply chain management systems. Your ideas of how to automate routine decisions can save much labor and time for your organization. Knowing how expert systems and geographic information systems work might stimulate fresh ideas in your mind for implementation of new ISs, which can not only save labor and time but also be a competitive tool for your organization.

DECISION SUPPORT SYSTEMS

To save time and effort in their decision making, knowledge workers use several types of decision support applications. One such type, a **decision support system (DSS)**, is a computer-based information system designed to help knowledge workers select one of many alternative solutions to a problem. DSSs can help corporations increase market share, reduce costs, increase profitability, and enhance product quality. By automating some of the decision-making process, the systems give knowledge workers access to previously unavailable analyses. Technically, certain analyses could be performed by managers, but it would be prohibitively time-consuming and would render late, and therefore bad, decisions. DSSs provide sophisticated and fast analysis of vast amounts of data and information. Although the use of DSSs typically increases with the level of management, the systems are used at all levels, and often by non-managerial staff.

The definition of a DSS has been changing over the years. The following sections discuss the components of stand-alone DSSs: either self-contained applications or applications that are designed to address a rather narrow decision-making domain. You should realize that some components of a computer-based decision aid, such as databases, might already be in place when a new DSS is developed. Therefore, consider the following discussion a general framework and not a rigid recipe for the development of all DSSs.

The majority of DSSs comprise three major components: a data management module, a model management module, and a dialog module (see Figure 10.3). Together, these modules (1) help the user enter a request in a convenient manner, (2) search vast amounts of data to focus on the relevant facts, (3) process the data through desired models, and (4) present the results in one or several formats so the output can be easily understood. These steps follow the decision-making sequence described by Herbert Simon.

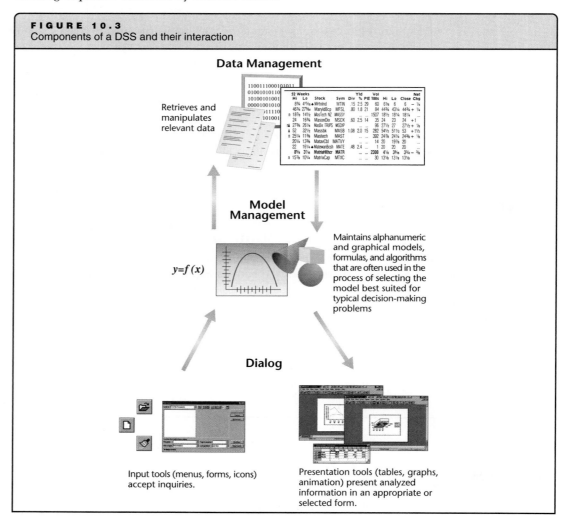

FIGURE 10.3
Components of a DSS and their interaction

Data Management

Retrieves and manipulates relevant data

Model Management

$y = f(x)$

Maintains alphanumeric and graphical models, formulas, and algorithms that are often used in the process of selecting the model best suited for typical decision-making problems

Dialog

Input tools (menus, forms, icons) accept inquiries.

Presentation tools (tables, graphs, animation) present analyzed information in an appropriate or selected form.

The Data Management Module

A DSS's **data management module** is a database or data warehouse that provides the data for the intelligence phase of decision making. For example, an investment consultant always needs access to current stock prices and those from at least the preceding few years. A data management module accesses the data and provides a means for the DSS to select data according to certain criteria: type of stock, range of years, and so on.

A DSS might use a database created specially for that system, but DSSs are usually linked to databases used for other purposes as well, such as purchasing, shipping, billing, and other daily transactions. When organizations use a supply chain management (SCM) or customer relationship management (CRM) system, the databases of such systems provide the data for the DSS. In fact, the DSS itself might be part of that system. Companies prefer their DSSs to access the data warehouse rather than the transactional database, to provide substantially more historical data than is available in transactional databases. This enables the DSS to consider data that covers a longer time period and/or a larger geographic area. Indeed, the major reason for building data warehouses is to enhance decision making.

Many DSSs are now closely intertwined with other organizational systems, including data warehouses, data marts, and ERP systems, from which they draw relevant data. For example, Rapt, Inc. offers a decision support application that helps optimize purchasing decisions of goods, especially for high-volume purchasers. The application, called Rapt Buy, is a Web-based application that captures business variables in data marts through an SAP or Oracle SCM system. The application's analytical software builds models that identify various elements of risk and then recommends purchasing strategies. It considers dozens of economic variables, including demand for the raw materials and yield (the percentage of the materials that are actually used in the final products). The system suggests how many units of each component the company should purchase to avoid carrying too much or too little inventory. In addition, the system provides multiperiod plans for optimizing procurement into the future. Within minutes, the application analyzes the potential effect of various procurement and negotiation strategies. At Sun Microsystems Corp., forecasting demand for new products can be off by up to 70 percent. Before implementing this system, procurement officers at this large manufacturer of servers spent many hours of manual analytical work. Now, they use this system to calculate forecasts faster and more accurately.

POINT OF INTEREST

So, What Will the Experts Do?

The increasing amount of decision making that is being automated might require less and less human decision making. However, this does not mean that we will need fewer experts. Automated decision making frees experts to deal with the unusual and more complex problems that information systems cannot solve. However, the increasing expertise that is captured in automated decision aids raises a question about the definition of an expert. If a physician uses IT to diagnose most of her patients, is she still considered an expert? If a veteran pilot uses IT to automatically take off, fly, and land airplanes, are the thousands of flight hours credited to him indicative of his expertise?

The Model Management Module

To turn data into useful information, the system utilizes its **model management module**, which offers a single fixed model, a dynamically modified model, or a collection of models from which either the DSS or the user selects the most appropriate one. A fixed variable model does not change. A dynamically modified model is one that is automatically adjusted based on changing relationships among variables.

A sequence of events or a pattern of behavior might become a useful model when the relationships among its inputs, outputs, and conditions can be established well enough that they can be used to analyze different parameters. Models are used to predict output on the basis of different input or different conditions or to estimate what combination of conditions and input might lead to a desired output. Models are often based on mathematical research or on experience. A model might be a widely used method to predict performance, such as best-fit linear analysis, or it might be built by the organization, using the experience that employees in

eg. Robot.

the firm have accumulated over time. Many companies will not divulge details of the models they have programmed because they view them as important trade secrets and valuable assets that could give them competitive advantages. Patterns or models might be unique to a certain industry or even to an individual business. For example:

- In trying to serve bank customers better, operations research experts create a model that predicts the most efficient positioning and scheduling of tellers.

- In the trucking business, models are developed to minimize the total mileage trucks must travel and maximize the trucks' loads, while maintaining satisfactory delivery times. Similar models are developed in the airline industry to maximize revenue.

- Another model for revenue maximization in the airline industry will automatically price tickets according to the parameters the user enters: date of the flight, day of the week of the flight, departure and destination points, and the length of stay if the ticket is for a round-trip flight.

- Car rental companies use similar models to price their services by car class, rental period, and drop-off options in different countries.

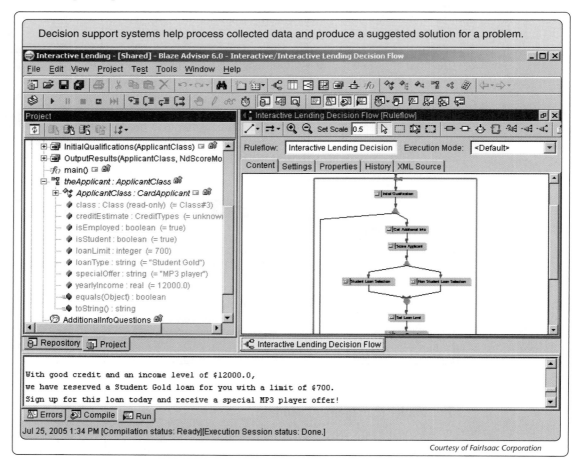

Decision support systems help process collected data and produce a suggested solution for a problem.

Courtesy of FairIsaac Corporation

Among the general statistical models, a linear regression model is the best-fit linear relationship between two variables, such as sales and the money spent on marketing. A private business might develop a linear regression model to estimate future sales based on past experience. For example, the marketing department of a shoe store chain might apply linear regression to the relationship between the dollar amount spent on search Web site advertising and change in sales volume. This linear relationship can be translated into a program in a DSS. Then the user can enter the total amount to be spent on search Web site advertising for the next year into the DSS, and the program will enter that figure into the model and find the estimated change in the sales volume. The relationship between the two variables can be plotted, as shown in Figure 10.4.

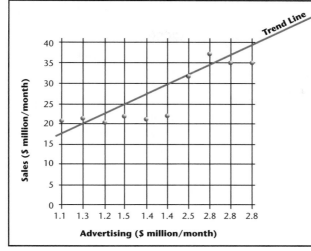

FIGURE 10.4
A linear regression model for predicting sales volume as a function of dollars spent on Web advertising

Advertising ($M/month)	Sales ($M/month)
1.1	20.3
1.3	21
1.2	20.1
1.5	22.7
1.4	21.9
1.4	22
2.5	32
2.8	36
2.8	35
2.8	34.8

Note that the actual data points rarely lie on the regression line produced from the data. This illustrates the uncertainty involved in many models. For instance, in Figure 10.4, if the marketing managers tried to estimate the sales volume resulting from spending $1.4 million per month on Web advertising, their estimates for both months plotted on the graph would be more than the actual sales. In spite of these discrepancies, the regression line might be adequate in general for modeling, with the understanding that results are not necessarily precise. Also note that models often describe relationships among more than two variables and that some models can be expressed as a curve, rather than a straight line.

Usually, models are not so simple. In this advertising and sales example, for instance, many more factors might play a role: the number of salespeople, the location of the stores, the types of shoes offered for sale, the search keywords with which the advertising is associated, and many more parameters. Therefore, before models are programmed to become part of a DSS, the environment in which the decision will be executed must be carefully considered.

Not all DSS models are business-oriented. In some areas, especially engineering, DSS models might simulate physical rather than business environments. For example, aeronautical engineers use computer models of wind tunnels to view how a computer model of an aircraft with a new wing design might behave. It is significantly less expensive to construct a software model than to build a physical model. The simulation provides valuable information on vibrations, drag, metal fatigue, and other factors in relation to various speeds and weather conditions. The output, in the form of both animated pictures and numerical tables, enables engineers to make important decisions before spending huge amounts of money to actually build aircraft—decisions such as the angle in which the aircraft wings are swept, the shape of the fuselage's cross section, the spreading of weight over different parts of the plane, and so forth. When using this type of model, engineers base part of their decision on visual examination of the behavior of the simulation model.

The Dialog Module

For the user to glean information from the DSS, the system must provide an easy way to interact with the program. The part of the DSS that allows the user to interact with it is called the **dialog module**. It prompts the user to select a model, allowing the user to access the database and select data for the decision process or to set criteria for selecting such data. It lets the user enter parameters and change them to see how the change affects the result of the analysis. The dialog might be in the form of commands, pull-down menus, icons, dialog boxes, or any other approach. In essence, the dialog module is not much different from the user interfaces of other types of applications. As an increasing number of DSSs are available for use through the Internet, some dialog modules are especially designed to be compatible with Web browsers. Many such DSSs are accessed through corporate intranets.

The dialog module is also responsible for displaying the results of the analysis. DSSs use various textual, tabular, and graphical displays from which the decision maker can choose. Take the previous advertising effort scenario, for example, where the company's marketing manager is trying to decide how to spend promotional dollars. The dialog component of the DSS presents a menu allowing the marketing executive to select Web search advertising from a variety of promotional choices and to choose the amount to be spent in that channel (see Figure 10.5). Now the dialog module calls up the part of the database that holds current data on advertising expenditures and sales volumes for the corresponding months. At this point, the system might either present a list of models for analyzing the data from which the user can choose or, if it is sophisticated enough, select a model automatically, based on the problem at hand. The model projects sales figures based on the data from the database, and the dialog component presents the results of the analysis. The output helps the executive make a decision by answering the question, "Will the proposed amount to be spent on Web ads yield a large enough boost in sales?"

Sensitivity Analysis

An outcome is almost always affected by more than one parameter; for instance, the sales volume of a product is affected by the number of salespeople, the number of regional sales representatives, the amount spent on national and local television advertising, price, competition, and so on. However, outcomes rarely respond in equal measure to changes in parameters. For instance, a small change in price per unit might result in a dramatic increase in sales, which means sales volume has a high sensitivity to product price. However, the same sales might increase only slightly in response to a huge investment in advertising dollars, which means that sales have a low sensitivity to advertising expenditure. It is important to pinpoint the parameters to which the outcome is highly sensitive, so that an organization can focus efforts where they are most effective. Sometimes the parameters to which an outcome is most sensitive also affect other parameters, so these interactions must be carefully tracked as well.

If a company wishes to maximize profit, managers must find the optimal combination of many factors. To equip a DSS to help achieve this goal, an approximate mathematical formula that expresses the relationship between each factor and the total profit is built into the DSS. Then a **sensitivity analysis** is conducted to test the degree to which the total profit grows or shrinks if one or more of the factors is increased or decreased. The results indicate the relative sensitivity of the profit to the changes. If the outcome is affected significantly even when the parameter is changed only a little, then the sensitivity of the outcome to the parameter is said to be high. The opposite is also true: if the outcome is affected only a little, even when the parameter is varied widely, the outcome is said to be insensitive to the parameter. For instance, a manager might ask, "What is the impact on total quarterly profits if television advertising is decreased by 10 percent and the number of commissioned sales representatives is increased by 5 percent?" Because questions typically are phrased in this format, sensitivity analysis is often referred to as **what-if analysis**. Note that you can use a DSS to perform what if analyses on multiple parameters at the same time.

Equally important is the use of sensitivity tests to learn which parameters do *not* make a difference. For example, based on data collected during a promotion via coupons, marketing analysts might learn that discounts did not increase sales and/or did not bring in new customers. The obvious decision would be not to conduct similar promotions in the near future.

You might be familiar with sensitivity analysis from using electronic spreadsheets. Spreadsheets enable you to enter both data and formulas in cells. Thus, they are an excellent tool for building both the data and the models that decision support systems need, and therefore they make excellent tools for building decision support software. Changing data in one or several cells will result in a different solution to a problem. This allows you to see the effect that a change in one parameter has on the calculated outcome.

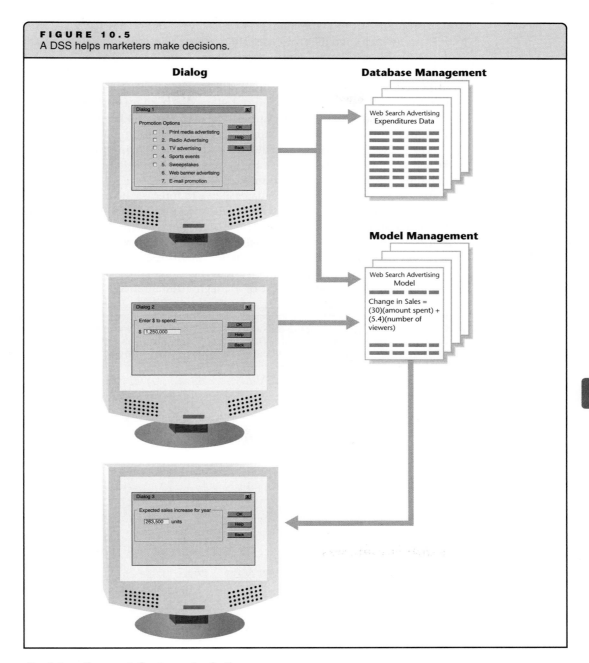

FIGURE 10.5
A DSS helps marketers make decisions.

Dialog

Dialog 1
Promotion Options
☐ 1. Print media advertising
☐ 2. Radio Advertising
☐ 3. TV advertising
☐ 4. Sports events
☐ 5. Sweepstakes
6. Web banner advertising
7. E-mail promotion
OK
Help
Back

Dialog 2
Enter $ to spend:
$ 1,250,000
OK
Help
Back

Dialog 3
Expected sales increase for year
263,500 units
OK
Help
Back

Database Management

Web Search Advertising
Expenditures Data

Model Management

Web Search Advertising
Model

Change in Sales =
(30)(amount spent) +
(5.4)(number of
viewers)

Decision Support Systems in Action

DSSs can be used on demand, when a manager needs help in making an occasional decision, or they might be integrated into a scheme that enforces corporate policy. In either case, DSSs help maintain standard criteria in decision making throughout the organization. A growing number of organizations implement software applications that produce decisions automatically and in real time. The only labor involved is the entry of relevant parameters, and when the DSS is linked to the organization's Web site, even this activity might not be performed by employees but by clients. Following are some examples of how DSSs are used for various purposes.

Food Production and Retailing

How much of each type of cookie should a cookie store produce today? Which ingredients should be taken out of the refrigerators, and how long before they are to be mixed and put in the oven? These are only some of the decisions that operators of a Mrs. Fields Cookies store would have to make. However, the decision making is done for them by a DSS. The system, installed in each store, decides for the managers which types of cookies to make, what quantities of each type, and which ingredients to use. The software considers numerous parameters, including the store's historical sales volumes, season, day of the week, time of day, whether it is a holiday, location, and the weather. The company—which operates 390 stores in the United States and more than 80 stores in other countries—has structured the business environment for the store managers. Each store manager must follow the system's instructions regarding which baked products to make and their quantities. The system prescribes the ingredients for each product, when to pull dough from refrigerators, and how to bake, including oven temperatures and time.

The software Mrs. Fields Cookies used was so effective in running food stores that the owners created a subsidiary, Fields Software Group. The firm sells its software to several companies, including Burger King Corporation.

In the restaurant industry, managers have to forecast the number of patrons and the amount of ingredients to purchase, as well as where to purchase it to minimize cost. FoodPro, a DSS developed and sold by Aurora Information Systems, helps make such decisions. Based on the historical data restaurants accumulate, the system helps with these decisions. A recipe database is used to propose decisions on ingredients, quantities, and consolidated purchases from specific vendors. Other components of the system include financial forecasting, invoicing, accounting, and practically every other aspect of managing restaurants.

Part of a spreadsheet that helps decide whether or not to make an investment. The original calculation (top) yields a positive net present value (NPV); a correction in the expenses of the first year (middle) changes the return to negative; a change in the interest rate of the loan taken for the project (bottom) makes it profitable again.

	A	B	C	D	E	F
1	Year	1	2	3		
17						
18	Net Cashflow	(10,756,900)	877,880	12,078,900		
19	Interest Rate	6.50%				
20	NPV	**673,122**				
21						
22						
23						
24						
25						
26						

Sheet1 / Sheet2 / Sheet3 /

	A	B	C	D	E	F
1	Year	1	2	3		
17						
18	Net Cashflow	(11,556,900)	877,880	12,078,900		
19	Interest Rate	6.50%				
20	NPV	**(78,052)**				
21						
22						
23						
24						
25						
26						

Sheet1 / Sheet2 / Sheet3 /

	A	B	C	D	E	F
1	Year	1	2	3		
17						
18	Net Cashflow	(11,556,900)	877,880	12,078,900		
19	Interest Rate	5.50%				
20	NPV	**120,882**				
21						
22						
23						
24						
25						
26						

Sheet1 / Sheet2 / Sheet3 /

Agriculture

The Canadian government has sponsored the development of a series of DSSs for farmers, one of which is Prairie Crop Protection Planner, a system that helps farmers make decisions regarding weed, insect, and disease control on the Canadian prairie. The Web-based system helps select crop protection products; calculate application rates and costs; identify weeds, insects, and diseases; identify herbicide injury symptoms, keep pesticide application records, and quickly reference crop protection products.

When a farmer enters basic information about a particular crop and the pest problem—weeds, insects, or diseases—the Planner outlines options such as chemicals and ways to apply them. Farmers can describe their spraying equipment, size of the field, and current chemical prices from local suppliers, and the Prairie Crop Protection Planner calculates application rates, costs per acre, the amount of product the farmer will need to use in the sprayer's tank, and the amount of chemical needed to spray the field.

Similarly, many other national and regional government agencies offer decision support systems on the Web. For example, the U.S. Department of Agriculture provides several decision tools online for farmers. One of them uses three decades of historical temperatures to help farmers decide in which regions of Nebraska to plant grapevines, and which types, to avoid crop spoilage due to extreme temperatures. Part of the output is in the form of a color-coded map.

Tax Planning

Some applications that people may not think of as DSSs actually are. TurboTax, TaxCut, and other tax-preparation applications have been developed over the years to do much more than help fill out forms. They come with sophisticated formulas to help taxpayers plan the best strategy in selecting options, with the final purpose of minimizing the tax paid. For example, the applications compare filing status and deduction options: which approach would result in a lower combined tax, filing as two individuals or joint filing as husband and wife? Itemizing deductions, or taking a standard deduction? Taking a smaller education credit, or a larger education deduction? Based on the taxable income and the combination of deductions taken, the applications warn users about their chances of being audited by the Internal Revenue Service and give them a chance to modify deductions. The applications also remind users of optional deductions, tell them what the deductions entail (e.g., if you take deduction X you may not take deduction Y or you increase the probability of being audited) and thereby make it easy for filers to make decisions. And when users complete their tax preparation for the past year, they can plan their tax for next year—based on their total income and type of income (wages, business, capital gains, and so forth)—and make decisions on how much to contribute to pension funds, charity, and other purposes that serve as tax shields to reduce the tax owed next year.

Web Site Planning and Adjustment

Because so many companies use the Web for marketing, selling, and customer support, decisions on how to design Web sites are extremely important. Some companies offer DSSs specifically designed to analyze shoppers' behavior at their sites based on captured data such as pages viewed, options clicked, and the sequence of pages shoppers view. For example, Datanautics, Inc. offers G2, a path analysis system that analyzes how visitors navigate through a site. The purpose is to convert shoppers into buyers, a process marketers call conversion. Managers might be impressed that 30 percent of shoppers who follow a certain sequence of Web pages purchase an item. However, the software might reveal that another, unexpected sequence ends up with 90 percent of the shoppers buying something. This can lead to a decision to enhance those pages or eliminate certain pages between the home page and the last page before the purchase takes place. Another company, WebTrends, sells WebPosition, a decision tool that helps decide which keywords to use for improved listing on search engines, how to optimize Web pages for greater conversion, submit URLs to search engines, and analyze conversions.

Yield Management

You might be surprised to learn that the passenger sitting next to you on an airplane paid a third of what you paid for the same flight. This is the result of recommendations the airline receives from a DSS whose purpose is to maximize revenue. The concept is called **yield management** or *revenue management*. For each flight, revenue managers enter a myriad of data, including departure point and time, destination point and time of arrival, the number of airports in which the airplane stops, the airplane capacity, and information on utilized capacity from previous operations of the particular flight. They change pricing, or let the system change prices, according to the time a ticket is purchased, and how long before the passenger flies back. The dilemma is between offering low prices to fill up the plane, or upping the price and risking flying with some empty seats.

The purpose of yield management DSSs is to find the proper pricing to maximize the overall revenue from selling seats for each flight. The result is often price discrimination, which is legal and a common practice in the airline industry: you might pay a different price depending on how far in advance you purchased the ticket, the fact that a companion flies with you, the number of days between departure and return, and several other variables. Typically, airlines double or triple the price of a ticket when it is purchased only a few days before departure, because usually the availability of seats on competitors' flights is limited. Also, late purchasers tend to be business people who are reimbursed for their travel and therefore are less price-sensitive. Airlines take advantage of these facts and the expectation that customers who make a late reservation usually do so because they have little flexibility in selecting the flight date. Other variables are less obvious, and therefore DSSs are used to model demand and sensitivity to prices.

Similar decision aids are used in the hospitality industry. For example, Harrah's Entertainment, the operators of a chain of hotels and casinos, uses such a system to set room rates for its hotels and for offering different rates to different levels of members in its customer loyalty programs. Room rates might be lower, or even free, for customers who regularly spend a great deal of money on gambling. Like many other companies, the chain has a customer loyalty program called Total Rewards. Harrah's data analysis program, called Revenue Management System, recognizes a Total Rewards member's telephone number and allows reservation agents to offer lower prices for rooms during a busy weekend for a high-value customer—one who usually spends a lot—or to raise the price of a Saturday night stay for customers who don't yield much profit.

Financial Services

Manually deciding how much money to loan to which customer at what interest rate could delay the decision process to a point of losing the potential customer. Loan applicants are reluctant to wait more than a day or even a few hours for the bank's response. Automated decision aids can produce offers within minutes after a customer enters data. The DSS combines this data with data retrieved from credit history databases and preprogrammed rule models to create a rapid response.

Consider DeepGreen Financial, an online home equity lender owned by Lightyear Capital. This bank is actually a computer program. To apply for a mortgage loan, customers go to the company's site (www.deepgreenfinancial.com) and fill out an application, which takes no longer than five minutes. A DSS retrieves the customers' credit report, engages a scoring formula, accesses an online valuation of the property to be mortgaged, examines fraud and flood insurance conditions, and produces a decision on the loan (such as full amount requested or a smaller amount, with what down payment, and at what interest rate). Eighty percent of applicants receive a response within two minutes. The system also selects a local notary public. All that is left for the applicant is to select a closing date. Although many loan applicants still find it strange to think of an online information system as an equity lender, between 2000 (its inception year) and 2007, DeepGreen extended more than 65,000 loans totaling $5 billion.

An increasing amount of decision making in the financial services and many other industries are made this way: automatically and in real time. This saves many hours of labor and ensures speedy service for customers.

Benefits Selection

Employers can save labor costs and provide convenience to their employees by helping them make decisions. One area where this can occur is selection of benefits, especially health care plans. ADP, Inc., a company known for its payroll processing services, teamed with Subimo LLC to offer an online application to help employees evaluate health care plan costs. The analysis is based on individual medical needs. The tool uses an individual's age, gender, geographic region, and health condition to estimate several costs of a family's health care. It predicts the health

Many banks use automated decision aids to determine a client's creditworthiness.

© Image Source Black/Getty Images

services the individual or family would most likely need. The tool's modeling is based on the demographic data and experience of over 60 million U.S. patients. Employees can use the predictions to select plans and to decide how to fund a flexible spending account, a health saving account, or health reimbursement arrangement.

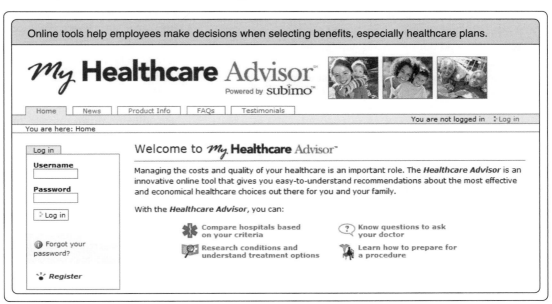

Online tools help employees make decisions when selecting benefits, especially healthcare plans.

Similar decision support tools are offered by other companies, including some of the largest health care management organizations (HMOs) in the United States. A tool called MyHealthcareAdvisor.com, also offered by Subimo, serves small businesses that wish to augment their employee coverage. It is estimated that about 60 million Americans have access to tools that help make decisions on health care plans and options within plans. Interestingly, about 95 percent of enrollment transactions are executed online.

Decisions by Machines

Every year thousands of people are denied credit not because they are bad credit risks, but because of errors that lending institutions make. For example, a newly married woman might be denied credit because her credit history is still only under her maiden name. Errors in recording monthly payments are also a common reason for denying credit; it's possible for one late payment to deduct many points from one's credit score. For some banks, if your credit score is in the 500s or even 600s on a scale of 350–850, you are not considered a good credit risk. Some banks might agree to give you a loan with an interest rate that is higher than most people get.

A single late credit-card or mortgage payment could lower your credit score severely, but you might never know that, because the law does not require anyone to notify you. Many people do not even know that a credit score system exists, and that it is shared by all banks and other lenders. Chances are you will find out about your too-low score only when you apply for a loan and only if you ask for the reason of denial or higher offered interest rate. And when you find out and try to explain that there was a mitigating circumstance for the late payment, your explanation might not help you much. That's because American banks use computers for credit decisions. It is often a computer program that decides who will or will not receive credit. And the decision is final.

Chapter 9, "Challenges of Global Information Systems," pointed out the major differences between U.S. and European privacy laws. One of the areas in which the policies differ is automated decision making that affects individuals. There is no U.S. law that even addresses the issue, while the European Union (EU) restricts the use of automated decision making that affects individuals in its data protection directive.

- **European Protection.** The full title of the EU's directive is *Directive 95/46/EC of the European Parliament and of the Council of 24 October 1995 on the protection of individuals with regard to the processing of personal data and on the free movement of such data*. EU directives are akin to the U.S. Constitution; each member state may formulate its own laws within the framework of the directives. Practically speaking, this means member states may restrict organizations more and afford more protection than the directive requires, but not less.

  Article 15 of the directive is titled "Automated Individual Decisions." It grants every person the right "not to be subject to a decision which produces legal effects concerning him or significantly affects him and which is based solely on automated processing of data intended to evaluate certain personal aspects relating to him, such as his performance at work, creditworthiness, reliability, conduct, etc."

- **Who Needs Protection?** To what extent should organizations rely on computer-based decision aids to make business decisions about individuals? Automated decision making is used routinely in the United States by banks, credit-card companies, mortgage companies, employers, and, to some extent, educational institutions. The affected individuals might be consumers, credit applicants, employees, job applicants, prospective students, applicants for membership in associations, and people who are evaluated by organizations in other capacities.

  Creditworthiness is determined by processing personal financial data in models that have been developed specifically to sort the good risks from the bad risks. A bad risk is a person or institution that is likely to default on a loan. Should credit-card companies, for instance, ask their officers to open a manila folder for every American adult and make a decision on creditworthiness only after leafing through the filed documents in it? Should they be banned from using an automated process that makes the decision for them based on the same criteria that the officers would use manually? And when employers sift through hundreds and thousands of digitized résumés of job applicants, should they be banned from using software that retrieves the résumés with keywords that suggest a good fit for the job, while eliminating those without them? Would it be practical for Google not to use software to sift through the 1.5 million job applications the company receives every year?

  The use of automated decision making offers not only added efficiency but also enhanced effectiveness. When using a DSS or an expert system, the user enjoys the knowledge and experience that have been accumulated by other people over many years. Thus, in addition to efficiency, automated decision making might be more effective than manual decision making.

- **Hidden Injustice.** On the other hand, shifting decision making to a machine might create injustices. Suppose your record is among several hundred records of applicants considered for a position. The records were obtained from a third party, a company that sells personal information. Your qualifications are excellent, but your record also indicates a law violation. The system removes you from the pool of eligible candidates, and you do not get the job. Had you seen your record before its processing, you could have told the company that this entry was an error: you were charged once but acquitted in court. If the company had contacted you, you could have ironed out the misunderstanding and possibly have gotten the job.

Is the EU directive excessive? Do American organizations overuse automated decision making? Is it practical, in the digital age, to give up the efficiency of automated decision making to determine an individual's creditworthiness or job performance? Does a little more justice in credit and employment justify giving up the greater efficiencies of automated decision making?

EXPERT SYSTEMS

It is not always possible to exploit expertise by coupling quantitative data from a database with decision models. In such cases, an expert system might be required. An **expert system (ES)** is developed to emulate the knowledge of an expert to solve problems and make decisions in a relatively narrow domain. A *domain* is a specific area of knowledge. For example, in medicine a domain is often a diagnosis of a specific disease or a family of related diseases, such as bacterial diseases. The purpose of ESs is to replicate the unstructured and undocumented knowledge of the few (the experts), and put it at the disposal of the many other people who need the knowledge, often novices or professionals in the same domain but with far less expertise. Advanced programs might include **neural networks** (computer programs that emulate the way the human brain works) which can learn from new situations and formulate new rules in their knowledge bases to address events not originally considered in their development. Expert systems and neural networks are two techniques researched and implemented in a field called **artificial intelligence (AI)**. The field is so called because it focuses on methods and technologies to emulate how humans learn and solve problems.

POINT OF INTEREST

The Turing Test

Scientists continue the quest for software that will be at least as smart as humans, so that expertise can be enhanced and delivered through information technology. In 1950, Alan Turing, a British mathematician, published an article titled "Can Machines Think?" His own answer was yes. Today the Turing test is this: An interrogator is connected to a person and to a machine via a terminal and cannot see either. The interrogator asks both the person and the machines questions and is to determine by their answers which is a human and which is a machine. If the machine can fool the interrogator, it is considered intelligent. In 1990, Hugh Loebner offered to grant a gold medal and $100,000 to the first person who could build such a machine. At the annual competition, judges present the same questions to computers and people, but cannot see either. The communication is by text, similar to online chat. Competitors try to build software whose answers would be indistinguishable from those of humans. So far nobody has won. You can find information about the competition at www.loebner.net/Prizef/loebner-prize.html.

As Figure 10.6 illustrates, the major difference between DSSs and ESs is in the "base" they use and how it is structured. DSSs use data from databases. An ES uses a **knowledge base**, which is a collection of facts and the relationships among them. An ES does not use a model module but an inference engine. The **inference engine** is software that combines data that is input by the user with the data relationships stored in the knowledge base. The result is a diagnosis or suggestion for the best course of action. In most ESs, the knowledge base is built as a series of IF-THEN rules.

Figure 10.7 provides a simple illustration of how such rules are used to conclude which disease is infecting a tree. If the humidity is low, the average air temperature is higher than 60 degrees Fahrenheit, the tree leaves are dark green, and the tree's age is 0-2 years, then the tree has

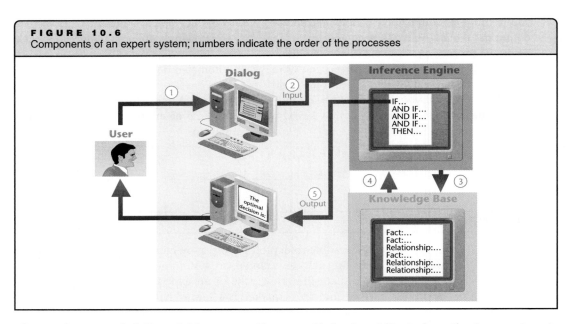

FIGURE 10.6
Components of an expert system; numbers indicate the order of the processes

disease A at a probability of 90 percent. However, if the humidity is low, the temperature is between 40 and 65 degrees Fahrenheit, the tree leaves are green, and the tree's age is 0-2, there is a 40 percent probability that the disease is A and 50 percent that the disease is B. The diagnosis helps reach a decision on proper treatment to stop the disease. A real expert system for such diagnosis would consist of many more rules, usually hundreds or thousands, because there are many more conditions—antecedents—and combinations of factors that may cause a disease; and there are more diseases that a particular tree may have.

In a mineral exploration, for example, such rules can be: IF the drilling depth is so many meters, AND IF the sample includes a certain percentage of carbon (and so forth), THEN there is a 90-percent probability that so many meters further down there is oil of commercial quality and quantity. Such rules are often not quantitative but qualitative, and therefore can only be stored as a knowledge base rather than a database.

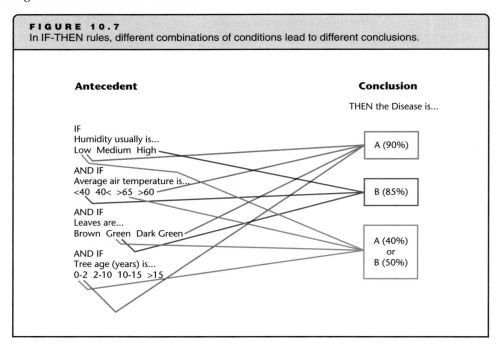

FIGURE 10.7
In IF-THEN rules, different combinations of conditions lead to different conclusions.

ES researchers continue to look for ways to better capture knowledge and represent it. They test the results of such efforts in highly unstructured problem-solving domains, including games. One such game that has intrigued both researchers and laypeople is chess. The game is a highly unstructured environment in which the number of possible moves is enormous, and hence, the player must be an expert to select the best move for every board configuration.

Rather than containing a set of IF-THEN rules, more sophisticated ESs use neural networks (neural nets), programs that are designed to mimic the way a human brain learns. An ES is constructed with a set of rules, but as data on real successes and failures of decisions is accumulated and fed into the system, the neural network refines the rules to accomplish a higher success rate.

Business applications have increasingly combined neural nets and ES technologies in software that monitors business processes and supply chain management. One example is an application called BizWorks, which was developed by InterBiz Solutions, a division of Computer Associates International. It was installed at Myers Industries, an international manufacturer of plastic and rubber products for industrial, agricultural, automotive, commercial, and consumer markets. The application uses past production data so it can predict when molding machines are likely to produce defective products. It monitors data coming from the machines, looking for conditions similar to those resulting in defective products in the past. Using another technology called intelligent agents, the software can alert customers. An **intelligent agent** is software that is "dormant" until it detects a certain event, at which time it performs a prescribed action. When BizWorks detects an imminent problem, it checks to see which customers ordered the products that the problematic machine is scheduled to make. If the problem threatens on-time delivery to a key customer, the sales and customer service managers receive an alert. They can then notify the customer.

POINT OF INTEREST

Deep Thoughts

In their attempts to create "thinking" software, researchers and engineers use the game of chess as a highly unstructured domain. In 1997, the then world chess champion, Garry Kasparov, lost a match to the computer program Deep Blue, which was developed by IBM. In 2003, Kasparov played against two more advanced computer programs, Deep Junior and X3D Fritz. Both matches ended in ties. While the software is becoming increasingly "smarter," experts say that eventually software will beat all chess grandmasters mainly because of speed, not sophistication. Computers calculate the best moves much faster than humans.

Source: McClain, D. L., "In Chess, Masters Again Fight Machines," *The New York Times* (www.nytimes.com), Technology Section, June 21, 2005.

Neural nets have been very effective in detecting fraud of many types. For example, 85 percent of credit-card issuers use a neural network product called Falcon from HNC Software, a subsidiary of the credit-rating firm Fair Isaac Corporation. The application uses large volumes of cardholder purchasing data and analyzes it to establish spending patterns. Deviations from these patterns trigger an investigation. Using mathematical algorithms, the software calculates, on a scale of 1–999, the likelihood that a transaction is fraudulent. For instance, if a cardholder historically has used his or her card once per week to purchase gasoline and groceries within a certain Pennsylvania zip code, the purchase of groceries in Vermont would trigger a low score of fraud likelihood. If the same card is suddenly used to purchase high-liquidity merchandise such as jewelry, the likelihood would be set at a much higher number. The low-risk pattern changes if the cardholder starts traveling often and makes purchases all over the country or in other countries.

Insurance companies use neural nets to detect fraudulent claims both from the insured party and from health-care providers. Empire Blue Cross Blue Shield has used the technology for many years and has saved millions of dollars. In one case, it caught a doctor who allegedly provided an annual respiratory test that normally is provided no more than twice per lifetime. In another case, it caught a doctor who filed a claim for a pregnancy test, but the software detected that the test was given to a man.

Another technique to use expertise is case-based reasoning, and the software that supports the technique is often called case-based ESs. The important parameters of a case to be analyzed are compared to many cases until one or a small number of them are found highly similar to the analyzed case. The system then brings up the decision made in those cases, and a successful decision is applied. The knowledge base is a database of cases. Instead of an inference engine, the system uses software that searches key parameters in the archival case reports.

Case-based reasoning is useful especially in medical decision making. Comparison with previous, similar cases helps to diagnose a symptom and recommend remedial action such as medication and other treatments.

Expert Systems in Action

ESs have been implemented to help professionals in many different industries, such as health care, telecommunications, financial services, and agriculture. The following is a small sample.

Medical Diagnosis

Because medicine is one of the most unstructured domains, it is not surprising that many of the early ESs were developed to help doctors with the diagnosis of symptoms and treatment advice, as mentioned earlier. MYCIN (diagnosis of bacterial diseases), CADUCEUS (internal medicine diagnostics), and PUFF (diagnosis of pulmonary diseases) are only a few of these systems. PUFF includes instrumentation that connects to the patient's body and feeds various data about the patient's condition into the ES to be analyzed for pulmonary diseases.

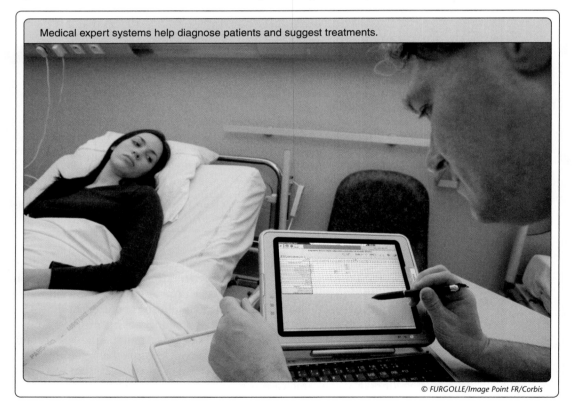

Medical expert systems help diagnose patients and suggest treatments.

© FURGOLLE/Image Point FR/Corbis

More recently, German scientists developed an ES that enhances the accuracy in diagnoses of Alzheimer's disease, which gradually destroys memory. The system examines positron emission tomography scans of the patient's brain. The scans provide images that can be reliably interpreted only by experienced physicians. Based on the expertise of such experts, the ES can detect Alzheimer's before the appearance of behavior typical of the disease. More than 4.5 million Americans suffer from Alzheimer's, and the proportion of the U.S. population that suffers from this yet incurable disease is growing. An early, accurate diagnosis helps patients and their families plan and gives them time to discuss care while the patient can still take part in decision making. When tested on 150 suspected patients, the ES performed as well as the experts. Now it can serve any doctor in early diagnosis of the disease.

Medical Management

In addition to diagnostic ESs, some hospitals use systems that help discern which tests or other initial treatment a patient should receive. Some of the decisions might be administrative. For example, The University of Illinois at Chicago Medical Center uses an application called Discern Expert. It monitors patient data and events and recommends action, such as admission, transfer to another ward or hospital, discharge, or order of tests or treatments. For instance, a set of rules in the system can look like this:

> IF: An order for a contrast-enhanced CAT SCAN is received
> AND: The patient's BUN level is HIGH
> AND: The patient's CREATININE level is HIGH
> THEN: Send a message to the patient's physician via electronic mail indicating a possible adverse effect of contrast agent use in this setting.

Blood urea nitrogen (BUN) is caused by the breakdown of blood, muscle, and protein. High levels of it might indicate kidney disease. Creatinine is a protein produced by muscle tissue and released into the blood. High levels of it might indicate kidney failure. Discern Expert helps staff members prevent complications or unnecessary testing. Once a medical ES is composed, it can be used anywhere, bringing expertise to poor regions of the world, where expert doctors are in dire shortage.

Telephone Network Maintenance

AT&T uses an ES to diagnose and fix network failures. The system consists of three parts: Monitor, Consultant, and Forecaster. Monitor constantly checks AT&T's telephone network for errors. When a problem is detected, the system uses a synthesized voice to warn network specialists, who can then use Consultant to walk them through recommended troubleshooting and repair procedures to correct the problem. Before the company started using the ES, a small number of highly trained specialists did the troubleshooting, which is now done by employees with less training. Forecaster, the third part of the ES, checks system files and notifies personnel of problems likely to occur, based on previous experience, allowing the staff to prevent problems from occurring.

POINT OF INTEREST

ES, You're Practicing Law!

Entrepreneur Henry Ihejirika offered his bankruptcy expert system at two Web sites. Clients could go to Ziinet.com or 700law.com, answer questions, and have the proper forms and petitions prepared for them by the system. A client who used the system testified that it was excellent, providing much better service than a clerk would. When a bankruptcy trustee noticed errors in one client's filing, the client blamed the system. Ihejirika was summoned to court. The judge decided that the system's knowledge and advice was as a good as a lawyer's, and therefore had practiced law without a license. The entrepreneur was ordered to remove the service from the Web. He appealed, but the appellate court upheld the decision. He used a human lawyer for his defense.

Source: Poulson, K., "AI Cited for Unlicensed Practice of Law," *Wired Blog Network*, March 5, 2007; posted by Robert Hudock.

Credit Evaluation

Holders of American Express (AmEx) credit cards can potentially charge the card for hundreds of thousands of dollars per purchase. Obviously, most retailers and restaurateurs will not process a charge before they contact AmEx for approval. The AmEx clerk who considers the request uses an ES. The system requests data such as account number, location of the establishment, and amount of the purchase. Coupled with information from a database that contains previous data on the account, and a knowledge base with criteria for approving or denying credit, the ES provides a response.

Another expert system called FAST (Financial Analysis Support Techniques) helps with credit analysis. The system is used by more than 30 of the top 100 U.S. and Canadian banks as well as some of the largest industrial and financial companies in the world. It gives a credit analyst access to the expertise of more experienced advisors, accelerating the training process and increasing productivity.

The system provides complex analysis of the data contained in applicants' financial reports. The expert system not only provides English-language interpretation of the historical financial output but also prepares the assumptions for annual projections and produces text output linkable to word-processing software. It eliminates much of the tedious writing of analytical reports, producing standard financial statement reviews.

Loan officers periodically update the knowledge base to customize it for a bank's current loan policy, as well as national and local economic forecasts and interest rate projections. The system consistently and reliably interprets the relationship of these variable factors and the levels of sensitivity that the loan officers associate with a particular financial statement.

Detection of Insider Securities Trading

Like other similar institutions, the American Stock Exchange (AMEX) has a special department to prevent insider trading of the securities under its supervision. Insider trading is the trading of stocks based on information available only to those affiliated with a company, not to the general public. This practice is a serious breach of U.S. federal law. To detect insider trading, the department receives information from several sources on unusual trading activity and uses this information to identify a stock it might want to investigate. Using an ES, the department's analysts access a large database of the stock's history and choose a time period of interest. The system provides questions that the analysts can answer with the information they received from the database. The questions are formulated to reflect the experience of expert investigators. After the analysts finish answering all the questions, the system provides two numbers: the probability that a further investigation is warranted, and the probability that it is not.

Detection of Common Metals

Metallurgists are experts, and their time is expensive. Also, they usually work in laboratories, which are expensive, too. General Electric Corp. developed an expert system that helps nonexperts to identify common metals and alloys outside laboratories. The user provides information on density, color, and hardness of the metal and results of simple chemical tests that can be performed by novices outside the laboratory setting. If the user provides sufficient information, the system will positively identify the metal or alloy. If the information is insufficient, the system will provide a list of possible metals in order of likelihood. Even such a list can be helpful in some situations, saving much time, labor cost, and the need to wait for lab testing.

Irrigation and Pest Management

Knowing the quantities of water and pesticides to use at different stages of peanut growing can save farmers millions of dollars. After much research, the National Peanut Research Laboratory of the U.S. Department of Agriculture developed an ES called EXNUT to help peanut growers make these decisions. Scientists produced a large knowledge base on plants, weather, soil, and other factors that affect the yield of peanut fields. Farmers feed EXNUT with data about the field throughout the growing season, such as minimum and maximum soil temperature and rainfall measures, and the program provides recommendations on irrigation, the application of fungicide, and the likelihood of pest conditions. It recommends that farmers withhold water during certain stages and that they use the highest and lowest soil temperatures as indicators of soil moisture and plant health.

The department further developed the system and changed the name to EXNUT Irrigator Pro. Farmers who are not considered experts were able to increase their yield to quantities greater than those harvested by expert farmers, while using less water and fungicide. It costs a farmer $3.71 per acre to use the system, but the results are impressive: growers have increased yield by 200–300 pounds per acre with the help of the ES. The total yield of an acre is usually 5,000–6,000 pounds. The $3.71 cost gained additional revenue of $65.90 per acre.

A related ES, HARVPRO, helps optimize peanut harvest time. Growers sample peanuts and give them to an analyst. The analyst blasts off the outer layers of the peanuts' hulls to check the ripeness of the middle hull. The darker the middle hull, the more progressive the peanut's ripeness. The analyst enters into the system the maturity profile and other data. The output is recommendations for an optimal harvest time. Since most growers have several fields, they can plan an optimal schedule of harvesting each field when its peanuts are at peak ripeness, neither too soon nor too late.

Diagnosis and Prediction of Mechanical Failure

Finding out what causes a failure in a system can be daunting. Therefore, it is not surprising that a great number of ESs help methodically diagnose what might cause a failure. For example, technicians at Cessna use the Cessna Diagnostic and Repair ES for the Citation X, an executive jet. If, say, the airplane's floodlight fails, the system starts the analysis with three possible situations. The technician clicks the menu item that describes the situation. In the bottom-left frame the ES lists the components that make up the failing assembly. When the technician clicks an item in the right frame, the ES displays a drawing of the relevant switches. The system contains many "maps" of switches as well as photos of different parts of the cockpit and drawings of electrical circuits. The final output is a set of instructions for fixing the problem.

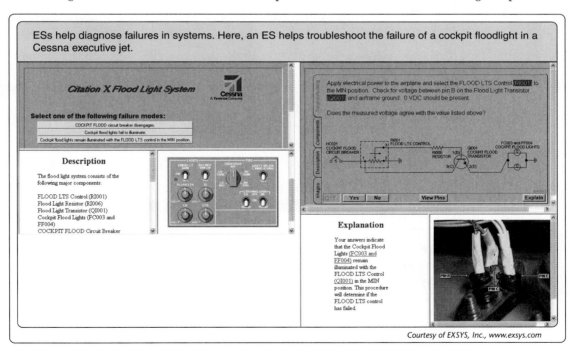

ESs help diagnose failures in systems. Here, an ES helps troubleshoot the failure of a cockpit floodlight in a Cessna executive jet.

Courtesy of EXSYS, Inc., www.exsys.com

A reliable way to predict the failure of diesel locomotive engines is to examine the oil from the engine. Experienced technicians at Canadian Pacific Railroad took many years to develop this expertise, which involves a technician analyzing a sample of lubrication oil for metal impurities, and a mechanic analyzing the data. The process not only takes years to learn, but is difficult to teach to novices, so Canadian Pacific decided to develop an ES for this purpose.

The system takes the spectrum data as input and uses the rules provided by the experts. A technician can use the output report, which details components requiring service and those likely to fail soon. The system has had remarkable success; analysis of more than 10,000 samples has yielded accurate predictions 98 percent of the time. The company saved money by replacing components before they failed. In some cases, the replacement of a single component saved more money than was spent on the development of the ES.

GROUP DECISION SUPPORT SYSTEMS

When a team of people are to make decisions, a **group decision support system (GDSS)** can be useful. The systems are often named group intelligence systems, collaborative systems, or simply group systems. Their purpose is to facilitate the contribution of ideas, brainstorming, and choosing promising solutions. Often, one person serves as a facilitator of the entire process. Typically, a session starts by defining a problem to be resolved or a decision to be made; followed by contribution of ideas, evaluation of the ideas (such as pros and cons of each idea), and some method of voting on the ideas. The voting determines the ranking of ideas to solve the problem or make a decision. The entire list of suggested decisions is then submitted to the final decision maker. If the group is authorized to make the decision, then the top-ranked offered decision is adopted. ThinkTank 2, an application offered by Group Systems, and FacilitatePro, offered by Facilitate.com, are typical examples of such systems.

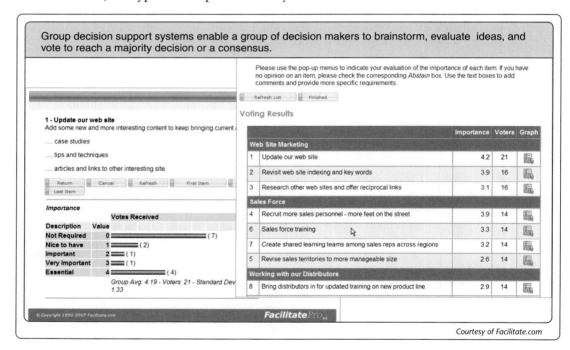

Group decision support systems enable a group of decision makers to brainstorm, evaluate ideas, and vote to reach a majority decision or a consensus.

Courtesy of Facilitate.com

The process not only helps in structuring the group decision-making process, it also creates an environment different from sitting around a table. GDSSs allow participants to maintain anonymity during the entire session or parts of it. This removes the fear of putting forth ideas that might be dismissed or ridiculed. It also puts all the participants on equal footing regardless of rank or seniority. Anonymity helps elicit more creative ideas and a more open and thorough decision-making process. It also results in a consensus or at least a decision by a majority. Such decisions are less politically motivated and therefore garner more support when implemented.

GEOGRAPHIC INFORMATION SYSTEMS

As mentioned in Chapter 5, "Business Software," some decisions can be made only when examining information on a map. Many business decisions concern geographic locations—as input, output, or both. For example, consider the process of choosing the best locations for new stores or determining how to deploy police forces optimally. For map-related decisions, **geographic information systems (GISs)** are often the best decision aids. GISs process location data and provide output. For instance, a GIS could be used to help a housing developer determine where to invest by tracking and displaying population changes on a map, highlighting in color increases of more than 10 percent over the past three years. With this information, a developer could easily decide where to invest on the basis of population growth trends. Other examples include the following:

- Delivery managers looking for the shortest distance a truck can travel to deliver ordered goods at the lowest cost.

- School district officials looking for the most efficient routes for busing school children to and from their homes.

- City planners looking to deploy services to better serve residents, which might include police officers deciding how to deploy their forces on the basis of precinct maps indicating levels of criminal activity.

- Health-care agencies analyzing which areas of a community need more or less attention and resources for treatment of certain diseases or injuries that result from criminal violence.

- Oil companies looking to determine drilling locations on the basis of geological tests.

- Hunters, fishers, hikers, and other people who enjoy outdoor recreation looking for suitable sites and trails for their activities based on their requirements, such as local fauna and trail length.

- Mapping concentrations of people at work and in shopping centers to help banks decide where to install new ATMs.

This ArcView screenshot describes the market penetration for a retailer in Nashville, TN. Zip code boundaries were used to calculate these percentages, with dark blue representing the highest penetration. Through this analysis method it is immediately clear that zip codes closest to the selected store have the highest market penetration and largest number of customers.

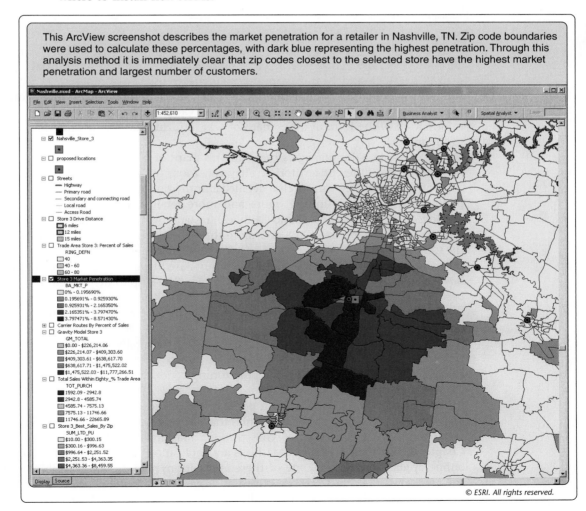

In Springfield, Massachusetts, for instance, health-care professionals integrated data collected for Hampden County with information from the area's two major medical centers and the city's health, planning, and police departments to use in combination with the region's map. They use models that help identify geographic areas and population groups that need health-care intervention in youth violence and late-stage breast cancer detection.

A typical GIS consists of (1) a database of quantitative and qualitative data from which information is extracted for display, (2) a database of maps, and (3) a program that displays the information on the maps. The digitized maps are produced from satellite and aerial photography. Displays might be in the form of easily understood symbols and colors or even moving images. For instance, an oil exploration map might show different concentrations of expected crude oil deposits in different hues of red. Or, population density might be similarly displayed on a map using different hues of blue. A more sophisticated GIS might display, in colors or icons, concentrations of specific consumer groups by age, income, and other characteristics.

POINT OF INTEREST

Maps, Please

ESRI is a leading developer of geographic information systems. Its Web site provides a list of GISs that you can use online. Go to www.esri.com/software/internetmaps/index.html and peruse the maps. Some are used for location of real estate for sale or lease; others for environmental protection; and others for water resource management, disaster handling (such as floods and fire), and many other purposes. Many of the sites are run by local governments for the use of their administrators and citizens.

Web technology helps promote the use of GISs by private organizations and governments alike. Intranets allow employees to bring up thousands of maps from a central repository on their own PCs. HTML and XML, the primary languages used to compose and retrieve Web pages, support the presentation of pictures with marked areas, which makes them ideal for retrieval of marked maps. Clicking different areas of a map can zoom in and out as well as bring up related information in the form of other maps or text, utilizing the multimedia capabilities of the Web to the fullest.

For example, sales managers can bring up maps of continents and see how past sales have performed over different territories. They can zoom in and zoom out on a territory. With the click of a mouse they can receive detailed information on who serves the territory and other pertinent information. Indeed, more and more retail chains are adopting GISs for decision making. Pollo Tropical, a Hispanic-Caribbean restaurant chain, operates 75 restaurants in Florida, New Jersey, and New York City, and more than 20 franchised restaurants in Latin America and the Caribbean. Its managers used "gut feeling" to determine where to open new restaurants. Now they use a GIS with geodemographic data purchased from MapInfo (now a division of Pitney Bowes), a leader in location intelligence software. The system helps them pinpoint where their best customers live and work in the company's effort to expand outside Florida.

In government work, a city clerk can bring up a map of the neighborhood of a resident, zoom in on the resident's house pictured on the map, click on the picture, and receive information such as real-estate taxes owed and paid over the past several years. Further information, such as whether a neighborhood uses septic tanks or a sewage system, might be rendered by different colors. The map can also show different zoning codes, such as land designated for residential, industrial, or commercial purposes.

You may be familiar with some popular Web-based GISs: Google Earth, Mapquest, Yahoo Maps, and others. Some of the companies that own these applications have opened the software for anyone to create specialized GISs through mashups. Recall our discussion of mashups in Chapter 5, "Business Software." Mashups are applications that combine features of two or more online applications. Often, one of these applications shows maps. Combining the maps with data such as the location of certain information—events, clubs, real estate for sale—creates a specialized GIS. For example, www.zillow.com mashes a map with up-to-date estimates of real estate prices. The Web has hundreds of similar mashups involving maps. One site where you can see over a hundred of such mashups is www.programmableweb.com/mashups.

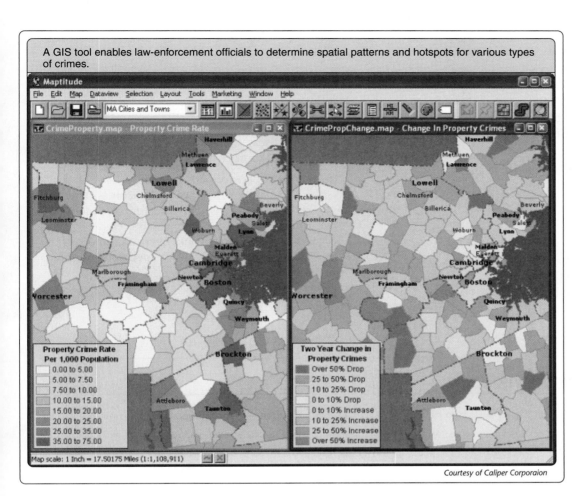

A GIS tool enables law-enforcement officials to determine spatial patterns and hotspots for various types of crimes.

Courtesy of Caliper Corporaion

SUMMARY

- Decision aids include decision support systems, expert systems, group decision support systems, geographic information systems, and any other software tool that helps with decision making automatically or on demand. Some are stand-alone systems, and others are part of larger systems. Most are accessible through a Web browser.

- The decision-making process combines three major phases: intelligence, design, and choice. In the first phase, data is collected from which relevant information will be gleaned. In the design phase, the data is organized into useful information and processed by models to analyze it and produce potential courses of action. In the final stage, the decision maker selects an alternative, that is, makes the decision.

- Problems span a continuum between two extremes: structured and unstructured. A structured problem is one for whose solution there is a proven algorithm.

An unstructured problem is one for which there are multiple potential solutions. A semistructured problem is one that is neither fully structured nor totally unstructured.

- Most DSSs have three components. The data management module gives the user access to databases from which relevant information can be retrieved. The model management module selects, or lets the user select, an appropriate model through which the data are analyzed. The dialog module serves as an interface between the user and the other two modules. It allows the user to enter queries and parameters, and then presents the result in an appropriate or selected form, such as tabular or graphical.

- DSSs provide a quick way to perform sensitivity analysis. The user can change one or several parameters in the model and answer "what if" questions, called what-if analysis.

- Powerful software tools such as electronic spreadsheets let users with little expertise in systems development create their own DSSs.

- Expert systems are developed to emulate the knowledge of an expert. Developers use artificial intelligence techniques.

- An expert system consists of a knowledge base, inference engine, and a dialog module.

- Neural network software is often integrated into an expert system to enable the system to learn and refine decision rules based on success or failure.

- Expert systems are used in narrow domains in which typical diagnosis and decisions are unstructured, such as health care, mineral exploration, stock investment, and weather forecasting.

- When decisions involve locations and routes, professionals can use geographic information systems (GISs). GISs provide maps with icons and colors to represent measurable variables such as population concentrations, potential natural resources, deployment of police forces, pinpointing concentrations of diseases, and other factors that involve locations and routes.

- Computerized decision aids practically leave decision making to machines. When machines determine whom to hire, whom to accept for higher education, or to whom to extend credit, the decision-making process could overlook important circumstances, in which case the decisions might not be accurate or fair.

DEBOER FARMS REVISITED

Carl DeBoer has been using an expert system to help manage pests on his farm. Such expert systems, as well as decision support systems, are increasingly used to assist the DeBoers and other farmers in their work.

replication of expertise, timely response, and consistent solutions. Cite some examples of these contributions that the PestPRO system has made to DeBoer Farms.

What Would You Do?

1. Dr. Wildes told Carl that the PestPRO system took years to refine. Explain how the historical data and comparison with scientists' results can help computer systems "learn" through feedback.

2. The chapter illustrated nine contributions of expert systems to organizations: planning, decision making, monitoring, diagnosis, training, incidental learning,

New Perspectives

1. Carl explored a decision support system that helps manage risk and an expert system to help handle infestations. What other types of decision support systems might help farm owners manage their businesses?

KEY TERMS

REVIEW QUESTIONS

1. What is a decision? When does a person have to make a decision? *P. 342*

2. Calculating a complex trajectory of a spaceship to Mars is a structured problem, whereas diagnosing the cause of a rash on a person's skin is often unstructured. How so? *P. 343-344*

3. DSSs use models to process data. Explain what a model is. Give an example that is not mentioned in the chapter. *P. 347 ATM. Pricing.*

4. Many DSSs are not stand-alone anymore, but are embedded in other ISs. What are those ISs? *P. 347*

5. What is a sensitivity test? Give three examples. *P. 350*

6. The airline and hospitality industries use DSSs for yield management. What is yield management, and what is the output of a yield management DSS? *Glossary - output is revenue.*

7. What is the purpose of an expert system? How can it serve as a competitive tool? *Slide 12*

8. Explain how expert systems can distribute expertise. *P. 357*

9. How could an ES be used to detect probable fraud committed by a bank employee? *By analyzing Fin reports & bank acct history.*

10. What is the advantage of combining ES and neural net technologies? *P. 357*

11. What is a GIS? What purpose does it serve? *Slide 14*

12. Name the three major elements that are combined to make up a GIS. *Slide 18*
   - marketing
   - Holiday cooking
   - add ingredients to unique dish.

DISCUSSION QUESTIONS

13. DSSs and ESs give structure to an often unstructured decision-making process. Explain this statement.

14. Bank officers use DSSs to make decisions on extending or denying credit. Universities might use DSSs to decide which applicant to admit and which applicant to reject. Would you agree to using DSSs for such decisions without human intervention?

15. Some companies (such as Mrs. Fields Cookies) use DSSs to make decisions for their knowledge workers. The decisions are based on previous experience and on corporate policy. Would you prefer to work for a company that requires you to use a DSS for the decisions you have to execute, or would you rather work for a company that lets you make decisions on your own? Explain.

16. Some managers say that you should never accept the output of any computer-based decision aid without scrutiny. Do you agree? Explain.

17. Give three examples of a business decision for which you would not use a decision support system or any other software. Give three non-business examples. Explain your choices.

18. Think of executives in human resources, finance, marketing, and information systems. In which of these areas could executives benefit the most from using GIS technologies? Explain.

19. Some DSSs allow the user to select a model for an analysis. Would you like to have the final say in determining which model would be used, or would you prefer to let the software use whatever model it chose for the analysis, and then produce a decision? Why?

20. Credit scoring firms such as Fair Isaac base their scoring only on electronic data collected from financial sources. A low credit score may significantly affect a person's or a family's life. What information that may be relevant is not considered in the scoring?

21. How could you use a GIS for scheduling your classes? What would be the likely sources of data for such a system?

22. Some GISs are used for very small geographic areas, such as a campus or a single building. Can you think of how a GIS could help the maintenance department of a university campus?

23. As an increasing amount of decision making is carried out by software, do you think the demand for college-educated workers will diminish?

24. You noticed that the family doctor and the specialists that you see consult their PC and handheld computers for practically every diagnosis they make and for every prescription they give you. Does this make you trust these physicians less than physicians who do not consult their computers?

25. Many software packages offer expertise in a multitude of areas, such as investment, nutrition, and writing your own will. Do you trust the advice of such software?

26. "The more professionals use ESs, the less expertise they accumulate, because the less actual hands-on experience they gain. This is akin to a pilot who spends most of his flying time watching an autopilot system rather than flying the airplane with his own hands." Do you think that ESs can *decrease* real expertise for its users? Why or why not?

APPLYING CONCEPTS

27. Make a list of six daily activities you perform. They might include preparing homework assignments, shopping, and other activities. Ensure that three of the activities call for decision making that is unstructured (or semistructured) and that three involve structured decision making. Prepare a one-page report listing the activities, stating what decision making is required, and explaining why the decision making is unstructured or structured.

28. Prepare a one-page report on the career you intend to pursue. Give at least four examples of activities involved in such a career that call for problem solving. Explain which problem solving is structured and which is unstructured.

29. You are the head of a medical team that wants to learn about the spread of a new disease in your state. You decide to engage a company that designs GISs. List the types of data that would be needed for the system and which agencies are likely to have collected the data. Suggest visual effects to make it easy to identify concentrations of sick people and the spread of the disease.

30. Use an electronic spreadsheet such as Excel to design a DSS for solving the following problem. A publisher makes and sells books that have different titles but have the same format and are made of the same materials (paper, ink, binding, and so forth). Three types of costs are involved in the process:

Risk Category	Points
Age	
20 < age < 25	0
25 < age < 60	10
60 < age	0
Income (annual $) per household member	
INCOME < 20,000	0
20,000 < income < 40,000	10
40,000 < income < 60,000	20
60,000 < income < 85,000	30
85,000 < income	40
Employment at current workplace	
self-employed	10
less than 2 years	5
more than 2 years	10
Net worth (in $)	
NW < 30,000	0
30,000 < NW < 50,000	10
50,000 < NW < 70,000	20
NW > 70,000	30

Fixed cost per title: $15,000 for setting up the press, regardless of how many pages a title has or how many copies will be made of that title.

Fixed cost per page: Setting the plates for printing costs $2.00 per page, regardless of the number of books. (Thus, for a book of 324 pages, the per-page fixed cost would be $648.)

Variable cost per page: The printing and binding of each page of each copy of a title costs 7 cents.

Assume that no other costs are involved (such as shipping and handling). Prepare a decision model that allows the publisher to decide the following:

a. For a given number of copies, what should be the book's minimum retail price per copy to break even?

b. For a given retail price, what is the break-even quantity of copies?

Test your decision tool for a title that has 250 pages. If the publisher intends to sell 40,000 copies, what is the break-even price? If the publisher decides to price the book at $18.00, what is the break-even quantity that must be sold?

E-mail your spreadsheet to your professor, attached to a message in which you answer the two test questions.

31. Use a spreadsheet or download a free ES shell for this assignment. Build a simple ES that determines the eligibility of an applicant for a bank loan of $50,000 for 30 years. Eligibility is determined by the number of points the applicant accumulates in several risk categories, based on the following:

Loan size

No points are deducted for a $10,000 loan. Henceforth, 2 points are deducted from the total score for each additional $10,000 of the requested loan.

Loan life

No points are deducted for the first 5 years of the loan's life (years to repay the loan). Henceforth, 1 point is deducted for every 5 years.

Rule

If the applicant's total number of points is equal to or exceeds 60, grant the loan.

32. To make decisions, decision makers must collect data. Consider this quote from *CIO Insight*: "The more personal information the company can gather, the more it can effectively target its ads—and charge advertisers premium prices. And Google's information-gathering techniques could present a compelling case for companies that want to direct their messages to finely targeted audiences. If a user specifies weather and searches in and around a particular geographic area, for instance, wouldn't local providers jump at the chance to push their services straight to that user's screen? And wouldn't the user want such services? To make decisions one needs much relevant data and proper tools." Google offers many online services to individuals, and it also collects information about them, through DoubleClick and other subsidiaries. Research Google and its subsidiaries. List the types of data the company is able to collect directly and indirectly about individuals. For each such application or service, say what can be collected and how this may threaten privacy. Then, consider the entire "package" of applications and services, and explain what risks to privacy this "package" poses.

33. Team up with two other students. Contact a local stockbroker. Ask the broker to give you a list of the most important points he or she considers when predicting appreciation of a stock. Ask how each point (such as last year's earnings per share, percentage of appreciation/ depreciation over the past six months, and the like) affects the net result, as a percentage. Use the input to formalize the model in a spread-sheet application. Select a portfolio of 100 units of 10 traded stocks. Use the model to predict the increase in the price of each stock and the value of the entire portfolio a year into the future. If you know how to use macros, embellish your new DSS with a user-friendly dialog module.

34. Team up with another student. Obtain a digi-tized map of your campus. If you cannot copy one, you can use a scanner to digitize a paper map. Use a Web design application (such as FrontPage) to create a simple GIS. On the map page, mark 10 areas so that when clicked, another page comes up with additional infor-mation about the building or part of building you marked. Examples of marked areas that you could include: registrar's office, school of busi-ness building, the student union, and the com-puter lab. E-mail your GIS to your professor.

FROM IDEAS TO APPLICATION: REAL CASES

Minimize Credit Risks

Companies that extend credit want to be able to assess the risk of not being paid by their clients, with many of whom they have had no previous experience. While large companies often have their own risk assessment experts or use the expensive services of consultants, medium-size companies with annual revenues of $50–200 million have traditionally found it difficult to assess the risk involved in doing business globally. One American company saw this as a business opportunity and decided to offer these firms the tools to reduce their exposure to credit risks, network security threats, ID theft, and other risks. The Web allowed it to offer expertise online.

American International Group, Inc. (AIG) is one of the world's largest insurance and financial services companies, with operations in 130 countries. The company offers its Web site TradeCredit.com (www. aigtradecredit.com), which provides up-to-the-minute credit information about businesses that intend to buy goods and services from AIG's clients. Clients log on and can assess and monitor the buying power and repayment capabilities of a customer. AIG's system tracks, in real time, the financial standing of companies around the globe. This helps the seller—AIG's client—to set down payments and payment timetables that take into consideration the risk of not being paid on time or at all.

In addition, the system tracks and assesses country-specific risk variables, such as political instabilities and changes in trade laws. It also updates risks related to specific industries rather than the country in which a buyer operates. The risks are determined by experts who update the systems frequently.

AIG developed the original system in-house and implemented it in 1999. However, every time risk algorithms changed, programmers had to modify the code. The programming was time-consuming and costly. The IT department had to modify the original system so it would contain a structured credit algorithm that could be adapted to individual countries based on parameters that characterized business conditions and credit practices in each country. The idea was to separate the algorithms from tables that contain changing parameters. This way, managers can change parameters in country and industry tables, and the changes take effect in the algorithms automatically. This keeps the need to reprogram to a minimum.

Paul Narayanan, the manager of AIG's eBusiness Risk Solution Division, said that the major challenge was to transform the business rules into software code and design the software so that fine-tuning of those rules could be done without too much hassle. The system contains risk models shared by all countries as well as special rules for each country and even regions within countries.

For the original DSS, experts developed complex risk models in Microsoft Excel. The models were eventually converted into Java for installation at AIG's Web site. The system became increasingly complex. After several years of operations, the IT professionals realized that developing rules for additional countries took too long, about two months for each country. AIG purchased a financial risk DSS called Blaze Advisor from Fair Isaac, a credit assessment firm.

With the integration of Blaze Advisor, managers can change decision rules without involving programmers. Within the first three months of using the integrated system, AIG added 14 countries to the mere two that it covered before. The rules for a country that is already in the system can be updated with new rules within eight hours. Risk models, rating criteria, and decision algorithms for new countries can be added in 5–14 days.

A subscriber to AIG TradeCredit.com logs on and enters the details identifying a company with which the subscriber wishes to do business. The subscriber receives real-time assessment of the potential buyer's ability to pay for goods and services. The subscriber can also receive the payment history of the company and make his or her own decisions. The system works so well that it is endorsed by the U.S. Chamber of Commerce for its members.

Sources: "Fair Isaac Blaze Advisor Business Rules Automate Risk Decisions for AIG TradeCredit.com," (www.fairisaac.com), 2005; (www.aigtradecredit.com), 2007.

Thinking About the Case

1. Why is a system such as TradeCredit so important, especially in the era of e-commerce?

2. What are the advantages of using an online DSS such as this?

3. Why was the original software (developed in-house) merged with Blaze Advisor?

Modeling Health Care

How much of the physiology of the human body can be embedded in a computer program? How many diseases develop in a pattern that can be simulated in a software application? The answer is quite a few. Archimedes Model, named after the famous Greek mathematician and discoverer, is a medical decision-making application. What are the chances that a 31-year-old female with cancer and who smokes will respond better to one treatment than another? Ask Archimedes.

The unique decision support system was developed by David Eddy, the son and grandson of surgeons. Although he is a medical doctor, halfway through his residency he heard a different calling. He wanted to help patients via digitized medical modeling instead of a surgeon's scalpel. He obtained a Ph.D. in mathematics and developed a career in mathematical modeling to improve medical tests and treatments.

Eddy was dissatisfied with the shortcomings of medical treatment. When he set out to design his new system, the generally accepted patterns of diagnosis and treatment were simple: if a patient had a certain symptom, such as a lump in the chest, doctors would prescribe a certain treatment. There was little regard for the many other factors that might cause the situation as well as have an impact on the success of the treatment. Doctors rarely considered the treatment's chances of benefits and harms.

Eddy's system was developed by the Biomathematics Unit of Kaiser Permanente's Care Management Institute. The unit's head, Leonard Schlessinger, and his team used differential equations and algorithms to describe the anatomy, physiology, and progress of diseases. For example, based on epidemiological studies and clinical trials, they created a model of coronary artery disease. The model calculates the location and extent of the blockage and occurrence of symptoms. It then calculates the outcomes of the disease. Algorithms model the actions of patients and caregivers: patient behavior in seeking care, the performance of tests, and treatments by physicians. Before prescribing a treatment, physicians can try different treatments and change the process of care for a disease and then explore the effect that the system reports.

The starting point of modeling a disease and its treatment is a session with experts who know all about the disease. They describe the symptoms, impact on patients, and suggested treatments. Eddy and his team map out the symptoms associated with the progression of the disease. They add to the model equations that represent the symptoms and progression as well as the patient demographics. In all, there might be 50–100 variables that impact the progression of a disease. Building a model for one disease takes over a year and costs $0.5–1 million. However, the rewards are impressive. Using Archimedes for proper drug prescriptions more than 10 years for just 10,000 patients with coronary artery disease can result in 4063 avoided heart attacks, 893 avoided deaths, and monetary savings of $44 million for Kaiser Permanente.

Physicians can use Archimedes Model to examine the impact of age, sex, race, income, employment, and eating habits—among other variables—on a person's health, as well as the effect of different treatment regimens. The application's accuracy is equal to that of doctors with tens of years of experience. It contains rules that researchers reach only after many years of clinical trials involving thousands of real patients. Medical trials that would take months or years can be simulated with Archimedes and take no more than 30 minutes.

The decisions doctors can make with Archimedes include the choice of tests, decision to hospitalize or not to hospitalize, medications, and the timing of treatments. The system also provides the cost of treatment and probable outcomes.

The system helps medical staffs and managers of medical centers plan better. It helps decide on which treatments to spend money. For example, it might show hospital managers that if they treat 1,000 patients who have angina pectoris (chest pains) with a certain treatment, the hospital can prevent 72 heart attacks and the cost of the treatment will be $6 million. Comparing the information with similar information about other treatments and alternative treatments of other diseases, managers can decide how best to allocate resources. Thus, Archimedes addresses two purposes: medical treatment and medical management.

The Archimedes software was originally designed specifically to make projections helpful to both clinicians who decide about medical care and to managers who make decisions about resource allocation. However, it can be used by a wide variety of organizations, including pharmaceutical companies, disease management organizations, health-care organizations, researchers, and medical associations to solve both clinical and administrative healthcare decisions.

Sources: Wallace, P., "The Care Management Institute: Making the Right Thing Easier to Do," *The Permanente Journal*, Spring 2005, Volume 9 No. 2; Southwick, K., " 'Doctor Data' Digitizes Medical Care," *CIO Insight* (www.cioinsight.com), April 17, 2003; *Care Management Institute*, "Archimedes," (www.kpcmi.org/media/archimedesqa.html), June 20, 2003; (http://archimedesmodel.com).

Thinking About the Case

1. What is the data on which the Archimedes Model system relies?

2. A physician familiar with Eddy and his work admires the logic embedded in Archimedes but says it would be difficult to sell a system like this to other hospitals, because health care is an emotionally charged issue. Do you agree with the observation?

3. Does the system replace doctors? Why or why not?

Mapping Risk

The purpose of purchasing insurance is to pass risks to an insurance firm. However, insurance firms do not like risks any more than the individuals and businesses they insure. Therefore, insurance executives try to pinpoint the locations, or entire regions, with greater risks and those with lower risk. Knowing this helps them price their insurance policies better, or decide if they want to insure a certain asset at all. In addition, many insurance firms reinsure, which means they pay premiums to reinsurers, companies that reimburse the insurers if the insurers have to compensate their own clients.

Early in the 20th century many insurance companies refused to insure small properties, especially in certain urban areas. Therefore, in 1907, 150 owners of tenement houses on the East Side of New York City formed a trade association to protect their interests. They named it Greater New York Taxpayers Association. Over the years, the association turned into the GNY insurance company. GNY consists of Greater New York Mutual Insurance Company and Insurance Company of Greater New York. A.M. Best, the premier rater of insurance companies, granted the firm an A+ rating.

Greater New York Insurance Companies (GNY) has $200 million of direct insurance premiums. The terrible events of September 11, 2001, cost insurance companies $19 billion. After the 9/11 terrorist attacks, GNY found out that reinsurers were less willing to do business in the city. It could not reinsure as much of its business as it could before. One way of coping with the problem was to know the relationship between a location and its risks. Insurers want to know how much money they would have to pay in case of total destruction of an asset. They call this total exposure. If the firm could show reinsurers total risks and total exposure relative to locations, they might be more comfortable taking the risks.

GNY executives wanted to be able to see all the risks they had by policy type and exposure within a certain radius from each pinpointed location, such as the Empire State Building. If the building were to be destroyed, what would be the total exposure: $2 billion? $5 billion? They decided to adopt mapping software. They considered products from MapInfo, CDS Business Mapping, Baseline Business Geographics, and Millbrook, Inc.

After product demonstrations at the company's offices in East Brunswick, New Jersey, GNY decided to purchase MapInfo's system. MapInfo's programmers adapted the system to GNY's requirements. They installed a Web server in the company's New York offices so that the system could be available through an intranet to all employees. The data for the GIS came from policy, claim, and accounting information, all of which resided on an IBM AS/400 computer. The data is processed weekly through geocoding, that is, mapping it onto maps. The process involves 50,000–60,000 records and takes about eight minutes. The result is an updated Access database that holds both the textual and numeric data and its relation to map locations.

Employees use Microsoft Internet Explorer to enter a location's address and a radius, usually 500 feet or greater. The system returns a map with icons, and a legend on the right-hand side lists the displayed content represented by each icon. When the search is for a product (that is, a certain type of insurance policy), a dot with a certain color represents each product. For example, if the search is by risk exposure (the maximum payment for a claim), red dots might represent $5 million and higher, blue dots might represent $2 million and higher, and so forth. Alternatively, the size of the dot can be changed for size of exposure.

If users so desire, they can obtain information such as agent code, construction code, product, protection class, and policy type for each insured asset on the map. Each information search can be performed by any three codes, including zip code. By clicking a link below the legend, search results can be exported into an Excel spreadsheet.

Management found that the investment was worthwhile. It paid $25,000 for the hardware and $64,000 for the software, training, and maintenance during the first year of operation. GNY has a maintenance contract with MapInfo, which costs $40,000 per year for the first two years. Now, underwriters do not need to use separate applications to associate risks with locations; they simply map the risks on a map. The underwriters are more comfortable with their decisions on whether to take a risk and insure an asset, and GNY's reinsurers are more comfortable with their own decisions when they see the risks drawn on a map.

Sources: (www.gisdevelopment.net), 2005; (www.gny.com), 2005; O'Donnell, A., "Mapping System Reassures Reinsurers," *Insurance & Technology* (www.insurancetech.com), March 3, 2003.

Thinking About the Case

1. Is the system described in the case really a GIS? If not, explain why. If it is a GIS, explain what makes it one.

2. Is the system a DSS? Explain why or why not.

3. If you were a reinsurer's underwriter looking at information on this system, give an example of why you might refuse to insure an asset.

11

ELEVEN

Business Intelligence and Knowledge Management

LEARNING OBJECTIVES

As more and more business operations are managed using information from information systems and sometimes automatically *by* information systems, large amounts of data are collected and stored electronically. With proper software tools, data stored in databases and data warehouses enables executives to glean business intelligence—information that helps them know more about customers and suppliers—and therefore helps them make better decisions. Information technology also makes it possible to organize stored knowledge and garner knowledge from vast amounts of unstructured data.

When you finish this chapter, you will be able to:

■ Explain the concepts of data mining and online analytical processing.

■ Explain the notion of business intelligence and its benefits to organizations.

■ Identify needs for knowledge storage and management in organizations.

■ Explain the challenges in knowledge management and its benefits to organizations.

■ Identify possible ethical and societal issues arising from the increasing globalization of information technology.

DEBOER FARMS:
Harvesting Technology's Benefits

Carl DeBoer looked out at the crowd. When Steve Janssen first asked him to talk about precision farming at a workshop for other farmers, Carl thought Steve was having a good joke at his expense. Carl was not exactly a technical genius. But Steve convinced Carl that he could serve as an example to others, since DeBoer Farms had successfully implemented the system.

A New Type of Data

Carl looked at his notes and began. "Precision farming is revolutionary not because it uses satellites and space-age technology, but because it's an entirely new way to collect information about your farm, to manage that information, and to use it to help you make better decisions about how to increase your yield, reduce your risk, and increase profits.

"I had always monitored information, as you all do, on a per-field basis. I took soil samples within a field, averaged the results, and then spread fertilizer uniformly over the field. I seeded the same way—evenly over the whole field. I measured my yield *per field*. Of course, I knew that there were differences within the field—I noticed that some parts of a field produced a better yield per acre than others and that some areas had more weeds. I knew there were differences *within* my fields, but until precision farming, there was no practical way to get meaningful information about these differences. There was no practical way to gather, store, or analyze this data, much less to customize the type and amount of fertilizer or seed to the different areas within a field. Precision farming has changed all that. It gathers and analyzes new data within a field and gives you detailed information on what to do about it."

Collecting Data for the Warehouse

"Okay, let's talk about how the system gathers this new data," Carl continued. "Precision farming uses a geographic information system, or GIS, to make maps of your farm. You use data collection devices in the fields and handheld computers to enter the data, and the GIS converts the data into yield, soil nutrients, topology, and weed maps.

"Other kinds of data collection devices include an on-the-go weed detection system that relies on an optical camera. I use that system, but I know other folks are using sensors that detect salt bands or grain protein.

"You can also enter data yourself. As you take soil samples, you use a handheld computer to record the exact location of each sample within the field. You can also record the patches of your field that have weeds. If you're taking insect counts, you can enter that information too.

"All this data you're collecting is used in two ways. First, it's used for all your current operations. If you've recorded a patch of weeds in a part of the field, you can program your tractor to spot-spray that area the following day. Second, the data is collected into a data warehouse that stores all your input and output over the years. Collecting and analyzing this historical data can really help to increase your yield."

Mining the Data

"Here's how it works," Carl continued. "Once you collect all this data into your warehouse, the program analyzes the data to find correlations. You might want to know what events lead to others. Has early planting led to insect infestation? Is crop rotation combating insect infestation? The software can mine the data and answer these questions.

"The program can also find clusters in your data, groups of data that you didn't know were related. For example, when the field floods, what part of the field will be nitrogen deficient as a result of runoff? Is there a pattern? If so, the software can detect it and you can program your tractor to fertilize those areas—without going through the trouble and expense of soil testing!

"The program also mines the data in the warehouse to make predictions. The program analyzes input factors like soil nutrients and seed density and crop production. You folks are probably all familiar with yield maps, maps that show crop production in a given season. This program generates a prescription map. After analyzing input and output correlations, the program generates recommendations regarding the types and amounts of fertilizers, seeds, and pesticides to use in each area of your field. The prescription map shows the amount of input factors—like fertilizer, pesticide, and seed—to use in different places. It shows the amount and type of input that will most likely produce the greatest output.

"A number of different software products can perform this analysis, but I know Sam Beatty is going to be speaking about the pros and cons of each of those systems later."

Implementing the System

"The next step is to implement the results of the analysis. For this, you need instruments that use variable rate technology, or VRT. You might have heard of a VRT sprayer or a VRT planter. These devices, mounted on a tractor, adjust the rate of application according to the prescription map that your software created. You can also use different seeds or fertilizer blends within a field. You can spot-spray for weeds or insects and enter that information into the system and then view it later on the map."

Carl looked up from his notes. "Is all that clear?"

Laughter erupted from the audience. Then he noticed Fred Halpern raising his hand. "Go ahead, Fred," Carl said.

"Are you really saving money with this system?"

"Last year I saved $11 per acre. My yield improved, and I also used less seed, less pesticide, and less fertilizer. And you know what else, Fred? When you use less fertilizer, you have less runoff and that's better for the environment—and our groundwater."

"I'd heard that," Fred said. "But it really does sound a bit complicated."

"Well, we had to purchase new computer hardware, software, telecommunications equipment, and related high-tech field machinery. We had to spend time and money training so that we could figure out how to use data-logging software, field-mapping software, and VRT planters and sprayers. But, Fred, you know why they asked me to speak, don't you?" Carl paused to see if anyone would respond. "Well, if I can do it, anyone can!"

DATA MINING AND ONLINE ANALYSIS

Recall from our discussion in Chapter 7, "Databases and Data Warehouses," that data warehouses are large databases containing historical transactions and other data. However, data warehouses in themselves are useless. To make data warehouses useful, organizations must use software tools to process data from these huge databases into meaningful information. Because executives can obtain significantly more information about their customers, suppliers, and their own organizations, they like to call information gleaned with such tools business intelligence (BI) or business analytics. The two main uses of these databases are data mining and online analytical processing. These terms are often used interchangeably by some people.

Data Mining

Data warehouses could be regarded as a type of mine, where the data is the ore, and new useful information is the precious find. **Data mining** is the process of selecting, exploring, and modeling large amounts of data to discover previously unknown relationships that can support decision making. Data-mining software searches through large amounts of data for meaningful patterns of information.

Some data-mining tools are complex statistical analysis applications, and others use additional tools which go beyond statistical analysis and hypothesis testing. While some tools help find predefined relationships and ratios, they do not answer the question that more powerful data-mining tools can answer: "What are the relationships we do not yet know?" This is because the investigator must determine which relationships the software should look for in the first place. To answer this question, other techniques are used in data mining, including artificial intelligence techniques, described in Chapter 10, "Decision Support and Expert Systems."

To illustrate the difference between traditional queries and data-mining queries, consider the following examples. A typical traditional query would be: "What is the relationship between the amount of product X and the amount of product Y that we sold over the past quarter?" A typical data-mining query would be: "Discover two products most likely to sell well together on a weekend." The latter query lets the software find patterns that would otherwise not be detected through observation. While data has traditionally been used to see whether this or that pattern exists, data mining allows you to ask *what* patterns exist. Thus, some experts say that in data mining you let the computer answer questions that you do not know to ask. The combination of data-warehousing techniques and data-mining software makes it easier to predict future outcomes based on patterns discovered within historical data.

Data mining has four main objectives:

- *Sequence* or *path analysis*: Finding patterns where one event leads to another, later event.
- *Classification*: Finding whether certain facts fall into predefined groups.
- *Clustering*: Finding groups of related facts not previously known.
- *Forecasting*: Discovering patterns in data that can lead to reasonable predictions.

These techniques can be used in marketing, fraud detection, and other areas (see Figure 11.1). Of the four types of analysis, you may be most familiar with clustering. When you search for a certain item, you often see a list of other items with a message similar to "Customers who purchased this item also bought...". Data mining is most often used by marketing managers, who are constantly analyzing purchasing patterns so that potential buyers can be targeted more efficiently through special sales, product displays, or direct mail and e-mail campaigns. Data mining is an especially powerful tool in an environment in which businesses are shifting from mass-marketing a product to targeting the individual consumer with a variety of products that are likely to satisfy that person. Some observers call this approach "marketing to one."

Why You Should Learn About BI and KM Tools

Information technology has advanced from fast calculation machines to systems that produce useful information using structured data and then to software that turns unstructured information into knowledge. Knowing how to use BI tools will help you to independently produce highly useful information from data warehouses and other large data sources. In your work you will also need to use other peoples' knowledge. Much of this knowledge exists in the recorded work and in the minds of coworkers and experts outside your organization. Knowing how to use these tools will help you as well as others perform better. As a knowledge worker you will be able not only to use your own, limited knowledge but also augment it with the experiences of other people.

FIGURE 11.1
Potential applications of data mining

DATA-MINING APPLICATION	DESCRIPTION
Consumer clustering	Identify the common characteristics of customers who tend to buy the same products and services from your company.
Customer churn	Identify the reason customers switch to competitors; predict which customers are likely to do so.
Fraud detection	Identify characteristics of transactions that are most likely to be fraudulent.
Direct marketing	Identify which prospective clients should be included in a mailing or e-mail list to obtain the highest response rate.
Interactive marketing	Predict what each individual accessing a Web site is most likely to be interested in seeing.
Market basket analysis	Understand what products or services are commonly purchased together, and on what days of the week.
Trend analysis	Reveal the difference between a typical customer this month and a typical customer last month.

Predicting Customer Behavior

In banking, data mining is employed to find profitable customers and patterns of fraud. It is also used to predict bankruptcies and loan payment defaults. For example, when Bank of America looked for new approaches to retain customers, it used data-mining techniques. It merged various behavior patterns into finely tuned customer profiles. The data was clustered into smaller groups of individuals who were using banking services that didn't best support their activities. Bank employees contacted these customers and offered advice on services that would serve them better. The result was greater customer loyalty (measured in fewer accounts closed and fewer moves to other banks).

POINT OF INTEREST

Did You Really Buy It?

Many retailers will accept a returned item even if you cannot show a receipt. Some people take advantage of this and "return" stolen goods for cash. To minimize the phenomenon, some retailers have joined the Return Exchange, a program run by a company by the same name in Irvine, California. Now, when you return merchandise without a receipt to Staples, The Sports Authority, Guess, or another member retailer, you are asked to present a driver's license or other identification. The information is recorded and communicated to Return Exchange. The company uses statistical models to determine if certain "customers" engage in fraud. If the software determines that a person has defrauded retailers, the stores will stop accepting the person's returns. However, neither the retail chains nor the company disclose the criteria used to make the decision. Return Exchange does not allow retailers to share customer information.

Companies selling mobile phone services face a growing challenge of *customer churn* (switching to a competitor). Some surveys show that more than 50 percent of mobile phone users consider switching to a competitor at any given time, and 15 percent plan to switch to a competitor as soon as their contract expires. Mobilcom GmbH, a German company with 4.56 million customers and 1100 employees, uses data mining to identify such customers and approach them with inducements to continue or renew their contract before they switch. The company uses an application called DB Intelligent Miner from IBM. The software periodically looks for patterns of customer churn and assigns each customer a score representing the likelihood of canceling the contract. The software considers many variables, among which are complaint history and the number of days to expiration. Customer loyalty is extremely important because the cost of obtaining a new customer far exceeds the cost of retaining an existing one, especially in a highly competitive market such as mobile telephones.

To ensure a steady flow of customer data into their data warehouses, companies in almost every industry—from airlines to lodging, dining, and gambling—operate customer loyalty programs similar to the original frequent-flier programs. Membership is often free, and customers leave a record every time they make a purchase even if they do not use a credit card to pay. In many cases, mining such data provides business intelligence to target individual customers.

A large U.S. airline collects every possible piece of data on passengers in a central data warehouse, from frequent-flyer numbers through reservations and flight details. The airline uses data-mining tools to extract information that helps retain frequent flyers. For example, the executives can query the data warehouse to see how many flight disruptions, cancellations, or delayed arrivals its best customers experience in a given month. This helps the airline to proactively contact these customers and offer them incentives to ensure their continued business.

POINT OF INTEREST

Chemical Mining

By some estimates, 85 percent of recorded corporate knowledge is in text files. Dow Chemical Co. maintains a BI center in Midland, Michigan. Using a software tool called ClearResearch from Clear Forest Corp. (a subsidiary of Reuters), Dow's research staff has extracted useful information from various unstructured sources, including abstracts of chemical patents registered over the past century, published research articles, and the company's own files. The software has reduced the time it takes Dow's researchers to decide what they need to read. You can try the software at http://sws.clearforest.com/SWS.htm and use it free of charge.

Sources: Robb, D., "Text Mining Tools Take on Unstructured Data," *Computerworld* (www.computerworld.com), June 21, 2004; (www.clearresearch.com), July 2007.

UPS has an organizational unit called Customer Intelligence Group. The group analyzes patterns of customer behavior so it can make predictions that help the company enhance services and retain customers. For example, the group is able to accurately predict customer defections by examining usage patterns and complaints. When the data of a specific customer indicates that the customer might defect, a salesperson contacts that customer to review and resolve any problems. The software helped to significantly reduce the loss of customers.

Identifying Profitable Customer Groups

Financial institutions, especially insurance companies, often dismiss high-risk customers. Better analysis of such customers can yield good business, as Progressive Casualty Insurance Company has proven. Progressive is the fourth largest U.S. insurance firm. The company uses proprietary analytical software and widely available insurance industry data. The company defines narrow groups or "cells" of customers, for example, college-educated motorcycle riders ages 35 and older whose credit scores are above 650 and who have no accidents recorded. For each cell, the company performs a statistical regression analysis to identify factors that most closely correlate with the losses that this particular group causes. For each cell, the company then sets premiums that should enable the company to earn a profit on its portfolio of customer groups. The company uses simulation software to test the financial implications of accepting the analyzed groups as customers. This way, Progressive can profitably insure customers in traditionally high-risk categories. Other insurance companies reject such applicants and refuse to renew the contracts of customers who became high-risk because of claims such as for car accidents. These companies do so without bothering to analyze the data more deeply.

Utilizing Loyalty Programs

Loyalty programs such as frequent flier and consumer clubs help organizations amass huge amounts of data about their customers. Some grocery chains, for example, issue discount coupons only to the most loyal customers. Harrah's Entertainment, Inc., the casino and hotel chain, uses its data warehouse to target individual customers, rather than groups. The technique—whose specifics the company refuses to disclose for obvious reasons—enables Harrah's to tailor lodging, dining, and gambling packages that are attractive to its customers. It helps Harrah's discern the small spender from the big spender and decide how to price those services according to individual spending patterns at the company's facilities. This is an example of yield management or revenue management, a concept introduced in Chapter 10, "Decision Support and Expert Systems." Harrah's relies heavily on its software applications to price-discriminate. It gives sales agents instructions to charge people who have a history of little spending on gambling higher per-night rates than they charge big gamblers. The case study Sure Bet at the end of this chapter provides more details on Harrah's utilization of BI.

Collecting data through a customer loyalty program and gleaning business intelligence from the data catapulted Harrah's to a leader status in the hotel-casino industry.

AP Photo/ Isaac Brekken

Inferring Demographics

Some companies use data-mining techniques to try to predict what customers are likely to purchase in the future. As mentioned in previous chapters, Amazon.com is a leader in exploiting customer data. The company registered U.S. Patent Number 6,865,546, titled "Methods and systems of assisting users in purchasing items." The software developed by Amazon determines the age of the recipient of an item purchased by a customer. The age range is estimated based at least in part on a customer order history of gifts purchased for the recipient. The first gift is associated with the first "age appropriateness designation." The second gift is associated with a second age appropriateness designation. An age range associated with the recipient is estimated. The software also captures and analyzes any data that may indicate the recipient's gender. The recipient's age progression is calculated, and the company uses it to offer the customer gifts for that person when the customer logs on to the site. So, if you purchase gifts from Amazon.com for your baby niece, do not be surprised if Amazon entices you to purchase items for a young girl, a young woman, and an older woman over the next few decades. Here is another example of what this data-mining tool can do: if you purchased perfume a week before Valentine's Day, it will infer that you bought the item as a Valentine's gift for a woman and offer certain colors for the wrapping paper.

Online Analytical Processing

Online analytical processing (OLAP) is another type of application used to exploit data warehouses. Although OLAP might not be as sophisticated in terms of the analysis conducted, it has extremely fast response time and enables executives to make timely decisions. Tables, even if joining data from several sources, limit the review of information. Often, executives need to view information in multiple combinations of two dimensions. For example, an executive might want to see a summary of the quantity of each product sold in each region. Then, she might want to view the total quantities of each product sold within each city of a region. And she might also want to view quantities sold of a specific product in all cities of all regions. OLAP is specially designed to answer queries such as these. OLAP applications let a user rotate virtual "cubes" of information, whereby each side of the cube provides another two dimensions of relevant information.

The Power of OLAP

Figure 11.2 shows the interface of a Web-based OLAP application whose purpose is to provide information about federal employees. You can go to www.fedscope.opm.gov and receive information about federal personnel in almost any imaginable dimension for several years. Dimensions include region of employment, level of service, occupation, salary range, and many more. The middle table shows number of employees by department and region. Clicking the Department of Defense link produces more detailed information for that department, using the same dimension as before, but only for this department. You could also receive similar data for a particular branch of the military such as the U.S. Navy (bottom table). This would be an example of **drilling down**, a process by which one starts with a table that shows broad information and successively retrieves tables of more specific information. The OLAP application lets you receive the information in numbers of employees or as their percentages in each region, department, or organizational units within the department.

OLAP applications operate on data organized especially for such use or process data from relational databases. A dynamic OLAP application responds to commands by composing tables "on the fly." To speed up response, databases can be organized in the first place as dimensional. In **dimensional databases**—also called **multidimensional databases**—the raw data is organized in tables that show information in summaries and ratios so that the inquirer does not have to wait for processing raw data. Many firms organize data in relational databases and data warehouses but also employ applications that automatically summarize that data and organize the information in dimensional databases for OLAP. Cognos, Hyperion (a subsidiary of Oracle), and many other companies sell multidimensional database packages and OLAP tools to use them.

OLAP applications can easily answer questions such as, "What products are selling well?" or "Where are my weakest-performing sales offices?" Note that although the word "cube" is used to illustrate the multidimensionality of OLAP tables, the number of tables is not limited to six, which is the number of sides of a real cube. It is possible to produce tables showing relationships of any two related variables contained in the database, as long as the data exists in the database. OLAP enables managers to see summaries and ratios of the intersection of any two dimensions. As mentioned in Chapter 7, "Databases and Data Warehouses," the data used by OLAP applications usually comes from a data warehouse.

OLAP applications are powerful tools for executives. For example, consider Figure 11.3. Executives of a manufacturing company want to know how the three models of their product have sold over the past quarter in three world regions. They can see sales in dollar terms (top table) and then in unit terms (second table). They can then drill down into summaries of a particular region, in this case North America, and see the number of units sold not only by model but by model and color, because each model is sold in three colors. This information might lead them to recommend to the dealer to stop selling Model 3 in blue in North America, because sales of blue units of this model are quite low in this region. While still investigating last quarter's sales in North America, the executives might want to examine the sales performance of each dealer in this region. It seems that Dealer 3 enjoyed brisk sales of Model 1, but not of Models 2 and 3. If the sales picture is the same for another quarter or two, they might decide to stop sales of these models through Dealer 3 and increase the number of Model 1 units they provide to that dealer.

In a similar manner, Ruby Tuesday, the restaurant chain, solved a problem at one of its restaurants. Managers who examined performance by location discovered that a restaurant in Knoxville, Tennessee, was performing well below the chain's average in terms of sales and profit. Analyzing the store's information revealed that customers were waiting longer than normally for tables, and for their food after they were seated. There could be many reasons for this: an inexperienced cook, understaffing, or slow waiters, to name a few.

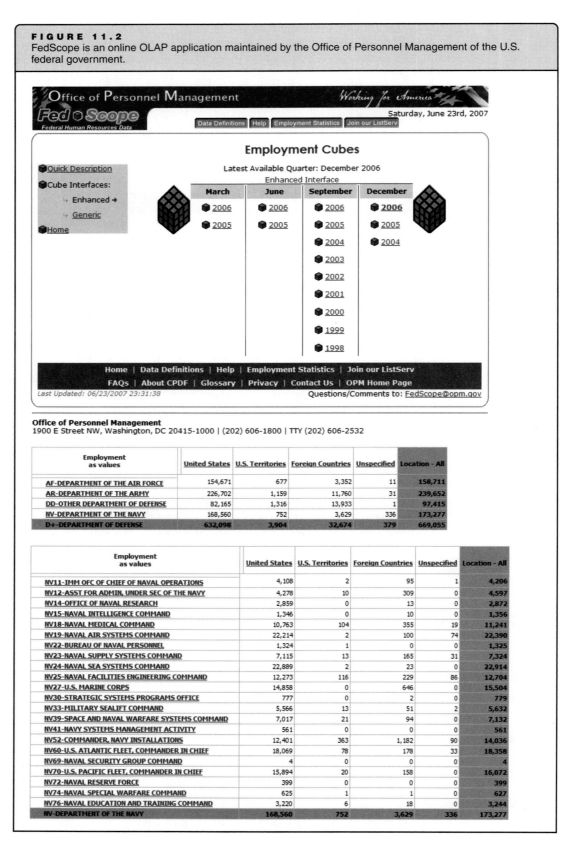

FIGURE 11.2
FedScope is an online OLAP application maintained by the Office of Personnel Management of the U.S. federal government.

Employment as values	United States	U.S. Territories	Foreign Countries	Unspecified	Location - All
AF-DEPARTMENT OF THE AIR FORCE	154,671	677	3,352	11	158,711
AR-DEPARTMENT OF THE ARMY	226,702	1,159	11,760	31	239,652
DD-OTHER DEPARTMENT OF DEFENSE	82,165	1,316	13,933	1	97,415
NV-DEPARTMENT OF THE NAVY	168,560	752	3,629	336	173,277
D+-DEPARTMENT OF DEFENSE	632,098	3,904	32,674	379	669,055

Employment as values	United States	U.S. Territories	Foreign Countries	Unspecified	Location - All
NV11-IMM OFC OF CHIEF OF NAVAL OPERATIONS	4,108	2	95	1	4,206
NV12-ASST FOR ADMIN, UNDER SEC OF THE NAVY	4,278	10	309	0	4,597
NV14-OFFICE OF NAVAL RESEARCH	2,859	0	13	0	2,872
NV15-NAVAL INTELLIGENCE COMMAND	1,346	0	10	0	1,356
NV18-NAVAL MEDICAL COMMAND	10,763	104	355	19	11,241
NV19-NAVAL AIR SYSTEMS COMMAND	22,214	2	100	74	22,390
NV22-BUREAU OF NAVAL PERSONNEL	1,324	1	0	0	1,325
NV23-NAVAL SUPPLY SYSTEMS COMMAND	7,115	13	165	31	7,324
NV24-NAVAL SEA SYSTEMS COMMAND	22,889	2	23	0	22,914
NV25-NAVAL FACILITIES ENGINEERING COMMAND	12,273	116	229	86	12,704
NV27-U.S. MARINE CORPS	14,858	0	646	0	15,504
NV30-STRATEGIC SYSTEMS PROGRAMS OFFICE	777	0	2	0	779
NV33-MILITARY SEALIFT COMMAND	5,566	13	51	2	5,632
NV39-SPACE AND NAVAL WARFARE SYSTEMS COMMAND	7,017	21	94	0	7,132
NV41-NAVY SYSTEMS MANAGEMENT ACTIVITY	561	0	0	0	561
NV52-COMMANDER, NAVY INSTALLATIONS	12,401	363	1,182	90	14,036
NV60-U.S. ATLANTIC FLEET, COMMANDER IN CHIEF	18,069	78	178	33	18,358
NV69-NAVAL SECURITY GROUP COMMAND	4	0	0	0	4
NV70-U.S. PACIFIC FLEET, COMMANDER IN CHIEF	15,894	20	158	0	16,072
NV72-NAVAL RESERVE FORCE	399	0	0	0	399
NV74-NAVAL SPECIAL WARFARE COMMAND	625	1	1	0	627
NV76-NAVAL EDUCATION AND TRAINING COMMAND	3,220	6	18	0	3,244
NV-DEPARTMENT OF THE NAVY	168,560	752	3,629	336	173,277

Managers at headquarters decided to take a look at the average time between when a check was opened at the cash register and the time the customer paid. In the restaurant industry this is an indication of an important factor: how long it takes to move from one party to another at

FIGURE 11.3
Using OLAP tables

Sales ($ 000)

	Model 1	Model 2	Model 3
North America	115800	136941	53550
South America	72550	63021	25236
Asia	65875	53781	17136
Total	**254225**	**253743**	**95922**

Sales (Units)

	Model 1	Model 2	Model 3
North America	4632	6521	2975
South America	2902	3001	1402
Asia	2635	2561	952
Total	**10169**	**12083**	**5329**

North America (Units)

	Model 1	Model 2	Model 3
Red	2401	1785	2512
Blue	1766	527	52
White	465	4209	411
Total	**4632**	**6521**	**2975**

North America Dealerships (Units)

	Model 1	Model 2	Model 3
Dealer 1	102	556	2011
Dealer 2	1578	2450	108
Dealer 3	2358	0	10
Dealer 4	20	520	57
Dealer 5	574	2995	789
Total	**4632**	**6521**	**2975**

a given table. The shorter the time, the better. The average time "to close a check" at Ruby Tuesday's restaurants is 45 minutes. At this particular location it was 55–60 minutes. Examining additional information, management concluded that the reason for the longer wait was increased demand thanks to an economic boom in the region. The company sent people to change the layout of the kitchen, positions of the cooks, and the placement of food. Cooking took less time, serving was faster, and the wait time decreased by 10 percent. More customers could be served, and revenue went up.

OLAP applications are usually installed on a special server that communicates with both the user's computer and the server or servers that contain a data warehouse or dimensional databases (although OLAP might also process data from a transactional database). Since OLAP applications are designed to process large amounts of records and produce summaries, they are usually significantly faster than relational applications such as those using SQL (Structured Query Language) queries. OLAP applications can process 20,000 records per second. As mentioned before, when using preorganized dimensional tables, the only processing involved is finding the table that corresponds to the dimensions and mode of presentation (such as values or percentages) that the user specified.

OLAP in Action

OLAP is increasingly used by corporations to gain efficiencies. For example, executives at Office Depot, Inc. wanted to know how successful sales clerks and stores were at cross-selling certain items. A store succeeds in cross-selling when it convinces customers to buy paper when they buy

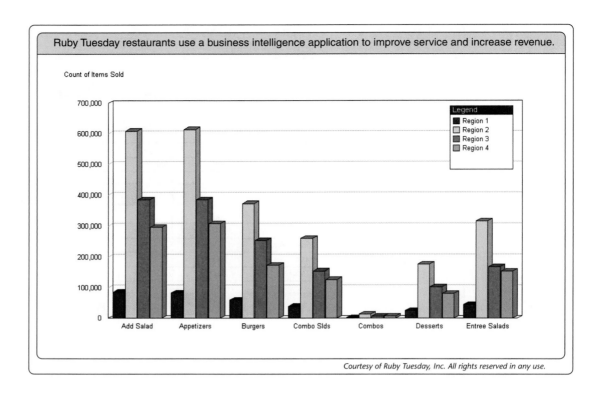

Ruby Tuesday restaurants use a business intelligence application to improve service and increase revenue.

Count of Items Sold

Legend
■ Region 1
□ Region 2
■ Region 3
■ Region 4

POINT OF INTEREST

Calories? Who Cares...

CKE Restaurants owns the fast-food chain Hardee's. The company makes extensive use of BI applications. Contrary to perceptions that customers would prefer low-calorie food, in 2004 CKE Restaurants recommended that Hardee's offer the Monster Thickburger, an artery-clogging "monument to decadence" of 1,420 calories, most of which come from the 107 grams of fat. It also contains 229 mg of cholesterol and 2,740 mg of sodium. The decision to offer the item came when management perused the results of analysis performed by the BI applications on data collected on sales of the experimental item: menu mixes, cost to produce a unit of the Monster, average units sold compared with other burgers, gross profits, and total sales per test store. The Monster Thickburger has been a great success and is still on the menu. Sales in Hardee's stores open for at least a year increased by 5.8 percent in the month after introduction, and the Monster directly contributed to much of the increase. The chain also sells a "Low Carb Thickburger" with a single rather than a double meat slab.

Sources: Levinson, M., "The Brain Behind the Big, Bad Burger and Other Tales of Business Intelligence," *CIO* (www.cio.com), March 15, 2005; (www.hardees.com), July 2007.

---

pens, or to purchase computer peripherals when they purchase a computer. The company used OLAP on its data warehouse, which saves transactions from 1,020 stores and by more than 60,000 employees in 10 countries. Making the proper conclusions helped the company to increase annual sales revenues by $117 million. Management now knows better which items cross-sell with other items and therefore makes better decisions on placing items on shelves in proximity.

Managers in some companies now track information about their products from the purchasing of raw materials to the receipt of payment, not only for operations but so they can learn more about their clients and their own business. For example, Ben & Jerry's, one of the largest U.S. ice cream makers, collects data about every container of ice cream it sells, starting with the ingredients. Each container is stamped with a tracking number, which is stored in a relational database. Using OLAP software, salespeople can track how fast new types of ice cream gain popularity, and which remain stagnant, on an hourly basis. Matching such information with about 200 telephone calls and e-mail messages the company receives weekly, managers can figure out which supplier's ingredients might have caused dissatisfaction with a certain product.

Employees who know very little about programming and report design are discovering that BI software is becoming easier to use. Intelligent interfaces allow them to enter questions in free form or close to free form. A part of the application that is called the semantic layer parses the question, which has been written as if you were speaking to a person, translates it into instructions to the computer to access the appropriate data mart or the proper columns of a data warehouse, and produces the answer, which is a number of charts showing trends. In a few seconds, a manager at Lands' End can find out which type of denim pants was the company's best-seller at Sears stores over the past six months. BI software has become so popular in large companies that Microsoft has integrated such software into its popular database management system, SQL Server.

More Customer Intelligence

We have discussed customer relationship management (CRM) in several previous chapters. A major effort of most businesses, especially retail businesses, in using such systems is to collect business intelligence about customers. Both data-mining and OLAP software are often integrated into CRM systems for this purpose. Since an increasing number of transactions are executed through the Web, managers can use data that is already in electronic form to analyze and strategize. The challenge is to address the right customer, at the right time, with the right offer, instead of spending millions of dollars in mass marketing or covering numerous Web sites with ads.

Clickstream software has become almost a standard feature of the server management software used by many companies. Clickstream software tracks and stores data about every visit to a site, including the site from which the visitor has come, which pages have been viewed, how long each page has been viewed, which items have been clicked, and so on. The data can then be analyzed to help redesign the site to make visits of all or certain demographic groups more attractive. The purpose may be to have the visitors spend more time at the site so its potential for advertising revenue from other companies grows, and/or to entice visitors to make more purchases.

Many companies find that using only the data collected directly from consumers does not provide a full picture. They approach third parties, companies that specialize in collection and analysis of consumer data. The companies, such as Avenue A | Razorfish, DoubleClick, and Engage Software, use cookies and spyware (explained in Chapter 8, "The Web-Enabled Enterprise") to collect consumer data, which can be combined with the individual's clickstream data.

By compiling billions of consumer clickstreams and creating behavioral models, these companies can determine individual consumers' interests from the sites they visited (what do they like?), the frequency of visits (are they loyal?), the times they surf (are they at work or at home?), and the number of times they click on ads or complete a transaction. Then, sites can display ads that match the typical interests at sites where the likely customers tend to visit. They can use software that will change the ad for each visitor by using cookies that identify the user.

Consider the challenge that was facing Drugstore.com, a Web-based drugstore headquartered in Bellevue, Washington. Management wanted to reach more customers who were likely to purchase its products, but they did not have the tools to discover who those people were. While Drugstore.com had plenty of information about customers—including name, address, and a list of past purchases—the company still did not know where exactly to find those customers on the Web or where to find more people who have the same buying habits. Management hired Avenue A | Razorfish, Inc., a firm that specializes in consumer profiling. Avenue A managers say they know where 100 million Web users visit, shop, and buy. This information comes from data they have collected for several years, not for any specific client. During a previous marketing campaign for Drugstore.com, Avenue A had compiled anonymous information about every Drugstore.com customer who made a purchase during the campaign. Avenue A knew what specific ad or promotion a given customer had responded to, what that customer had browsed for on the Drugstore.com site, whether the customer had made a purchase, and how many times the customer had returned to purchase.

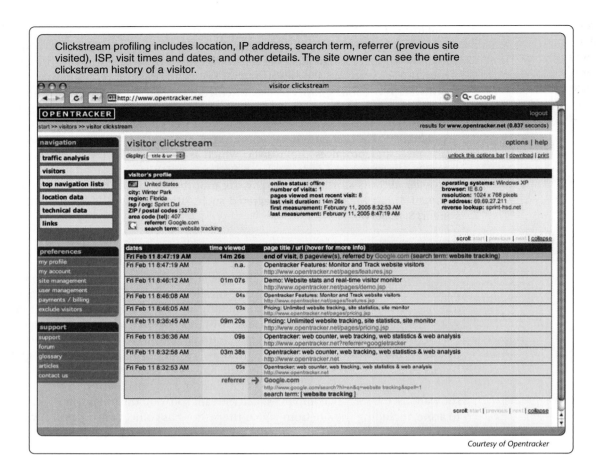

Clickstream profiling includes location, IP address, search term, referrer (previous site visited), ISP, visit times and dates, and other details. The site owner can see the entire clickstream history of a visitor.

Courtesy of Opentracker

Using its Web Affinity Analysis software, Avenue A could track Drugstore.com's individual customers across more than 3000 Web sites. Avenue A then constructed common themes in the customers' online behavior, such as the general Web sites they visited, visits to competing online drug retailers, and the likelihood that those individuals would click on ads. The company gave Drugstore.com a list of 1.45 million "high-quality prospects," shoppers with a high potential of purchasing from Drustore.com. Drugstore.com managers used the information to build a marketing strategy, assuming that those common characteristics and habits would be shared by as-yet-unconverted customers. (A converted customer is a shopper that is convinced to buy.) Using similar software helped Eddie Bauer, Inc. to decrease its marketing cost per sale by 74 percent over three months, and the Expedia, Inc. travel site to cut its cost per sale by 91 percent over eight months.

Dashboards

To make the use of BI tools convenient for executives, companies that develop BI tools create interfaces that help the executives and other employees to quickly grasp business situations. The popular name of such an interface is **dashboard**, because it looks something like a car dashboard. Car dashboards provide information in the form of clock-like indicators and scales. BI dashboards use similar visual images. They include speedometer-like indicators for periodic revenues, profits, and other financial information; plus bar charts, line graphs, and other graphical presentations whenever the information can be presented graphically. Figure 11.4 shows dashboards from Business Objects and XeoMatrix, providers of BI software. Similar dashboards are parts of BI tools offered by other vendors, including Cognos and SAS. ERP vendors, such as SAP and Oracle, also include dashboards in their applications. Dashboards are often designed to quickly present predefined business metrics such as occupancy ratios in hotels and hospitals, or inventory turns in retail.

FIGURE 11.4
BI dashboards help executives quickly receive metrics, ratios, and trends in mostly graphic format.

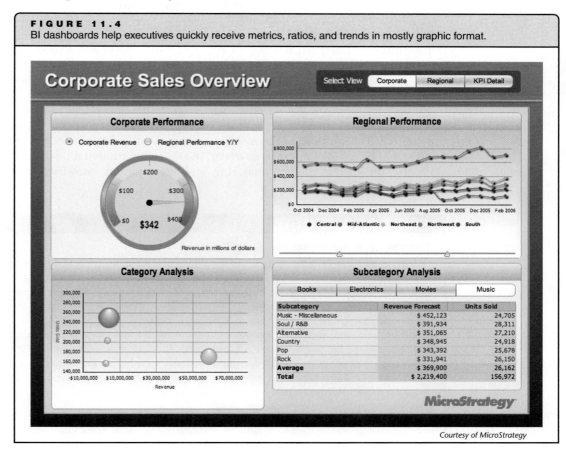

Courtesy of MicroStrategy

At TruServ, a member-owned hardware cooperative of 7000 retailers worldwide, executives use dashboards to monitor revenue and sales of individual items. The cooperative operates stores under the names True Value, Grand Rental Station, Taylor Rental, Party Central, Home & Garden Showplace, and Induserve Supply. Using the dashboard to conduct analyses, managers can pinpoint trends and changes over time and receive alerts to help monitor, interpret, and make decisions. They can better track inventory. In the past, 20 percent of the cooperative's inventory was in the "red zone." Red zone inventory is either liquidated or sold for a loss after a promotion is ended. The dashboard was instrumental in helping TruServ reduce this loss inventory to 5 percent. Employees can also track the relationship between a new promotion and sales of the promoted item, and decide to stop or continue the promotion based on the trends they see.

KNOWLEDGE MANAGEMENT

Imagine you work for a consulting firm. Your supervisor assigns you to a new client. As a smart professional, the first thing you want to check is whether your firm has had previous experience with this client and what knowledge has been accumulated about the experience. You heard that two ex-employees had contact with this client several years ago. It would be great to discuss it with them, but they are gone. Their knowledge is no longer available to the firm, because it is not recorded anywhere. The data recorded about the financial transactions with this client cannot provide the knowledge you are seeking: How easy or difficult was the interaction with the client's executives? What are the strengths and weaknesses of that organization? In engineering companies, engineers might want to see if colleagues have already encountered a problem they are trying to solve, and what the solution to that problem was. IT professionals might want to know if their colleagues have encountered a similar repeating failure with a network management system.

An organization can learn much about its customers, sellers, and itself by mining data warehouses and using OLAP software, but such techniques still do not satisfy another important challenge: how to manage knowledge, expertise that is created within and outside the organization. As discussed in Chapter 10, "Decision Support and Expert Systems," expertise in narrow domains can be programmed in expert systems. However, organizations would like to garner and manage much more knowledge. Effective management of knowledge can help both employees and customers.

Samuel Johnson, the author of an early English dictionary, said that one type of knowledge is what we know about a subject, and the other type is knowing where to find information about the subject. The purpose of knowledge management is mainly to gain the second type of knowledge. **Knowledge management (KM)** is the combination of activities involved in gathering, organizing, sharing, analyzing, and disseminating knowledge to improve an organization's performance.

POINT OF INTEREST

Knowledge Management and Baby Boomers

Baby boomers are Americans born between 1946 and 1964. Most of these people have worked many years for the same organization, accumulating an enormous amount of expertise. Generation X, which came after the baby boomers, is much smaller. This means that fewer people must absorb and use more knowledge. Boomers started to retire in the late 1990s, taking with them much expertise. For example, in 1997 Northrop Grumman saw 12,000 workers retire. Many of them were experts in various systems of the B-2 bomber, the most complex aircraft ever built. Since the productive life of these aircraft will continue for another 40 years or so, the company now suffers from a shortage of expert knowledge. Exit interviewers took video interviews of these retirees, but much of the knowledge could not be captured. Learning from this experience, Northrop Grumman now uses KM tools to capture and organize knowledge by using software tools years before engineers retire.

Source: Patton, S., "Knowledge Management (KM)–How to Beat the Baby Boomer Retirement Blues," *CIO*, January 15, 2006.

Information that can be gleaned from stored data is knowledge, but there is much more knowledge that organizations would like to store that they currently do not. The knowledge that is not maintained in information systems is typically of the type that cannot be extracted from readily captured data at Web sites or other electronic means of transactions. It is accumulated through experience. Much of it is kept in people's minds, on paper notes, on discussion

Top Results (148)

Industry: RadioPress Releases
... will be airing an exclusive live recording of the band Rockfour. ... Segment is a...
Web-Yahoo emediawire.com/new...

EQ Magazine - For audio and home recording st...
... successor to the Grateful Dead — a band more noted for its musically eclectic, ...
Web-MSN eqmag.com/issueIndex...

O'Connor Piano, MIDI Keyboard and Organ S...
... and was cast in the Busby Berkeley film Gold ... albums contained hits suc...
Web-MSN oconnormusic.org/mon...

Going home - musician Shirley Eikhard Perform...
... me that songwriting's a bit like mining for gold-you have to move a lot of ... ...
Web-Yahoo findarticles.com/p/ar...

NetRhythms: A to Z Album and Gig reviews
However, he just doesn't mine the past for his gold, Jessica Drive and ...... psy...
Web-Dogpile netrhythms.com/re...

EQ Magazine - Session File : Phish
... super-improvisational, free form live shows than for its gold-status studio ...
Web-Yahoo eqmag.com/story.as...

Rock On The Net: Bee ...
... world-wide with "New York Mining ... Gees Number Ones was certified gold. Love Songs was released.
Web-MSN rockonthenet.com/arti...

the Joshua Tree: Information from Answ...
Red Hill Mining Town" had originally been slated as a

Web (125)
News (0)
Blogs (7)
Premium (10)
Shopping (6)

transcripts, and in other places that are not readily accessible to a company's employees. Therefore, knowledge management is a great challenge. Knowledge management is the attempt by organizations to put procedures and technologies in place to do the following:

- Transfer individual knowledge into databases.
- Filter and separate the most relevant knowledge.
- Organize that knowledge in databases that allow employees easy access to it or that "push" specific knowledge to employees based on prespecified needs.

Knowledge management software facilitates these activities. As the cost of storage media continues to decrease and database management packages are increasingly more sophisticated and affordable, storage and organization of unstructured information have been less of a challenge. The more difficult issue is development of tools that address the third challenge: quickly finding the most relevant information for solving problems.

Capturing and Sorting Organizational Knowledge

The research company IDC argues that almost half of the work that **knowledge workers** do in organizations has already been done, at least partially. This work includes researching a certain subject, preparing a report, and providing information as part of a consulting contract. It estimated that labor worth $3,000–5,000 per knowledge worker is wasted annually because workers try to solve the same problem that other workers have already solved. Organizations could save this duplication, or replication, by collecting and organizing knowledge that is gained by members of the organization.

To transfer knowledge into manageable online resources, some companies require workers to create reports of their finding. Others, especially consulting firms, require their employees to create reports about sessions with clients. However organizations collect information, the results might be several terabytes of potential knowledge, but the challenge for employees is to know how to find answers to specific questions. Some software tools have been developed to help.

Electronic Data Systems Corp. (EDS), an IT consulting firm, requires all of its 130,000 employees to fill out an online questionnaire about their activities once per year. With 20,000 of these employees, EDS conducts surveys three times per year. Some of the questions provide multiple-choice answers, which make the input structured and easy to sort and analyze, but some of the most valuable input is in the form of free text. In the past, this part was forwarded to managers who learned and drew conclusions from it. Now the company uses an automated system, PolyAnalyst from Megaputer Intelligence, Inc., to sort the text information and create links between topics.

Motorola, the giant manufacturer of communications equipment, has 4 terabytes of data managed by a knowledge management application. The application enabled engineers to query this huge resource. Still, unless a worker knew exactly where the proper data was located or the names of people who were on a team that had solved the problem at hand, the worker could not find a proper answer. Motorola decided to implement Watson, an application developed by Intellext (now a unit of MediaRiver). Watson is installed on employees' PCs. It can be embedded in Microsoft Word, PowerPoint, and Outlook as well as the Web browsers Internet Explorer and Firefox. When using Word, it analyzes a user's document as it is being written, creates an automatic query about the subject, reaches out into the KM program, and pulls information that might be applicable to the task at hand from an individual PC or online resources that the user has designated. It works similarly when one uses Outlook or a Web browser.

Employee Knowledge Networks

While some tools build knowledge bases and help employees access them, others put the emphasis on directing employees to other employees who have a certain expertise. The advantage of this approach is that an expert can provide expertise that has not been captured in information systems (see Figure 11.5). Large companies, especially multisite ones, often waste money because employees in one organizational unit are not aware of the experience of employees in another unit. For example, one energy company spent $1 million on a product designed to work on oil rigs to prevent sediment from falling into wells. When the equipment was installed, it failed. The executives of another unit decided to purchase the same equipment, which, not surprisingly, failed in the other location. Then a third unit, elsewhere, purchased the equipment, which also failed. While one can justify the loss of the first $1 million as legitimate business expense in the course of trying a product, the other $2 million was lost because decision makers did not know that the equipment had already been tried and failed. To alleviate similar problems, some software companies, such as Tacit Systems, Inc., AskMe Corporation, Participate Systems, Inc., and Entopia, Inc., have developed **employee knowledge networks**, tools that facilitate knowledge sharing through intranets. Recall that an intranet uses Web technologies to link employees of the same organization.

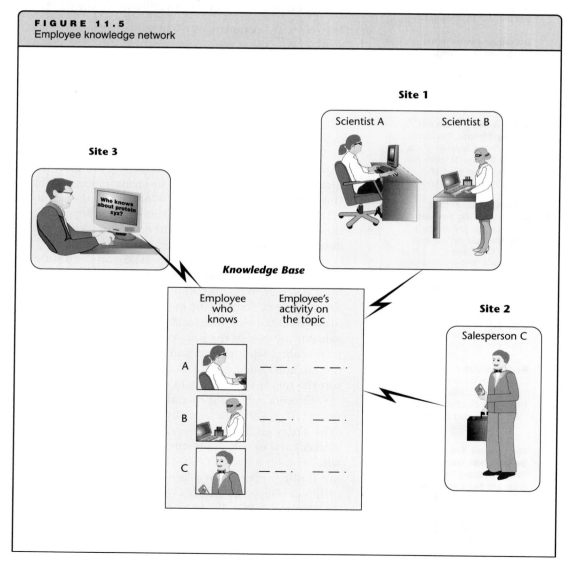

FIGURE 11.5
Employee knowledge network

Knowledge and Globalization

In the Middle Ages, Venice considered its expertise in making glassware not only a business trade secret but also a state secret. Divulging glassmaking knowledge to anyone outside the republic of Venice was punishable by death, because much of the state's economy depended on excluding other states and countries from such knowledge. Venice, like other states in that era, would never "offshore" any of the work to another country. Nowadays, matters are completely different. What was expertise a year ago has become routine work this year and will become automated next year. At that point, the expertise value in the product will have diminished, and to make a profit the organization that once had a comparative advantage in producing the product will have to use the least expensive labor available. It will offshore manufacturing to a factory in a country where labor is cheaper. The industry in the original country will lose jobs.

Information technology helps create knowledge but also expedites the turning of knowledge into routine, automated processes that can be carried out elsewhere. IT also expedites the transfer of knowledge from countries that created it to countries that can quickly use it. Software used to be developed almost exclusively in the United States. Much of the software that the world uses now has been developed in Germany, India, Ireland, Israel, and Russia. A growing amount of the software developed for U.S. companies is created in India and China. The programmers' expertise is similar, but the wages earned by programmers in those two countries are a fraction of what American programmers would be paid for the same work. This is a pivotal element in what is called *globalization*—moving from national economies to a global economy. Is this bad for countries such as the United States and good only for countries such as India and China?

BI and KM software is developed mainly in the United States, Germany, and the United Kingdom. However, these systems are sold anywhere and can help companies in other countries compete with companies in those "developed" countries. This puts developing countries in position to gain knowledge much faster than before and compete better. Now, the competition is not only in the manufacturing and service areas but also in R&D.

In the United States, some observers view the issue in the following light: America used to be a world leader in manufacturing, but other countries now have a comparative advantage in manufacturing, and their workers have taken the jobs that American laborers used to perform. For some time Americans had an advantage in providing services, but many of these services are now provided over the Internet and telephone lines by workers in other countries, so the service sector's advantage has diminished. The United States is still ahead in terms of innovation and creation of know-how, they say, but we are starting to see this advantage slipping away, too. And when other countries beat us in creation of knowledge, they ask, what's left with which to compete?

Should governments take measures—legal or otherwise—that protect their economic advantages? Should they penalize companies that offshore manufacturing jobs? Should they forbid the sale of know-how to other countries? Should they adopt the Venetian model? Or, should we look at the world as one large economy where each worker and each organization should compete for a piece of the pie regardless of national borders, so that consumers everywhere can enjoy products of the highest quality for the lowest price possible?

Tacit Systems' ActiveNet tool continuously processes e-mail, documents, and other business communications and automatically "discovers" each employee's work focus, expertise, and business relationships. The tool "mines" this unstructured data to build a profile of each employee in terms of topics and interests. The goal is to ensure that two people who might benefit from creating a connection in a workplace do so, so that one can learn from the experience of another about a specific issue. By analyzing e-mail and documents, the tool extracts the employee's interests and solutions to problems, and that information is added to the employee's profile. Other employees who seek advice can access the profile, but they cannot see the original e-mail or document created by the employee. This ensures uninhibited brainstorming and communication.

AskMe's software also detects and captures keywords from e-mail and documents created by employees. It creates a knowledge base that holds the names of employees and their interests. An employee can access a Web page at which the employee enters a free-form question. The software responds by listing the names of other employees who have created e-mail, text documents, or presentations on the subject, and the topics of their work. The employee can view the activity profiles of these people, and then contact them via the Web site, e-mail, instant messages, or paging. The responder can use the same Web site to respond and attach documents that might help the inquirer. AskMe's tool captures the communication, including attached documents, and adds them to the knowledge base. (Note that in this context the knowledge base is not organized as the knowledge bases in expert systems are.)

Knowledge from the Web

Consumers keep posting their opinions on products and services on the Web. Some do so at the site of the seller, others at general product evaluation sites such as epinions.com, and some on blogs. By some estimates, consumer opinions are expressed in more than 550 billion Web pages. This information is difficult to locate and highly unstructured. If organizations could distill knowledge from it they could learn much more than they do from conducting market research studies, such as focus groups, both about their own products and those sold by competitors.

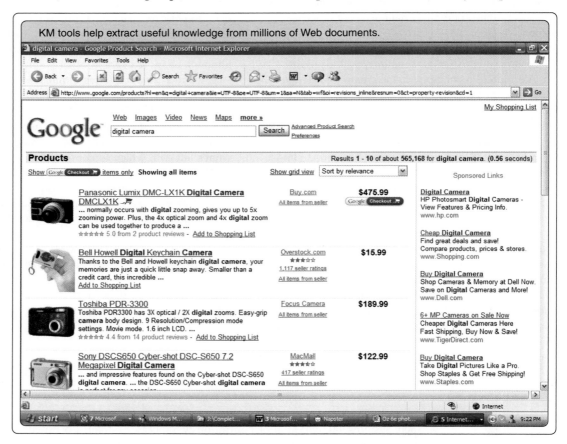

KM tools help extract useful knowledge from millions of Web documents.

Some companies have developed software tools that search for such information and derive valuable business knowledge from it. For example, Accenture Technology Labs, the technology research and development unit of the consulting firm Accenture, developed Online Audience Analysis. The tool searches thousands of Web sites daily and retrieves predetermined information about specific products and services. It then uses data-mining techniques to help organizations understand what consumers are saying about corporate brands and products.

Factiva, a subsidiary of Dow Jones, promotes a software tool by the same name. Factiva is accessible through a Web site and gathers information online from over 10,000 sources—newspapers, journals, market data, and newswires—information that amounts to millions of documents. About 60 percent of the information is not accessible to the general public. It screens

POINT OF INTEREST

Webwise

Several companies offer BI services based on data gleaned from the Web. One is Hitwise Pty., Ltd., which claims to be the leading online competitive intelligence service. Every day, the company monitors how more than 25 million Internet users interact with more than 800,000 Web sites in more than 160 industries. It provides marketing managers with insights on how their organizations' Web presence compares to competitive Web sites. The managers use the information in efforts such as affiliate programs, search advertising, and Web site content development.

Source: (www.hitwise.com), July 2007.

every piece of new information that is posted at any of these Web sites for information specified by a subscribing organization. The search can be more tailored and specific than searches performed through free search engines such as Google or Yahoo!. The software helps organizations add to their knowledge base, especially in terms of what others say about their products and services. The tool takes into account factors such as the industry and context in which an inquirer works to select and deliver the proper information. For example, a key word such as "apple" means one thing to an employee of a hardware or software organization and something completely different to an employee in agriculture or a supermarket chain.

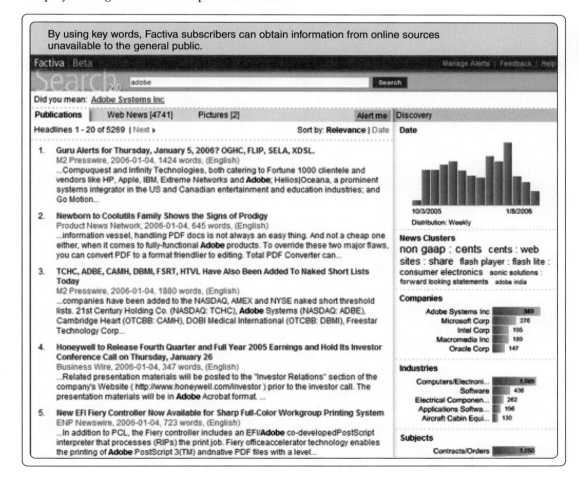

By using key words, Factiva subscribers can obtain information from online sources unavailable to the general public.

Autocategorization

To categorize knowledge into manageable data, companies use autocategorization software. **Autocategorization** or **automatic taxonomy** automates the classification (taxonomy) of data into categories for future retrieval. Practically all search engine sites, such as Google and Yahoo!, use autocategorization software, and continue to improve the software to provide more precise and faster responses to queries. Many companies have installed such software at their corporate Web sites.

For example, U.S. Robotics (USR), a large manufacturer of networking devices, operates in a market with narrow profit margins, and thus one call to the support personnel about a purchased item might wipe out the profit on that sale. Therefore, reducing support personnel labor is important. The firm's surveys showed that 90 percent of clients calling technical support had visited the USR Web site before calling. USR purchased autocategorization software from iPhrase Technologies, Inc. (a subsidiary of IBM) to help customers help themselves in searching for answers to their questions at the Web site so that customers would not have to telephone the support staff. The software can help interpret customer queries even if the queries are misspelled. The software improved the accuracy and responsiveness of the support database at USR's Web site. Consequently, support calls decreased by a third, saving the company more than $135,000 monthly.

Google, Yahoo!, and other companies in the search engine industry have developed applications that sift through documents both online and offline, categorize them, and help users bring up only links to the most relevant documents. These companies sell their products to corporations to use at their Web sites, intranets, and extranets.

- Business intelligence (BI) or business analytics is any information about the organization, its customers, and its suppliers that can help firms make decisions. In recent years organizations have implemented a growing number of increasingly sophisticated BI software tools.

- Data mining is the process of selecting, exploring, and modeling large amounts of data to discover previously unknown relationships that can support decision making. Data mining helps sequence analysis, classification, clustering, and forecasting.

- Data mining is useful in such activities as predicting customer behavior and detecting fraud.

- Online analytical processing (OLAP) helps users peruse two-dimensional tables created from data that is usually stored in data warehouses. OLAP applications are said to provide a virtual cube that the user can rotate from one table to another.

- OLAP either uses dimensional databases or calculates desired tables on the fly.

- OLAP facilitates drilling down, moving from a broad view of information to increasingly detailed information about a narrow aspect of the business.

- Dashboards interface with BI software tools to help users quickly receive information such as business metrics.

- Knowledge management involves gathering, organizing, sharing, analyzing, and disseminating knowledge that can improve an organization's performance.

- The main challenge in knowledge management is identifying and classifying useful information to be gleaned from unstructured sources.

- Most unstructured knowledge is textual, both inside an organization and in files available to the public on the Web.

- Employee knowledge networks are software tools that help employees find other employees who have expertise in certain areas of inquiry.

- Autocategorization (automatic taxonomy), the automatic classification of information, is one important element of knowledge management. Autocategorization has been used in online customer support Web pages to reduce the labor involved in helping customers solve problems.

DEBOER FARMS REVISITED

Carl DeBoer has been using precision farming, a new knowledge management system, on his farm. Let's review some aspects of systems they use and issues Carl faces in implementing the system.

What Would You Do?

1. After Carl finished speaking about his precision farming system, one member of the audience raised the concern that this emerging technology might be evolving so rapidly that the hardware and software she bought today to implement the system might be outdated in a year or two. If you were this farm owner, how would you deal with this concern? What kind of research could you do to determine the best system to buy or whether to wait for another year or two?

2. Precision farming allows farmers to apply different types and amounts of seeds, fertilizers, and pesti-

cides to different areas of the field. Five years after this knowledge system has been put into place, how would you use data mining to determine what decisions and techniques were most successful in varying these parameters within a field?

New Perspectives

1. Knowledge management systems are being implemented in the farming industry for a variety of tasks. They have been used by farmers who grow a variety of different crops as well as by dairy farmers and cattle ranchers. Search the Web for current and potential uses of knowledge management on farms. How prevalent do you think they will become? Prepare to discuss your findings in class.

KEY TERMS

autocategorization, 396
automatic taxonomy, 396
business analytics, 378
business intelligence (BI), 378
clickstream software, 387
dashboard, 389

data mining, 379
dimensional database, 383
drilling down, 383
employee knowledge
   network, 392

knowledge management
   (KM), 390
multidimensional
   database, 383
online analytical processing
   (OLAP), 382

REVIEW QUESTIONS

1. What is business intelligence?
2. What is OLAP, and why is it often associated with visual cubes?
3. What is the advantage of using a dimensional database rather than on-the-fly processing in OLAP?

4. Why is online analytical processing usually conducted on warehoused data or dimensional databases rather than on data in transactional databases?
5. What is "drilling down"?

6. What are data-mining techniques expected to find in the huge data warehouses that they scour?

7. Explain dimensional databases and the rationale behind their name. What is their use?

8. What is knowledge, and how does it differ from other information?

9. In general, what is the purpose of knowledge management in organizations?

10. What is the purpose of employee knowledge networks?

11. What is the benefit of tools that direct employees to experts rather than to stored knowledge?

12. What is autocategorization (automatic taxonomy)? How can autocategorization software help companies to serve customers and employees?

13. Context is a major factor when using tools to glean knowledge from Web sources. How so?

14. Data mining helps mainly in four ways: sequence analysis, classification, clustering, and forecasting. Data mining helps determine whether a person has committed fraud. Which of the four types of analysis help do that? Explain why.

15. The Web is a huge resource from which almost any organization could derive knowledge, yet few do. What is the major challenge?

DISCUSSION QUESTIONS

16. What does intelligence mean? Do you accept the use of the word in "business intelligence software tools," or do you think the use of this word is exaggerated compared with what these tools provide?

17. You are an executive for a large retail chain. Your IT professionals use data-mining software. They tell you of the following relationship the software found: middle-aged single men tend to purchase personal grooming products and light bulbs together. Should you assign employees to research the reason for this? How will you use this information?

18. Employee knowledge network software keeps tabs of much of what employees create on their computers and all the e-mail they send. As an employee, would you be comfortable with such software? Why or why not?

19. The term "business intelligence" has been used by IT professionals to mean many different things. What might be the reason for this?

20. Can businesses use free search engines, such as those provided by Google and Yahoo!, to efficiently gather useful knowledge for better decision making?

21. Recall the discussion of expert systems in Chapter 10, "Decision Support and Expert Systems." In what sense are employee knowledge networks similar to expert systems and in what sense are they different?

22. Consider Amazon.com's data-mining software, which infers demographic information about the recipients of gifts. Is letting software infer demographics less of an invasion of privacy than questionnaires or other forms of direct questioning? Is such inference more effective in obtaining customer information? Why or why not?

23. Some consumer advocates argue that using services such as Return Exchange (see the sidebar "Did You Really Buy It?") violate privacy, because legitimate customers who make multiple returns without receipt might be denied a legitimate right to return merchandise. The sore point is over disclosing the criteria used to determine who is probably engaged in fraud. Do you agree with this resentment, or do you think it is legitimate to use such a practice? What could be the motivation not to disclose the criteria?

24. Consider the discussion of autocategorization and smart search engines, like the one used byUSR. Suppose you purchased an electronic device and have a problem with it. You turn to the company's Web site and find either a FAQ (frequently asked questions) section or a Web page that invites you to enter a free-form question about your problem. Which option do you prefer, and why?

25. Companies would like to have systems that would allow them to store all the business-related knowledge that their employees have accumulated. Do you expect that such systems will exist in your lifetime? Why or why not?

26. Suppose a company developed a device that connects to your head to automatically debrief you at the end of each working day and store all that you have learned and experienced during the day in an information system. Would you be comfortable with this? Why or why not?

27. Search the Web for a story on an organization that successfully used data-mining techniques. Write a four- to five-page paper in which you describe what the software was able to do for the organization and how the results benefited the organization. Emphasize what information the organization now has that could not be obtained without data-mining techniques.

28. Write a one-page report explaining what one can do with an OLAP application that cannot be done with the same data in a spreadsheet or a relational database. Give at least two examples.

29. Go to www.fedscope.opm.gov. Produce the following tables for the latest year for which data is available:
    a. Number of U.S. federal employees by category size (large, medium, small) by country of service.
    b. Number of U.S. federal employees by category size, only for employees working in Australia.
    c. Number of U.S. federal employees by department within the *large* agency category, only for employees working in Australia.
    Is the sequence in which you produced the three tables considered drilling down? Explain why or why not.

30. Select an organization and research what it produces and where and how it operates. Using a pen and paper or graphical software, design three to six dashboards that the senior executives (vice president level) could use to receive up-to-date information to improve the organization's performance. Explain the purpose of each indicator.

31. Team up with another student. Select a specific company and write a report about its knowledge management needs. Start with a description of the activities that take place. List the types of employees who could benefit from access to documented expertise. Say which expertise you would take from internal sources and which from external sources. Give examples of knowledge that the company could use.

32. Team up with two other students. The team is to prepare a plan for The Researcher Connection, an employee knowledge network, to help the professors at your institution develop research ideas and conduct their research. List the elements of your proposed system and convince potential users in your report how it will help them (1) find relevant literature on a research subject, (2) learn who in the institution has done a similar study, (3) learn who in the institution would be interested in collaborating in the research, and (4) perform any other activity that is involved in conducting research and publishing the resultant article. Explain which of the resources to be used in the system already exist and will only need to be tapped (and how).

FROM IDEAS TO APPLICATION: REAL CASES

Sure Bet

In the casino-hotel industry, Harrah's name is well known for taking advantage of information technology. The chain makes exemplary use of loyalty programs, which helps it amass huge amounts of personally identifiable data. Executives use OLAP and data mining techniques to improve decisions.

Founded in 1937, Harrah's License Company operates 51 casinos in seven countries and employs over 80,000 people. Twenty-six of them are located in 13 states in the United States. Most of the casinos operate under the names Harrah's, Caesars, Bally's, and Horseshoe. The company is based in—surprise!—Las Vegas, Nevada. Until 1998, the company lagged behind competitors. It did not have the amount of cash its competitors had to invest in opulent hotels and attractive shows. Then a new chief operating officer (COO) was hired. Gary Loveman decided to use the only asset he could: customer data. Instead of investing in new properties, he spent money on building a data warehouse, purchasing BI software, and hiring business analysis specialists.

When Loveman (now president and CEO) took over, Harrah's already had a customer loyalty program called Total Reward. It offered a Total Reward card that is used in slot machines and other facilities. Members receive credit points every time they visit a Harrah's casino and gamble or spend a night at its hotels.

However, until the late 1990s, the cards did not help the company retain customers. Analysis showed that 65 percent of cardholders gambled elsewhere. Further analysis revealed that 82 percent of revenues came from 26 percent of customers. With further analysis, management found that these 26 percent were middle-age and older people with small incomes. While the company wants the loyalty of these gamblers, it also wanted to lure the bigger spenders.

Points are stored in member cards similar to credit cards and can be exchanged for rewards, cash, coupons, and complimentary services. The points are also tallied to place a cardholder in one of three categories: gold, platinum, or diamond. The value of services for which credit points can be traded increases with the category level. The cards were first introduced in 1997, but the data collected from them has been analyzed with growing sophistication since Loveman joined the company.

At every transaction—such as purchasing chips, buying food and drinks, or staying nights at a hotel—data is collected. The data is accumulated in a data warehouse built by Teradata, a division of NCR that specializes in data warehouses. Harrah's uses BI software from two companies, SAS and Tibco, to discover disloyal customers and turn them into loyal ones. The SAS application is called Customer Intelligence. In early 2007, the reward program had data on 41 million participants.

To increase loyalty, the company had to know what rewards would keep customers in its hotels and casinos and encourage them to spend more within its own chain rather than at competitors. Therefore, management conducted several experiments and analyzed the data collected. The analysis discovered that gift shop discounts were not attractive, but that discounts on hotel rooms were. Many customers did not stay in a hotel but actually lived near a casino. These customers valued casino chips more than any other incentive. Analyzing data of the same customers over time revealed that those who enjoyed their visit in one year spent more the next year, and vice versa. Management decided to change the basis for employee incentives: instead of paying bonuses based on generating income it does so based on customer satisfaction.

The SAS software enabled the company to divide members into 80 different segments. It uses a different marketing campaign for each segment. The company knows that customer behavior depends on where the customers live and where they gamble. Customers who live near a casino behave differently from those who live far. Those who gamble in Las Vegas have different gambling characteristics from those who gamble in New Orleans.

The huge amount of data on each individual and the great (and growing) number of participants enable the company to learn how customers interact with Harrah's. The BI applications discovered some interesting facts. Customers spent an average of 36 percent of their gambling budget at Harrah's. Much of the casino's gambling income came from what is known in the industry as low rollers, people who spend no more than $50 per day on gambling. While low rollers do not spend much daily, they visit the casino 30 times per year. By segmenting customers and following their interaction with Harrah's, the company developed and keeps adjusting typical patterns.

The patterns help the company predict customer behavior by demographic characteristics and spending history. The company designs offers for each customer. Some of the offers are made before the customer leaves the hotel. For example, the company knows that each gambler has a "pain threshold," a sum of dollars lost above which they will stop gambling. The SAS software calculates individual pain thresholds. The Tibco software tracks individual activities in real time, and responds to some of the events in real time. The Tibco software continuously compares individual gains and losses. When a Total Reward member is about to reach his or her pain threshold, the casino offers a free meal or a ticket to a show to keep her happy and keep her inside the building. Many people believe that their winning probabilities are affected by previous winning or losing. Therefore, a staff member may also offer to lock the machine until she is back from the free meal.

Harrah's does everything it can to accommodate high rollers, that is, big spenders. In the holiday season its hotels are fully booked. While this may sound good to any hotelier, it may not be ideal for casino hotels. The BI software revealed that high rollers tend to book late. Therefore, management decided to reserve some rooms for these customers. Even if they book late, they will find an available room.

Based on business data analysis, Harrah's offers different incentives to different customers. When this idea was first applied it broke a long-held rule. Other casino chains were not quick to follow. The time at which the incentive is offered and the value of the incentive depend on the value of the individual customer to Harrah's.

Harrah's income at its Las Vegas properties rose 10 percent since it started using the BI software, and income from operations increased 26.6 percent. Customers spent 45 percent more on gambling. The company's market share in the gaming industry grew significantly, and its stock price went up from $14 to over $85. Over the years 2003–2006, it spent $22 million on the data warehouse, BI software, and training. Nuclear Research, a business research firm, estimates that Harrah's reaped $208 million in measurable benefits from this investment. The company purchased Caesars' Palace and other competitors. In 2000, Harrah's value was $3 billion. In 2007, investors offered to acquire it for $17 billion. BI seems to have helped Harrah's make gold, platinum, and diamonds.

Sources: "Nothing Left to Chance," *InformationAge* (www.information-age.com), January 18, 2007; Arellano, N.E., "Harrah's bets on BI to gain customer loyalty," *Computerworld*, November 16, 2006; (www.harrahs.com), July 2007; (www.teradata.com), July 2007.

Thinking About the Case

1. Both the customer loyalty program (Total Reward) and data warehouse were in place when the new COO (now CEO) joined Harrah's. What, apparently, was not done until that point in time which is done now?

2. What is Harrah's purpose of using the BI software?

3. Tibco's software responds to events. Give an example.

4. How can segmentation of customers into 80 different groups save the chain money?

5. Why does the company not target only high rollers (big spenders)? Why does it also care so much about low rollers?

6. Is incentive discrimination ethical? Is it ethical to pander to a losing gambler's superstition to lock a machine until the gambler returns to it from a free meal?

One Step Ahead of the Criminal

If data mining can predict behavior, can it help stop criminals before they carry out their crime? This is what one police force in Virginia tries to do. The Richmond Police Department (RPD) uses statistical analysis tools from SPSS and predictive analysis tools from Information Builders to pinpoint locations and individuals that may be involved in crimes. The force can more rapidly respond to a crime by knowing where to deploy squads. In more successful cases, increased police presence prevents the crime.

Richmond has a population of over 192,000. RPD consists of 750 police officers in four precincts. Each precinct is divided into 12 sectors. RPD assigns 18–30 officers to each sector.

The two software companies combined their tools with those of RTI International, a research company specializing in "turning knowledge into practice." One of its senior researchers, Colleen McCue, showed RPD that with ample data, data mining techniques can predict when and where a crime is likely to occur. She helped the force deploy officers on New Year's Eve of 2005. During that period, gunfire complaints were reduced by almost 49 percent from the previous year. During the year, the number of guns seized grew by 246 percent. The same patterns helped RPD to deploy officers on July 4.

In the period of 2000–2004, McCue actually worked for RPD as supervisor of the crime analysis unit. At the start of the effort, she used a simple record management system to analyze crime. After the events of 9/11 she decided to adopt advanced statistical and data modeling tools.

Based on ample data collected over long enough time, any human behavior can be predicted. Thus, criminal behavior, too, can be predicted. Criminal activity of an individual is associated with variables such as month, week, and time of day; geographic location; event (such as a fair or party); and weather conditions.

McCue started using Clementine, an application from SPSS. The application helped perform the analysis she wanted to do. She first examined the threat assessment in investigative support work, linking crime to motive. She also looked for patterns of aggravated assault involving firearms. When analyzing historical data, relationships emerged between violent crimes and their time and location at certain probabilities.

Proper deployment of resources is extremely important not only because of crime prevention but also because the resources are limited and must be efficiently allocated. RPD receives 500,000 calls annually from citizens. Only 30,000 of these calls are about criminal activities. The others come from people who ask for police presence in their neighborhood because they perceive problems. However, these perceptions are not necessarily congruent with what the data mining results predict. Whenever RPD responds to a call, two officers are sent to the location for 20–30 minutes. The officers are the department's most expensive resource. When they were deployed on New Year's Eve, the smart assignment saved RPD $15,000 in overtime pay.

The successful implementation on the two holidays encouraged RPD to augment the system and link it to a map. Information Builders worked with ESRI, a leader in development of geographic information systems (GIS), to link the predictive information to a map. The maps are photos of the city taken from space. Police officers can zoom in on a city district, block, or individual addresses, including doors and windows. Information can be displayed in tables or on the maps.

To help the officers, Information Builders created icons, such as guns and needles, representing different crime risks. Crime probabilities are represented by colors: green for low probability, yellow for medium probability, and red for high probability. Crime probability and other information can now be grasped much faster than before, when it was on long printouts.

Information is available to officers anywhere because it is accessible through a Web browser and the Internet. Every day, data is fed into the predictive model, and the system outputs an eight-hour window for crime in specific locations. Officers are assigned to those locations appropriately.

Chief Rodney Monroe, who heads RPD, is satisfied with the system. However, he would like to extend the time frame from eight hours to 60–90 days. SPSS specialists believe this can be done. As an increasing amount of data is collected over time, the predictive models will be able to more accurately forecast crime time and location.

Sources: Haber, L., "BI Helps Police Predict Crime," *eWeek*, January 30, 2006; "New Book Details Analytic Methods Used to Forecast, Prevent Crime," (www.rti.com), September 21, 2006; Chun, L., "Policing via Predictions," Richmond.com, March 28, 2007.

Thinking About the Case

1. What are the similarities between predicting customer behavior and criminal behavior? Give an example.

2. What are the two most important benefits of the BI/GIS system that the RPD uses?

3. How can time help improve the system?

4. Are you concerned about the attempt to forestall crime at the individual level? At the neighborhood level? If you see an ethical issue in this, explain why. If you do not see an ethical issue, explain why some people may have such concerns.

(Still) Made in America

Most people are familiar with Gore-Tex, the fabrics developed and manufactured by W. L. Gore. The fabrics, used for sports and outdoor casual clothes, are water and wind resistant yet breathable. Gore, a private firm headquartered in Newark, Delaware, is one of a handful of American companies that make fabric in the United States. Because the wage of a textile worker overseas is a fraction of the wage in the United States, the country is no match for Asia and other areas involved in textiles. Michael Jennings, president of Michael Wesetly Clothing, says that 97 percent of Americans do not own a single piece of clothing manufactured in the United States.

Gore's main strength is innovation, and the main contributor to this strength is the democratic nature of the firm's organization. The company refers to its employees as associates. To a large extent, they are allowed to work in whatever area they believe they can contribute the most. Associates create project teams depending on their interest in working on certain ideas. By some management analysts, this is the closest a commercial organization has come to a democracy.

The privately held company, whose annual revenue is estimated at $1.5 billion, has allowed associates to devote a certain percentage of their time to work on whatever new ideas they might have. Over the years, the approach turned ideas into revenue, such as using fibers originally developed for fabric in dental floss. It is no wonder Gore has been listed on the "best companies to work for" list by *Fortune* magazine since 1998.

Gore consists of four divisions: fabric, medical, industrial, and electronics. It operates 17 development and production plants in Delaware and Maryland as well as several manufacturing plants in Asia and Europe. The products range from fabric laminates to industrial sealants. All the company's products are the result of innovative development and production of synthetic polymers, known to laypeople as various types of plastic.

Production processes at Gore are complex. For example, Gore-Tex is the product of a careful process. The fabric is laminated by gluing a polymer membrane onto fabric. There must be enough glue to ensure that the laminate bonds with the fabric, but if too much glue is applied, breathability is reduced.

To maintain its competitive edge Gore must foster collaboration among its workers. In Gore's case, this is not an easy task, especially because of the fluid organizational style. Rather than be assigned to departments for the long haul, scientists and many of the engineers create *ad hoc* project teams. When an assignment has been accomplished, a team is disbanded, and its members join other teams or create new ones to pursue new R&D ideas.

Scientists and engineers are well-versed in a number of software tools, from spreadsheets to statistical process control and capability analysis tools. However, it was practically impossible for workers in one plant to learn from the experience of workers in other plants. Engineers always want to know which chemical processes work and which do not. Gore engineers have used multiple software applications for the analysis.

Management decided to implement a system that would enable engineers to share their knowledge. IT professionals implemented Process Health Assessment Tool, which uses a custom-built Web service that enables workers to access reports detailing the state of Gore's production processes. The software is linked to various sources in which useful data resides: spreadsheets, archival databases, and ERP systems. The data is pulled into a single database.

An analytical software suite developed by SAS—a privately held company specializing in development of business intelligence—is then employed. It consists of a suite of tools that mine the data, model processes, and statistically analyze the data to suggest quality improvement. The suite prepares reports that help Gore engineers manage manufacturing operations. An engineer can obtain a report on the state of a specific process: materials applied, how they are applied, what has worked well, and what has not. In the textile area, reports detail production processes of fabric laminates such as Gore-Tex.

Gore's customers—apparel makers—come up with different specifications quite often. Engineers try varying amounts of materials, such as glues, in production, to meet those specifications. The ultimate purpose of the new software tool is twofold: to meet customer specifications and to predict the results of a process. The tool analyzes a myriad of variables and determines what the company calls "the health of the process." Predictability analysis determines whether a process will yield the same result consistently.

Predictability has a significant impact on much of the company's supply chain because it helps answer questions such as which materials to purchase and in what amounts, how to schedule orders from suppliers, and how to apportion the raw materials to the various plants. Issues such as storage in various conditions must be planned, too. Timing purchases of the materials is extremely important, because ski jackets and other seasonal products made by Gore's clients must be ready within a timetable that has little tolerance for errors.

With the new software, engineers from two, three, or even all four divisions can collaborate and share knowledge not only on details of chemical processes, but other operations as well. For example, they can compare suppliers and make decisions on which one to select based on quality of the raw material and experience with on-time deliveries.

Over the years, scientists and engineers at Gore developed their local lingo. Jose Ramirez, an industrial statistician at Gore, noted that the Process Health Assessment Tool has one major impact on how scientists and engineers communicate now. While the data they analyze might be very different, all now speak the same language when sharing information and expertise, because all communication through the tool is done along its two goals: satisfying customer specifications and predictability of processes.

Sources: Moore, J., "W. L. Gore: Dry Goods," *Baseline* (www.baselinemag.com), May 23, 2005; (www.gore-tex.com), 2005.

Thinking About the Case

1. Does the Process Health Assessment Tool produce knowledge or is it an employee knowledge sharing tool?

2. Why is this tool so important, especially at a company such as Gore?

3. How does the software tool help Gore streamline business operations?

© Bob Krist/CORBIS

PART FIVE

Planning, Acquisition, and Controls

CASE V: WORLDWIDE HOST

"Your home away from home." This motto captures the mission of Worldwide Host, a global hotel chain with premium properties in many European, Asian, and North and South American countries. The upscale chain has always prided itself on providing personalized service and making its guests feel welcome, wherever they travel. No detail is overlooked for guests' comfort—from a friendly morning wake-up call and fresh-squeezed orange juice with breakfast, to on-site gym and spa service, to fluffy towels and robes and mints on the down pillows in the evening. But Worldwide Host employees' travel experiences haven't always run as smoothly as its guests'. In fact, the chain's travel division needed upgrading, as its chief information officer, Michael Lloyd, knew all too well.

Outdated Travel Division, New Ideas

As Worldwide Host's CIO, Michael Lloyd often traveled to the hotel chain's properties throughout the globe, overseeing its reservations and other information systems. When he needed to make travel arrangements, Michael used the corporate travel division, composed of approximately 90 full-time employees. The staff members worked with airlines and rental-car companies—as well as Worldwide's individual hotel staff—to arrange the business travel for all of Worldwide Host's employees. When Michael needed to travel to London, Tokyo, São Paulo, or anyplace in the United States, he would call or e-mail the travel division with details of his trip. Travel staff would arrange the flights, nights at hotel properties, and rental cars, and then send him the information. But despite everyone's best efforts, he'd experienced miscommunications on dates, repeated calls and e-mails for clarification, and on more than one occasion, a missed airline connection because of a change in the airline schedule that didn't get passed along to him. If he was just one of the company's business travelers, what was happening with his coworkers? Hundreds were scattered across the globe. The travel system always frustrated Michael because it was inefficient. It had been around for decades and was no longer meeting Worldwide's needs.

One day while traveling, Michael had an idea: what if Worldwide Host moved its travel services exclusively to the Web? Allowing employees to enter their travel needs directly into a Web-based system would streamline the reservation process and eliminate errors. Such a system could save the company thousands of dollars in lost time. Michael took his idea a step further. Other hotel chains were creating alliances and opening their services to the public via the Web. As a leading hotel chain, Worldwide Host could do the same to remain competitive—revamping its corporate travel division into a full-fledged, e-commerce site, which employees and the general public could use. To accomplish this goal, the company would need to tie its existing travel reservation system into a Web-based interface—that much was certain. But it would also need to ensure that any new system was compatible with its airline and rental-car partners' systems. This would be a monumental effort because of the size of the companies. Each one logged thousands of reservations a day.

Still, the idea had many positive points. The higher public profile could generate additional revenues for Worldwide Host, the airlines, and rental-car companies. Currently, any unoccupied rooms, unsold plane seats, or unrented cars at the end of the day were lost revenue that couldn't be recouped. All companies needed to run their operations as close to full capacity as possible. So, offering the public access to those unfilled rooms, seats, and cars in a combined Web site could boost reservations, generating additional, much-needed revenue. He also knew the amount of time it took to arrange his own business travel, with many calls and e-mails to the travel division. Scheduling vacations through a travel agent meant additional time for the public, too. With today's fast-paced lifestyles, streamlining the process and putting control in the traveler's hands made sense.

Under the current system, Worldwide Host maintained a travel division used only by its own employees. Michael reasoned that with an e-commerce site, the travel division's employees could now serve the general public, spreading the

cost of the division over a broader revenue base. Maybe Worldwide Host could move them into a separate subsidiary. The revenues from the subsidiary could also help offset the cost of developing the new system. Michael took stock of the pros and cons of the Web site in his notebook. The prospect excited him. As soon as he reached his hotel room, he called the chief executive officer, Nathan Plummer, to discuss his idea. The new system would need top executive backing to become a reality.

High-Level Negotiations

Nathan Plummer was no newcomer to the hospitality field. His grandfather had started Worldwide Host in the 1930s, and Nathan had grown up in the business, along with his sister and brother. He and Michael met several times to map out a strategy for the proposed Web site. After lengthy internal discussions with Worldwide Host's management staff, Nathan decided to proceed with the e-commerce effort.

To gather initial support for the e-commerce site, Nathan and Michael began negotiating with Worldwide's current airline and rental-car business partners. All of the companies had a depth of experience in reservations systems, but they had never linked them into one all-encompassing system. Despite the difficulties that establishing a unified system could entail, the promise of increased bookings was very tempting. The partners were willing to listen. Nathan and Michael presented additional arguments: the companies could trim the transaction fees they now paid to travel agents for each reservation, saving out-of-pocket expenses. Customers could go online at their convenience to arrange their travel plans. The final point that persuaded the companies to participate was the advantage of partnering to develop the site. Each of the companies had been considering developing its own e-commerce site, but the cost to a single company was a major hurdle. A new combined site would allow each company to share the cost of setup with its partners, plus draw more Web traffic than single sites. Also, down the road, other travel organizations could join the alliance.

Meetings were long and involved, but eventually the partners struck a deal—the new Web site development effort, TripExpert.com, was born. To gather even more support for TripExpert, Nathan assured his partners that he planned to meet with other travel companies, such as additional airlines and hotel chains, cruise lines, and theme parks, to enlist their participation in the site once it was up and running.

Information Systems: A Critical Link

With the agreement of the travel companies to participate, Michael now turned to the big issue of the Web site itself. Key to TripExpert's development was the underlying information technologies. He jotted down some notes from his earlier meetings and brainstorming sessions and set up a meeting with General Data Systems (GDS), Worldwide Host's IS consulting firm. The firm had developed Worldwide Host's computerized reservation system and regularly assisted the hotel chain with upgrades and maintenance. Over the years, Worldwide Host had established a strong relationship with GDS and relied on its staff's expertise in systems development, network administration, and troubleshooting. Investigating development of the new Web site would require even closer collaboration between Worldwide and GDS.

GDS's director of new business assigned Judith Kozak, his most experienced systems analyst, to the TripExpert project. She had worked on Worldwide Host's systems many times and knew their capabilities well. Also, Michael respected her abilities, and the two had worked well together in the past.

Michael explained his vision of the new system to Judith. "So what we're talking about is a system based on Internet technology that would handle not only our internal travel needs but extend the reservations capabilities to the general public. Nathan, our CEO, and I spoke with the heads of North Trans and Blue Sky airlines as well as A-1 and Bargain Rent-a-Car companies. It took some persuading initially, but they realized the advantage of offering a full-service travel site. We'll need GDS's help, Judith, to explore the various options we have. Is a system currently available that we could buy and modify to fit our needs? We need to tie all our companies' systems together in some way. Or do we need to start from scratch and build a customized system?"

Judith replied, "Off the top of my head, I know of pieces that exist—some Web-based reservation systems and databases, and there is a mainframe-based system an airline has used for decades. But I'd need to investigate each of your partners' systems to see what their capabilities are. A speedy search engine is critical—customers don't like to wait more than a few seconds for responses when they shop on the Web. Delays were a problem with early e-commerce sites, but technology has come a long way since then. We'll see whether we can purchase a search engine component. Availability of the system will be key, too. Web traffic can spike unexpectedly, so we need to plan for peak demand."

"I'm also concerned about security," said Michael. "We need to assure the public that their financial information is absolutely safe with us or they won't use the site. Trust in our security will be crucial."

Forming a Development Team

"Since we're on the subject of requirements for the new system, have you discussed setting up a development team to investigate the possibilities?" asked Judith. "We need members from your travel division, hotel management and reservations staff, and in-house IS staff, to start with. User input is important to make sure we address all the business functions, and your IS staff can help with technical issues. Anyone else?"

Michael responded, "I think we need to interview staff from the airlines and car companies to get their perspective as users of the system, too, and we'll need you and some of GDS's best systems analysts on the team."

"Of course—I'll look into our staffing immediately. What about oversight and approval?"

"Nathan will head the executive steering committee, which will also include the chairman of our board, the chief financial officer, VP of hotel management, your director of new business, your CEO, and me. Our top execs need to know how we're proceeding and what plans we are making. This is such a large undertaking that we need management's oversight of our progress."

"Sure," Judith said. "We can work essentially the way we did when developing your new reservations system a few years ago. That system was critical to your business, and this new one will be just as important."

"So to start with, we'll need to explore the feasibility of each of the options we have," Michael said. "Can you begin looking at the technical aspects—investigate in detail any existing systems? In the meantime, I'll look into scheduling and staffing. Then we'll need to tackle the big issue—economics. Nathan will need a realistic budget for every option."

"When do you want to meet next?" asked Judith.

"How does your schedule look two weeks from now? Say, on Tuesday morning?"

"Sounds fine. I'll talk with my supervisor to get four of our best analysts lined up for the team. I'll e-mail to let you know."

"Great. We have a lot of work to do to get Trip-Expert off the ground."

BUSINESS CHALLENGES

After studying the next three chapters, you will know the basics of systems development efforts. You'll explore the issues of planning, systems development, alternatives for acquiring systems, and security and disaster recovery.

- In Chapter 12, *"Systems Planning and Development,"* you learn the steps to creating a plan for a new system, such as the e-commerce travel Web site for Worldwide Host, and what steps to follow to develop an e-commerce Web site, including feasibility studies and defining the essential functions of the new system.

- In Chapter 13, *"Choices in Systems Acquisition,"* you learn how Worldwide Host can evaluate the benefits and risks of alternative methods of acquiring an information system, including purchasing or leasing a program to create an integrated system.

- In Chapter 14, *"Risks, Security, and Disaster Recovery,"* you learn about the risks threatening information systems, especially those dealing with financial transactions on the Web, and ways to protect systems against attack.

© Bob Krist/CORBIS

12

TWELVE

Systems Planning and Development

LEARNING OBJECTIVES

Planning and developing new information systems can be complex. Systems planning often requires creating or adjusting strategic plans because of the great impact of IT on business models and operations. Those involved in development have to translate a business opportunity, a solution to a problem, or a directive into a working set of hardware, software, and networking components. Once a development project is under way, many people from different disciplines are usually involved in the effort. Communications skills are extremely important for successful results.

When you finish this chapter, you will able to:

- Explain the importance of and steps in IT planning.

- Describe the systems development life cycle, which is the traditional approach to systems development.

- Explain the challenges involved in systems development.

- List the advantages and disadvantages of different system conversion strategies.

- Enumerate and explain the principles of agile systems development methods.

- Be able to contribute a meaningful set of requirements when serving on a project development team for a new or modified IS.

- Explain the concept of systems integration.

- Discuss whether IT professionals should be certified.

WORLDWIDE HOST:
A Vision for the Future

Michael Lloyd, Worldwide Host's CIO, was convening a meeting with his TripExpert.com project staff. A few months had passed since he had been given the go-ahead to investigate development of the company's new travel Web site. He pointed to Worldwide's mission statement, which was posted on the boardroom wall:

Satisfying our guests' and employees' needs is the key to maintaining Worldwide Host's leading position in the global hospitality industry. We are dedicated to superior customer service and continued employee growth.

Michael spoke to the assembled team, composed of Worldwide Host and General Data Systems (GDS) employees. "As I've mentioned before, this statement is our guiding principle. Worldwide Host recognizes that information systems are central to its continued success. That's why we're here. We talked earlier about our need to look beyond our day-to-day IS issues, and I urged you all to rethink processes as we move ahead. We have an opportunity to reshape this company and its future—to look beyond our immediate technology needs to new technologies and processes that will allow future expansion and efficiency." With these initial words of encouragement, the team began to review the information they had gathered since their last meeting.

Investigating Existing Systems: Capabilities and Needs

Judith Kozak, GDS's top systems analyst, and her coworkers had been looking into the compatibility of Worldwide Host's existing reservations system with their airline and car-rental partners' systems. She noted that capabilities and systems differed: the airlines' reservation software needed to update flight information continually due to weather changes and equipment malfunctions, and airlines

and car-rental companies both needed to track the locations of their planes and cars to be sure that they were available where needed. In contrast, hotel systems are more static; properties themselves don't move, and cancellations occur much less frequently than in the airline industry. But the hotel systems needed information on each property that was more detailed—location of room, general appearance, type of beds, availability of meeting rooms, exercise facilities, and other details. In short, Worldwide Host's existing system didn't serve quite the same functions, or run on the same platform, as the other two systems. This made the prospect of tying the existing systems into a single system dubious.

Searching for System Alternatives

Corey Lee, another team member from GDS, had been investigating online reservations systems such as Hotwire, Expedia, and Travelocity. The best ones were not available for sale or were too expensive. One developed by an airline had been used by travel agents for decades. But it couldn't easily be linked to the Web, and its use was not intuitive—the average Web surfer would have trouble with its user interface. Also, the system ran on an old, outdated mainframe system.

Finally, Corey located one new Web-based global reservations system whose owners, Reservations Technologies, licensed their product. Since it didn't have all the components TripExpert.com would need, GDS explored the possibility of tying it to part of Worldwide Host's existing reservations systems and working with the owners to develop additional functionalities. In the meantime, GDS hired Alana Pritchett, who had the expertise to migrate the data from their existing system into this new one.

Stalling Out on Database Development

Alana Pritchett had been heading up the database team. She and her two coworkers had been talking with Worldwide Host employees to gather information about their existing reservations system and to develop lists of new business needs for the TripExpert.com site. They thought they had covered every new function that the database needed to handle for Web site connections: aside from increased capacity, they noted that they'd need to display picture files to show Web shoppers sample rooms. The travel staff hadn't previously worried about graphics capabilities. They had also listed a requirement for the system to track TripExpert site reservations so that a small Web service fee would be added to the room rate. But what they and the Worldwide Host staff forgot to consider were the discounted room rates that were to be offered to last-minute Web shoppers. Within three weeks of a reservation date, the system needed to release a block of rooms to the Web site—rooms that were not already reserved by traditional means. Those rooms were to be made available at a discount to Web shoppers, enticing them to book last-minute stays. Worldwide Host hoped to fill more of its hotels by offering the Web discount and thus boosting profits.

The database team members were already making preliminary design plans when their omission came to light. They reported the slipup to Michael Lloyd with trepidation. They believed that incorporating the new features into the system's design could mean a delay of about six weeks, throwing the whole project behind schedule.

Getting Back on Track

Michael called in Judith Kozak, GDS's lead analyst, to find a solution—and quickly. Delays in the Trip-Expert system would mean that Worldwide Host would lose competitive ground, since other sites were coming online. Judith had worked on Worldwide Host's existing reservations system and knew its capabilities well. She also knew GDS's analysts' skills.

"Michael, let me check Corey Lee's schedule. He's been working on licensing the Web airline-reservation system for us. I think he is ahead of schedule in his investigation of the system's capabilities and his negotiations, so maybe we can borrow him for a while. Is that what your schedule shows? Good. Corey has worked under deadline pressure before and is an experienced analyst. If he and I put our heads together on this problem, I think we can straighten it out."

"Thanks, Judith. Let me know how it's looking as soon as you can. The database is critical to the site's functioning. If we don't get that component in place, we'll jeopardize the whole system. What good is the front end without the back?"

"Not much," she answered. "I'll call Corey right now to see whether he can meet with Alana's team and me tomorrow. In the meantime, try not to worry. We've hit bad stretches before but come out OK in the end."

PLANNING INFORMATION SYSTEMS

In recent years, a growing number of corporations have implemented enterprise ISs such as ERP systems, SCM and CRM systems, or other systems that serve the entire organization or many of its units. The investment of resources in such systems, both in financial and other terms, is substantial, as is the risk in implementing such large systems. If the implementation is successful, the new system can significantly change the manner in which the organization conducts business and even the products or services it sells. For all these reasons it is necessary to plan the implementation of information systems, whether they are developed in-house, made to order by another company, or purchased and adapted for the organization. When planning, it is important to align IT strategies with the overall strategies of the organization. (In this discussion the terms "IT planning" and "IS planning" are used interchangeably.)

Steps in Planning Information Systems

IT planning includes a few key steps that are a part of any successful planning process:

- Creating a corporate and IT mission statement.
- Articulating the vision for IT within the organization.
- Creating IT strategic and tactical plans.
- Creating a plan for operations to achieve the mission and vision.
- Creating a budget to ensure that resources are available to achieve the mission and vision (see Figure 12.1).

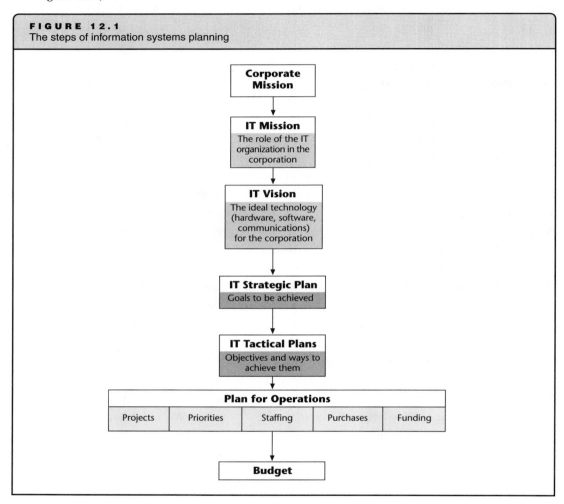

FIGURE 12.1
The steps of information systems planning

A *mission statement* is a paragraph that communicates the most important overarching goal of the organization for the next few years. Although the ultimate mission of any organization is to survive and—if it is a for-profit organization—to produce profit for its owners, mission statements are rarely limited to these points. Rather, they say how the organization intends to survive and thrive. For example, in Amazon.com's early years, its mission was to brand itself as the most recognized retailer on the Web and to create the largest possible market share. Management pursued this mission, though it resulted in years of financial loss.

An important part of an organization's overall mission statement is an IT mission statement that is compatible with the larger mission. It is usually a paragraph, or several paragraphs, describing the role of IT in the organization. Often, the IT mission and IT vision are combined into one statement. The IT vision includes the ideal combination of hardware, software, and networking to support the overarching mission. For example, Amazon.com's management continues to recognize that innovative IT, especially Web and fulfillment technologies, is the most important resource for the organization's success.

The CIO, with cooperation of senior management as well as managers in the IT unit, devises a strategic plan for implementation of IT in the organization. The plan addresses what technology will be used and how employees, customers, and suppliers will use it over the next several years. Since IT advances so quickly, strategic IT plans are typically prepared for no longer than five years.

The goals laid out in the strategic plan are broken down into objectives, which are concrete details of how to accomplish those goals. The objectives typically include resources to be purchased or developed; timetables for purchasing, developing, adapting, and implementing those resources; training of employees to use the new resources; and other details to ensure timely implementation and transition.

The objectives are further broken down into specific operational details. For each project, management assigns a project manager and a team; vendors are selected from whom available components of hardware, software, and services will be purchased; and funding is requested. When the financial requests are approved, the corporate budget includes the money to be spent over several months or years on these projects.

IT planning is not much different from planning any other acquisition of resources—starting with a vision of how the resources will be used to accomplish goals and breaking those ideas down into projects and the resources to be allocated to carry the projects to successful completion. In recent years, a growing proportion of IT funds have been spent on software, with most of the funds going to purchase and adapt software, rather than developing it in-house or assigning development to another company.

Inspiration for development and/or implementation of new information technologies come from several sources, including users like you. Figure 12.2 displays typical sources for new systems. Competitive pressure inspires many members of the organization, not only senior managers, to come up with creative ideas. Often, such ideas are included in IT planning. In many organizations a steering committee oversees IT planning and the execution of IT projects. Steering committees are composed of users, IT professionals, and senior managers. There is high probability that in your career you will find yourself on such a committee.

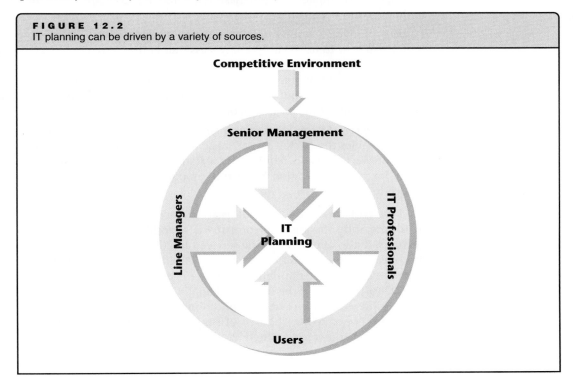

FIGURE 12.2
IT planning can be driven by a variety of sources.

Understand the Principles of Systems Development

By and large, organizations have recognized the need to let non-IT professionals play major roles in systems development. You might be called on to participate in this process, not just to provide input here and there but as a member of a development team. The IT professionals on the team need your insight into the business processes in which you participate. They need your advice on ways to improve these processes through the use of new or improved ISs. One approach to development, agile methods, actually views the users as sharing at least half of the responsibility for the effort.

Software developers count on you and your coworkers to provide them with proper requirements and feedback. You should be knowledgeable, active, and assertive in software development projects, because you will have to live with the products of these efforts. Also, when your organization decides to discard one IS and adopt a new one, your understanding of the implementation process and your proper cooperation will be highly valuable. Your knowledge will be solicited regularly and will play a valuable role in decision making if you work for a small organization.

The Benefits of Standardization in Planning

One major goal—and advantage—of planning is standardization. When management decides to adopt a certain IT resource for all its units, regardless of function or location, it standardizes its IT. Standardization results in several benefits:

- *Cost savings.* When the organization decides to purchase the same hardware or software for all its units, it has better bargaining power and therefore can obtain lower prices from vendors. This applies to purchasing or leasing computers of all classes—mainframe, midrange, and personal computers—as well as licensing software.

- *Efficient training.* It is easier to train employees how to use a small variety of software than to train them how to use a large variety. Less trainer time is required, and—more importantly— employees spend less time on training and more time on their regular assignments. This also saves cost in the form of fewer labor hours spent on training. Even if each employee only uses a single application, but the organization maintains several applications for the same purpose, training time is extended.

- *Efficient support.* Standardizing for a small number of computer models and software applications enables the IT staff to specialize in the hardware and software they have to support. The more focused skills required for a standard suite of hardware and applications make it easier for the organization to recruit support personnel, and results in more satisfactory service to users.

Hewlett-Packard, the IT giant, is an example of how a company attempts to standardize hardware and software. The company is a world leader in production of PCs, servers, and printers. It also offers consulting services. In 2003, HP embarked on a five-year program that it hoped would cut its IT costs from $3.04 billion to $2.11 billion. Among consolidation of data centers and other initiatives, management decided to reduce the number of servers from 19,000 to 10,000, and the number of applications from 5,000 to 1,500. For example, instead of 84 different procurement applications, the company planned to use only five. The company hoped to cut the cost of supporting applications alone from 80 percent of total spending on software to a mere 20 percent.

From Planning to Development

After planning a new IS or a set of ISs, management decides how to obtain the systems. In a great majority of cases, "systems" means software. For example, CRM and SCM systems rarely require specialized hardware (although they may require more, or more powerful, hardware). An increasing number of new systems are purchased and adapted for an organization's needs rather than developed in-house, although in-house development still takes place in many organizations. The approaches to systems development are the same regardless of who develops the system—the organization or its vendor.

Systems development generally is conducted in two approaches: the systems development life cycle (SDLC) and nontraditional methods, among which are many gathered under the umbrella of agile methods. SDLC is the more traditional approach and has been used for several decades. In certain circumstances, it should still be used. Agile methods developed out of prototyping, an application development approach that emerged in the 1980s aimed at cutting costs and time. **Prototyping** involves fast development of an application based on initial user requirements and several cycles of user input and developer improvements. Practicing the philosophy of prototyping—that coding should start as soon as possible and that users should be involved throughout the process—led to several methods of software development called agile methods. The following sections discuss both approaches.

THE SYSTEMS DEVELOPMENT LIFE CYCLE

Large ISs that address structured problems, such as accounting and payroll systems and enterprise software applications, are usually conceived, planned, developed, and maintained within a framework called the **systems development life cycle (SDLC)**. The approach is also called "waterfall" development, because it consists of several distinct phases that are followed methodically, and the developers complete the phases sequentially. Described graphically, the phases look like a waterfall from the side. The developers do not deliver pieces of the systems before the entire system is fully completed. Although textbooks might refer to the various phases and subphases of the SDLC by different names, or organize them slightly differently, in general, the process follows the same steps. While the SDLC is a powerful methodology for systems development, organizations are sometimes forced to take shortcuts, skipping a step here or there. Occasionally, time pressures, funding constraints, or other factors lead developers to use different approaches to systems development.

POINT OF INTEREST

Pulling the Plug

Sometimes the business needs of an organization change so fast that "freezing the code" of software is infeasible. WestJet Airlines, a Canadian discount airline based in Calgary, announced in July 2007 that it was pulling out of a project to develop a new reservation system. The company was willing to sacrifice its investment of $30 million. Travelport, the company developing the software for several companies but using WestJet as a partner and advisor, did not do anything wrong. WestJet simply grew very fast. As the project progressed, management wanted to add features such as the ability to partner with U.S. and international carriers. The original features of the aiRES—short for Airline Reservation Systems—fit a small discount airline, but not a larger, international one. A system that was excellent for 2005 was far too limited for 2008. WestJet suspended the work of 150 internal IT specialists and 50 outside consultants on the project.

Source: Duvall, M., "Airline Reservation System Hits Turbulence," *Baseline*, July 25, 2007.

The SDLC approach assumes that the life of an IS starts with a need, followed by an assessment of the functions that a system must have to fulfill that need, and ends when the benefits of the system no longer outweigh its maintenance costs, or when the net benefit of a new system would exceed the net benefits of the current system. At this point the life of a new system begins. Hence, the process is called a *life cycle*. After the planning phase, the SDLC includes four major phases: analysis, design, implementation, and support. Figure 12.3 depicts the cycle and the conditions that can trigger a return to a previous phase. The analysis and design phases are broken down into several steps, as described in the following discussion.

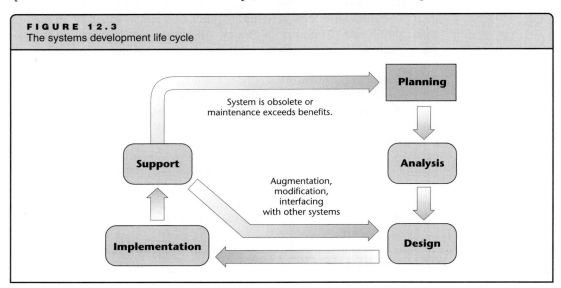

FIGURE 12.3
The systems development life cycle

Analysis

The **systems analysis** phase is a five-step process (summarized in Figure 12.4) that is designed to answer these questions:

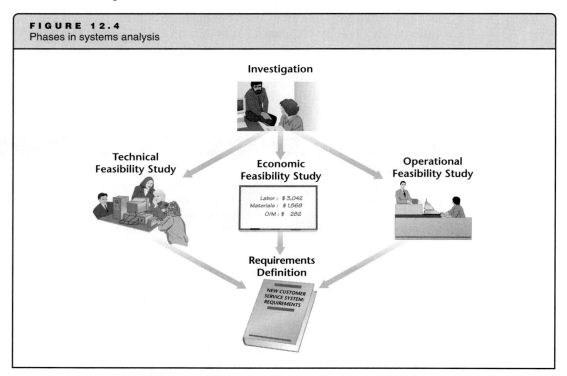

FIGURE 12.4
Phases in systems analysis

Investigation

- What is the business process that the system is to support?
- What business opportunity do you want the system to seize, what problems do you want it to solve, or what directive must you fulfill?

Technical Feasibility Study

- Is technology available to create the system you want?
- Which of the available technologies should we use?

Economic Feasibility Study

- What resources do you need to implement the system?
- Will the system's benefits outweigh its costs?

Operational Feasibility Study

- Will the system be used appropriately by its intended users (employees, customers, suppliers)?
- Will the system be used to its full capacity?

Requirements Definition

- What features do you want the system to have?
- What interfaces will the system have with other systems, and how should the systems interface?

Investigation

The first step in systems analysis is investigation, which determines whether there is a real need for a system and whether the system as conceived is feasible. Usually, a small *ad hoc* team—consisting of a representative of the sponsoring executive, one or two systems analysts, and representatives of business units that would use the new system or be affected by it—is put together to perform a quick preliminary investigation.

The team spends time with employees at their workstations to learn firsthand about the way they currently carry out their duties, and interviews the workers about problems with the current system. This direct contact with users gives workers the opportunity to express their ideas about the way they would like a new IS to function and to improve their work. The investigative team prepares a written report summarizing the information gathered. The team members also forward their own opinions on the need for a new system. They will not necessarily agree that a new system is justified.

If the preliminary report concludes that the business situation warrants investment in a new IS, a more comprehensive investigation might be authorized. The sponsoring executive selects members for a larger analysis team. Usually, members of the original team are included in this augmented group to conduct **feasibility studies**. The objective of the larger investigation team is to determine whether the proposed system is feasible technically, economically, and operationally.

The Technical Feasibility Study

A new IS is technically feasible if its components exist or can be developed with available tools. The team must also consider the organization's existing investment in hardware, software, and telecommunications equipment. For example, if the company recently purchased hundreds of units of a certain computer, it is unlikely that management will approve the purchase of computers of another model for a single new application. Thus, the investigators must find out whether the proposed system can run properly on existing hardware.

The Economic Feasibility Study

Like any project, the development of a new IS must be economically justified, so organizations conduct an economic feasibility study. That is, over the life of the system, the benefits must outweigh the costs. To this end, the analysts prepare a **cost/benefit analysis**, which can be a spreadsheet showing all the costs to be incurred by the system and all the benefits that are expected from its operation.

The most accurate method of economic analysis is the fully quantitative **return on investment (ROI)**, which is a calculation of the difference between the stream of benefits and the stream of costs over the life of the system, discounted by the applicable interest rate, as shown in Figure 12.5. To find the ROI, the net present value of the system is calculated by combining the net present value of the costs of the system with the net present value of the benefits of the system, using calculations based on annual costs and benefits and using the appropriate interest rate. If the ROI is positive, the system is economically feasible, or cost justified. Remember that during the time the system is developed, which might be several years, there are no benefits, only development costs. Operational costs during the system's life include software license fees, maintenance personnel, telecommunications, power, and computer-related supplies (such as hardware replacement, software upgrades, and paper and toner purchases). If the system involves a Web site, the cost of revising and enhancing the site by Webmasters and other professionals must also be included.

Figure 12.5 presents a simplified example of a cost/benefit spreadsheet and analysis for a small system. Since the net present value of the system is positive ($43,152,000), and therefore the benefits exceed the investment, the development effort is economically justified. In the figure, in the year 2015, the net present value starts to diminish. As this value continues to diminish, the organization should consider creating a new system. If the system is not replaced or significantly upgraded, the existing system will become a drain on the organization over time.

FIGURE 12.5
Estimated benefits and costs of an IS ($000)

Year	2009	2010	2011	2012	2013	2014
Benefits						
Increase in sales			56,000	45,000	30,000	10,000
Reduction in staff			20,000	20,000	20,000	20,000
Total Benefits	0	0	76,000	65,000	50,000	30,000
Costs						
Analysis	15,000					
Design	37,500					
Implementation	0	56,000				
Hardware	0	20,000				
Operation and maintenance	0	0	5,000	5,000	5,000	5,000
Total Costs	52,000	76,000	5,000	5,000	5,000	5,000
Difference	(-52,000)	(-76,000)	71,000	60,000	45,000	25,000
Discounted at 5%	(-49,524)	(-68,934)	61,332	49,362	32,259	18,657
Net present value for six years	43,152					

Often, it is difficult to justify the cost of a new IS because too many of the benefits are *intangible*, that is, they cannot be quantified in dollar terms. Improved customer service, better decision making, and a more enjoyable workplace are all benefits that might eventually increase profit but are very difficult to estimate in dollar amounts. This inability to measure benefits is especially true when the new IS is intended not merely to automate a manual process but to support a new business initiative or improve intellectual activities such as decision making. For example, it is difficult to quantify the benefits of business intelligence (BI) and knowledge management (KM) systems. Software vendors often promote "fast ROI" as a selling point, and express it in terms of the short period of time over which the adopting organization can recoup the investment. Still, such claims are difficult, if not impossible, to demonstrate. Therefore, the economic incentive for investing in a new IS is often "we must use it because our competitors use it" and a general expectation that the new IS will benefit the organization in at least one way.

When laws or regulations dictate the implementation of a new IS, no ROI analysis is carried out. For example, when companies implement software to comply with the record keeping and financial procedures of the Sarbanes-Oxley Act, the question is not whether to implement the system. The economic analysis becomes which software is the least expensive and which personnel—internal or hired from consulting firms—would implement the system for the least cost while maintaining the required standards.

POINT OF INTEREST

Ready, Set... Too Late

To meet their IT plans and complete projects on time, U.S. companies often hire foreign IT professionals. The United States allows foreign nationals with special skills to work in the country for a limited number of years. The U.S. Citizenship and Immigration Services (USCIS) issues H-1B visas for this purpose. Employers apply for the visas predominantly to hire software developers. The number of visas granted annually is limited to 65,000. In 2006, employers reached the number of approved applications for 2007 within two months. In 2007, the USCIS announced that applications for 2008 would be considered only if submitted on or after April 2, 2007. Employers submitted 195,000 applications on that same day, which also became the last day for applications. The USCIS randomly approved 65,000 and rejected the rest.

Source: Perelman, D., "Year's Supply of H-1B Visas Tapped Out on Day One," *CIO Insight*, April 4, 2007.

The Operational Feasibility Study

The purpose of the operational feasibility study is to determine whether the new system will be used as intended. More specifically, this analysis answers the following questions:

- Will the system fit into the culture of this organization?
- Will all the intended users use the system to its full capacity?
- Will the system interfere with company policies or statutory laws?

Organizational culture is an umbrella term referring to the general tone of the corporate environment. This includes issues such as tendency to share or not to share information among units and people, willingness to team-play, and the proclivity of employees to experiment with new ideas and technologies. The development team must consider culture to ensure that the new system will fit the organization. For example, if the system will be used by telecommuters, the organization must be open to telecommunications via the Internet. The analysts must find out whether this need would compromise information security and confidentiality, and implement the proper security measures.

Another point the team considers is compliance with statutory regulations and company policy. For example, the record-keeping system the staff wants to use might violate customer privacy or risk the confidentiality of government contracts with the company. If these issues cannot be overcome at the outset, then the proposed system is not operationally feasible.

Requirements Definition

When the analysts determine that the proposed system is feasible, the project team is assembled. Management or the consulting firm nominates a project leader who puts together a project team to develop the system until it is ready for delivery. The team includes systems analysts, programmers, and, often, representatives from the prospective groups of users.

One of the first pieces of information the analysts need to know is the system requirements. **System requirements** are the functions that the system is expected to fulfill and the features through which it will perform its tasks. In other words, system requirements are what the system should be able to do and the means by which it will fulfill its stated goal. This can be done through interviews, questionnaires, examination of documents, and on-the-job observations. Once facts are gathered, they are organized into a document detailing the system requirements.

The managers of the business unit, or business units, for which the system is to be developed often sign the document as a contract between them and the developers. This formal sign-off is a crucial milestone in the analysis process; if the requirements are not well defined, resources will be wasted or underbudgeted, and the completion of the project will be delayed.

Design

With a comprehensive list of requirements, the project team can begin the next step in systems development, designing the new system. The purpose of this phase is to devise the means to meet all the business requirements detailed in the requirements report. As indicated in Figure 12.6, **systems design** comprises three steps: a description of the components and how they will work, construction, and testing. If the decision is to purchase ready-made software, the description of components details how certain components will be adapted for the particular needs of the purchasing organization, and construction is the actual changes in programming code.

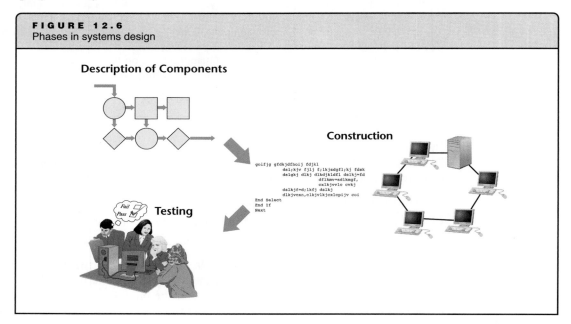

FIGURE 12.6
Phases in systems design

To communicate ideas about data, processes, and information gleaned from data, systems analysts and programmers use conventional symbols. The advantage of such conventions is that visual information can be grasped much faster and more accurately than text, much as a blueprint for a building conveys specifications more efficiently than the equivalent text. One such convention is the data flow diagram.

Data Flow Diagrams

A **data flow diagram (DFD)** is used to describe the flow of data in a business operation, using only four symbols for these elements: external entities, processes, data stores, and the direction in which data flows (see Figure 12.7). *External entities* include individuals and groups of people who are external to the system, such as customers, employees, other departments in the organization, or other organizations. A *process* is any event or sequence of events in which data is either changed or acted on, such as the processing of data into information or the application of data to decision making. A *data store* is any form of data at rest, such as a filing cabinet or a database. Data flows from an external entity to a process, from a process to a data store, from a data store to a process, and so on. Thus, a carefully drawn DFD can provide a useful representation of a system, whether existing or planned.

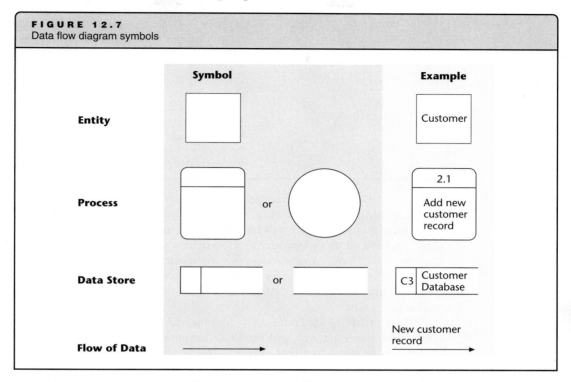

FIGURE 12.7
Data flow diagram symbols

The use of only four symbols and the simplicity of DFDs are their great advantage. Often, systems analysts produce several levels of DFDs for a system. The highest level contains the least number of symbols and is the least detailed. A lower level is more detailed; what might be represented only as a general process in the higher level is exploded into several subprocesses and several databases. The lowest-level diagram explodes some processes further and is the most detailed; it shows every possible process, data store, and entity involved. Usually, the first- and second-level diagrams are presented to non-IS executives, and the lowest-level DFD is considered by the IS professionals while they analyze or develop the system.

The DFD in Figure 12.8 shows a process of calculating a sales bonus. A salesclerk is an entity entering data (in this case, salespeople's ID numbers), which flows into a process, namely, the bonus calculation, which also receives data from the salespeople database (in this case, the dollar amount each salesperson sold over the past year). The result of the process, the bonus amount for each salesperson, is information that flows into a bonus file. Later, the company's controller will use the information to generate bonus checks.

DFD symbols are suitable for describing any IS, even if it is not computer-based. A DFD of the existing system helps pinpoint its weaknesses by describing the flow of data graphically and allowing analysts to pinpoint which processes and databases can be automated, shared by different processes, or otherwise changed to strengthen the IS. If a new IS is needed, a DFD of the conceptualized new system is drawn to provide the logical blueprint for its construction.

While DFDs are easy to learn and use, they have shortcomings—like any diagramming method—and cannot describe a system completely. For example, they do not specify computations within a process or timing relationships among data flows. A payroll DFD, for instance, cannot specify whether employee time sheets are checked as they are submitted or at the end of the week. Such details usually accompany DFDs as text comments.

Unified Modeling Language (UML)

As an increasing number of developed applications became object oriented, a new way to describe desired software was needed. Several diagramming sets were developed by the 1970s, but

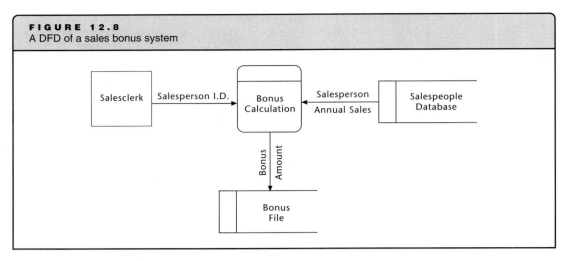

FIGURE 12.8
A DFD of a sales bonus system

in the late 1990s a *de facto* standard emerged: **Unified Modeling Language (UML)**. UML is a graphical standard for visualizing, specifying, and documenting software. It helps developers to communicate and logically validate desired features in the design phases of software development projects. It is independent of particular programming languages, but it does provide standard visual symbols and notations for specifying object-oriented elements, such as classes and procedures. It also provides symbols to communicate software that is used for constructing Web sites and Web-based activities, such as selecting items from an online catalog and executing online payments.

UML consists of diagrams that describe the following types of software: use case, class, interaction, state, activity, and physical components. A *use case* is an activity that the system executes in response to a user. A user is referred to as an *actor*. Use case diagrams communicate the relationships between actors and use cases. *Class diagrams* describe class structure and contents and use the three-part symbol for class: name, attributes, and methods (see the example in Chapter 5, Figure 5.4). *Interaction diagrams* describe interactions of objects and the sequence of their activities. *State charts* communicate the states through which objects pass, as well as the objects' responses to signals (called stimuli) they receive. *Activity diagrams* represent highly active states that are triggered by completion of the actions of other states; therefore, they focus on internal processing. *Physical diagrams* are high-level descriptions of software modules. They consist of components diagrams, which describe the software, including source code, compilation, and execution, and deployment diagrams, which describe the configuration of software components when they are executed. Figure 12.9 shows an example of modeling in UML.

Construction

Once the software development tools are chosen, construction of the system begins. System construction is predominantly programming. Professional programmers translate input, output, and processes, as described in data flow diagrams, into programs. The effort often takes months or even years (in which case the users might not be served well due to changes in business needs). When a program module is completed, it is tested. Testing is performed by way of walk-through and simulation.

In a walk-through, the systems analysts and programmers follow the logic of the program, conduct processes that the system is programmed to execute when running, produce output, and compare output with what they know the results should be. In simulation, the team actually runs the program with the data. When all the modules of the application are completed and successfully tested, the modules are integrated into one coherent program.

System Testing

Although simulation with each module provides some testing, it is important to test the entire integrated system. The system is checked against the system requirements originally defined in the analysis phase by running typical data through the system. The quality of the output is examined, and processing times are measured to ensure that the original requirements are met.

FIGURE 12.9
A sample UML model and its explanation

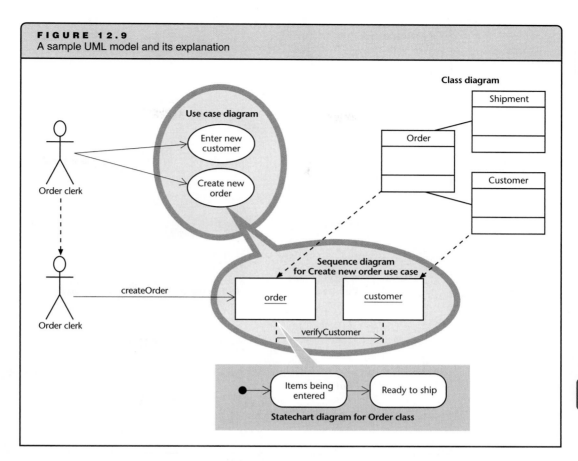

Testing should include attempts to get the system to fail, by violating processing and security controls. The testers should try to "outsmart" the system, entering unreasonable data and trying to access files that should not be accessed directly by some users or—under certain circumstances—by any user. This violation of typical operating rules is a crucial step in the development effort, because many unforeseen snags can be discovered and fixed before the system is introduced for daily use. If the new system passes the tests, it is ready for implementation in the business units that will use it.

Testing tends to be the least respected phase in systems development. Too often project managers who are under time pressure to deliver a new IS either hasten testing or forgo it altogether. Because it is the last phase before delivery of the new system, it is the natural "victim" when time and budget have run out. This rush has caused many failures and, eventually, longer delays than if the system had undergone comprehensive testing. For example, the delayed delivery of Microsoft Windows Vista was the result of extensive testing. A thorough testing phase might delay delivery, but it drastically reduces the probability that flaws will be discovered only after the new system is delivered.

Implementation

The **implementation** of a new IS, also called delivery, consists of two steps: conversion and training. Although training might precede conversion, if training is done on the job it can occur after conversion. **Conversion** takes place when an operation switches from using an old system to using a new system. Conversion can be a difficult time for an organization. Operators need to get used to new systems, and even though the system might have been thoroughly tested, conversion can hold some unpleasant surprises if bugs or problems have not been discovered earlier. Services to other departments and to customers might be delayed, and data might be lost. Four basic conversion strategies can be employed to manage the transition (see Figure 12.10).

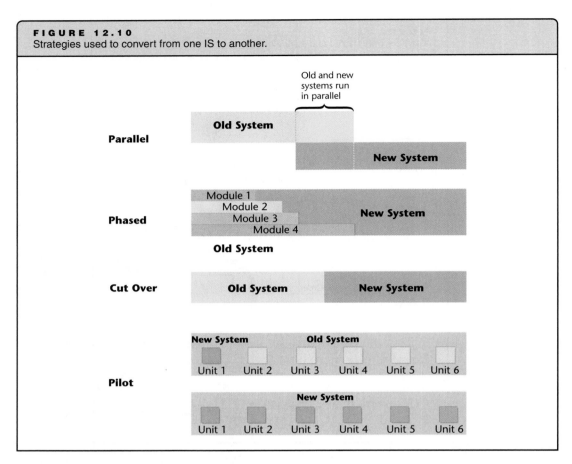

FIGURE 12.10
Strategies used to convert from one IS to another.

Parallel Conversion

In **parallel conversion**, the old system is used along with the new system for a predetermined period of time. This duplication minimizes risk because if the new system fails, operations are not stopped and no damage is caused to the organization. However, parallel conversion is costly because of the expenses, especially labor costs, associated with running two systems. It is also labor intensive because users must enter data twice, into the new and old systems. By and large, parallel conversion is rarely used nowadays. When it is, parallel conversion is used for internal applications used only by employees, not customer or business partners.

Phased Conversion

ISs, especially large ones, can often be broken into functional modules and phased into operation one at a time, a process called **phased conversion**. For example, conversion of an accounting IS can be phased, with the accounts receivable module converted first, then the accounts payable, then the general ledger, and so on. A supply chain management system might be implemented one module at a time: first, the customer order module, then the shipment module, then the inventory control module, and so on, up to the collection module. This phased approach also reduces risk, although the benefits of using the entire integrated system are delayed. Also, users can learn how to use one module at a time, which is easier than learning the entire system at once. However, when parts of both systems are used, there might be data inconsistencies between the two.

Cut-Over Conversion

In a **cut-over conversion**—also called **flash cut conversion** or direct conversion, or direct cut-over—the old system is discarded and the new one takes over the entire business operation for which it was developed. This strategy is highly risky, but it can be inexpensive, if successful, because no resources are spent on running two systems in parallel, and the benefits of the entire new system are immediately realized.

Pilot Conversion

If the new system is to be used in more than one business unit, it might first be introduced for a period of time in a single unit, where problems can be addressed and the system can be polished before implementing it in the other business units. This trial conversion is also possible for systems shared by many departments and disparate sites, as is increasingly the case due to the growing popularity of intranets and extranets. Obviously, **piloting** reduces risks because it confines any problems to fewer units. It is especially useful for determining how comfortable staff members and other users, such as suppliers and customers, are with a new system—a lesson that can be applied to the later units. As with the parallel strategy, the pilot strategy means that benefits of the full implementation of the system are delayed.

When a system is developed by a software vendor for a wide market rather than for a specific client, conversion often takes place at beta sites. A **beta site** is an organization whose management agrees to test the new system for several months and provide feedback. (In the Greek alphabet, beta is the second letter. Alpha, the first letter, is used for an Alpha site, the developing organization.)

POINT OF INTEREST

Who Will Guard the Guards?

With so much money invested in IT, some executives find the temptation to defraud their employers irresistible. The Association of Certified Fraud Examiners (ACFE) reported that in 2004 U.S. corporations lost $660 billion—6 percent of revenue—to "occupational fraud." Most of the loss was the result of procurement fraud, in which insiders collaborate with outside providers of services to defraud their own employers. According to ACFE, 67.8 percent of this fraud is committed by managers and executives. CIOs and other IT executives are often the culprits. Buca, a restaurant chain, found out in 2001 that its CFO and CIO used their employer's money to establish an IT consulting firm that served Buca, and received kickbacks worth at least $50,000. The New York City Office of Chief Medical Examiner discovered that between 2000 and 2005 its director of the MIS department and the director of records defrauded the city by awarding technology contracts to companies that did not deliver services. The amount involved: $8 million. The CIO of the Electric Reliability Council of Texas (ERCOT), four other senior IT managers, and an outside contractor funneled nearly $1 million to a company for work that was never done. IT executives are in a unique position of trust because their organizations count on their technical authority and because the budgets they handle often reach $1 billion per year.

Source: "Procurement Fraud: How Tech Insiders Cheat Their Employers," *Baseline* (www.baselinemag.com), June 7, 2006.

Support

The role of IT professionals does not end with delivery of the new system. They must support the system and ensure that users can operate it satisfactorily. **Support** includes two main responsibilities: maintenance and user help. Maintenance consists of postimplementation debugging and updating (making changes and additions), including adding features that were originally desired but later postponed so budget and time limits could be met. Usually, updating is the greater effort.

Debugging is the correction of bugs or problems in programs that were not discovered during tests. Updating is revising the system to comply with changing business needs that occur after the implementation phase. For example, if a company collects personal data for market analysis, managers might want to use the new IS to collect more data, which might require new fields in the databases.

Although maintenance is viewed by IS professionals as lacking in glamour, it should not be taken lightly or left to less-experienced professionals. Company surveys show that up to 80 percent of IS budgets is spent on maintenance, the cost of which varies widely from system to system. The major reason for this huge proportion is that support is the longest phase in a system's life cycle. While development takes several months to about three years, the system is expected to yield benefits over many years.

Efficient and effective system maintenance is possible only if good documentation is written while the system is being developed, and if the code is written in a structured, easy-to-follow manner. Documentation consists of three main types: paper books, electronic documents, and in-program documentation. The latter covers nonexecutable comments in the code, seen only when reviewing the application's source code. You can see this type of documentation when you retrieve the source code of many Web pages. In-program documentation briefly describes what each module of the program does and sometimes who developed it. Printed and electronic documentation is prepared both for programmers, who can better understand how to revise code, and for users who want to learn about the various features of the application.

AGILE METHODS

While the full approach of the SDLC or similar waterfall methods are used to develop ISs, it is widely recognized that these methods are lengthy, expensive, and inflexible. Systems developed on the SDLC model are often unable to adapt to vague or rapidly changing user requirements. To overcome these challenges, alternative methods have emerged that are collectively called **agile methods**. As Figure 12.11 illustrates, agile methods treat software development as a series of contacts with users, with the goal of fast development of software to satisfy user requirements, and then improving the software shortly after users request modifications. Agile methods make extensive use of iterative programming, involving users often, and keeping programmers open to modifications while development is still under way. The better known methods are Extreme Programming (XP), Adaptive Software Development (ASD), Lean Development (LD), Rational Unified Process (RUP), Feature Driven Development (FDD), Dynamic Systems Development Method (DSDM), Scrum, and Crystal. XP (not to be confused with the Microsoft operating system of the same name) is by far the most documented and best known of these methods. FDD and DSDM are more structured than other agile methods.

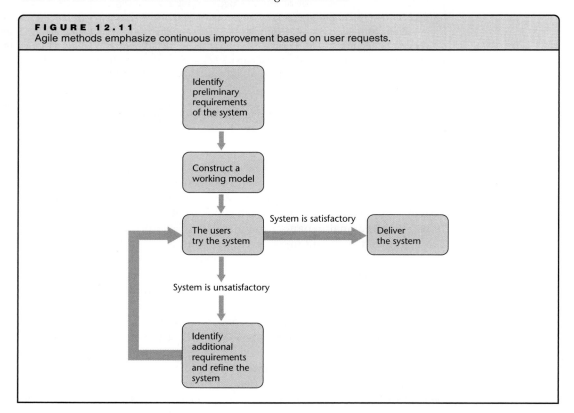

FIGURE 12.11
Agile methods emphasize continuous improvement based on user requests.

The differences among the methods are outside the scope of this discussion. However, the major advantage of all agile methods is that they result in fast development of applications so that users can have them within weeks rather than months or years. Users do not have to wait long for system modifications, whether they are required because of programmer errors or because users have second thoughts about some features.

However, the benefits of agile methods do not come without risks. First, the analysis phase is minimal or is sometimes eliminated completely. Reducing or skipping a thorough formal analysis increases the risk of incompatibilities and other unforeseen mishaps. Also, the developers devote most of their time to construction and little time to documentation, so modification at a later date can be extremely time consuming, if not impossible. Because of the inherent risks, there are times when agile methods are appropriate and others when they are not (see the discussion later in this section).

Software developers who espouse the approach usually subscribe to the *Manifesto for Agile Software Development*, which expresses the following priorities: individuals and interactions over processes and tools; working software over comprehensive documentation; customer collaboration over contract negotiation; and responding to change over following a plan. You can find the full Manifesto at *http://agilemanifesto.org/principles.html*. The software developed should primarily satisfy users, not business processes, because users must be satisfied with the applications they use even if that means changing processes. While program documentation is important, it should not come at the expense of well-functioning software, especially when time is limited and the programmers must decide how to allot their time—on better software or on better documentation. The customers of software development, the users, are not an adverse party and should not be negotiated with but regarded as codevelopers and co-owners of the software. Plans are good but might stand in the way of necessary changes. Responding to changing user requirements is more important than following a plan. If there is a development plan at all, it is fine to change it often.

All agile methods aim to have "light but sufficient" development processes. Therefore, project teams avoid use of formal project management plans, financial spreadsheets for budgeting, task lists, or any other activity that does not directly contribute to development of a functioning application.

While the SDLC or any other waterfall approach requires users to sign off on their requirements and then wait for the system to be completed, agile methods encourage users' involvement throughout the process and encourage developers to change requirements in response to user input if needed. The purpose of agile methods is not to conform to a static contract with the users but to ensure that the users receive an application with which they are happy. To avoid costly redesign, agile methods encourage developers to test each module as soon as it is complete.

For example, Extreme Programming (XP) includes the following principles: produce the initial software within weeks (rather than months) to receive rapid feedback from users; invent simple solutions so there is less to change and necessary changes are easy to make; improve design quality continually, so that the next "story" is less costly to implement; and test constantly to accomplish earlier, less expensive defect detection. (A *story* is a requirement or set of requirements delivered by the users.) Instead of formal requirements, developers encourage the users to give examples of how they would like to use the application in a certain business situation. Communication with users is highly informal and takes place on a daily basis.

Unlike more traditional methods, XP encourages two programmers to work on the same module of code on the same computer. This fosters constructive criticism and feedback. The constant communication between the two coders is meant to ensure cross-fertilization of ideas and high-quality software. The idea is that two minds working on the same code create synergy, and that two pairs of eyes are more likely to spot bugs than a single pair.

Critics of agile programming in general and XP in particular argue that the relaxed approach to planning as well as ceding decision making and accountability to clients (users) might result in disasters, especially if such methods are applied to large, complex projects. The critics cite the DaimlerChrysler payroll system (Chrysler Comprehensive Compensation, known as C3). C3 was the first large application developed with the XP method and was eventually canceled. The software never delivered more than one-fourth of the features it was supposed to have.

On the other hand, supporters give examples of success. One is a system developed for Domino's Pizza. When the company needed a new sales-tracking system, its CIO realized the project was too big and the time allotted for it—nine months—too short. He hired the services of experienced XP consultants and programmers. The clients described each feature—a story in XP parlance—on an index card. Each feature was coded in less than a week. The software was ready on time and with full functionality. It links point-of-sale registers with applications that track orders at the chain's 7,000 stores, which sell 400 million pizzas annually. Management now can analyze sales by toppings, crusts, sizes, delivery address, and soda sold with the pies.

When to Use Agile Methods

Agile methods are an efficient approach to development when a system is small, when it deals with unstructured problems, and when the users cannot specify all the requirements at the start of the project. They are also useful when developing a user interface: the developers can save time by quickly developing the screens, icons, and menus for users to evaluate instead of forcing the users to provide specifications.

When a system to be developed is small in scale, the risk involved in the lack of thorough analysis is minimal, partly because the investment of resources is small. (A small system is one that serves one person or a small group of employees. A large system is one that serves many employees, who might be accessing the system via a network from different sites.) If the small-system development takes longer than planned, the overall cost is still likely to be smaller than if a full SDLC were performed.

When users cannot communicate their requirements, either because they are not familiar with technological developments or because they find it hard to conceptualize the system's input and output files, processes, and user interface, developers have no choice but to use agile methods. In this case the users are often able to communicate their requirements as the development proceeds. For example, it is easier for marketing personnel to evaluate Web pages designed for a new electronic catalog and promotion site than to describe in detail what they want before seeing anything. Without being shown actual examples, users often can offer little guidance beyond "I will know it when I see it." It is easier for future users to respond to screens, menus, procedures, and other features developed by IT professionals than to provide a list of requirements for them.

When Not to Use Agile Methods

Agile methods might not be appropriate for all systems development. If a system is large or complex, or if it is designed to interface with other systems, using agile methods might pose too great a risk because the methods skip feasibility studies. Some experts do not recommend the use of agile methods for large systems (with the possible exception of Crystal, which accommodates scalable software development) because such systems require a significant investment of resources; therefore, system failure could entail considerable financial loss. The systematic approach of the SDLC is recommended if the system is complex and consists of many modules, because extra care must be applied in documenting requirements and the manner in which components will be integrated, to ensure smooth and successful development.

For the same reasons, use of agile methods should be avoided when a system is to be interfaced with other systems. The system requirements and integration must be analyzed carefully, documented, and carried out according to a plan agreed on by the users and developers before the design and construction phases start. This early consensus reduces the risk of incompatibility and damage to other, existing systems. Therefore, accounting ISs, large order-entry systems, and payroll systems as whole systems are rarely developed under agile methods. Other factors that should encourage use of waterfall methods are the size of the development team, how often the application is expected to be modified, how critical it is in terms of affecting people's lives and key organizational goals, and how tight the development budget is.

An additional risk with any type of prototyping is the difference between visible and nonvisible features of the software. Users tend to judge software by its visible elements and be less concerned about features such as database integrity, security measures, and other invisible but important elements. They may discover too late that some features are either missing or do not function to their satisfaction.

Figure 12.12 summarizes factors in deciding when and when not to use agile methods.

FIGURE 12.12
When and when not to use agile methods

When to use agile methods	When not to use agile methods
Small-scale system	Large-scale system
System solving unstructured problems	Complex system
When it's difficult for users to specify system requirements	System with interfaces to other systems
When the development team is small and co-located	When the team is large or distributed in multiple sites
System requirements are dynamic	System requirements are fairly static
System will not put people and critical organization goals at risk	System will significantly affect people's well-being and critical organizational goals
Development project budget is tight	Development is well-funded

PROJECT PLANNING AND MANAGEMENT TOOLS

Several tools exist to help plan and manage development projects. Some of the tools encompass planning and managing the development of many ISs. IBM's Rational Portfolio Manager is software that helps organizations plan investment in a new system, and then plan and manage the development project and delivery. All who are involved in the projects can track the progress and the sums spent on resources. Similarly, Primavera's ProSight helps plan and manage application portfolios, resources, budgets, and compliance with government regulations. eProject, offered by a company with the same name, is a Web-based application that helps team members and customers define tasks and manage projects by remotely accessing information about the project, such as personnel assignments, expenses, milestones, and completion times.

While some tools, such as Rational Portfolio Manager, are geared to help with software development projects, others, such as ProSight, eProject, @Task, Microsoft Project, and many others, are designed to accommodate planning and management of any type of project.

Project planning and management tools help plan and control the assignment and costs of resources as well as budgeting and completion time of milestones.

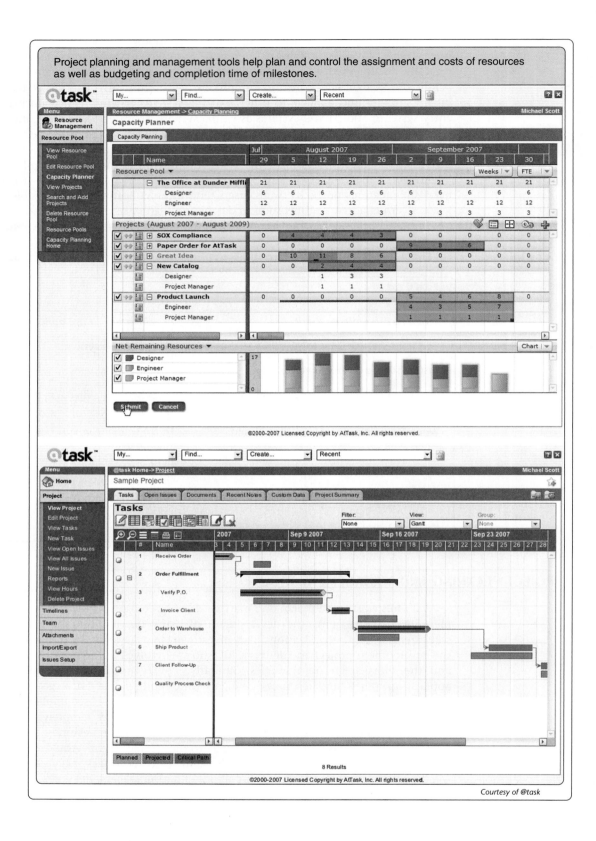

Courtesy of @task

POINT OF INTEREST

Just Checking

British Telecom (BT) has a subsidiary called BT Exact that is responsible for research, technology, and IT operations for its giant parent company. BT Exact manages several thousand IT development projects concurrently every year. To monitor the projects, the CIO meets with each project team every 90 days. Performance is measured by several metrics, including customer satisfaction (for systems or parts of systems that have been delivered) and maintaining cost levels that ensure the planned return on investment (ROI). If all goals for the 90 days are met, the project team is awarded bonuses. An extra incentive is transparency: the CIO publishes a calendar of delivery dates of all the IT projects.

Source: D'Agostino, D., "British Telecom's Tech Transformation," *CIO Insight*, February 6, 2007.

SYSTEMS INTEGRATION

Firms often must wrestle with highly distributed, heterogeneous environments populated with applications for special tasks, which cannot be accessed by systems used for other tasks. Often, the disparate systems cannot "talk to each other" because they run on different operating systems (or, as IS professionals say, on different platforms).

Much of what IT professionals do is systems integration, rather than analysis and development of a stand-alone IS. **Systems integration** looks at the information needs of an entire organization, or at least of a major division of it. The analysts consider the existing, but often disparate, ISs and then produce a plan to integrate them so that data can flow more easily among different units of the organization and users can access different types of data via a single interface. Consequently, many IS service companies call themselves systems integrators. Systems integration has become increasingly important mainly because more and more ISs are linked to Web sites, because more legacy databases are integrated into new enterprise applications such as SCM and ERP systems, and because of the growing linking of ISs between organizations (see Figure 12.13). *Legacy systems* are older systems that organizations decide to continue to use because the investment in a new system would not justify the improved features, or because the old systems have some advantage that cannot be obtained from newer systems.

FIGURE 12.13
Situations calling for systems integration

- Linking existing ISs to Web sites
- Linking databases to Web sites
- Interfacing legacy systems with new systems
- Linking legacy databases with enterprise applications
- Sharing information systems among organizations

Systems integration is often more challenging than systems development. In fact, some IT professionals regard systems development as a subspecialty of systems integration because the integrator must develop systems with an understanding of how data maintained in disparate systems can be efficiently retrieved and used for effective business processes, and because legacy systems must often be interfaced with recently acquired systems.

For example, marketing managers can have richer information for decision making if they have easy access to accounting and financial data through their own marketing IS. The better the integration, the better they can incorporate this information into their marketing information.

Systems integrators must also be well-versed in hardware and software issues, because different ISs often use incompatible hardware and software. Often, overcoming incompatibility issues is one of the most difficult aspects of integration. Consider business intelligence systems, which were discussed in Chapter 11, "Business Intelligence and Knowledge Management." The concept of extracting business intelligence from large data warehouses often involves integration of several ISs. The challenges are significant, and by some estimates more than half of all BI projects are never completed or fail to deliver all the expected features and benefits.

Systems integration has become increasingly complex because it now involves the ISs not only of a single organization but of several organizations. In the era of extranets, the challenge is many times more difficult because IT professionals must integrate systems of several different companies so that they can communicate and work well using telecommunications. Imagine how difficult it is to integrate disparate legacy systems of several companies. For this reason, companies often contract with highly experienced experts for such projects.

Should IS Professionals Be Certified?

When organizations commit millions of dollars to developing systems, they count on IT professionals to provide high-quality systems that will fulfill their intended purposes without harming their businesses, their employees, or their consumers. But the products of IT professionals often fail and cause serious damage. Some people argue that because of the high investment and high risk usually associated with systems development and operation, IT professionals, like other professionals, should be certified. These people argue that certification would minimize problems caused by ISs. Others argue that certification might stifle free competition and innovation, or even create a profession whose members will make it difficult to pass certification examinations so that current members can continue to enjoy high income.

Certification is meant to guarantee that the experts have passed tests that ensure their skills. The government or other authorized bodies are expected to license experts, thereby certifying which people have knowledge and skills in a particular discipline that are significantly greater than those of a layperson. Proponents of the measure argue that certification could reduce the malfunctioning of ISs.

Certification Pros. Some experts say certification could minimize the number and severity of IS malfunctions. Civil engineers must be certified to plan buildings and bridges. Doctors pass rigorous exams before they receive their licenses and begin to practice without supervision. Public accountants must be licensed to perform audits. Lawyers must pass the bar exams to practice. Why, these people ask, should IS professionals be allowed to practice without licensing?

Software experts do possess all the characteristics of professionals. They work in a field that requires expertise, and the public and their clients usually are not qualified to evaluate their skills. Certification could help the following groups in their relationships with IT specialists:

- *Employers* often hire software professionals without knowing what they are getting. They count on the information included in the candidate's résumé and, sometimes, on letters of recommendation. Mandatory certification might protect potential employers against charlatans. Also, certification would provide potential employers with information on a candidate's suitability for different levels of performance. For example, a professional might be qualified to participate in a systems development team but not to head the project team.

- *Clients* could realize even greater benefit from mandatory certification. While employers can learn, in time, of the real capabilities of their personnel, businesses that hire consultants have no previous employment experience on which to rely.

- *Society* might enjoy fewer software-related failures. Only those who are qualified would be allowed to engage in development and maintenance of information systems, thereby improving the overall integrity of ISs. Certification is especially needed for those holding key development positions for systems whose impact on society is significant, such as medical ISs and software embedded in weapons systems.

Certification Cons. Two arguments are raised against mandatory certification:

- It is difficult, if not impossible, to devise a way to measure software development competence. For instance, there are many different methods for developing applications, and there is no proven advantage of one over another. A computer professional might be very experienced in one method but not in others. It would be unfair to disqualify that individual merely on this basis.

- Some argue that mandatory certification might create a "closed shop" by using a single entry exam designed to admit very few people. In such a scenario, the status and income of those admitted would be enhanced at the expense of those excluded. With little fear of competition within the closed group, there is often little incentive to improve skills.

- IT progresses very fast, faster than many other fields. Ensuring that a specialist know how to use certain information technologies and methods today does not ensure that this person will know how to use technologies and methods two years from now. This person's knowledge may be obsolete and render the certificate useless.

Where We Operate Now. Mandatory certification or licensing of IT professionals is rare. Only Texas, British Columbia, and Ontario require licensing of software development professionals. In fact, the industry cannot reach agreement about who should be considered an IT professional. Some organizations, such as the Institute for Certification of Computer Professionals (ICCP), test and certify people who voluntarily take their tests. (About 50,000 people have been certified by ICCP globally, out of millions who consider themselves IT professionals.) Some software companies certify analysts and programmers to install their companies' tools. However, there are no certification regulations for IT professionals in the United States or anywhere else that are similar to those for many other professions.

SUMMARY

- IT planning is important especially because investing in IT is typically substantial and because of the high risk in implementing enterprise applications.

- Standardization is often an important part of IT planning. Standardization helps save costs, provides efficient training, and results in efficient support.

- The systems development life cycle (SDLC) and other waterfall methods consist of well-defined and carefully followed phases: analysis, design, implementation, and support.

- The purpose of systems analysis is to determine what needs the system will satisfy.

- Feasibility studies determine whether developing the system is possible and desirable from a number of viewpoints. The technical feasibility study examines the technical state of the art to ensure that the hardware and software exist to build the system. The economic feasibility study weighs the benefits of the system against its cost. The operational feasibility study determines whether the system will fit the organizational culture and be used to full capacity.

- System requirements detail the features the users need in the new system.

- In systems design, developers outline the systems components graphically and construct the software. Tools such as data flow diagrams and the Unified Modeling Language (UML) are used to create a model of the desired system.

- When the system is completed, it is implemented. Implementation includes training and conversion

from the old system to the new system. Conversion can take place by one of several strategies: parallel, phased, cut-over, or piloting.

- The systems life cycle continues in the form of support. The system is maintained to ensure operability without fault and satisfaction of changing business needs.

- Agile methods are a popular alternative to the traditional systems development life cycle. Agile methods place considerable emphasis on flexible requirements and frequent interaction with users. These methods skip detailed systems analysis and aim at delivering a new application in the shortest possible time.

- Several applications help plan and manage development projects. Some are geared toward planning and management of software development. Some are Web-based, allowing remote access.

- Systems integration is often much more complicated than systems development, because it requires the IT professionals to make different applications communicate with each other seamlessly. The complexity is multiplied when integrating ISs of several organizations that must work together over the Web.

- Because of the major responsibility of IS professionals, the question of whether certification is needed has come up. If doctors, civil engineers, lawyers, and public accountants are subject to mandatory certification, many people argue that IS professionals should be, too.

WORLDWIDE HOST REVISITED

Worldwide Host's TripExpert.com project team has been busy investigating opportunities for development of the new Web site. They are keeping strategic planning issues in mind as they learn more about options for the site.

What Would You Do?

1. The case at the beginning of the chapter lists Worldwide Host's corporate mission statement. From that statement, information given in the opening case, and the examples in the chapter, write a possible IS mission statement for Worldwide Host. Be sure to include information on IS's place in the organization and its chief contributions.

2. In what ways has the TripExpert.com team been dealing with systems integration, instead of systems development? Cite examples of systems that Worldwide Host is trying to integrate.

New Perspectives

1. The TripExpert.com site is facing a time crunch. The chapter discussed agile methods as ways to speed development of information systems. Would these techniques work for the TripExpert.com project? Why or why not?

2. The database team overlooked a key new requirement for the database system. Could this mistake have been prevented? If so, how and at what stage of the systems development life cycle?

KEY TERMS

agile methods, 426
beta site, 425
conversion, 423
cost/benefit analysis, 418
cut-over conversion (flash cut conversion), 424
data flow diagram (DFD), 420
feasibility studies, 417

implementation, 423
organizational culture, 419
parallel conversion, 424
phased conversion, 424
piloting, 425
prototyping, 415
return on investment (ROI), 418

support, 425
system requirements, 420
systems analysis, 416
systems design, 420
systems development life cycle (SDLC), 415
systems integration, 431
Unified Modeling Language (UML), 422

REVIEW QUESTIONS

1. Why is IT planning so important?
2. As part of their IT planning, many organizations decide to standardize. What does standardization mean in this context, and what are its potential benefits?

3. Why is traditional systems development referred to as a "cycle"? What determines the cycle's end?
4. Systems developers often use the term "application development" rather than "systems development." Why?

5. What are the benefits of using data flow diagrams? Who benefits from DFDs?

6. SDLC is usually recommended for developing an IS that will be interfaced to other ISs. Give two examples of an IS that is interfaced with at least two other ISs.

7. Recall the discussion of IT professionals in Chapter 1, "Business Information Systems: An Overview". Of the following professionals, who does the majority of the systems construction job: the CIO, systems analyst, database administrator (DBA), or programmer? Why?

8. What are the advantages of agile methods over waterfall development methods, such as the traditional SDLC? What are the risks?

9. Why are agile methods so helpful when users cannot define system requirements?

10. An increasing number of IS professionals prefer to call the end users of their creations "customers," even if the developers and users are employees of the same organization. Why?

11. What is systems integration?

12. Why is systems integration more complicated when the systems involve the Web than when they do not?

13. The emergence of the Web as a vehicle for business increased the need for systems integration. How so?

DISCUSSION QUESTIONS

14. The modern view of systems development is that it should be a continuation of IS planning. Why?

15. Consider a new chain of shoe stores. The marketing department of the corporation would like to know the customers and their preferences. What questions would you ask before developing an IS for data collection and analysis?

16. The analysis phase of systems development includes fact finding. Suggest ways to find facts, other than the ways mentioned in this chapter.

17. In data flow diagrams, a process is always labeled with an action, while entities and data stores are labeled with nouns. Why? Give two examples for each of these elements.

18. You are asked to recommend a conversion strategy for a new accounts receivable system. The system will be used only by the controller's office. Which strategy will you recommend, and why?

19. You are asked to recommend a conversion strategy for a new ERP system that includes accounting, sales, purchasing, and payroll modules. Which strategy will you recommend, and why?

20. What are the elements that make the responsibilities of IT professionals similar to those of other professionals, such as engineers and financial analysts?

21. Do you support mandatory certification of IT professionals? Why or why not? If you do, which IT professionals (listed in Chapter 1, "Business Information Systems: An Overview") would you require to pass tests? Why?

22. Many IT professionals say that trying to certify all specialists in this field is impractical. Why?

23. Many software companies (such as Microsoft, Oracle, and SAP) certify people as consultants for their products. For instance, you might become a certified SAP R/3 Technical Consultant. Is this type of certification the same, in principle, as the certification of a physician, lawyer, or certified public accountant (CPA)? Explain.

24. Suppose you are the IT director for a hospital. You have a small crew that helps the medical and administrative staffs with their computers and applications, but when a new system must be developed, you must hire IT professionals. How would you conduct your search for reliable IS developers? Whom would you contact, and what questions would you ask?

25. You are the CIO for a large university hospital. The medical staff of the oncology ward would like to build an expert system for diagnosis. Your preliminary review shows that the financial investment would be considerable. What questions do you ask (of both the doctors and your staff) to decide whether to use a thorough SDLC or agile methods to develop the system? List and explain your questions.

26. You are trying to explain to your supervisor the general details of a proposed IS. The IS involves a server connecting many PCs. Your supervisor is not an IS professional and has no idea what a DFD is. How would you prefer to communicate your ideas: verbally; in writing, but without diagrams; with a DFD; or with a combination of some or all of these means? Explain your choice.

27. During development of a new IS, professional jargon might facilitate communication among IS professionals, but it might be detrimental when used to communicate with users. Explain.

APPLYING CONCEPTS

28. Prepare a 10-minute software-based presentation (use PowerPoint or another application) to make a presentation on the topic: "Factors that have made IS planning difficult over the past five years." Include in your presentation developments in hardware, software, and telecommunications; globalization; the Internet; the IT labor force; and any other area that has had an impact on IT planning.

29. You were hired as an IS consultant by a small chain of stores that rents domestic appliances. Partly because operations are run with paper records, one store does not know what is going on in the other stores. The president of this small company thinks that the chain doesn't utilize its inventory efficiently. For example, if a customer needs a lawnmower and the appliance is not available in store A, the salespeople cannot tell the customer if the mower is available at another outlet or offer to bring it for the customer from another outlet. The president would like an IS that would allow the chain to serve the customers better and that would help with tracking and billing, too. She would like to take advantage of the Web to help both employees and customers. Both should know what is available for rent and at which store at any given time. List the questions you would ask in your fact-finding effort and indicate who in the organization would be asked each question.

30. Assume you are the leader of a team that has just completed construction of a Web site that provides information but also allows online purchasing of your company's products. Enumerate and explain the steps you would take to test the system. Prepare a software-based presentation (using PowerPoint or a similar application) to explain all the testing steps and why each must be taken. (*Hint*: Keep in mind different operating systems, Web browsers, screen sizes, and so forth.)

31. Prepare a DFD that describes the following application: Gadgets, Inc., sells its items through traveling salespeople. When a salesperson receives a signed contract from a client, he or she enters the details into a notebook computer. The salesperson later transmits the record to the company's mainframe computer at its headquarters. The program records the details in four files: sales, shipping, accounts receivable, and commissions. If the buyer is a new customer (one who is not yet in the customer database), the program enters the customer's record into the customer database and generates a thank-you letter. The program also calculates the 5 percent commission, which is recorded in the commission file with the salesperson's code. At the end of the month, the program produces a paper report with the records of all the new customers. In addition, if the total monthly sales of the salesperson exceed $100,000, the program generates a congratulatory letter showing that total. If the total is less than $5,000, the program produces a letter showing the total and the sentence: "Try harder next month."

32. Prepare a DFD that communicates the following business walk-up car rental scenario: When a customer approaches the counter at Buggy Car Rental, a serviceperson asks the customer for the details of the desired car. He or she then checks in a computerized database to see whether a car with these features is available. If a car is available, the serviceperson collects pertinent information from the customer (including an imprint of the customer's credit card), fills out a contract, and has the customer sign the contract. The customer is then given a key and is told where to find the car in the parking lot. The serviceperson indicates in the database that the car is no longer available. If a car with the desired specifications is not available, the serviceperson offers a car of a higher category at no extra charge. If such a car is not available either, the service person offers an available car of a lower category. The customer either rents it or refuses to rent it. When the car is returned, the customer pays by check or by charging the credit card and returns the keys. The serviceperson gives the customer a copy of the signed contract, indicates in the database that the car is now available, and records its new mileage.

33. Team up with another student. Each of you should select a different agile method from the list appearing in this chapter. Each should write a one-page summary of the principles, benefits, and shortcomings of the method. Then, sit together and write a one-page summary of the differences between the two methods along the three points.

34. Team up with another student to search the Web for tools that facilitate software development, and choose three tools. List the features provided in each of the tools. Assume that the vendors' claims are true. Which phases and activities of the systems development life cycle does each tool support? Which would you prefer to use in systems development? Why? Prepare a 5-minute software-based presentation (using PowerPoint or a similar application) to present your findings and explain your recommendations.

FROM IDEAS TO APPLICATION: REAL CASES

Standardization at Standard Pacific

Standard Pacific Homes is one of the largest home-builders in the United States. It is a Fortune 500 company that prides itself of having built more than 97,000 homes in California, Florida, Arizona, the Carolinas, Texas, Colorado, Nevada, and Illinois over the past four decades. Its revenue in 2005 was $5 billion. Over the years, the company enjoyed growth in business as a result of acquiring smaller companies.

In 2005, when Rob Kelle joined the company as CIO, he found out that the organization used many different technologies, too many to make him and his staff comfortable. Standard Pacific was incorporated in California in 1961. Kelle realized that the organization's Southern California culture would not allow him to dictate policy. Acquired companies had to be given as much freedom as possible. They were forced to use software applications selected by Standard Pacific, but were left the choice of how to use these applications. This approach gave the acquired business managers a measure of comfort. They felt good that nobody dictated to them how to use applications, and that they could continue to operate as they had been used to.

A year and a half later, the CIO regretted this free-hand policy. The policy prevented the company from consolidating data and applying business intelligence (BI) analysis across units. The various databases were organized differently. This caused Kelle's staff "a maintenance nightmare." Whenever his IT professionals wanted to pull data, they had to do so from 26 different systems. Often, different types of data had the same names in different units, which caused confusion and often resulted in comparing apples to oranges. This was the main reason why Standard Pacific did not purchase a BI application; there was none that could accommodate this inconsistent array of databases.

Sales data were not tied electronically to the company's ERP system. The data had to be manually extracted from the various databases and fed into the ERP system. While the CIO realized that the organizational culture did not allow him to shove standardized systems down the throats of employees, he met with user groups to convince them to accept his suggestions. He knew that people usually reverted to their old applications if the new ones did not work fast enough for their taste. He convinced user groups to use standardized applications, and use them often to learn them quickly.

The CIO started implementing several changes. He moved the ERP system from an AS/400 computer to Windows servers. Newly recruited software engineers were more comfortable with Windows than the operating system on the AS/400. The change came at a good time, because the storage capacity of the AS/400 was reaching its limit.

As a construction company, the organization uses computer-aided design (CAD). The CIO decided to subscribe to a hosted extranet application called BuzzSaw, provided by CAD company AutoDesk. BuzzSaw enables workers in the field to access a wide array of construction documents, including drawings, specifications, budgets, and schedules. Since all the data is on the same server, whenever a construction engineer makes a change, the change is accessible and visible to all who need the information. The system sends business managers alerts when a change in a project plan alters a budget item or impacts the schedule.

For a 2700-employee company, Kelle's staff of 90 is relatively small. The annual IT budget of $20 million is quite small, too. But this is even more of an incentive to move the organization into shared and consistent applications. The CIO moved all networking to Cisco equipment, including VoIP telephony. All units use Microsoft's Outlook Exchange for e-mail. Many of the new applications are developed using .NET software development tools to enable Web-based applications that all units can use through browsers.

Before taking the position of CIO at Standard Pacific, Kelle had served as a consultant with Ernst & Young. He is an experienced professional. He is moving the company into more standardized technologies, but he is doing it gradually, and—so far—successfully.

Sources: Schuman, E., "Home Builder Crafts Consistency," *eWeek*, April 19, 2006; (www.standardpacifichomes.com), July 2007; (usa.autodesk.com), July 2007.

Thinking About the Case

1. The CIO allowed acquired businesses to use business applications as they saw fit. He later regretted the decision. Why?

2. Standardizing IT at Standard Pacific may be more difficult than in some other organizations, not necessarily for technical reasons. What makes the implementation of this company's IT plan more challenging?

3. Standard Pacific's CIO is moving the company into using more consistent information technologies through sharing applications. Give examples.

A Sick Medicaid System

Medicaid is a program that states administer to help people with low income obtain health care and related services. The program is sponsored and partially funded by the U.S. federal government. In recent years, many states had to modify their Medicaid information systems to comply with the federal Health Insurance Portability and Accountability Act (HIPAA) of 1996. Most people know HIPAA for its privacy requirements. Indeed, HIPAA requires that states (and other health care providers and administrators) secure Medicaid records. However, many states used the occasion of implementing HIPAA to implement additional changes in their Medicaid systems, such as adding the capability to submit claims online.

Maine's Department of Human Services (DHS) serves 262,000 Medicaid recipients and processes more than 120,000 Medicaid claims per week. To adapt their Medicaid claims systems to HIPAA, most states decided to leave their old ("legacy") systems as is and integrate them into a Web site. DHS decided to build a totally new system. The results were disastrous. In 2006, Maine was the only state that did not comply with HIPAA.

Maine's old Medicaid system was implemented in the 1970s on a Honeywell mainframe computer. DHS managers reasoned that building a new system would be easier and less expensive to maintain than upgrading and maintaining the old system. The department's IT staff was assigned to build the system. Some service providers with experience in the field, such as Electronic Data Systems (EDS), offered to outsource state claim processing. Maine declined the offers. The IT staff believed that building a new system would allow more flexibility when updates were needed. Such updates have been required over the years, and were added to the software. They include adding Medicaid services, adding new health care providers, and changing service rates. The staff also believed it could manage the system better than having it outsourced to a service provider.

The old system checked three information elements: whether the provider's record was in the database; whether the patient was eligible; and whether the service rendered was covered. The new system was to check 13 information elements. For example, the system would also ensure that the provider was authorized to perform the service on the date the service was provided, and see if the provider's license was valid.

In April 2001, DHS issued a request for proposal (RFP). The requirements called for a system that was more secure and would clear claims faster, track costs better, and give health care providers more accurate information on their claim status. DHS received two

offers: one from Keane, Inc. for $30 million, and the other from CNSI for $15 million. Keane had some experience with medical systems. The firm also worked on the Maine Medicaid eligibility software. CNSI had no experience with medical systems. The state's head of procurement decided to hire the low bidder.

The development team included 50 employees from CNSI and the 15 members of the DHS IT department. The team had to include in the system code pieces that were modified numerous times over the legacy system's life of 25 years. The team also had to interview a dozen Medicaid experts in the Bureau of Medical Services, a unit of DHS, so the developers could receive accurate information about Medicaid rules. These rules were needed for proper programming. Often, the experts could not spare the time for these meetings, and CNSI programmers used what they believed were the proper rules. In many cases these rules were wrong, the contractor's programmers had to speak with the experts, and then they had to fix the code. This delayed completion times.

The department head knew that the team had to consist of 60–70 members, but was reluctant to ask for more resources. The project was to be completed in two years. The federally mandated deadline for implementing HIPAA was October 1, 2003, which ostensibly gave the team enough time to complete the project. However, over the next two years Medicaid rules changed, and the team had to change the program accordingly. Despite the long hours of programming, errors kept popping up, and the team had to recode. Also, the team had to add storage capacity and computing power to accommodate the increase in information resulting from the new rules. This further delayed the completion of assignments.

In January 2003, a new governor took office. John Baldacci promised to save costs. One way to do so was by merging the Department of Behavioral and Developmental Services with the Department of Human Services into the new Department of Health and Human Services (DHHS). The new department had to merge its information systems. Merging the systems took away IT staff members from the Medicaid project effort. It also diverted the attention of executives from the project. The extra work and the delays increased the cost of the project from $15 million to $22 million.

The team missed the HIPAA deadline. Time pressure changed some original project requirements. The project leaders decided to test only some components of the system rather than all of them. A sample of 10 health care providers was pilot-tested with some claims, but the claims were not processed as expected. DHHS staff members were not adequately trained to use the system. And except for an announcement about new provider ID codes, the providers were not given training on how to use the new system.

The CIO decided to cut-over to the new system in January 2005. The cut-over was such that the system could not revert back to the legacy system. The rationale was that to meet HIPAA requirements, the new system could not use the old identifying numbers: tax ID numbers for providers, and social security numbers for patients.

On January 21, 2005, the new system started its productive life. The CNSI team members reported that all was going well. After three days, the DHHS head decided to check statistical data. He knew something was wrong. The system sent 24,000 claims—about 50 percent of all claims—into a "suspended" file. Usually, suspended claims were those determined rejected, or with an error that was not significant enough to reject but required further investigation for payment. The error rate seemed too high. The old system suspended only 20 percent of the claims. Instead of payments, many doctors received zero payment. When they resubmitted the claims, the system automatically rejected them again because it was programmed to reject any claim it had already rejected.

The CIO and programmers examined the code and found errors. For example, it turned out that without asking Medicaid experts, the programmers limited each claim to 1,000 lines, while some contained up to 10,000 lines. Any claim with more than 1,000 lines was rejected. The staff of MaineCare, the new name of the Bureau of Medical Services, asked providers to limit claims to 1,000 lines while the team was working to fix the code. The claim backlog reached 100,000, and the state owed health care providers $50 million. Despite their efforts, the team members could not fix the problems. The system continued to reject legitimate claims.

The governor ordered the DHHS to fix the system by March. March passed, and the system still did not function properly. The state hired XWave, a firm specializing in systems integration and IT consulting to try to fix the problems. A new department head was installed. By the end of summer 2005 more than 647,000 claims representing payments of $310 million were suspended in the system's database.

XWave consultants soon found out that a major part of the problem was lack of communication among programmers. The state's programmers and CNSI programmers worked on different parts of the software without telling each other what they were doing. The DHHS director decided to do something that should have been done from the start. He nominated a Medicaid expert, a physician who headed MaineCare, to work full time with the software development team. She clarified all the intricacies of the rules, and helped programmers to implement them in the software.

Throughout this ordeal, MaineCare personnel worked hard to serve providers by phone and by manual work. However, they could only process 1,000 claims per week. In October 2005, the DHHS director announced that the system was working as well as the old one. His explanation: the new system now suspended only 20 percent of claims, the same rate as the old one. However, doctors argued that the system still rejected legitimate claims. They also wondered why the new system was compared to the old one. What was the purpose of spending millions of dollars on a new system that would be, at best, as good as the old system?

The new system's failure caused much damage to Maine's health care industry. Doctors, dentists, hospitals, clinics, and nursing homes were not paid for their services for weeks. Some dentists and therapists were forced to shut down their businesses. Other caregivers had to take out loans to pay their bills and stay in business.

Over six years of effort, Maine spent $70 million on the system, $24 million of which was paid to CNSI. CNSI's vice president of technology claimed in May 2007 that the rate of clean claims—claims that are processed in the first pass through the system—rose to 96 percent. Nonetheless, DHHS officials recognized that the system was flawed. In December 2006 the state decided to outsource its Medicaid claims management to a private firm. The firm would use its own system and would update it as the federal government's Medicaid demands increased. The contract was expected to be awarded in 2010. In the meantime, the state contracted with CNSI to maintain the existing system.

Sources: Holmes, A., "Maine's Medcaid Mistakes," April 15, 2006; Mehta, V., "CNSI Inks New Contract to Provide Ongoing Operations and Maintenance Services for Maine's Medicaid Claims Management System," (www.cns-inc.com), May 14, 2007; (http://blogs.govexec.com/techinsider/archives/2007/05/mainecare.php), May 21, 2007.

Thinking About the Case

1. Were there any factors that contributed to the project failure which were not the fault of the project team and its leaders?

2. Some critics said that the fact that only two bidders made offers, and that the price quotes were so different, should have alerted DHS that either the RFP was unrealistic or that the low bidder could not develop the system properly. What is your opinion?

3. Some states decided to continue to use their old (legacy) Medicaid claim systems, update them, and provide a Web link to the system, instead of developing a new system. What are the advantages of this approach?

4. In many cases organizations decide to convert to a new IS by cutting over. In what sense was the cutover in this case riskier than in other such conversions?

13

THIRTEEN

Choices in Systems Acquisition

LEARNING OBJECTIVES

Developing systems in-house or commissioning a software development firm is the most expensive way to acquire ISs. Other alternatives might be less expensive and offer different benefits. Some of the alternatives have been mentioned in previous chapters, but they are discussed in more depth here and will provide a deeper understanding of systems acquisition.

When you finish this chapter, you will able to:

- Explain the differences among the alternatives to tailored system development, which include outsourcing, licensing ready-made software, using software as a service, and encouraging users to develop their own applications.

- List the business trade-offs inherent in the various methods of acquiring systems.

- Describe which systems acquisition approach is appropriate for a particular set of circumstances.

- Discuss organizational policies on employee computer use.

WORLDWIDE HOST:
Tapping Others' Expertise

Worldwide Host is a leader in the hotel industry, not in software development. CIO Michael Lloyd convinced his executive team long ago that it made better financial sense for the hotel chain to contract with a software firm to develop or upgrade its information systems while he and his staff concentrated on hardware and day-to-day support. It took time, many interviews, and the review of several proposals to find a firm that fit well with Worldwide's unique needs, but General Data Systems (GDS) fit the bill. The firm was a top-notch software developer with a long track record in the industry, and it provided the technical expertise that Worldwide needed to keep abreast of ever changing technology. The partnership between the two firms allowed Worldwide to maintain a low IT staffing level and use GDS for help with new business needs and problems. The TripExpert.com Web site project was just the latest in the two firms' collaboration.

Adding Another Firm to the Mix

Michael was meeting with GDS analysts Judith Kozak and Corey Johnson to go over some decisions for the TripExpert.com Web site. Corey began reporting on his latest information concerning their plans to purchase a license from Reservations Technologies for an existing reservation system developed for the Web.

"We ran benchmark tests on the system's performance, and it did provide the transaction response time we need on the Web. The system also has good scalability—it can handle projected peak customer demand. We repeated the tests several times with different sets of data, and it performed well. So, the system seems to be a good option for us," he said.

Michael interrupted. "How long has this company been in business, and who else has used their system? I want to be sure they're reliable, since we're staking a big part of Worldwide's future on the TripExpert site."

"They've been around for about nine years—not long for a software company overall—but pretty old for a Web software firm. I checked the background of some of their technical staff, and they received advanced degrees in computer science, artificial intelligence, and electrical engineering from Stanford, MIT, and the University of Illinois. Plus, they gained practical experience at other companies before launching their firm. They are well respected in the field," reported Corey.

"What about their clients?" Michael persisted.

"They've worked with quite a few airlines—GlobalAir, Svenska, Universal Airlines. North Trans, one of your airline partners, recommended them to me. We could set up a time to review some of their operations on-site, if that would make you feel more comfortable."

"Great. I'd like to hear firsthand from their customers. Let's set that up in the next couple of weeks. I have some travel coming up, so the sooner, the better."

Fitting It All Together

Judith asked, "What about the additional functionalities that we need for TripExpert? Will GDS have to hire programmers for that purpose, or will Reservations Technologies take care of it?"

"Their system was designed to allow easy modifications. They'll create the additional functionalities. The company also offers technical assistance as part of a licensing agreement, so that would cover us if we need their help in modifying the system to tie into our existing reservations system," Corey responded.

"Speaking of the hotel reservation component, Judith, how are we coming on Worldwide's new system? Are we back on track after the database glitch?" Michael asked.

"We lost five weeks overall after we pulled additional staff in to work on the room rate discounting component. We'll keep trying to gain back a day or two wherever we can, but we need to maintain our quality standards. Also, we need to begin planning our training sessions for your travel division. I've been putting some materials together as we go."

Michael laughed. "Another task to add to my list—can't wait."

OPTIONS AND PRIORITIES

In Chapter 12, "Systems Planning and Development," you learned about software development and that few companies develop their own ISs in-house. Recall, also, that "systems" almost always means "applications," and therefore the terms will be used interchangeably in this chapter, as in Chapter 12. The four alternatives to in-house development by IT specialists, as illustrated in Figure 13.1, are outsourcing, licensing, using software as a service (SaaS), and having users develop the system. If an application of the desired features and quality can be obtained from more than one of these sources, then the major factor left to be considered is usually cost. The preference then would be to license, because of immediate availability and low cost. If the application cannot be licensed, the next choice would usually be to obtain use of the system as a service from an application service provider (ASP) because the system is immediately available for use and the organization does not have to lay out a large sum up front for such use. If ASPs do not offer the desired IS and it can be developed by non-IT employees, then this would usually be the chosen alternative. If non-IT employees cannot develop the IS, the choice might then be to outsource IS development. However, as you will see, outsourcing is a concept that might encompass more than just commissioning the development of an application.

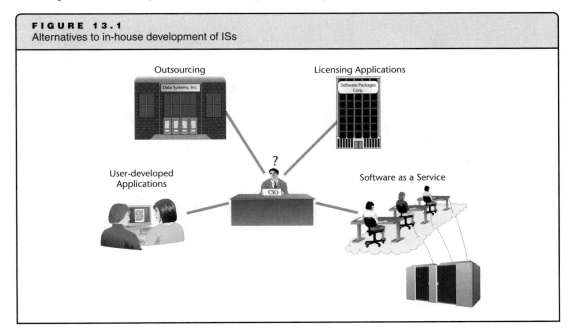

FIGURE 13.1
Alternatives to in-house development of ISs

Many factors must be considered in addition to quality and cost. Therefore, these alternatives are not fully comparable and often cannot be prioritized as simply as they have been here. The purpose of this discussion is to clarify the advantages and disadvantages of these options. As you will see, a variety of factors drive organizations to decide how they acquire ISs and the service that supports the maintenance and use of the systems.

OUTSOURCING

Outsourcing in general means hiring the services of another organization or individual to perform some of the work that otherwise would be performed by you or your employees. In the IT arena **outsourcing** has two meanings. One is to commission the development of an application to another organization, usually a company that specializes in the development of this type of application. The other is to hire the services of another company to manage all or parts of the services that otherwise would be rendered by an IT unit of the organization. The latter concept might not include development of new applications.

POINT OF INTEREST

World Interactive IT Outsourcing Map

CIO magazine provides a world map of IT outsourcing at: www.cio.com/article/123711/Click_and_Save_with_Our_Interactive_Global_Outsourcing_ Map . The map has several "pins," each for a major country in which IT outsourced work is performed. Clicking a pin provides a summary of relevant facts: overall ranking as an outsourcing destination, average second year programmer salary, rank of geopolitical risk, English proficiency, pros, and cons. All facts are for 2006, but most will probably apply for several years into the future.

Source: Goldberg, M., & Overby, S., "Click and Save with Our Interactive Global Outsourcing Map," *CIO*, July 16, 2007.

Outsourcing Custom-Designed Applications

Often, an organization has a need that no existing software can satisfy. For example, if the cost-accounting procedures of a particular company are so specific that no commercially available software can perform them, the company must develop **custom-designed**, or **tailored**, **software**. In recent years, the number of companies developing applications in-house has declined. The majority of custom-designed applications are developed by companies that specialize in providing consulting and software development services to other businesses.

Why You Should

Understand Alternative Avenues for the Acquisition of Information Systems

As an increasing number of business activities are supported and enhanced by ISs, it is extremely important for organizations to acquire systems that best fit their needs and are available as soon as possible, and to minimize the cost of systems acquisition and maintenance. As explained in Chapter 12, "Systems Planning and Development," employees should involve themselves in the process of deciding which ISs will be introduced into their business units and what features they will have. Since there are several ways to obtain ISs, professionals like you must understand the advantages and disadvantages of each. If you have a concern with a certain approach to acquire the system you need, you should voice it and be able to propose other options.

While custom-designed applications are more expensive than purchased ones, they have several advantages (see Figure 13.2).

- *Good fit to need*: The organization enjoys an application that meets its needs exactly, rather than settling for the near fit of a ready-made program.

- *Good fit to culture*: When custom-developing a system, developers are more sensitive to the organizational culture. Employees enjoy an application that fits their work. When licensing a packaged application, employees sometimes must change their work to accommodate the software.

- *Dedicated maintenance*: Because the programmers are easily accessible to the company, they are familiar with the programs and can provide customized software maintenance. Maintenance includes modification for business changes (including mergers with and acquisition of other organizations) and upgrading of the software when new technologies become available.

- *Smooth interface*: When a system is custom-made for an organization, special care can be taken to ensure that it has proper interfaces with other systems. The new system can communicate smoothly with those systems.

- *Specialized security*: Special security measures can be integrated into the application. Because the program is custom designed, security measures are known only to the organization.

- *Potential for strategic advantage*: Recall from the discussion in Chapter 2, "Strategic Uses of Information Systems," that companies gain a strategic advantage when they can employ an IS that their competitors do not have. A unique application might give a business a strategic advantage because it is the only business that can enjoy the application. For example, no CRM application can do for a business what an enterprise application that was developed specifically to serve its customers in a unique way can do.

FIGURE 13.2
Advantages and disadvantages of custom-designed applications

Advantages

◆ Good fit of features to business needs

◆ Good fit of features to organizational culture

◆ Personnel available for maintenance

◆ Smooth interfaces with other information systems

◆ Availability of special security measures

◆ Potential for a strategic advantage

Disadvantages

◆ High cost

◆ Long wait for development if IS personnel are busy with other projects

◆ Application may be too organization-specific to interface with systems of other organizations

The greatest disadvantage of tailored applications is their high cost. Tailored software development requires an organization to fund all development costs; in contrast, costs of developing off-the-shelf and other ready applications are distributed over a larger number of expected purchasers. Another disadvantage of custom-designed development is that the production schedule can be delayed because IS personnel might not be available for long periods. Another important downside is that custom-designed software is less likely to be compatible with other organizations' systems. If organizations with different tailor-made systems decide to link their systems, they might incur significant cost to modify one or both of the systems.

Clients of outsourced software development should also be aware of an inherent conflict of this option: on one hand they want the developing firm to conform to a contract that includes specific requirements of the software. On the other hand, specific requirements may make the development effort inflexible and potentially costly: if the client company needs to change requirements as the development progresses, the developers might either refuse to deviate from the original requirements or might agree to make the changes for hefty additional charges. Contracts for outsourced software development might also be incompatible with some development methods, such as agile methods, discussed in Chapter 12, "Systems Planning and Development". The essence of such methods is the clients' ability to request modified or new features as the development moves forward, which might stand in stark contrast to the contract.

Many North American and European countries have outsourced development of well-defined applications to professionals in other countries, an act often referred to as **offshoring**. Programmers in India, China, and the Philippines earn a fraction of their colleagues in Western countries while often mastering the same level of skills. Hiring these programmers might reduce the cost of development significantly. Offshoring has caused layoffs of programmers in Western countries and created much bitterness among those professionals and supporters of local labor. However, this is apparently an inevitable result of the growing scope of economic globalization.

POINT OF INTEREST

Offshoring Back

Bharti Tele-Ventures is India's largest provider of mobile phone services. Its main rivals are Tata and Reliance, and competition is fierce. Demand for its services has been evidenced in revenue growth: from $509 million in 2003 to over $4 billion in 2006. Bharti had 3 million subscribers in 2003, but knew that to sustain profits it would have to grow its subscribers to 25 million within just a few years. This meant the company would have to do the impossible: hire 10,000-20,000 workers within two years. CEO Sunil Mittal decided to do what many other CEOs do: focus on what his company did best, and outsource all other activities. In 2004, he signed $400 million contracts with Sweden's Ericsson, Germany's Siemens, and Finland's Nokia to maintain Bharti's network. He also outsourced most of Bharti's information systems services to IBM for a period of 10 years for $750 million. Bharti would focus on marketing, devising new services, and developing new business opportunities.

Source: Chandler, C., "Wireless Wonder," *Fortune*, January 22, 2007, pp. 131-136.

Outsourcing IT Services

A large number of businesses have turned to IT companies for long-term services: purchasing and maintaining hardware; developing, licensing, and maintaining software; installing and maintaining communications networks; developing, maintaining, and operating Web sites; staffing help desks; running IT daily operations; managing customer and supplier relations; and so on. An organization might use a combination of in-house and outsourced services. It might outsource the development of an IS, but then put its own employees in charge of system operation, or it might outsource both the development and operation of the system. When a business outsources only routine business processes, such as customer order entry or human resource transactions, the practice is sometimes called *business process outsourcing*. Note, however, that this term refers to the outsourcing of many activities, whereas this discussion is limited to only IT services.

In considering whether to outsource IT services, management should ask the following questions:

- What are our core business competencies? Of the business we conduct, what specialties should we continue to practice ourselves?

- What do we do outside our specialties that could be done better for us by organizations specializing in that area?

- Which of our activities could be improved if we created an alliance with IT organizations?
- Which of our activities should we work to improve internally?

Many companies have come to realize that IT is not their core competency and should not be a focus of their efforts. In addition, the pace of developments in IT might require more expertise than is available within many organizations.

POINT OF INTEREST

Outsourcing and Tact

In 2006, Bank of America announced that it would outsource the jobs of 100 help desk technicians from the San Francisco Bay area to India. The bank ordered the soon-to-be-fired American workers to train their replacements as a condition for receiving their severance payments.

Source: Horowitz, A., Jacobson, D., McNichol T., & Thomas, O., "101 Dumbest Moments in Business," *Business 2.0*, January/February 2007, p. 107.

A growing portion of corporate IS budgets is allocated for purchased (outsourced) services. IT companies that made their reputation by providing hardware and software, such as IBM and Unisys, have seen revenue from the outsourcing service portion of their business grow faster than the revenue from hardware and software sales. Among the largest IT service providers are IBM, EDS, Accenture, Computer Sciences Corp. (CSC), Unisys, First Data, AT&T, Capgemini, Perot Systems, and Hewlett-Packard. For the sake of simplicity and clarity here, such companies are called vendors, and the organizations to which they outsource are called clients. (Note that some trade journals refer to vendors as outsourcers.) Outsourcing is typically a long-term contractual relationship in which the vendor takes over some or all of the client's IT functions. Typical outsourced functions are listed in Figure 13.3.

FIGURE 13.3
Typical outsourced IT services

- ◆ Application development and software maintenance
- ◆ Hardware purchasing and hardware maintenance
- ◆ Telecommunications installation and maintenance
- ◆ Help desk services
- ◆ Web site design and maintenance
- ◆ Staff training

IT outsourcing contracts are typically signed for long periods of time, usually for 7 to 10 years. The sums of money involved are very large, some reaching billions of dollars. For example, in March 2003 Motorola signed a 10-year, $1.6 billion contract with CSC to handle its IT infrastructure. Until May 1, 2003, when the contract became effective, Motorola handled all of its IT needs in-house. CSC now handles Motorola's worldwide midrange computers, desktop computers, telecommunications, and data centers. IBM signed a 10-year, $2.5 billion outsourcing contract with Deutsche Bank to take care of the bank's IT needs in eight European countries. In January 2003, IBM signed a 7-year, $5 billion contract to satisfy most of JPMorgan Chase's IT needs. In July 2005, Perot Systems took over several IT services of Metaldyne for a period of 10 years. Metaldyne is a designer and supplier of automotive components. Perot Systems provides network management, service desk help, and data center operations. Although the number of IT outsourcing contracts worth $1 billion or more has declined over time, the sums involved are still large. For example, in 2006, Abu Dhabi's Water and Electricity Authority signed a 10-year, $110 million contract with Injazat Data Systems, a joint venture of Mubadala Development Company and EDS; and in 2007, insurance company Allianz of America signed a 7.5-year, $330 million contract with IBM. In August of 2007, the U.S. federal government signed a 10-year outsourcing contract with Accenture for $50 billion. This was the largest single IT contract the U.S. government had ever signed.

There is a peculiar—and paradoxical—aspect to IT outsourcing: while contracts are signed for long periods of time, they typically involve rapidly changing technologies. Vendors often agree to sign outsourcing contracts only if the period is at least five years because of the human resource commitment they have to make, but strategic IT plans—as discussed in Chapter 12, "Systems Planning and Development"—are for only 3–5 years. As a result, clients sometimes find themselves bound by contracts that no longer satisfy their needs. They then try to renegotiate the contract. For example, in July 2001, Tenet Healthcare Corporation extended its outsourcing relationship with Perot Systems before its contract expired. The companies signed the original contract in 1995 for seven years. This time, the operator of 114 acute-care hospitals nationwide wanted the vendor to focus on enhancing the infrastructure and applications for a wide array of Web-based applications to support communication among employees and with insurance companies. It therefore asked to renegotiate the contract before the original termination date of 2002. The new contract was for 10 years and was worth $550 million.

Renegotiation of outsourcing contracts is not unusual. Several companies that signed long-term contracts have found that the financial burden was too heavy or that the expected benefits had not materialized. In April 2004, Sears, Roebuck and Co. signed a 10-year, $1.6 billion outsourcing contract with CSC. Eleven months later, in May 2005, Sears terminated the agreement, claiming that CSC failed to perform some of its obligations. At this writing, the case is being litigated in court.

Making educated decisions on outsourcing IT services has become a major success factor for organizations. In its 2007 report titled "Does IT Matter? Hackett Concludes the Answer is Yes," the consulting firm Hackett Group lists five ingredients for successful use of ISs: standardization of data and consolidation of applications; focusing on high-return opportunities; not minimizing IT costs single-mindedly; maximizing the value of information assets; and outsourcing selectively with a focus on effectiveness. The research results clearly showed that among the 2100 companies it used for the analysis, those that achieved higher efficiency and effectiveness spent 7 percent more per end-user on IT operations than typical companies, but, on average, earned that amount back fivefold in lower operational costs. Good decisions on which IT services to outsource and how to manage the outsourcing contracts were a major contributor to the corporations' success.

POINT OF INTEREST

Outsourcing SOX

After several corporate scandals, the U.S. Congress passed the Sarbanes-Oxley Corporate Governance Act in 2005. Corporations were given ample time to take the necessary steps to comply, some of which involve software that helps to prevent fraud and ensure accurate financial filings. The law, named after its authors, is popularly known as SOX. U.S. corporations spent $6 billion in 2006, and were expected to spend another $6 billion in 2007, to comply with the law. About a quarter of this sum was spent on software development and modification. Indian software companies have benefited tremendously. Apparently, much of the software engineering work is outsourced to companies in India, which specialize in SOX-related software.

Source: Bellman, E., "A Cost of Sarbanes-Oxley: Outsourcing to India," *Wall Street Journal*, July 14, 2005; Sullivan, L., "Compliance Spending To Reach $28 Billion By 2007," *ChannelWeb* (www.crn.com/government).

Advantages of Outsourcing IT Services

Clients contract for IT services to offload in-house responsibility and to better manage risks. When a client outsources, management knows how much the outsourced services will cost; thus, the risk of miscalculation is eliminated. Additional advantages make the contracting option attractive:

- *Improved financial planning*: Outsourcing allows a client to know exactly what the cost of its IS functions will be over the period of the contract, which is usually several years. This allows for better financial planning.

- *Reduced license and maintenance fees*: Professional IS firms often pay discounted prices for CASE (computer-aided software engineering) tools and other resources, based on volume purchases; they can pass these savings on to their clients.

- *Increased attention to core business*: Letting outside experts manage IT frees executives from managing it. They can thus concentrate on the company's core business—including developing and marketing new products.

- *Shorter implementation cycles*: IT vendors can usually complete a new application project in less time than an in-house development team can, thanks to their experience with development projects of similar systems for other clients. (However, they are not likely to use less time if they lack experience with such systems, or if they insist on a waterfall development process rather than an agile method.)

- *Reduction of personnel and fixed costs*: In-house IS salaries and benefits and expensive capital expenditures for items such as CASE tools are paid whether or not the IS staff is productive. IS firms, on the other hand, spread their fixed and overhead costs (office space, furnishings, systems development software, and the like) over many projects and clients, thereby decreasing the expense absorbed by any single client.

- *Increased access to highly qualified know-how*: Outsourcing allows clients to tap into one of the greatest assets of an IT vendor: experience gained through work with many clients in different environments.

- *Availability of ongoing consulting as part of standard support*: Most outsourcing contracts allow client companies to consult the vendor for all types of IT advice, which would otherwise be unavailable (or only available from a highly paid consultant). Such advice might include guidance on how to use a feature of a recently purchased application or on how to move data from one application to another.

As you can see, cost savings is only one reason to outsource IS functions. In fact, studies show that saving money is not the most common reason for outsourcing. Surveys have shown that executives expect several benefits from an outsourcing relationship. Figure 13.4 shows the most cited expectations, such as access to technological skills and industry expertise. To many executives, these anticipated benefits are more important than cost savings, especially in light of reports that in many cases outsourcing did not save the client money.

Risks of Outsourcing IT Services

Despite its popularity, outsourcing is not a panacea and should be considered carefully before it is adopted. In some situations, organizations should avoid outsourcing. The major risks are as follows:

- *Loss of control*: A company that outsources a major part of its IT operations will probably be unable to regain control for a long time. The organization must evaluate the nature of the industry in which it operates. While outsourcing can be a good option in a relatively stable industry, it is highly risky in one that is quickly changing. Although the personnel of an IT service company might have the necessary IS technical skills, they might jeopardize the

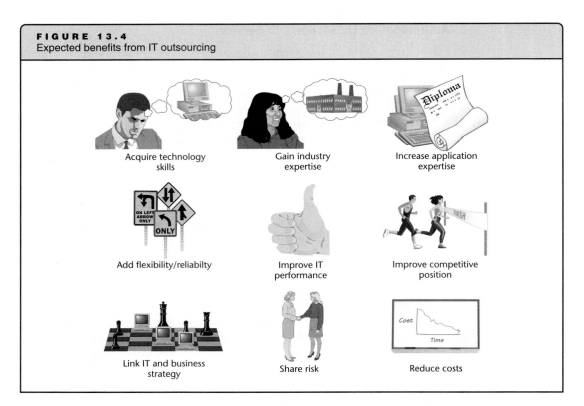

FIGURE 13.4
Expected benefits from IT outsourcing

Acquire technology skills

Gain industry expertise

Increase application expertise

Add flexibility/reliabilty

Improve IT performance

Improve competitive position

Link IT and business strategy

Share risk

Reduce costs

client's core business in the long run if they cannot adapt to constantly changing business realities in the client's industry. Sometimes when this problem becomes clear, the client might have disposed of all personnel who could react properly to such developments. Moreover, even if the client organization still employs qualified IT professionals, the vendor might object to their involvement in activities that, according to the outsourcing contract, are outside their jurisdiction.

- *Loss of experienced employees*: Outsourcing often involves transferring hundreds, or even thousands, of the organization's employees to the IS vendor. For example, as part of the outsourcing contract between Motorola and CSC in 2003, CSC absorbed 1300 of Motorola's IT employees, and when the Wall Street company JPMorgan outsourced its IT functions to IBM in 2003, IBM hired its client's 4000 IT employees. The organization that absorbs the workers can usually employ them with lower overhead expenses than their former employer and use their skills more productively. The client eliminates this overhead cost, but it also gives up well-trained personnel. In addition, if most of the vendor's personnel serving the client are the same employees that the client maintained until the outsourcing contract was signed, the company's ability to gain new expertise from outsourcing could be compromised.

- *Risks of losing a competitive advantage*: Innovative ISs, especially those intended to give their owners a competitive advantage, should not be outsourced. Outsourcing the development of strategic systems is a way of disclosing trade secrets. Confidentiality agreements can reduce, but never completely eliminate, the risk. A competitor might hire the same vendor to build an IS for the same purpose, thereby potentially eliminating the first client's advantage. In addition, assuming that these systems incorporate new business or technical concepts, vendors will bring less than their usual level of experience—and therefore fewer benefits—to the project. Outsourcing strategic or core business ISs incurs more risk than outsourcing the routine tasks of operational ISs (see Figure 13.5).

- *High price*: Despite careful precontract calculations, some companies find out that outsourcing costs them significantly more than they would have spent had they taken care of their own ISs or related services. Several clients have pressured vendors to renegotiate their outsourcing contracts or have found a way to terminate the contract because executives believed they could enjoy the same level of service, or higher-quality service, by maintaining a corporate IT staff. To minimize such unpleasant discoveries, the negotiating team must clearly define every service to be included in the arrangement, including the quality of personnel, service hours, and the scope and quality of services rendered when new hardware and software are adopted or when the client company decides to embark on new ventures, such as e-commerce initiatives or establishment of an intranet.

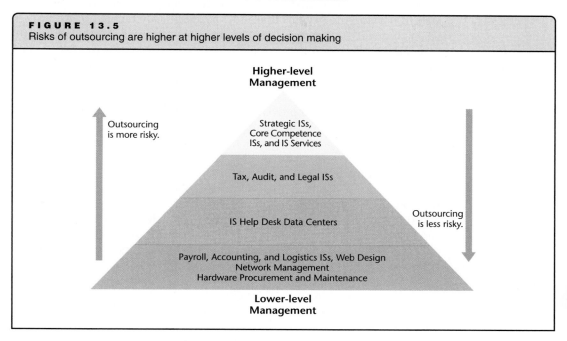

FIGURE 13.5
Risks of outsourcing are higher at higher levels of decision making

Higher-level Management

Outsourcing is more risky.

Strategic ISs, Core Competence ISs, and IS Services

Tax, Audit, and Legal ISs

IS Help Desk Data Centers

Outsourcing is less risky.

Payroll, Accounting, and Logistics ISs, Web Design
Network Management
Hardware Procurement and Maintenance

Lower-level Management

The most important element of an outsourcing agreement for both parties, but mostly for the client, is what professionals call the **service-level agreement**. The negotiators for the client must carefully list all the types of services expected of the vendor as well as the metrics to be used to measure the degree to which the vendor has met the level of promised services. Clients should not expect vendors to list the service level and metrics; the *clients* must do it. It is in the client's interest to have as specific a contract as possible, because any service that is not included in the contract, or is mentioned only in general terms, leaves the door open for the vendor not to render it, or not to render it to a level expected by the client.

LICENSING APPLICATIONS

Businesses can select from a growing list of high-quality packaged software, from office applications that fit on a CD to large enterprise applications. Therefore, purchasing prepackaged software should be the first alternative considered when a company needs to acquire a new system. Recall that "purchased" software is almost always *licensed* software. The purchaser actually purchases a license to use the software, not the software itself. Thus, here the term "licensing" means purchasing a license to use. Unless an IS must be tailored to unique needs in an organization, licensing a prepackaged system might well be the best option.

Ready-made software can be classified into two groups: one is the relatively inexpensive software that helps in the workplace, such as Microsoft Office and similar suites, including software that supports more specific tasks such as project management and tax preparation. Such software usually costs tens of dollars to several hundred dollars for a single user or thousands of dollars for a company with many employees. The other group includes large software applications that support entire organizational functions, such as human resource management and financial management, or enterprise applications that span the entire organization. Such packages include ERP, SCM, and CRM applications and typically cost millions of dollars.

Software Licensing Benefits

When licensing a software package, the buyer gains several benefits: immediate system availability, high quality, low price (license fee), and available support. Immediate availability helps shorten the time from the decision to implement a new system and the actual implementation. If the company maintains an IT staff that develops applications, purchasing software frees the staff to develop the systems that must be specifically tailored to its business needs.

High-quality software is guaranteed through purchase partly because the software company specializes in developing its products and partly because its products would not survive on the market if they were not of high quality. Large developers often distribute prerelease versions, called **beta versions**, or simply betas, of software to be tested by companies (called beta sites) that agree to use the application with actual data for several months. The beta sites then report problems and propose improvements in return for receiving the fully developed software free or for a reduced license fee. By the time the software is released to the general market, it has been well tested.

Because software companies spread product development costs over many units, the price to a single customer is a fraction of what it would cost to develop a similar application in-house or to hire an outside company to develop it. Also, instead of devoting its own personnel to maintain the software, the buyer can usually contract for long-term service and be notified of new, advanced versions of the application. All software development companies provide after-the-sale support. Often, buyers enjoy a period of three months to one year of free service.

Even large companies that could afford to develop ISs on their own often elect to purchase when they can find suitable software. For example, CMS Energy, a $9 billion energy producer in Jackson, Michigan, decided to install a Web-based supply chain management system to link with the company's equipment suppliers. The company's information technologists wanted to build the system themselves. The executive vice president and chief financial and administrative officer, whose professional background is in IT, nixed the idea. He estimated that the cost of a homegrown system—about $20 million—would be greater than the savings it would deliver in its first few years of operation. Instead, he suggested the company use packaged software. The alternative cut the cost of the system in half.

You might be more familiar with off-the-shelf applications than with larger, more complex packaged applications. However, in recent years, enterprise applications have constituted a far larger part of IT expenditures on packaged software. As mentioned earlier, enterprise applications are complex applications that serve many parts of an organization, often several departments. They consist of several modules, each of which can be interfaced with another module from the same vendor.

Organizations cannot simply purchase such large applications and install them; they must employ professionals who specialize in the installation of the software, which might take months. Within limits, the providers of these large applications agree to customize part of the applications to the specific needs of a client. However, such customization is very expensive and is often risky; in some cases, customization has taken significantly longer than planned and was not completed to the full satisfaction of the client.

Software Licensing Risks

Although licensing a ready-made application is attractive, it has its risks:

- *Loose fit between needs and features*: Ready-made software is developed for the widest common denominator of potential user organizations. It might be useful to many, but it will be optimal for few. Companies must take extra care to ensure that ready-made software truly complies with company needs, including organizational culture. Obtaining input from many potential users in the selection process reduces this risk.

- *Difficulties in modifications*: Many companies find that they must have packaged software such as ERP and SCM applications modified to meet their specific needs, and too many of them find that the vendor does a poor job. For example, Nike spent $400 million to have i2 Technologies implement i2's SCM software. Nike claimed that the software did not work properly, causing shortages of high-demand products and overstocks of less popular items. Nike's management said that the software, which was supposed to lower operating costs and streamline communication with suppliers and buyers, failed both in performance and functionality. i2 blamed difficulties on customizing the software and Nike's inappropriate implementation of the software according to i2's suggested methods. Nike found the methods too rigid and did not implement them. Apparently, the mishap reduced Nike's sales in the first quarter after implementation by $100 million. Apparently, "just do it" did not suffice. "Do it right" would have been a better approach.

- *Dissolution of the vendor*: If the vendor goes out of business, the purchaser is left without support, maintenance service, and the opportunity to purchase upgrades to an application to which it is committed. Except for checking the financial strength of potential vendors, there is not much the purchaser can do to reduce this risk.

- *High turnover of vendor personnel*: Turnover among IS professionals is significantly higher than in other occupations. If a substantial number of employees involved in application development and upgrading leave a vendor, support is likely to deteriorate, and upgrades will be of poor quality. Purchasers can do little to reduce this risk.

Steps in Licensing Ready-Made Software

When selecting a particular software package, companies invest a lot of money and make a long-term commitment to conducting their business in a particular manner. Factors such as the complexity of installation, cost of training, and quality and cost of after-sale service must be considered in addition to the demonstrable quality of the software. Once a company decides that it will purchase a ready-made application, a project management team is formed to oversee system implementation and handle all vendor contact. The project management team has the following responsibilities (see Figure 13.6):

- *Identifying the problem or opportunity*: This step is similar to the initial inquiry and fact-finding step in the systems development life cycle (SDLC), discussed in Chapter 12, "Systems Planning and Development". The inquiry results in the identification of gross functional requirements and key integration points with other systems. The report generated often serves as a basis for a request for information from potential vendors.

- *Identifying potential vendors*: On the basis of information in trade journals (printed and on the Web) and previously received promotional material, as well as client references, vendors who offer applications in the domain at hand are identified. In addition to these sources, IS people might gather information at trade shows, from other organizations that have used similar technology, and from colleagues.

- *Soliciting vendor information*: The project manager sends a **request for information (RFI)** to the vendors identified, requesting general, somewhat informal information about the product.

- *Defining system requirements*: The project manager lists a set of functional and technical requirements and identifies the functional and technical capabilities of all vendors, highlighting the items that are common to both lists as well as those that are not. The project management team involves the users in defining system requirements to ensure that the chosen application will integrate well with existing and planned systems.

- *Requesting vendor proposals*: The team prepares a **request for proposal (RFP)**, a document specifying all the system requirements and soliciting a proposal from each vendor contacted. The response should include not only technical requirements but also a detailed description of the implementation process as well as a timetable and budget that can be easily transformed into a contractual agreement. The team should strive to provide enough detail and vision to limit the amount of precontract clarification and negotiation.

- *Reviewing proposals and screening vendors*: The team reviews the proposals and identifies the most qualified vendors. Vendor selection criteria include functionality, architectural fit, price, services, and support.

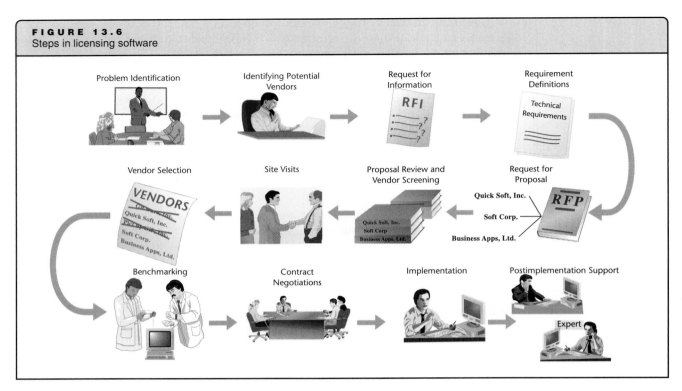

- *Visiting sites*: The complexity of the RFP responses might make evaluation impossible without a visit to a client site where a copy of the application is in use. The team should discuss with other clients the pros and cons of the application.

- *Selecting the vendor*: The team ranks the remaining vendors. The selection factors are weighted, and the vendor with the highest total points is chosen for contract negotiation. Sometimes make-or-break factors are identified early in the process to eliminate vendors that cannot provide the essential service. By now, the team has gathered enough information on the functionality of the various systems.

- *Benchmarking*: Before finalizing the purchasing decision, the system should be tested using **benchmarking**, which is comparing actual performance against specific quantifiable criteria. If all other conditions are the same for all the bidders, the vendor whose application best meets or exceeds the benchmarks is selected.

- *Negotiating a contract*: The contract should clearly define performance expectations and include penalties if requirements are not met. Special attention should be given to the schedule, budget, responsibility for system support, and support response times. Some clients include a clause on keeping the source code in escrow. If the vendor goes out of business, the client will receive the source code, without which the system cannot be maintained. The client should tie all payments to completion of milestones by the vendor and acceptance of deliverables.

- *Implementing the new system*: The new system is introduced in the business units it will serve, and user training is conducted.

- *Managing postimplementation support*: Vendors expect buyers of their large applications to request extensive on-site postimplementation support. Unexpected lapses or unfamiliarity with the system might require fine-tuning, additional training, and modification of the software. It is best to develop an ongoing relationship with the vendor because a solid relationship will foster timely service and support.

When choosing a vendor, organizations look for the quality and reliability of the product, but several additional factors, such as quality of service and support, vendor's support for industry standards, and vendor financial soundness, are extremely important. In surveys, IS managers have almost invariably revealed the importance of factors considered in selecting a vendor, as shown in Figure 13.7 (in descending order). Product quality and reliability stand well ahead of the price/performance ratio.

FIGURE 13.7
How IT managers rank the importance of product purchase factors

FACTOR	RATING
Quality and reliability	_____
Product performance	_____
Quality of after-sale service and support	_____
Trustworthiness of vendor	_____
Price/performance ratio	_____
Ease of doing business with vendor	_____
Vendor's support for industry standards	_____
Openness of future strategies and plans	_____
Vendor financial stability	_____

SOFTWARE AS A SERVICE

In an age when virtually every employee knows how to use a Web browser and access applications online, why should companies install applications at all? Salesforce.com was founded by Mark Benioff and three partners in March 1999 in a one-bedroom apartment in San Francisco. The concept was simple: offer intuitive, easy-to-use customer relationship management (CRM) software on demand via the Internet, and charge customers per use, per month for access to the software. Today, the company has 29,800 customers and 646,000 subscribers, the employees of these customers. Salesforce.com offers many more types of on-demand software than just CRM.

An organization that offers the use of software through communication lines is called an **application service provider (ASP)**. The concept is called **software as a service (SaaS)** or *software on demand*. Salesforce.com, CSC, IBM Global Services, NetSuite, Inc., Oracle Corp., Microsoft Corp., and RightNow Technologies, Inc. are among the better known players in this industry, but many other companies offer such services. According to the IT research firm Gartner, SaaS revenues totaled $6.3 billion in 2006 and were expected to reach more than $19 billion by 2011. The firm estimated that SaaS spending would constitute 25 percent of the total software market by that year.

An ASP does not install any software on a client's computers. Rather, the application is installed at the ASP's location, along with the databases and other files that the application processes for the client. However, clients can choose to save all the files produced by the application on their own local storage devices. The clients' employees access the application through the Web. They call up the application, enter data, process the data, produce reports online and on paper, and in general use the application the same way they would had it been installed at their location. SaaS interface software enables thousands of users from multiple corporate clients to use the same application simultaneously.

ASPs do not necessarily offer their own software packages. They often offer software developed by other companies. For example, USinternetworking (a subsidiary of AT&T) provides on-demand software by Oracle, Microsoft, and Ariba. On-demand service may cost several tens or hundreds of dollars per month per user, depending on the software rented.

POINT OF INTEREST

IBM: Not Only Computers

IBM is the world's largest information technology services company. That's "services," not "computers." For several years the main stream of revenue for the company have been its software development, consulting services, and outsourcing services rather than sales of computers. Consider the company's revenue for the second quarter of 2007 by activity: Global Technology Services (outsourcing and software development): $8.7 billion; Global Business Services (business consulting): $4.34 billion; Systems and Technology (including sales of mainframe computers, servers, and microchips): $5.1 billion. IBM signs outsourcing contracts worth over $40 billion every year.

Source: Gollner, P., "IBM Net Jumps 12 Percent as Revenue Gains on Software," Reuters, July 19, 2007.

As Figure 13.8 shows, renting and using software through the Web has benefits as well as risks. As in any time-limited rental, the client does not have to commit large sums of money up front. No employees have to devote time to learning how to maintain the software, nor to maintaining it once it is installed. No storage hardware is required for the applications and associated data, because the vendor uses its own hardware. And the software is usually available significantly sooner than if installed at the client's location; while it might take years to install and test enterprise applications on-site, an online renter can use the same application within days after signing a contract. And even if an organization is willing to pay for the software, it might not find skilled personnel to install and maintain the software.

FIGURE 13.8
Benefits and risks of Software as a Service (SaaS)

Benefits

- No need to learn to maintain the application

- No need to maintain the application

- No need to allocate hardware for the installation

- No need to hire experts for installation and maintenance

- Timely availability

Risks

- Possible long transaction response time on the Internet

- Security risks, such as interception by competitors

For many small companies this option is clearly the best. Holden Humphrey Co. is a lumber wholesaler in Chicopee, Massachusetts. It has 24 employees. The company's president decided it made no financial sense to hire IT personnel or pay for licensed software. The company pays $1,000 per month to an ASP, which enables nine of the employees to remotely access inventory management, accounting, and CRM applications.

The "software on demand" approach is attracting a growing clientele. Clients are mainly small and medium corporations, but some large organizations also prefer this option. The obvious risk is that the client cedes control of the systems, the application, and possibly its related data to another party. Although some vendors are willing to make minor changes to suit the client's needs, they will not make all requested changes. Some experts argue that by using SaaS, clients have less control over their systems, and that it is better to retain the ability to modify applications in-house. Response time might become a problem as well, because neither the ASP nor the client has full control over traffic on the Internet. Also, as with all activities through a public network, there are security risks, such as interception of information by a competitor.

For this reason, some clients prefer to use a leased line rather than the Internet to connect to the ASP. For instance, Simpson Industries, a manufacturer of auto parts in Plymouth, Michigan, uses an ERP system offered by IBM Global Services. But employees use the application through a leased line (a line that only Simpson can use) to connect directly to IBM's service center in Rochester, New York. When considering using a leased line, IT managers should consider the cost. While a DSL, cable, or optical fiber link costs $30 to $50 per month, a leased line with the same capacity costs $1000 to $2000 per month. Organizations should also consider the type of application and data their company is about to use.

Caveat Emptor

In recent years, faster links to the Internet and a more stable ASP industry have made SaaS an attractive option. However, even with reputable providers, some subscribers were disappointed because the scope of services and level of reliability were not what they had expected when they signed the contract. Managers in organizations considering ASPs should heed the following "commandments":

1. *Check the ASP's history.* Ask the provider for a list of references, and contact these customers to ask about their experience. Ask how soon the provider switched to a new version of the application they rented.
2. *Check the ASP's financial strength.* Request copies of the ASP's financial reports. Ensure that it has enough funds or secured funding to stay in business for the duration of your planned contract.
3. *Ensure you understand the price scheme.* Ask whether the price changes when you decide to switch to another application. Ask whether the price includes help desk services.
4. *Get a list of the provider's infrastructure.* Ask to see a list of the ASP's hardware, software, and telecommunication facilities. Ask the ASP to identify its business partners for hardware, software, and telecommunication services. Ask how data, including sensitive data such as credit-card account numbers, are stored and protected. Ask about security measures.
5. *Craft the service contract carefully.* Ensure that the contract includes penalties the ASP will pay if services are not rendered fully. Ensure that your organization will not have to pay penalties for early termination.

One important point to check when examining the list of facilities is uptime. **Uptime** is the proportion of time that the ASP's systems and communication links are up and running. Since no provider can guarantee 100 percent uptime, ASPs often promise 99.9 percent ("three nines," in professional lingo) uptime, which sounds satisfactory, but it might not be. Three nines mean that downtime might reach 500 minutes per year. This is usually acceptable for customer relationship management systems. Human resource managers or sales representatives, who typically use ISs less than 50 hours per week, might settle even for two nines (99 percent guaranteed uptime). However, experts recommend that organizations look for ASPs that can guarantee five nines—99.999 percent uptime—for critical applications. This high percentage of uptime ensures downtime of no more than five minutes per year. Some firms specialize in monitoring the uptime of ASPs. One such company is Towers Perrin, a management consulting firm that monitors the uptime of 200 Web-based applications.

Who hires the services of ASPs? Although you will find a variety of companies among ASP clients, the majority of clients fall into four categories:

- Companies that are growing fast and rely on software for deployment of their operations.
- Small companies that do not have the cash to pay up front, but who must use office, telecommunications, and basic business operations applications.

- Medium-sized companies that need expensive software, such as enterprise applications, for their operations but cannot afford the immediate payment of large sums (examples are ERP applications from companies such as SAP and Oracle).

- Organizational units at geographical sites where it is difficult to obtain desired software or personnel to install and maintain the software. These sites are typically located far away from a regional headquarters in a less-developed country. The office at that site can then use applications from a more developed country.

Another type of service provider, similar to an ASP, started to catch the attention of businesses in need of IT services: the **storage service provider (SSP)**. An SSP does not rent software applications, but rents storage space. Instead of spending money on the purchase of magnetic disks, a company can contract with an SSP and have all or some of its files stored remotely on the SSP's storage devices. The storage and retrieval are executed through communication lines, in most cases the Internet. SSPs charge by the number of terabytes used per month. Some of the leading SSPs are StorageNetworks, ManagedStorage International (a subsidiary of Incentra Solution, Inc.), and Amazon.com. We discussed Amazon.com's S3 (Simple Storage Service) initiative in the case study "As Vast as the Amazon" in Chapter 2, "Strategic Uses of Information Systems."

USER APPLICATION DEVELOPMENT

If an adequate application is not available on the market, or if an organization does not wish to take the risks discussed earlier with purchasing or renting, and if the application is not too complex, another alternative to software development is available. In **user application development**, nonprogrammer users write their own business applications. Typically, user-developed software is fairly simple and limited in scope; it is unlikely that users could develop complex applications such as ERP systems. If end users do have the necessary skills, they should be management should allow to develop small applications for immediate needs, and when they do, such applications can be maintained by the end users. They should be encouraged to develop applications that will be used for a brief time and then discarded. End users should not develop large or complex applications, applications that interface with other systems, or applications that are vital for the survival of the organization. They should also be discouraged from developing applications that might survive their own tenure in the organization (Figure 13.9).

FIGURE 13.9
Guidelines for end-user development of applications

End users should develop if...	*End users should not develop if...*
End users have the necessary skills	The application is large or complex
The application is small	The application interfaces with other systems
The application is needed immediately	The application is vital for the organization's survival
The application can be maintained by the users	The application will survive the user-developer tenure
The application will be used briefly and discarded	

Managing User-Developed Applications

The proliferation of user-developed applications poses challenges to managers, both in IT units and other business units. In addition to the rules outlined in Figure 13.9, management must cope with the following challenges:

- *Managing the reaction of IT professionals*: IT professionals often react negatively to user development because they perceive it as undermining their own duties and authority. To solve this problem, management must set clear guidelines delineating what types of applications end users may and may not develop.

- *Providing support*: To encourage users to develop applications, IS managers must designate a single technical contact for users. It is difficult to provide IT support for user-developed applications, because the IT staff members are usually unfamiliar with an application developed without their involvement. Yet, IT staff should help solve problems or enhance such applications when end users think their own skills are not adequate.

- *Compatibility*: To ensure compatibility with other applications within an organization, the organization's IT professionals should adopt and supply standard development tools to interested users. Users should not be allowed to use nonstandard tools. Note that compatibility in this context is for the purpose of transferring data among end users; interfacing user-developed applications with other organizational systems should be discouraged.

- *Managing access*: Sometimes, users need to copy data from organizational databases to their own developed spreadsheets or databases. If access to organizational databases is granted for such a purpose, access should be tightly controlled by the IT staff to maintain data integrity and security. Users should be forewarned not to rely on such access when developing their own applications if this is against the organization's policy.

The proliferation of desktop computers and easier-to-use development tools in the workplace has been a major stimulus in application development by users.

© George Doyal/Getty Images

Advantages and Risks

User development of applications offers several important advantages :

- *Shortened lead times*: Users almost always develop applications more quickly than IS personnel, because they are highly motivated (they will benefit from the new system); their systems are usually simpler in design; and they have a head start by being totally familiar with the business domain for which they are developing the application.

- *Good fit to needs*: Nobody knows the users' specific business needs better than the users themselves. Thus, they are apt to develop an application that will satisfy all their needs.

- *Compliance with culture*: User-developed software closely conforms to an individual unit's subculture, which makes the transition to a new system easier for employees.

- *Efficient utilization of resources*: Developing software on computers that are already being used for many other purposes is an efficient use of IT resources.

- *Acquisition of skills*: The more employees there are who know how to develop applications, the greater an organization's skills inventory.

- *Freeing up IS staff time*: User-developers free IS staff to develop and maintain an organization's more complex and sophisticated systems.

However, with all the advantages, application development by users also has some drawbacks. They must be considered seriously. The risks are as follows:

- *Poorly developed applications*: User-developers are not as skilled as IS personnel. On average, the applications they develop are of lower quality than systems developed by professionals. Users are often tempted to develop applications that are too complex for their skills and tools, resulting in systems that are difficult to use and maintain.

- *Islands of information*: An organization that relies on user development runs the risk of creating islands of information and "private" databases not under the control of the organization's IS managers. This lack of control might make it difficult to achieve the benefits of integrated ISs.

- *Duplication*: User-developers often waste resources developing applications that are identical or similar to systems that already exist elsewhere within the organization.

- *Security problems*: Giving end users access to organizational databases for the purpose of creating systems might result in violations of security policies. This risk is especially true in client/server environments. The creation of "private databases" known only to the individual user is risky. The user might not be aware that the information he or she produces from the data is "classified" under an organization's policy.

- *Poor documentation*: Practically speaking, "poor documentation" might be a misnomer. Usually, users do not create any documentation at all because (1) they do not know how to write documentation, and (2) they develop the application on their own to have it ready as soon as possible, and they don't want to take the time to document it. Lack of documentation makes system maintenance difficult at best and impossible at worst. Often, applications are patched together by new users, and pretty soon nobody knows how to iron out bugs or modify programs.

Computer Use Policies for Employees

The increasing numbers of PCs and the pervasive use of e-mail and the Web in businesses have exposed more people to ISs. The U.S. Census Bureau found that over 60 percent of workers' jobs involved computer use for all or part of the workday. This enables workers to be more productive, but computers are often used for unproductive, or even destructive, activities. If an employee uses a company car without permission, the act is obviously wrong. But if an employee uses a company computer to store private files, is this wrong? Accessing a company's intranet is legitimate and encouraged. Accessing another employee's file might be wrong. However, some employees might not be aware of the differences. What are the appropriate personal uses of company computers? Is the answer to this question already covered in existing laws? Should companies have policies that define the appropriate uses of their IT resources? Do we need new laws to ensure a law-abiding workforce? The answers to these questions vary.

- **When There Is No Corporate Policy.** Although unauthorized use of computers might be considered theft, authorities usually do not deal with it as such. Perhaps this is why most state statutes do not specifically address unauthorized use of computers. There is, however, one exception: California law states that an employee might use an employer's computer services for his or her own purpose without permission if no damage is caused and if the value of supplies and computer services does not exceed $100.

  If someone from outside a company accessed the company's computer without authorization and used it for any purpose whatsoever, the act would clearly be criminal under the laws of many countries and of every state in the United States. However, if an *employee* uses the same company computer after hours to prepare a homework assignment for a college class, the act might not be considered unethical, let alone criminal, unless the organization has a clear policy against such activity. What about creating a résumé or writing a letter as part of a job search? Without a company policy, the answer to this question is not clear.

  Widespread access to the Web makes the issues even more complicated. Employees have been fired for surfing the Web for their own personal purposes during work time. Some have been fired for surfing the Web during lunch breaks or after work hours; while they did not waste company-paid time, management objected to the specific sites they accessed, mostly those displaying pornographic images.

- **Company Policies Work.** To avoid misunderstanding, employers should provide clear guidelines, stating that any computer use not for the company's direct benefit, without the prior approval of the company, is forbidden. One simple measure that some organizations have taken is to have a written policy that is conspicuously posted, signed by employees upon hiring, or both. The notice could read as follows:

  "Company policy forbids any employee, without prior authorization of the employee's supervisor, to (a) access or use any equipment or data unless such access is work-related and required to fulfill that employee's duties, or (b) alter, damage, or destroy any company computer resource or property, including any computer equipment, system, terminal, network, software, data, or documentation, including individual employee computer files. Any such act by an employee might result in civil and criminal liability under federal, state, and local laws."

  Many companies do not object to recreational or educational use of their computers by employees outside of company time. If this is the case, the policy should say so. Without a policy, companies should not be surprised when their employees' interpretation of reasonable personal use differs from their employers'. However, if there is no clear policy, employees should always remember that a PC is a work tool that their employer put at their disposal for responsible use as part of their job. It is not there to help their own business, shop on the Web, or entertain them either during or outside of paid time. Thus, for example, they should not use e-mail or instant messaging to chat with their friends or browse the Web for their enjoyment. Yet, is sending a personal e-mail message during lunch break really much different from using a company pen to write a personal note during lunch break? Perhaps the best way to avoid misunderstanding is to simply ask your employer if what you intend to do is objectionable.

  To enforce their policies—explicit or implicit—many companies resort to means that are, in some minds, questionable. They use surreptitious surveillance software to monitor employee use of IT. Consider BeAware, an application offered by Ascentive LLC. The software tracks all employee PC activity with live, real-time monitoring of e-mails, Web surfing, chats, and program usage, recording screen shots, time used, and content.

Everything that you do with your computer can, at the same time, be tracked and recorded, and everything that you see on your monitor is also seen on the monitor of the person who tracks you. The same software is often used to monitor children's Web surfing activities and spouses suspected of infidelity. The question is, should employees be treated as undisciplined children or cheating spouses? An equally fair question is, should employers not have the right to monitor what's done with their equipment, especially on paid time?

- The alternatives to having applications developed in-house are outsourcing, licensing ready-made software, using software as a service, and user application development.

- Outsourcing has two meanings in IT: commissioning the development of a tailored application to an IT company, and assigning all or some of the IT services of the organization to a vendor of IT services.

- Outsourcing custom-designed applications might afford the organization good fit of the software to need, good fit to culture, dedicated maintenance, smooth interface, specialized security, and potential for strategic advantage.

- The potential advantages of outsourcing IT services include improving cost clarity and reducing license and maintenance fees, freeing the client to concentrate on its core businesses, shortening the time needed to implement new technologies, reducing personnel and fixed costs, gaining access to highly qualified know-how, and receiving ongoing consulting as part of standard support. However, outsourcing IT services has some potential risks: loss of control, loss of experienced employees, loss of competitive advantage, and high price. To ensure that the client enjoys all the expected services and their quality, a detailed service-level agreement must be signed with the IT service vendor.

- When an organization purchases a license to use ready-made software, it enjoys high-quality software that is immediately available at low price (license fee). However, licensed ready-made software has some potential risks: loose fit between needs and the software features, difficulties in modifications, bankruptcy of the vendor, and high turnover of the vendor's employees.

- Using software as a service has become popular. The client pays monthly fees based on the type of application used and the number of users, and its employees use the applications via a network, mostly through the Internet. ASP clients enjoy availability of applications, avoid the costs of storage hardware and large IT staffs, and do not have to make a long-term commitment of capital to software that might become obsolete in two or three years. The downsides of using an ASP are the loss of control over applications, the potentially low speed of interaction, and the security risks associated with using an IS via a public network.

- The advantages to user application development include a short lead time, good fit of application capabilities to business needs, good compliance with organizational culture, efficient utilization of computing resources, acquisition of skills by users, and the freeing of IS staff to deal with the more complex challenges of the systems. Disadvantages of user-developed applications include the risk of poorly developed applications, undesirable islands of information and private databases, duplications of effort, security problems, and poor documentation. Thus, user development of applications needs to be managed. IS managers need to determine the applications that users should and should not develop and dictate the tools that should be used.

- Well over half of America's office workers now have rich computer resources at the tips of their fingers. Often, employees do not know which activities are welcomed and which are not. If an organization lacks a clear policy, employees are not discouraged from abusing computers. This abuse is especially true when employees access Web sites that are objectionable to their employer or when employees use e-mail for purposes not intended by the employer. If no policy has been established, the simple rule is that employees should not use their computers for anything but work.

WORLDWIDE HOST REVISITED

Worldwide Host has developed a solid partnership with General Data Systems, using the software firm to develop new systems or upgrade existing systems. Take a closer look at some of the relationships firms establish in today's marketplace—the ways they are established and their advantages and disadvantages.

What Would You Do?

1. Worldwide Host has outsourced its software development to General Data Systems. What are some advantages that the hotel chain receives from this arrangement? What are some possible risks? Make a list of the pros and cons.

2. In the opening case, Michael Lloyd seems concerned about the qualifications of Reservations Technologies, the firm from which Worldwide plans to license the reservations system. Prepare a set of questions for him to ask both the company and its clients.

New Perspectives

1. The chapter mentions software as a service (SaaS) as an option to acquire software. Do some research about SaaS and application service providers (ASPs) to see whether any of them offer systems such as the ones Worldwide Host needs to use. Could Worldwide Host use an ASP for any of its needs? If so, which specific needs could be satisfied?

2. Several years in the future, Worldwide Host replaces its reservations system with one customized for it by a software development firm. A competitor approaches Worldwide's management and offers to pay an attractive annual fee for a license to use the system. Should management agree? Why or why not?

KEY TERMS

application service provider (ASP), 457
benchmarking, 456
beta versions, 454
custom-designed (tailored) software, 446

offshoring, 448
outsourcing, 446
request for information (RFI), 455
request for proposal (RFP), 455
service-level agreement, 453

software as a service (SaaS), 457
storage service provider (SSP), 460
uptime, 459
user application development, 460

REVIEW QUESTIONS

1. List and explain all the various options now available for an organization to enjoy the services of an IS.

2. Few organizations would develop an application in-house or pay another company to develop it if a similar application can be licensed. Why?

3. What are the benefits and risks of outsourcing IT services?

4. The major hardware and software makers, such as IBM and Hewlett-Packard, derive an increasing portion of their revenue from outsourcing contracts. Analyze and explain why they focus more of their efforts in this direction.

5. What might cause a client to ask to renegotiate a long-term outsourcing contract?

6. You are the CIO of a large manufacturing company. A software vendor approaches you with an offer to have your company serve as a beta site for a new human resource application. What would you consider before making a decision?

7. What is an RFI? What is the difference between an RFI and an RFP? The ideal response to an RFP is one that can be easily transformed into a contract. Why?

8. What is the purpose of benchmarking? Often, benchmarking involves visiting other organizations that have applied the system under consideration. Why?

9. What would you benchmark in a system whose purpose is to enter customer orders and accept customer credit-card account numbers for payment at your Web-based site?

10. When purchasing an off-the-shelf application, to which phase of the SDLC is the postimplementation support and service equivalent?

11. Some organizations charge the purchase price of an application that serves only a particular organizational unit back to the unit. Why does the existence of a charge-back arrangement create an incentive to have users develop their own applications?

12. Why don't users commonly document the applications they develop? Why is poor documentation a problem?

13. List and explain the benefits and risks of using the services of an on-demand software provider.

14. Some companies use software as a service (SaaS) because they want to concentrate on core competencies. What is a core competency?

15. What is a storage service provider (SSP)? How is it different from an ASP?

DISCUSSION QUESTIONS

16. Some outsourcing clients have devised contracts that incentivize their vendors to develop new, innovative ISs for the client. What elements would you include in a contract like this if you were a manager for an outsourcing (client) company?

17. Vendors like to market themselves as "partners" with their outsourcing clients. Why?

18. Do you think that development of ISs by end users should be encouraged? Do the benefits of the practice outweigh its risks?

19. Will ready-made software applications ever meet all the needs of all businesses? Explain.

20. The volume of the software as a service (SaaS) market is growing and predicted to reach one-fourth of the software market. In addition to the benefits listed in the chapter, what are the technological developments that cause this growth?

21. One of Salesforce.com's slogans is *The End of Software*. Do you agree that the term describes the concept of SaaS? Why or why not? Do you think that SaaS will eventually become the only way in which corporations use software? Why or why not?

22. A CIO said that while he would not use a public network such as the Internet with an ASP for some types of ISs, he would allow employees to use the Web for other types, such as an accounting application. Give three examples of applications that you would recommend and three that you would not recommend be used through the Web. Explain your choices.

23. Except for Salesforce.com, the industry leaders in the SaaS market are companies that have been software leaders before the concept emerged. What do you think is the reason for this?

24. Explain why you agree or disagree with the following statement: "Employees are smart enough to know what they should and should not do with their computers. A conduct policy will not prevent wrongdoing."

25. When using the services of an ASP, the client gains an important element that is often overlooked: support service. Considering only support service, would you prefer to have the rented software installed on your own company's hardware or leave it installed at the vendor's site? Explain.

26. Should employees be allowed to use their employers' e-mail for private communication at all? Should they be allowed to do so outside of paid time?

27. Assume that you are the CEO of a company that provides computers and access to e-mail and the Web to almost all of its employees. You are about to circulate a new IT use policy in the company. List and explain your "ten commandments" (or fewer, or more) for employees' use of software, e-mail, and the Web.

APPLYING CONCEPTS

28. You are a manager for a new company that is about to start selling textbooks to college bookstores via the Web. Several firms specialize in software that supports transactions and data collection on the Web. Prepare an RFI for an application to support your new company's effort on the Web, including posted catalogs, orders, shipment tracking, payment, and data collection for future marketing.

    Submit the list of questions you want prospective bidders to answer, and be ready to provide an explanation for including each of the questions.

29. A small company considering three storage service providers (SSPs) has asked for your advice. Research the SSPs mentioned in this chapter, and make your recommendations to the company. Consider cost, flexibility, length of client commitment, communication speed, security, reputation, and any other relevant information. Base your recommendation on your analysis of all the relevant strengths and weaknesses of each provider.

HANDS-ON ACTIVITIES

30. In recent years several companies experienced either a total failure or major mishaps when trying to have vendors implement enterprise applications. Consider "failure" as inability to complete the project or a project that ended up costing the client significantly more than expected, including lost revenue. Find at least three sources about such a case. Synthesize, list, and explain what happened and why. Conclude with your own recommendations of what could be done to avoid or minimize the damage. The recommendations should be written so that potential clients of such projects could take proper precautions. Your report should be about 1,500 words long.

31. In recent years a growing amount of software has been outsourced by U.S. companies to other countries, such as India and China. Research the Web and write a two-page paper that lists and explains the benefits and the disadvantages of offshoring application development. If you conclude that there are some benefits or advantages to offshoring of specific types of applications, say so and explain your conclusion.

32. Every online retailer (e-tailer) uses a virtual shopping cart application. Team up with another student, research the Web for companies that sell such applications, and write a 1,500 to 2,000-word report that summarizes the following points: (a) Who are the major companies that sell these applications? (b) What are the prices of these packages? (c) How long does it take to install the applications? (d) How is the relationship with the bank processing the credit-card payments established? (e) Are there ASPs that offer the use of such systems over the Internet? (Namely, the e-tailer uses a shopping cart application installed at the ASP's location.) (f) If there are ASPs that rent such systems, what are the payment schemes? *Note*: Many Web-hosting companies offer shopping carts as part of their packages. You may use these companies, but if you do, report whether the shopping cart service must be tied to the purchasing of other services.

33. Throughout this book, many organizations that provide information and advice on IT have been mentioned. Explore the Web sites of these organizations and the Web sites of IT-related magazines to find the latest statistics on the different alternatives for obtaining business applications or the use of business applications. Create a PowerPoint presentation to answer the following questions, and express your answers in pie charts:
    a. What was the dollar amount spent on IT in your country in each of the past three years, and what percentage of this amount was spent on software acquisition?
    b. How was the amount spent on software distributed among in-house development, purchased (or licensed) ready-made software, and outsourced development?

FROM IDEAS TO APPLICATION: REAL CASES

Wawa Flying High

Like so many executives, Wawa's CIO Neil McCarthy wanted his organization to have an enterprise application that helps make effective and timely decisions. He wanted all departments to have access to pertinent data in the same format, so they would not erroneously perceive different information from the same data, as often happened. But Wawa did not have the software tools to ensure that until 2004.

Headquartered in Pennsylvania, Wawa was once just a dairy. After some years, the privately held firm established convenience stores in which it sold its dairy products. The variety of items sold kept growing. Wawa's coffee became famous ("over 125 million cups sold every day!"). The chain grew from one location in Pennsylvania to 550 stores in Pennsylvania, Delaware, New Jersey, Maryland, and Virginia. The stores are operated by 13,000 "associates." In the early 2000s, Wawa started opening its own gas stations, and, of course, each gas station has a Wawa store. The company's logo, a Canada goose (Wawa, as the Native Americans called it), appears on more convenience stores as the chain grows. Wawa's 2006 revenue was $3.9 billion.

McCarthy realized that unit managers could not slice and dice a given set of data in the same way, because they were using different applications. Business units saw data in different formats, which often led to decisions based on different information. McCarthy wanted the entire organization to have a unified view of any data, and therefore the same information for decision making, whether the information was produced by the marketing, accounting, or real-estate department. McCarthy wanted to give all units access to information that would enable managers to decide which products to continue to carry and which to discontinue as well as which prices to increase or decrease depending on demand trends. He needed a good retail enterprise software package.

To meet the new goal, Wawa decided to implement an ERP system. Management approached SAP and Retek. At the time, SAP and Oracle were engaged in a fierce bid to acquire Retek, a company that developed and sold an ERP and business intelligence software package called ProfitLogic. Eventually, Oracle won the bid and purchased Retek. Unrelated to Oracle's new acquisition, Wawa selected SAP as its supplier of ERP software.

Wawa believed that SAP's upgrades would be less complicated than Retek's. McCarthy's opinion was that with Retek, every upgrade is like starting implementation of the software from scratch, and that Retek approached each upgrade as if it were a customized installation. McCarthy found that SAP's "out of the box" software met all of Wawa's initial needs, so very little customization would be required, if at all. Unlike some retailers who carry hundreds of thousands of different items, Wawa sells only 5,000. Thus, the IT staff believed that Wawa could pretty well "grow with the system," namely scale and modify the software relatively easily if the need arose.

Wawa also uses Oracle software. Shouldn't that have tipped the choice toward Oracle? McCarthy did not think so. He cited the company's use of Microsoft operating systems, but this did not mean that the company should also adopt other Microsoft products, such as SQL Server.

The company agreed to pay SAP between $5 million and $9 million for the license. (Wawa disclosed only the amount's range.) McCarthy hired a consulting firm, The Lakewest Group, to prepare a business case, including an ROI (return on investment) calculation. He presented it to his fellow corporate executives. They liked the business case for the system.

The IT department did not rush to begin the system's implementation. The plan was to deploy the modules in phases over four to five years. First, the financial modules were to be deployed. The human resource and data warehousing modules would be last to be implemented. The long deployment was one reason why McCarthy preferred the SAP system, which would not require much modification and upgrading. His 18-year experience taught him that such upgrades are a distraction. Since the SAP applications were not expected to require radical upgrades, he believed that the SAP system would result in less disruptive modifications to the software.

Apparently, Retek's software is richer in features, and this is why it needs to be more customized for each client. However, when customized, it fits business needs better than SAP's "plain vanilla" retail software package. Joe Polonski, Retek's vice president of enterprise strategy, was disappointed to lose Wawa's business. He said that Wawa's decision to adopt software that required fewer upgrades made business sense if the company settled for what SAP offered and was inclined to change business processes to fit the software, rather than the other way around.

Sources: Schuman, E., "Wawa CIO: Upgrade Fear Dictated Multimillion-Dollar SAP Purchase," *CIO Insight* (www.cioinsight.com), May 20, 2005; (www.wawa.com), 2007.

Thinking About the Case

1. McCarthy could probably prepare an ROI calculation with his own staff, or with other staff from Wawa. How could hiring an external consulting firm help his case convincing management to acquire the system?

2. What is the risk in frequently modifying licensed software?

3. If you were Wawa's CIO, which alternative would you prefer: software with more features that fit more business needs but requires more modifications, or software that provides fewer features and requires fewer modifications?

Outsourcing Vendor, SaaS Client

Optimizing the allocation of resources for IT services sometimes creates unexpected situations. This is a story of a leading vendor of IT outsourcing that decided to outsource one of its own IT functions. More accurately, it relies on software as a service.

Siemens AG is a German-based global corporation and one of the world's largest firms. It specializes in electronics and engineering. It employs 475,000 people in 190 countries, and reported revenue of $107.4 billion in 2006. One of its subsidiaries, Siemens IT Solutions and Services, Inc. (formerly Siemens Business Services, Inc.), is headquartered in Norwalk, Connecticut. It employs more than 43,000 people worldwide and reported revenue in 2006 of $6.4 billion. The company provides IT-related services, from consulting to systems integration to software development and management of IT infrastructures.

Many companies, including some large U.S. corporations, outsource IT services to Siemens IT Solutions and Services. The Siemens division manages their data centers, call centers, and other functions. To provide the services, the company employs more than 700 servers and 2800 PCs. The large databases that the company manages for its clients have increasingly attracted criminals who try to break into them and steal data. Many of the attacks have been on applications, rather than the databases themselves.

Management could not afford to have clients' data compromised. The company was using a set of vulnerability scanning tools licensed from the parent company, Siemens, to protect against such attacks. However, the chief security officer (CSO) realized that this was not an adequate solution. The software did not provide answers to all the types of attacks. It had to be updated frequently. The small staff had to constantly test information systems for vulnerabilities and patch security holes—they had to find or develop and then install the proper software to eliminate the vulnerabilities. His staff was too small to cope with the growing challenges.

The CSO decided to explore solutions outside the company. He looked for a security testing service that could be put to use immediately. iQwest Technologies, a company specializing in security, helped him and his team select a SaaS vendor. He selected Qualys, a company based in Redwood Shores, California. Qualys specializes in providing data and application security services on a SaaS basis. Siemens IT Solutions and Services agreed to try the service for six months before it made a long-term commitment.

To ensure seamless operations between Qualys's and the company's applications, the CSO hired iQwest Technology. iQwest helped with installation of Qualys's appliances and Web-based management controls. The appliances (hardware) and Web-management controls provide a seamless link for Siemens Business Service employees to use Qualys applications over the Internet.

In the past, staff members had to physically go to every location where equipment was installed to conduct testing or patch security holes. Now, Qualys does the work remotely through communication lines. The CSO's staff can set up the company's security appliances—hardware to block unauthorized access—anywhere in the world and start scanning new operations without sending anyone to the field to manually configure new servers or PCs.

Qualys took over the security responsibility for the 700 servers and 2800 PCs. The CSO believed that Siemens IT Solutions and Services handed over a system that had zero security holes. Yet, Qualys's testing applications found 70 problems of which he and his staff were not aware. After six months, the company decided to continue with Qualys.

Since Qualys took over the IT security work, Siemens IT Solutions and Services business grew, both in the number of clients and in business requested by existing clients. The CSO sees a direct connection between the two developments. Apparently, using Qualys as a SaaS provider was timed well, because clients have increased their demand for security. In the past, clients used to require a meeting or two per year and posed only a few questions about security measures. Now, they often show up with a 500-item questionnaire. They are highly inquisitive about how their data is protected.

Qualys' CEO said that an increasing number of businesses use his company's services, as well as those of competitors. One reason for this, he said, is that companies can offload much of the burden of developing and using their own security testing applications, which results in lower cost. The other reason is the enforcement of increasingly stricter security standards by government.

Siemens IT Solutions and Services accomplished its goals. It can assure clients that their data is well protected, and it protects the data without increasing staff or spending money on development and constant updating of software. It relies on an expert in this field.

Source: Hines, M., "Security at Your Service," *eWeek.com*, January 22, 2007; (www.it-solutions.usa.siemens.com), August 2007.

Thinking About the Case

1. What are the reasons the Siemens division turned to another company for using security testing software? Why is this a better solution than using in-house software and staff?

2. Siemens IT Solutions and Services is itself a vendor of IT outsourcing. Do you think that its use of SaaS or outsourcing some of its IT functions might hurt the company's reputation as an expert IT outsourcing vendor?

3. From the details provided in the case, do you think this is a SaaS arrangement or an outsourcing arrangement with Qualys? Is this a combination of the two?

4. Siemens IT Solutions and Services used a consultant—iQwest Technology—to select the SaaS vendor and to ensure seamless links between its own applications and the vendor's. What are the benefits of using a consultant for this purpose?

14

FOURTEEN

Risks, Security, and Disaster Recovery

LEARNING OBJECTIVES

As the use of computer-based information systems has spread, so has the threat to the integrity of data and the reliability of information. Organizations must deal seriously with the risks of both natural and human menaces. A computer expert once noted: "The only truly secure system is powered off, cast in a block of concrete, and sealed in a lead room with armed guards. And even then I have my doubts." Indeed, there is no way to fully secure an information system against every potential mishap, but there are ways to significantly reduce risks and recover losses.

When you finish this chapter, you will be able to:

- Describe the primary goals of information security.

- Enumerate the main types of risks to information systems.

- List the various types of attacks on networked systems.

- Describe the types of controls required to ensure the integrity of data entry and processing and uninterrupted e-commerce.

- Describe the various kinds of security measures that can be implemented to protect data and ISs.

- Improve the security of your personal information system and the information it stores.

- Recognize online scams.

- Outline the principles of developing a recovery plan.

- Explain the economic aspects of information security.

WORLDWIDE HOST:
Battling Back from Attacks

Worldwide Host's TripExpert.com site was fulfilling its promise. The site had been up for 10 months. CIO Michael Lloyd thought back to the end of the Web site's development project, when he had congratulated the General Data Systems staff for their hard work in getting the site up and running. At the time, he did not pay much attention to the fact that the system was linked to a public network. Soon, he realized that such a link requires special consideration. The first event to draw his attention was a prank.

Web Site Defaced

Michael's Webmaster, Susan O'Donnell, rushed into his office one day to report that someone had defaced the TripExpert home page. Since the site was so important to Worldwide, she checked it daily to ensure that it was running smoothly. She was shocked at what she saw on the site when she logged on in the morning. Someone had placed offensive images and language on the page, so she immediately took the site offline.

"It'll probably take us until early afternoon to get the site up again," she said. "Fixing the home page takes only a few minutes, but ensuring that this doesn't happen again will take longer. We're working on it right now as our top priority."

Michael responded, "I'm coming down now to see for myself. We get transactions of about $90,000 an hour from that site. We've got to get it back online as soon as possible."

Michael and Susan hurried down to the information center, waved their badges in front of the radio ID scanner, entered their codes, and opened the door. Inside, two IS staff members were working to clear the Web page of the intruder's messages. All four worked furiously to find the hole that allowed the intruder to deface the home page. They quickly replaced the page with a backup copy. After several hours of hard work, they patched the server software with code that they believed would eliminate the security hole and brought the site back online.

Attacks Continue

Five weeks after the defacement incident, Worldwide Host received a second blow: its site was among several that were hit with a denial-of-service attack. Requests were swamping Worldwide's servers. Michael and his IS security chief decided to disconnect the servers from the Net. Michael knew this meant a loss of profit of thousands of dollars, but there was nothing else he could do.

They turned the site on after an hour. Apparently, the attack had subsided, but Michael was worried that the attacker had also tried to damage databases or steal information.

"What is the extent of the problem, Jason? Did they breach any of our internal systems?"

"Doesn't look like it, but they were trying to get into our database. The secure servers that we use for transactions withstood the attempt to penetrate the customer database."

Michael sighed with relief. "Our security firm said that secure servers were critical to keeping our transactions private. Our system depends on the safety of our customers' information."

"Well, the security people were right. I'll keep running my diagnostics on the damaged software," said Jason. "I'll probably have to reformat the hard drives, reinstall the operating systems, and then the applications. That'll take some time."

"Susan is preparing a statement to post on the Web site to reassure our customers. We need to get that up fast. What else can we do to prevent denial-of-service attacks?"

"Since traffic load is the critical issue, we could add more servers to handle both the attacks and legitimate transactions. But the cost of those extra servers might be high."

"Let's look into it. We need to keep our site online."

GOALS OF INFORMATION SECURITY

In March 2006, a database of Florida International University was invaded by unauthorized visitors. The database contained the records of thousands of students and applicants. In May 2007, hackers accessed the names and Social Security numbers of 22,396 students at the University of Missouri. Two weeks later, the records of 4000 students and applicants to Northwestern University discovered that their records were posted online by a hacker who copied them from a university database. A few hours earlier, the names and Social Security numbers of all 64,000 Ohio state employees were stolen from a state agency intern who left a storage device containing a backup copy of the records in his car. The device also contained the personal details of 53,797 participants in the state's pharmacy benefits management program, and the Social Security numbers of over 75,500 dependents of these people. In July 2007, Fidelity National Information Services revealed that the personal information of more than 2.3 million people had been stolen from its database by a disgruntled employee. This is a small sample of what may happen to information systems and the data stored on them.

As you have already seen, the development, implementation, and maintenance of ISs constitute a large and growing part of the cost of doing business; protecting these resources is a primary concern. The increasing reliance on ISs—combined with their connection to the outside world through a public network, the Internet—makes securing corporate ISs increasingly challenging. Adding to the challenges is the increased tendency to store data on portable devices. The role of computer controls and security is to protect systems against accidental mishaps and intentional theft and corruption of data and applications. They also help organizations ensure that their IT operations comply with the law and with expectations of employees and customers for privacy. The major goals of information security are to:

- Reduce the risk of systems and organizations ceasing operations.
- Maintain information confidentiality.
- Ensure the integrity and reliability of data resources.
- Ensure the uninterrupted availability of data resources and online operations.
- Ensure compliance with policies and laws regarding security and privacy.

To plan measures to support these goals, organizations first must be aware of the possible risks to their information resources, which include hardware, applications, data, and networks; then, they must execute security measures to defend against those risks.

In recent years, the U.S. Congress passed several laws that set standards for the protection of patient, student, and customer privacy and compliance with corporate internal controls. They include the Health Insurance Portability and Accountability Act (HIPAA) and Sarbanes-Oxley Act (SOX). These laws have an important effect on securing information and, therefore, on securing information systems. Other countries have similar laws that have similar implications for information security. However, corporate concern should not be focused only on complying with the law. It should ensure that information resources are secure in order to minimize situations that might practically take them out of business.

Understand Risks, Security, and Disaster Recovery Planning

Some time ago you started working for a small company. You love your new job. It's Monday 10:00 a.m., and the IS you use is down. Apparently, there's a blackout, and in a freak accident the backup generators don't work. Do you know what *you* are supposed to do? As explained and demonstrated throughout this book, information is the life-blood of any modern organization. Practically every aspect of business depends on the currency of processed data and the timely provision of information. This fluent process can be achieved only if information systems are protected against threats. As a professional, you must be aware of what might happen to the ISs upon which you and your colleagues or subordinates depend. You must protect the systems against events that threaten their operation and make it impossible to carry out critical business activities. When a new system is developed, you should ask the developers to provide a system that not only supports the functions of your business unit but also incorporates controls that will minimize any potential system compromises. You also need to be prepared for a disaster, and should know how to implement your part of the business recovery plan to help restore operations as soon as possible.

RISKS TO INFORMATION SYSTEMS

On a Monday evening in July 2007, Netflix's online rental system went down and remained unavailable until Tuesday afternoon. The popular video rental company refused to disclose the reasons for the failure, but its customers were not happy for being locked out of service for 18 hours. In recent years, especially because of the growth of online business, corporations have considered protection of their IS resources an increasingly important issue, for good reasons. **Downtime**, the time during which ISs or data are not available in the course of conducting business, has become a dreaded situation for almost every business worldwide. By some estimates, U.S. businesses lose $4 billion annually because of downtime. An online airline reservation business can lose $90,000 per hour of downtime; an online retail business loses an average of $900,000; a credit-card company could lose $2.6 million; and an online brokerage house might lose up to $6.5 million per hour of downtime. The estimates by type of application for all industries are also mind-boggling. According to the Standish Group, the cost per minute of CRM applications not being available to a company is typically about $2,500. E-commerce applications typically have a downtime cost per minute of about $7,800. The costs are averages and depend on industry, the size of the company, and other factors. These costs may be much larger in many cases. The following section discusses the most pervasive risks to IS operations. In addition to the threats we discuss here, it is important to remember that terrorism has posed a serious threat to all aspects of ISs: hardware, software, and data.

Risks to Hardware

While stories about damage to ISs by malicious Internet attacks grab headlines, the truth about risks to ISs is simply this: the number one cause of systems downtime is hardware failure. Risks to hardware involve physical damage to computers, peripheral equipment, and communications media. The major causes of such damage are natural disasters, blackouts and brownouts, and vandalism.

Natural Disasters
Natural disasters that pose a risk to ISs include fires, floods, earthquakes, hurricanes, tornadoes, and lightning, which can destroy hardware, software, or both and cause total or partial paralysis of systems or communications lines. Floodwater can ruin storage media and cause short circuits

that burn delicate components such as microchips. Floods, in this context, include those occurring as a result of plumbing problems inside buildings. Lightning and voltage surges cause tiny wires to melt and destroy circuitry. In addition, wildlife and human error occasionally destroy communications lines; animals gnaw cables, and farmers and construction workers occasionally cut cables inadvertently.

Blackouts and Brownouts

Computers run on electricity. If power is disrupted, the computer and its peripheral devices cannot function, and the change in power supply can be very damaging to computer processes and storage. **Blackouts** are total losses of electrical power. In **brownouts**, the voltage of the power decreases, or very short interruptions occur in the flow of power. Power failure might not only disrupt operations, but it can also cause irreparable damage to hardware. Occasional surges in voltage are equally harmful, because their impact on equipment is similar to that of lightning.

The popular way of handling brownouts is to connect a voltage regulator between computers and the electric network. A voltage regulator boosts or decreases voltage to smooth out drops or surges and maintains voltage within an acceptable tolerance.

To ensure against interruptions in power supply, organizations use **uninterruptible power supply (UPS)** systems, which provide an alternative power supply for a short time, as soon as a power network fails. The only practical measure against prolonged blackouts in a public electrical network is to maintain an alternative source of power, such as a generator that uses diesel or another fuel. Once the main power stops, backup batteries provide power until the generator starts, gets up to speed, and produces the power needed for the computer system.

UPS units are a solution to extreme changes in voltage and can provide several minutes to several hours of backup battery power.

Courtesy American Power Conversion Corp. (APC)

Vandalism

Vandalism occurs when human beings deliberately destroy computer systems. Bitter customers might damage ATMs, or disgruntled employees might destroy computer equipment out of fear that it will eliminate their jobs or simply to get even with their superiors. It is difficult to defend computers against vandalism. ATMs and other equipment that are accessible to the public are often encased in metal boxes, but someone with persistence can still cause severe damage. In the workplace, the best measure against vandalism is to allow access only to those who have a real need for the system. Sensitive equipment, such as servers, should be locked in a special room. Such rooms usually are well equipped with fire-suppression systems and are air-conditioned, and thus protect also against environmental risks.

Risks to Data and Applications

The primary concern of any organization should be its data, because it is often a unique resource. Data collected over time can almost never be recollected the same way, and even when it can, the process would be too expensive and time consuming. The concern for applications, especially if the applications are not tailor-made, should come second. All data and applications are susceptible to disruption, damage, and theft. While the culprit in the destruction of hardware is often a natural disaster or power spike, the culprit in damage to software is almost always human.

Theft of Information and Identity Theft

Sometimes the negligence of corporations and the careless use of technology, especially on public links to the Internet, create security "holes" or vulnerabilities. In one case, a young man installed a program called Invisible KeyLogger Stealth on public-use computers in 14 Kinko's stores, where customers can access the Internet. (Such Internet-connected PCs are also available in public libraries and airports.) **Keystroke logging**—or simply keylogging—software records individual keystrokes. For one year, his software secretly recorded more than 450 usernames and passwords, which he used to access existing bank accounts and create new ones. He was caught when he used an application called GoToMyPC. Subscribers to the GoToMyPC service can use an

Natural disasters are a serious threat to hardware.

application by the same name to link to a PC from another PC and fully control the remote one as if they were sitting in front of it. Using the application, he remotely accessed and used one of his victims' PCs. Using the PC at home, this person noticed that the cursor was moving "by itself." The cursor opened files and subscribed to an online payment transfer service. The perpetrator pled guilty in court.

In 2005, massive-scale keylogging was put to work online by a criminal ring. As discussed in Chapter 8, "The Web-Enabled Enterprise," spyware software is used for several purposes. This time spyware was used to install a keylogging application that recorded communication with the victims' bank, insurance company, or other financial institutions. The collected data included credit-card details, Social Security numbers, usernames, passwords, instant-messaging chat sessions, and search terms. Some of the data was then saved in a file hosted on a server in the United States that had an offshore-registered domain name. Sunbelt, a company that develops and sells antispam and security software, managed to obtain access to a victim's computer and track what the spyware did. The company reported that the online thieves obtained confidential financial details of customers of 50 international banks. The keystroke logging software was small (26 KB), and took advantage of Internet Explorer browsers. For example, it accessed the browser's Protected Storage area, in which users often save their usernames and passwords for convenient automatic logins. Sunbelt recommended disabling this feature.

In some cases it is employees who unwittingly give away important information such as access codes. Con artists use tricks known as **social engineering**. They telephone an employee who has a password to access an application or a database, introduce themselves as service people from a telephone company or the organization's own IT unit, and say they must have the employee's password to fix a problem. Employees are often tempted to provide their password. The "social engineers" then steal valuable information.

Once criminals have a person's identifying details, such as a Social Security number, driver's license number, or credit-card number, they can pretend to be this person. This crime is called **identity theft**. The imposter can easily withdraw money from the victim's bank accounts, put charges on the victim's credit card, and apply for new credit cards. Since an increasing number of applications for such instruments as well as financial transactions are executed online, identity theft has become a serious problem. According to research firm Javelin Strategy & Research, 8.4 million U.S. adults were identity-theft victims in 2007, and their combined financial loss was $49.3 billion.

Both social engineering and breaking access codes to steal data from online databases have caused huge damage to corporations. Connecting databases to the Internet is necessary for proper operation of multisite organizations and organizations that must share data remotely with business partners. The only way to minimize hacking into such systems is to improve security measures.

POINT OF INTEREST

Like Handing Your Home Key to a Burglar

InTechnology.com published the most common passwords that people use to access their computers, in descending order of frequency: "password," "123456," "qwerty," "abc123," "letmein," "monkey," "myspace1," "password1," "blink182," and one's first name. A strong password is one that takes a long time to figure out by an unauthorized party, even if that party uses sophisticated software to crack it. Stronger passwords are longer, and contain a mix of letters, digits, and special characters (e.g., the characters +*&^%$#@!). Thus, a password such as Tbird4me&u would be a strong password. Several Web sites provide tools to measure the strength of your password. Try *www.securitystats.com/tools/password.php* and *www.microsoft.com/athome/security/privacy/password_checker.mspx*.

Source: InTechnology.com, 2007.

In recent years, identity theft has been more prevalent as part of *phishing*, a crime discussed in Chapter 8, "The Web-Enabled Enterprise." Crooks spam millions of recipients with bogus messages, supposedly from legitimate companies, directing them to a site where they are requested to "update" their personal data, including passwords. The sites are ones constructed by the criminals who steal the personal data and use it to charge the victim's credit account, apply for new credit cards, or—in the worst situations—also apply for other documents such as driver's licenses and apply for loans online.

In a more recent form of phishing, *spear phishing*, criminals use personal information to attack organizational systems. For example, after fraudulently obtaining the personal information of an employee, crooks use the information to obtain entry codes to systems to which the employee has access. They then exploit the systems to obtain more information or cause damage. According to the Anti-Phishing Working Group in Cambridge, Massachusetts, businesses had more than 55,000 phishing incidents in April of 2007. Not surprisingly, financial institutions are the most targeted sector. Spear phishers strive to steal money from online accounts.

Data Alteration, Data Destruction, and Web Defacement

Alteration or destruction of data is often an act of mischief. Data alteration is not a new phenomenon. In 1983, a group of Milwaukee teenagers accessed a computer system at Sloan-Kettering Cancer Center in New York via a modem and altered patients' records just for "fun." An alert nurse noticed a double—and lethal—dose of a medication in a patient's record and called a doctor. She saved the patient's life.

Phishers send e-mails with links to sites that look legitimate, but are not. If a site has been identified as fraudulent, a Web browser may post a warning when the recipient clicks the link. In this case, a day after the warning the phishers removed the materials from their server.

POINT OF INTEREST

The PhishTank

The PhishTank (*www.phishtank.com*) invites you to submit suspicious e-mail messages for vetting by the public. It is "a collaborative clearing house for data and information about phishing on the Internet." The site published its Top 10 organizations whose names were used to phish in June 2007, with the number of submissions: eBay: 4,371; PayPal: 3,342; Regions Bank: 318; National City: 282; Bank of America: 195: Poste Italiane: 136; JPMorgan Chase: 55; Amazon. com: 47; Wachovia: 47; Washington Mutual: 41.

Source: PhishTank report, June 2007.

As mentioned before, an organization's data is often the most important asset it owns, even more important than its hardware and applications. Even if data is altered or destroyed as a prank, the damage to the organization is considerable. The effort to reinstate missing or altered records from a backup copy might entail expensive labor. Even if the actual damage is not substantial, IT staff must spend a lot of time scanning the data pools to ascertain the integrity of the entire resource, and they must also figure out how the perpetrator managed to circumvent security controls. This activity itself wastes the time of high-salaried employees.

Often, the target of online vandals is not data but the organization's Web site. Each day, some organizations find their Web sites have been defaced. Defacement causes several types of damage: first-time visitors are not likely to stay around long enough or revisit to learn about the true nature of the site, and they might associate the offensive material with the organization; frequent visitors might never come back; and shoppers who have had a good experience with the site might leave it forever because they no longer trust its security measures.

To deface a Web site, an intruder needs to know the site's access code or codes that enable the Webmaster and other authorized people to work on the site's server and update its pages. The intruder might either obtain the codes from someone who knows them or use special "brute force" software that tries different codes until it succeeds in accessing the pages.

The best measure against defacement, of course, is software that protects against unauthorized access, or as it is more commonly known, hacking. However, since such software might fail, the public damage can be minimized by ensuring that members of the organization monitor the home page and other essential pages frequently. When the defacement is detected shortly after it occurs, the defaced pages can be replaced with backups before too many visitors have seen the rogue pages. An increasing number of Web sites are restored within hours or even minutes from the defacement.

The cure to any unauthorized entry to an IS is for the organization to find the hole in its security software and fix it with the appropriate software. Such software is often called a "patch." Software companies that sell server management applications often produce patches and invite clients to download and install them.

To combat hackers, organizations use honeytokens. A **honeytoken** is a bogus record in a networked database that neither employees nor business partners would ever access for legitimate purposes. When the intruder copies the database or the part of the database that contains that record, a simple program alerts security personnel, who can start an investigation. The program that detects the incident might also reside on a router or another communications device that is programmed to send an alert as soon as it detects the honeytoken. To entice the intruder to retrieve the honeytoken when only searching for individual records, the honeytoken might be a bogus record of a famous person, such as a medical record of a celebrity in a medical database or the salary of the CEO in a payroll database.

To learn of security holes and methods of unauthorized access, organizations can establish honeypots. A **honeypot** is a server that contains a mirrored copy of a production database (a database that is used for business operations), or one with invalid records. It is set up to make intruders think they have accessed a production database. The traces they leave educate information security officers of vulnerable points in the configuration of servers that perform valid work. In some cases, security people have followed an intruder's "roaming" in the honeypot in real time. Note, however, that different sources have different definitions of the terms honeypot and honeytoken. For example, some define honeypot as any trap set for abusers, including a physical computer, and a honeytoken as a special case where the trap is only data.

POINT OF INTEREST

Vishing You Were Hear

Vishing—voice phishing—is stealing VoIP (Voice over Internet Protocol) services for financial gain. Criminals manipulate the VoIP software of organizations or individuals so they can sell calling minutes to others. The ploy may sound theoretical, but it is not. In 2006, two people were arrested and accused of a scam in which they hacked into the networks of several unnamed companies and hijacked their VoIP bandwidth for resale. Further details of the case were not disclosed, but this is probably not an isolated case. Stealth Communications, Inc., a New York data communications company, estimated that by mid-2007 VoIP thieves were stealing 200 million minutes monthly. The value of these calls is about $26 million.

The term vishing also refers to harvesting private data when individuals use their VoIP phones. In many cases, credit card holders are asked to dial their credit-card account number before speaking to a representative. Criminals who intercept the call harvest the number. In most cases, vishing starts when individuals receive a call from an 800-number with a fake caller ID and are asked to call a certain telephone number supposedly because there is a problem with their card.

Source: Prince, B., "Experts: Enterprises Must Focus on VoIP Security," *eWeek*, July 3, 2007; Jaques, R., "Cybercriminals switch to VoIP 'vishing'," (vnunet.com), July 10, 2006.

Computer Viruses, Worms, and Logic Bombs

Computer **viruses** are so named because they act on programs and data in a fashion similar to the way viruses act on living tissue: computer viruses easily spread from computer to computer. Because so many computers are now connected to one another and many people share files, people unknowingly transmit to other computers viruses that have infected their own files. Once a virus reaches a computer, it damages applications and data files. In addition to destroying legitimate applications and data files, viruses might disrupt data communications: the presence of viruses causes data communications applications to process huge numbers of messages and files for no useful purpose, which detracts from the efficiency of transmitting and receiving legitimate messages and files. The only difference between a computer virus and a worm is that a **worm** spreads in a network without human intervention. A worm attacks computers without the need to send e-mail or open any received files. Most people refer to both types of rogue code as viruses, as does this book.

Almost as soon as e-mail became widespread, criminal minds used it to launch viruses. The Melissa virus of 1999 was an early demonstration of why you should be suspicious of e-mail messages even when they seem to come from people or organizations you know. In the Melissa case, an innocent-looking e-mail message contained an attached Microsoft Word document that, when opened, activated a macro that sent an infected message to the first 50 entries in the victim's Microsoft Outlook address book. Many other viruses spread in a similar way: the recipient is tempted to open—and thereby activate—a file that is attached to a message. The program in that file then destroys files, slows down operations, or does both, and uses vulnerabilities in the operating system and other applications to launch copies of itself to other computers linked to the Internet. Since Melissa, there have been thousands of virus and worm attacks, and millions of computers continue to be infected.

There are many more viruses waiting for victims. CERT/CC (Computer Emergency Response Team/Coordination Center), operated by Carnegie Mellon University, works for the U.S. government and is one of the major distributors of information on new viruses, worms, and other threats to computer security. It estimated that there are at least 30,000 computer viruses somewhere on public networks at any given time; other sources estimate the number at 40,000. CERT says that about 300 new ones are created each month. In 2006, there were 2.5 times more new viruses than in 2005. Totalvirus.com is a site that uses 30 antivirus products to search files for viruses. On a single day in March 2006, over 9,400 infected files were submitted to the site. Only 28 of the files were detected with known viruses. All other files contained new viruses that had not been identified and logged in the database.

One way to protect against viruses is to use **antivirus software**, which is readily available on the market from companies that specialize in developing this kind of software, such as Kaspersky, Symantec, and McAfee, or the free AVG and free open source ClamWin. Subscribers can regularly update the software with code that identifies and deletes or quarantines new viruses, or choose automatic updates, in which virus definitions are updated automatically when the computer is connected to the Internet. However, if a new virus is designed to operate in a way not yet known, the software is unlikely to detect it. Most virus-detection applications allow the user to automatically or selectively destroy suspect programs. Another way to minimize virus threats is to program network software, especially e-mail server software, to reject any messages that come with executable files that might be or contain viruses. Some e-mail applications, such as Microsoft Outlook, are programmed to reject such files.

Some viruses are called **Trojan horses**, analogous to the destructive gift given to the ancient Trojans, as described in Greek mythology. In their war against Troy, the Greeks pretended they were abandoning the city's outskirts and left behind a big wooden horse as a present. The Trojans pulled the horse into the city. When night fell, Greek soldiers hidden within the horse jumped out and opened the gates for thousands of their comrades, who conquered the city. In computer terms, a Trojan horse is any virus disguised as legitimate software or useful software that contains a virus. Many people also refer to spyware that comes with useful software as Trojan horse software.

A growing number of viruses and worms take advantage of vulnerable features of operating systems, most notably Microsoft Windows. Most attack this company's operating systems because the large majority of organizations worldwide use Microsoft operating systems to run their servers and computers. Software vendors provide patches against direct intrusion into computer systems and distribute security patches against viruses and worms. However, it is up to security professionals and network administrators to implement the patches as soon as they become available.

Some rogue computer programs do not spread immediately like a virus but are often significantly more damaging to the individual organization that is victimized. A **logic bomb** is software that is programmed to cause damage at a specified time to specific applications and data files. It lies dormant until a certain event takes place in the computer or until the computer's

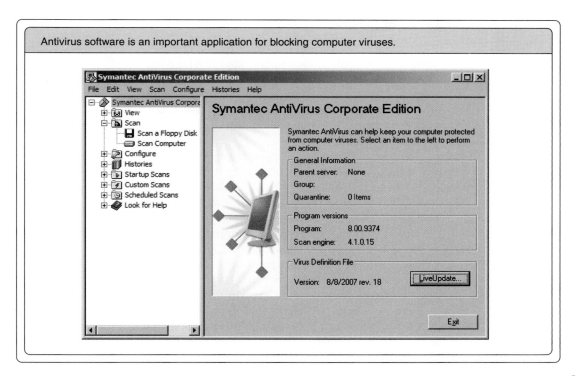

Antivirus software is an important application for blocking computer viruses.

inner clock reaches the specified time; the event or time triggers the virus to start causing damage. Logic bombs are usually planted by insiders, that is, employees of the victimized organization. In one case, a man named Timothy Lloyd was convicted of planting a logic bomb on Omega Engineering's computer system after he found out he was about to be fired. Lloyd, who had worked for the company for 11 years, planted six destructive lines of code on a company network server. He had tested the bomb and then reconstructed all the files. Twenty days after he left the company, the bomb erased all of the company's contracts, designs, and production programs, as well as proprietary software used by the company's manufacturing machines. The 31-year-old man's act cost the company an estimated $12 million, led to laying off 80 employees, and diminished its competitive position in the electronic manufacturing market. A plant manager for the company who testified at Lloyd's trial said the company would never recover from this sabotage. Lloyd was sentenced to 41 months in prison.

Nonmalicious Mishaps

Unintentional damage to software occurs because of poor training, lack of adherence to backup procedures, or simple human error. Although unintentional damage rarely occurs in robust applications, poor training might result in inappropriate use of an application so that it ruins data, unbeknownst to the user. For instance, when faced with an instruction that might change or delete data, a robust application will pose a question such as: "Are you sure you want to delete the record?" or issue a warning such as "This might destroy the file." More common damage is caused by the failure to save all work and create a backup copy. Destruction of data often happens when using a word-processing program to create or change text files and when updating databases.

Unauthorized downloading and installation of software that might cause damage can be controlled by limiting administration rights to employees. Many organizations instruct operating systems to deny such rights to most employees. They program ISs to accept new software installation only when the proper access codes are entered.

RISKS TO ONLINE OPERATIONS

The massive movement of operations to the Internet has attracted hackers who try to interrupt such operations daily. In addition to unauthorized access, data theft, and defacing of Web pages, there has been a surge in denial-of-service attacks and hijacking of computers.

Denial of Service

In February 2005, people who tried to access the Web sites of the Japanese prime minister and cabinet office could not do so because the sites were victims of a denial-of-service attack. **Denial of service (DoS)** occurs when a Web site receives an overwhelming number of information requests, such as merely logging on to a site. The intention of high-volume log-on requests is to slow down legitimate traffic on the site's server; business can slow to a halt. The server's frantic efforts to handle the massive amount of traffic denies legitimate visitors and business partners access to the site.

In most such attacks, the perpetrator launches software that uses other people's computers for the attack—unbeknownst to them; thus the attacks are sometimes known as distributed denial-of-service (DDoS) attacks. Professionals call the computers used in these attacks "zombies." Zombie computers not only exacerbate the volume of traffic but also make it impossible to track down the generator of the attack.

DoS attacks continue to be a major problem. In April and May 2007, the Web sites of Estonia's prime minister's office, banks, schools, and other organizations were simultaneously attacked. Experts from the North Atlantic Treaty Organization (NATO) helped the small country investigate the attack. They found out that computers from the United States, Canada, Brazil, Vietnam, and other countries were involved in the DoS attack. In mid-May 2007, the attacks stopped.

Because it is impossible to stop anyone from trying to log on to a Web site, there is no full cure for a DoS attack, but equipment is available that can filter most illegitimate traffic targeting a site. The equipment detects repeated requests that come from the same IP addresses at an abnormal frequency and blocks them, and it can be programmed to block all incoming communication from suspected servers. The equipment can filter about 99 percent of false requests, but using the equipment slows down communication, so the site's response is slowed. In addition, blocking requests might also deny access to legitimate visitors from suspected servers, especially if the server is used by an ISP that provides Internet access to thousands of people and organizations. One way to mitigate DoS attacks is for an organization to use multiple servers, which is a good idea anyway to handle periods of legitimate traffic increases.

No organization is immune to DoS. Some of the most visible Web sites have been attacked, including those of eBay, Amazon, CNN, and the U.S. White House. All had to shut down their sites for several hours. Amazon, eBay, and other commercial sites have lost revenue as a result. Even CERT has been forced to shut down its site because a DoS attack sent information into its Web site at rates several hundred times higher than normal. Symantec Corp., an antivirus software vendor, recorded an average of 5213 DoS attacks per day in the second half of 2006. This was actually a decline from an average of 6110 attacks in the first half of that year.

Computer Hijacking

You might not be aware of it, but there is a good chance your networked computer has been hijacked. No, nobody would remove it from your desk, but if it is connected to the Internet, it is used by other people. **Hijacking** a computer means using some or all of the resources of a computer linked to a public network without the consent of its owner. As you have seen, this has been done for DoS, but it is also done for other purposes.

Hijacking is carried out by surreptitiously installing a small program called a *bot* on a computer. Like many viruses, these programs are downloaded unwittingly by people who use chat rooms and file-sharing networks. When your computer is hijacked, your Internet connection might slow to a crawl. The damage to corporations in the form of reduced productivity can be great. The main purpose of hijacking computers is spamming: using hijacked computers to send unsolicited commercial e-mail to large numbers of people, often millions of addresses. Spammers do so for two reasons: they hide the real source of the e-mail so that they cannot be identified and pursued, and they take advantage of the hijacked machines' computer resources—CPU time, memory, and communications link—to expedite the distribution of spam.

Nobody knows how many computers are infected with bots, but estimates for 2007 ranged from 10 million to 150 million. Symantec counted 6.7 million active bots in one scan the company conducted. Since not all bots are active at the same time, the number of bots is probably much higher.

To hijack computers, spammers exploit security holes in operating systems and communications software, and then surreptitiously install e-mail forwarding software, much as one would install a virus. Most users do not notice the extra work their computers do. One precaution is to check why a computer continues activity (such as hard disk work) when the owner does not use it. Computer owners can also install special software that detects e-mail forwarding applications.

Computer hijacking is also done to turn computers into zombies to help a DoS. Instead of exploiting the computers to send e-mail, they are used to send repeated service requests to Web servers.

CONTROLS

Controls are constraints and other restrictions imposed on a user or a system, and they can be used to secure systems against the risks just discussed or to reduce damage caused to systems, applications, and data. Figure 14.1 lists the most common controls. Controls are implemented not only for access but also to implement policies and ensure that nonsensical data is not entered into corporate databases.

FIGURE 14.1
Common controls to protect systems from risks

◆ Program robustness and data entry controls

◆ Backup

◆ Access controls

◆ Atomic transactions

◆ Audit trail

Application Reliability and Data Entry Controls

Apart from performing programmed functions, reliable applications can resist inappropriate usage, such as incorrect data entry or processing. The most reliable programs consider every possible misuse or abuse. A highly reliable program includes code that promptly produces a clear message if a user either makes an error or tries to circumvent a process. For example, a Web site invites users to select a username and password, and the operators demand passwords that are not easy to guess. The application should be programmed to reject any password that has fewer than a certain number of characters or does not include numerals. A clear message then must be presented, inviting the user to follow the guidelines.

Controls also translate business policies into system features. For example, Blockbuster Video uses its IS to implement a policy limiting debt for each customer to a certain level. When a renter reaches the debt limit and tries to rent another DVD, a message appears on the cash register screen: "Do not rent!" Thus, the policy is implemented by using a control at the point of sale. Similar systems do not allow any expenditures to be committed unless a certain budgetary item is first checked to ensure sufficient allocation. A spending policy has been implemented through the proper software.

Backup

Probably the easiest way to protect against loss of data is to automatically duplicate all data periodically, a process referred to as data **backup**. Storage media suitable for routine backup were discussed in Chapter 4, "Business Hardware." Many systems have built-in automatic backup programs. The data might be duplicated on inexpensive storage devices such as magnetic tapes. Manufacturers of storage devices also offer Redundant Arrays of Independent Disks (RAID) for this purpose. As explained in Chapter 4, **RAID** is a set of disks that is programmed to replicate stored data, providing a higher degree of reliability.

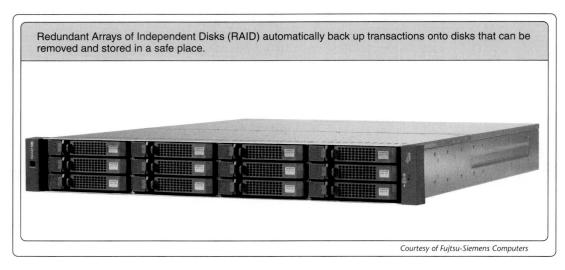

Redundant Arrays of Independent Disks (RAID) automatically back up transactions onto disks that can be removed and stored in a safe place.

Courtesy of Fujitsu-Siemens Computers

Of course, backing up data is not enough. The disks or tapes containing backed-up data must be routinely transported off-site, so that if a business site is damaged by a disaster, the remote storage can be used since it is likely to be spared. In the past, many companies had a truck haul backup disks and tapes to the storage location at the end of every business day, and some might still do so. However, due to developments in telecommunications in recent years, most corporations prefer to back up data at a remote site through communications lines. In fact, this approach is safer than transporting physical disks and tapes; on several occasions such media fell off vans or were stolen. Often, the backup disks or tapes reside thousands of miles away from the organization's business offices. For additional protection, backup disks or tapes are locked in safes that can withstand fire and floods.

Companies can also use the services of firms that specialize in providing backup facilities. The vendor maintains a site with huge amounts of disk space linked to the Internet. The online data backup service typically provides client organizations with an application that copies designated files from the client's systems to the remote disks. For obvious reasons, some professionals call this type of service "e-vaulting." One company that provides the service is AmeriVault (*www.amerivault.com*).

Access Controls

Unauthorized access to information systems, usually via public networks such as the Internet, does not always damage IT resources. However, it is regarded as one of the most serious threats to security because it is often the prelude to the destruction of Web sites, databases, and other resources, or theft of valuable information.

Access controls are measures taken to ensure that only those who are authorized have access to a computer or network, or to certain applications or data. One way to block access to a computer is by physically locking it in a facility to which only authorized users have a key or by locking the computer itself with a physical key. However, in the age of networked computers, this solution is practical only for a limited number of servers and other computers. Therefore, these organizations must use other access controls, most of which rely on software.

Experts like to classify access controls into three groups: what you know, what you have, and who you are. "What you know" includes access codes such as user IDs, account numbers, and passwords. "What you have" is some kind of a device, such as a security card, which you use directly or which continuously changes coordinated access codes and displays them for you. "Who you are" includes your unique physical characteristics.

The most common way to control access is through the combination of a user ID and a password. While user IDs are usually not secret, passwords are. IS managers encourage users to change their passwords frequently, which most systems easily allow, so that others do not have time to figure them out and to limit the usefulness of stolen passwords. Some organizations have

Companies can subscribe to online backup services that automatically create a remote backup of designated data.

systems that force users to change their passwords at preset intervals, such as once a month or once every three months. Some systems also prevent users from selecting a password that they have used in the past, to minimize the chance that someone else might guess it, and many require a minimum length and mix of characters and numerals. Access codes and their related passwords are maintained either in a special list that becomes part of the operating system or in a database that the system searches to determine whether a user is authorized to access the requested resource.

A more secure measure than passwords is security cards, such as RSA's SecureID. The device is distributed to employees who need access to confidential databases, usually remotely. Employees receive a small device that displays a 6-digit number. Special circuitry changes the number both at the server and the device to the same new number every minute. To gain access, employees enter at least one access code and the current number. The device is small enough to be carried on a key chain or in a wallet. This two-factor access control increases the probability that only authorized people gain access. This is an example of using both what you know and what you have.

In recent years, some companies have adopted physical access controls called biometrics. A **biometric** characteristic is a unique physical, measurable characteristic of a human being that is used to identify a person. Characteristics such as fingerprints, retinal scans, or voiceprints can be used in biometrics. They are in the class of "who you are." When a fingerprint is used, the user presses a finger on a scanner or puts it before a digital camera. The fingerprint is compared against a database of digitized fingerprints of people with authorized access. A growing number of laptop computers have a built-in fingerprint scanner for the same purpose. The procedure is similar when the image of a person's retina is scanned. With voice recognition, the user is instructed to utter a word or several words. The intonation and accent are digitized and compared with a list of digitized voice samples.

Several manufacturers of computer equipment offer individual keyboard-embedded and mouse-embedded fingerprint devices. For example, SecuGen Corporation offers EyeD Mouse, a mouse that includes a fingerprint reader on the thumb side of the device. It verifies a fingerprint in less than a second. Using biometric access devices is the best way not only to prevent unauthorized access to computers but also to reduce the workload of Help desk personnel. Up to 50 percent of the calls Help desk personnel receive come from employees who have forgotten their passwords.

Access controls include (top to bottom) what you know (password), what you have (a device), and what you are (biometric characteristics).

Courtesy of RSA Security, Inc / istickphoto.com

Atomic Transactions

As you know, in an efficient IS, a user enters data only once, and the data is recorded in different files for different purposes, according to the system's programmed instructions. For instance, in a typical order system, a sale is recorded in several files: the shipping file (so that the warehouse knows what to pack and ship), the invoice file (to produce an invoice and keep a copy in the system), the accounts receivable file (for accounting purposes), and the commission file (so that the salesperson can be compensated with the appropriate commission fee at the end of the month). As indicated in Figure 14.2, a system supports atomic transactions when its code only allows the recording of data if they successfully reach all their many destinations. An **atomic transaction** (from the Greek *atomos*, indivisible) is a set of indivisible transactions; either all transactions are executed or none are—never only some. Using atomic transactions ensures that only full entry occurs in all the appropriate files.

For instance, suppose the different files just mentioned reside on more than one disk, one of which is malfunctioning. When the clerk enters the sale transaction, the system tries to automatically record the appropriate data from the entry into each of the files. The shipping, accounts receivable, and invoice files are updated, but the malfunctioning commission file cannot accept the data. Without controls, the sale would be recorded, but unknown to anyone, the commission would not be updated, and the salesperson would be deprived of the commission on this deal. However, an atomic transaction control mechanism detects that not all four files have been updated with the transaction, and it doesn't update any of the files. The system might try to update again later, but if the update does not go through, the application produces an appropriate error message for the clerk, and remedial action can be taken.

Note that this is a control not only against a malfunction but also against fraud. Suppose the salesperson collaborates with the clerk to enter the sale only in the commission file, so he or she can be rewarded for a sale that has never taken place—and then plans to split the fee with the clerk. The atomic transaction control would not let this happen. Recall our discussion of relational database management systems. Virtually all current relational DBMSs have atomicity—the ability to make transactions atomic—as a required feature.

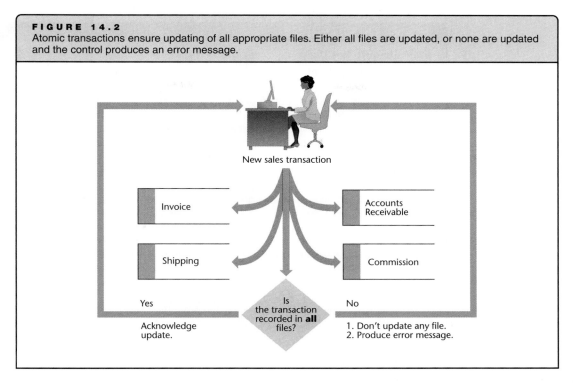

FIGURE 14.2
Atomic transactions ensure updating of all appropriate files. Either all files are updated, or none are updated and the control produces an error message.

New sales transaction

Invoice

Accounts Receivable

Shipping

Commission

Yes — Acknowledge update.

Is the transaction recorded in **all** files?

No — 1. Don't update any file.
2. Produce error message.

Audit Trail

In spite of the many steps taken to prevent system abuse, it nonetheless occurs. Consequently, further steps are needed to track transactions so that (1) when abuses are found, they can be traced, and (2) fear of detection indirectly discourages abuse. One popular tracking tool is the **audit trail**: a series of documented facts that help detect who recorded which transactions, at what time, and under whose approval. Whenever an employee records a transaction, the system prompts the employee to provide certain information: an invoice number, account number, salesperson ID number, and the like. Sometimes an audit trail is automatically created using data, such as the date and time of a transaction or the name or password of the user updating the file. This data is recorded directly from the computer—often unbeknownst to the user—and attached to the record of the transaction.

The laws and regulations of many countries require certain policy and audit trail controls, and since so many operations are performed using ISs, the controls must be programmed into software. In the United States, the Sarbanes-Oxley Act of 2002 requires corporations to implement audit trails and other measures in their systems.

Audit trail information helps uncover undesirable acts, from innocent mistakes to premeditated fraud. The information helps determine who authorized and who made the entries, the date and time of the transactions, and other identifying data that is essential in correcting mistakes or recovering losses. The audit trail is the most important tool of the **information systems auditor** (formerly known as the electronic data processing auditor), the professional whose job it is to find erroneous or fraudulent cases and investigate them.

SECURITY MEASURES

As you've seen so far in this chapter, the increase in the number of people and organizations using the Internet has provided fertile ground for unauthorized and destructive activity. This section describes several ways that organizations can protect themselves against such attacks, including using firewalls, authentication and encryption, digital signatures, and digital certificates.

Firewalls and Proxy Servers

The best defense against unauthorized access to systems over the Internet is a **firewall**, which is hardware and software that blocks access to computing resources. Firewalls are routinely integrated into the circuitry of routers, as discussed in Chapter 6, "Business Networks and Telecommunications." Firewall software screens the activities of a person who logs on to a Web site; it allows retrieval and viewing of certain material, but blocks attempts to change the information or to access other resources that reside on the same computer or computers connected to it.

It is important to note that while firewalls are used to keep unauthorized users out, they are also used to restrict unauthorized software or instructions, such as computer viruses and other rogue software. When an employee uses a company computer to access external Web sites, the firewall screens for viruses and active attempts to invade company resources through the open communications line. It might also be programmed to block employee access to sites that are suspected of launching rogue programs, or to sites that provide no useful resources. The firewall then prohibits the user from logging on to those sites.

As Figure 14.3 illustrates, a firewall controls communication between a trusted network and the "untrusted" Internet. The firewall can be installed on a server or a router. Network professionals use firewall software to check which applications can access the Internet and which servers might be accessed from the organization's network.

To increase security, some companies implement the **DMZ** (demilitarized zone) approach. The DMZ is a network of computers that are connected to the company's trusted network (such as an intranet) at one end and the untrusted network—the public Internet—at the other end. The DMZ includes resources to which the organization allows direct access from the Internet. It might include a Web site and computers from which people can download files. A DMZ provides a barrier between the Internet and a company's organizational network, which is usually an intranet. The connection between the DMZ and the organization's trusted network is established by using a proxy server.

A **proxy server** "represents" another server for all information requests from resources inside the trusted network. However, a proxy server can also be placed between the Internet and the organization's trusted network when there is no DMZ. For example, this might be the arrangement when the organization establishes its Web site as part of its trusted network. The proxy server then retrieves Web pages for computers requesting them remotely through the Internet. Thus, external computers requesting Web pages never come in direct contact with the computer

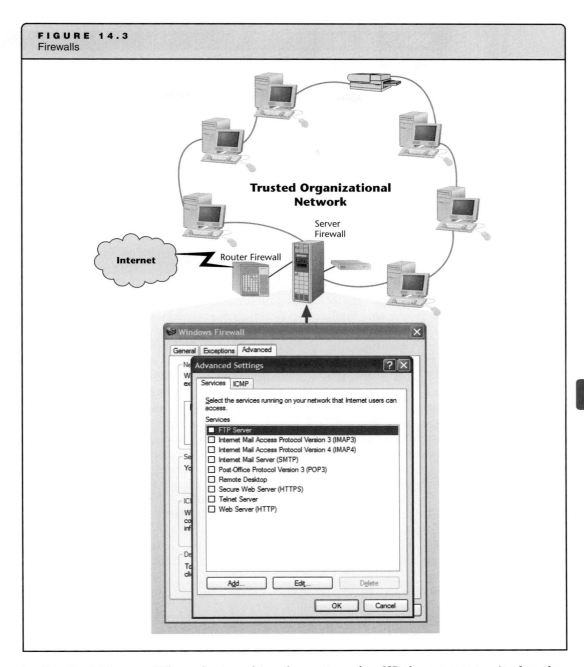

FIGURE 14.3
Firewalls

hosting the Web pages. When a business hires the services of an ISP, the proxy server is often the one operated by the ISP.

Both the organizational network server and proxy server employ firewalls. In Figure 14.3, the firewalls would be installed on the server of the organizational network and the router. The router is often called a "boundary router." The double firewall architecture adds an extra measure of security for an intranet.

Authentication and Encryption

With so much Web-based commerce and other communication on the Internet, businesses and individuals must be able to authenticate messages. That is, they must be able to tell whether certain information, plain or encrypted, was sent to them by the party that was supposed to send it. Note that the word "message" is used here for any type of information, not only text. It might be images, sounds, or any other information in digital form.

Authentication is the process of ensuring that the person who sends a message to or receives a message from you is indeed that person. Authentication can be accomplished by senders and receivers exchanging codes known only to them. Once authentication is established, keeping a message secret can be accomplished by transforming it into a form that cannot be read by anyone who intercepts it. Coding a message into a form unreadable to an interceptor is called **encryption**. Authentication also often occurs when an encrypted message is received, because the recipient needs to ensure that the message was indeed encrypted and sent by a certain party.

Both authentication and secrecy are important when communicating confidential information such as financial and medical records. They are also essential when transacting business through a public network. For example, millions of people now buy and sell shares of stock and other financial products on the Web, businesses and individuals make purchases through the Web and use credit-card account numbers for payment, and medical clinics use the Web to transmit patient records to insurance companies and prescriptions to pharmacies. All must authenticate the recipient and keep the entire communication confidential.

To authenticate the users and maintain secrecy, the parties can use encryption programs. Encryption programs scramble information transmitted over the network so that an interceptor only receives unintelligible data. The original message is called **plaintext**; the coded message is called **ciphertext**. Encryption uses a mathematical algorithm, which is a formula, combined with a key. The key is a unique combination of bits that must be used in the formula to decipher the ciphertext. As indicated in Figure 14.4, the receiving computer uses the key to decipher the ciphertext back into plaintext.

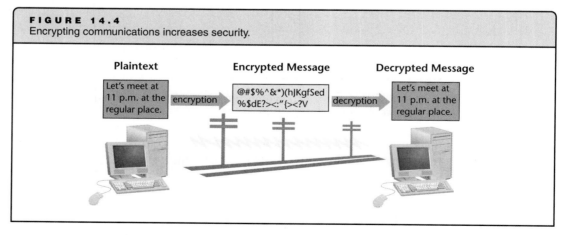

FIGURE 14.4
Encrypting communications increases security.

Plaintext — Let's meet at 11 p.m. at the regular place. — encryption → Encrypted Message — @#$%^&*)(hJKgfSed%$dE?><:"{><?V — decryption → Decrypted Message — Let's meet at 11 p.m. at the regular place.

To illustrate the use of encryption algorithms and keys, here is a simple example. Suppose you send a secret message that you want the recipient to decipher. Remember that each character in your digital message is represented by a byte, which is a combination of eight bits. The byte can be expressed as a numeric value. For instance, the character represented by 00010101 has a decimal value of 21. So, each character in your message has a numeric value. To decipher it, you might devise the following algorithm:

$$y = x + k,$$

where x is the original value of the character, k is the key, and y is the new (encrypted) value of the byte. The value of k is secret and known only to you and to the recipient. Suppose you and the recipient agree that the key's value is 00101000 (decimal 40). Now, each original character is first manipulated through the algorithm before transmittal. For example, the byte 00010101 (decimal 21) will now be transmitted as 00111101 (decimal 61), and 10111001 (decimal 185) will be transmitted as 11100001 (decimal 225). The result of any manipulation of text by this algorithm is a string of characters that makes no sense to anyone who cannot figure out the algorithm and the key. In reality, the algorithms are usually known to many people, or can be figured out relatively easily; it is the key that cannot be so easily deduced. In this case, the key is an addition of 40 in decimal (expressed as 00101000 in binary) to each byte (character). Since the recipient knows the key, he or she can use the algorithm, along with the key, to decipher your message into readable text. Note that this is an extremely simple example. In reality, encryption algorithms are significantly more complex. Also note that the key used in this example is a combination of eight bits, which is quite easy to figure out. It would be significantly more difficult to figure out a key consisting of 128, 256, 512, or 1024 bits, which are commonly used in Internet communication. When keys that long are used, even with the latest hardware and most sophisticated code-breaking software the average time to decipher an encrypted message is so long that the probability of success is extremely small.

Public-Key Encryption

As Figure 14.5 indicates, when both the sender and recipient use the same secret key (which is the case in the earlier example), the technique is called **symmetric encryption**. However, symmetric encryption requires that the recipient have the key before the encrypted text is received. Therefore, the key is referred to simply as a *secret key* or *private key*. While it is fairly simple to keep the secrecy of a message when the sender and recipient have the same key beforehand, it is impractical in daily transactions on the Internet. For example, a retail Web site would not be able to function if every buyer would require a secret key with each transaction to ensure confidentiality. Therefore, in such communication, there must be a way for the sender to communicate the key to the recipient before the message is sent. To this end, the parties use an **asymmetric encryption** comprising two keys: one is public, and the other is private. It is clear why this type of encryption is also called "public-key" encryption.

A public key is distributed widely and might be known to everyone; a private key is secret and known only to the recipient of the message. When the sender wants to send a secure message to the recipient, he uses the recipient's public key to encrypt the message. The recipient then uses her own private key to decrypt it. A mathematical relationship exists between the public and private keys. The public and private keys are related in such a way that only the public key can be used to encrypt messages, and only the corresponding private key can be used to decrypt them. It is virtually impossible to deduce the private key from the public key. All applications that use public keys and private keys use the same principles. What differentiates them from one another is the different encryption algorithm each uses.

Online businesses often switch site visitors to a secure server when they are asked to provide secret information such as credit-card account numbers or other personal data. The secure server provides the visitor's Web browser with the site's public key. The browser uses it to encrypt the credit-card number and any other personal information. The secure server uses the private key to decrypt the information. Once an encrypted exchange is established, the server can send the visitor's browser a secret key that both can use. Moreover, the server can change the key often during the session to make decryption more difficult.

Transport Layer Security

A protocol called **Transport Layer Security (TLS)** is used for transactions on the Web. TLS is the successor of Secure Socket Layer (SSL) and works following the same principles as SSL, with some additional improvements that are outside the scope of this discussion. TLS is part of

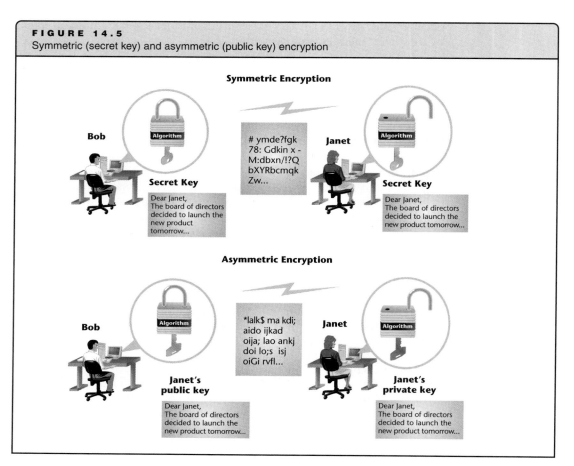

FIGURE 14.5
Symmetric (secret key) and asymmetric (public key) encryption

Symmetric Encryption

Bob — Algorithm — Secret Key

ymde?fgk 78: Gdkin x - M:dbxn/!?Q bXYRbcmqk Zw...

Janet — Algorithm — Secret Key

Dear Janet,
The board of directors decided to launch the new product tomorrow...

Dear Janet,
The board of directors decided to launch the new product tomorrow...

Asymmetric Encryption

Bob — Algorithm — Janet's public key

*lalk$ ma kdi; aido ijkad oija; lao ankj doi lo;s isj oiGi rvfl...

Janet — Algorithm — Janet's private key

Dear Janet,
The board of directors decided to launch the new product tomorrow...

Dear Janet,
The board of directors decided to launch the new product tomorrow...

virtually all current Web browsers. Current versions of browsers use TLS with a 128-bit key. TLS uses a combination of public key and symmetric key encryption. It works as follows:

1. When a visitor connects to an online site, the site's server sends the visitor's browser its public key.
2. The visitor's browser creates a temporary symmetric (secret) key of 128 bits. The key is transmitted to the site's server encrypted by using the site's public key. Now both the visitor's browser and the site's server know the same secret key and can use it for encryption.
3. The visitor can now safely transmit confidential information.

How safe is a 128-bit key? It would take 250 PCs working simultaneously around the clock an estimated average of 9 trillion times the age of the universe just to decrypt a single message. This is the reason why practically all financial institutions use 128-bit encryption, and if you want to bank online, you must use a browser that supports this key length. However, how long it takes an interceptor to decipher depends on current speed of hardware and sophistication of code-breaking software. As hardware becomes faster and software becomes more sophisticated, standard keys usually are set longer.

When you log on to secure servers you might notice that the "HTTP://" in the URL box at the top of the browser turns into an HTTPS:// (or https://), and a security icon, usually a little closed padlock, appears in the browser. It is advisable not to transfer any confidential information through the Web if you don't see both https:// and the padlock icon. **HTTPS** is the secure version of HTTP, discussed in Chapter 8, "The Web-Enabled Enterprise." HTTPS encrypts communication using SSL or TLS. Fortunately, all this encryption and decryption is done by the browser. When you access a secure area of a Web site, the communication between the site's server and your Web browser is encrypted. The information you view on your screen was encrypted by the software installed on the site's server and then decrypted by your browser.

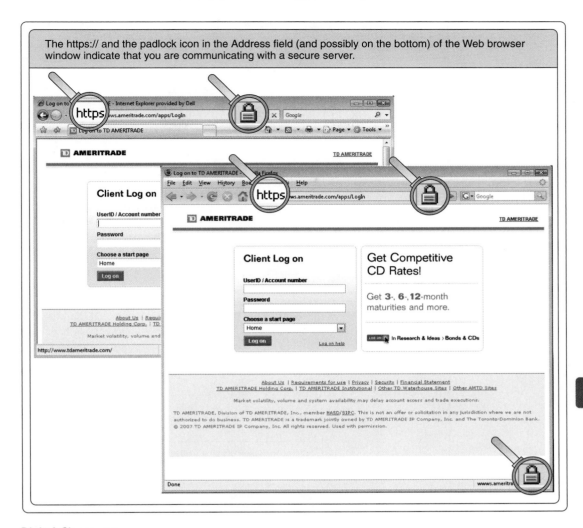

The https:// and the padlock icon in the Address field (and possibly on the bottom) of the Web browser window indicate that you are communicating with a secure server.

Digital Signatures

A **digital signature** is a way to authenticate online messages, analogous to a physical signature on a piece of paper, but implemented with public-key cryptography. The digital signature authenticates the identity of the sender of a message and also guarantees that no one has altered the sent document; it is as if the message were carried in an electronically sealed envelope.

When you send an encrypted message, two phases are involved in creating a digital signature. First, the encryption software uses a hashing algorithm (a mathematical formula) to create a message digest from the file you wish to transmit. A **message digest** is akin to the unique fingerprint of a file. Then, the software uses your private (secret) key to encrypt the message digest. The result is a digital signature for that specific file.

How does it work? Follow the flowchart in Figure 14.6. Suppose you want to send the draft of a detailed price proposal to your business partner. You want to be certain that the document you intend to send is indeed the one she receives. She wants the assurance that the document she receives is really from you.

1. You attach the price proposal file to an e-mail message. The entire communication is essentially one message, indicated as "Plain message" in Figure 14.6.
2. Using the hashing software, your computer creates a message hash, the message digest, which is a mathematically manipulated file of the message and is not readily readable by a human.
3. You then use a private key that you have previously obtained from the public-key issuer, such as a certificate authority, to encrypt the message digest. Your computer uses your private key to turn the message digest into a digital signature.
4. The computer also uses your private key to encrypt the message in its plain (unhashed) form. Your computer sends off both files.

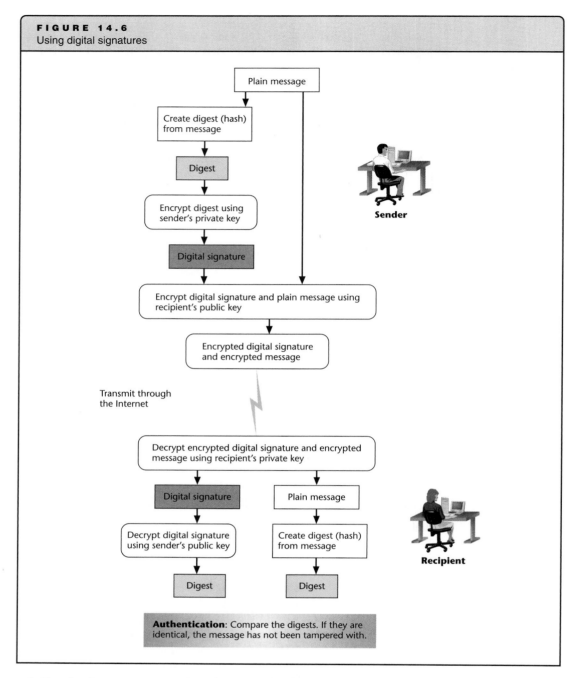

FIGURE 14.6
Using digital signatures

Plain message

Create digest (hash) from message

Digest

Encrypt digest using sender's private key

Digital signature

Encrypt digital signature and plain message using recipient's public key

Encrypted digital signature and encrypted message

Sender

Transmit through the Internet

Decrypt encrypted digital signature and encrypted message using recipient's private key

Digital signature

Plain message

Decrypt digital signature using sender's public key

Create digest (hash) from message

Recipient

Digest

Digest

Authentication: Compare the digests. If they are identical, the message has not been tampered with.

5. Your business partner receives the encrypted files: the digital signature (which is an encrypted message digest) and the encrypted message, which usually come as one file.

6. Your business partner's computer uses her private key (which is mathematically related to her public key, which you used) to decrypt both your digital signature and your encrypted unhashed message.

7. The decrypted digital signature becomes the message digest. Hashing the decrypted unhashed message turns this message into a digest, too.

8. If the two message digests are identical, the message received is, apparently, the one you sent, unchanged.

Since the message digest is different for every message, your digital signature is different each time you send a message. As described here, senders of encrypted messages obtain the public key of the recipient from an issuer of such keys. In most cases, the issuer is a certificate authority, and the recipient's public key is included in the recipient's digital certificate, which is discussed next.

Digital Certificates

To authenticate a digital signature, both buyers and sellers must use digital certificates (also known as digital IDs). **Digital certificates** are computer files that serve as the equivalent of ID cards by associating one's identity with one's public key. An issuer of digital certificates is called a **certificate authority (CA)**, an organization that serves as a trusted third party. A CA certifies the identity of anyone who inquires about a party communicating on the Internet. Some CAs are subsidiaries of banks and credit-card companies, and others are independent. American Express CA, Digital Signature Trust Co., VeriSign, Inc., and GlobalSign NV are just a few of the numerous companies that sell digital certificates. To view a long list of CAs you can go to *www.pki-page.org*. A CA issues the public (and private) keys associated with a certificate.

A digital certificate contains its holder's name, a serial number, expiration dates, and a copy of the certificate holder's public key (used to encrypt messages and digital signatures). It also contains the digital signature of the certificate authority so that a recipient can verify that the certificate is real. To view the digital certificate of a secure online business, click the padlock icon in the address bar or status bar of your browser. Figure 14.7 shows the same certificate presented through two different Web browsers.

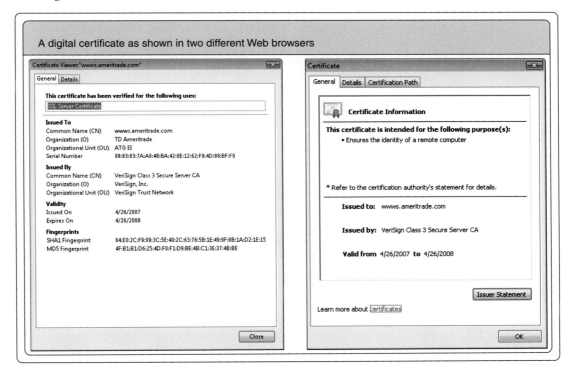

A digital certificate as shown in two different Web browsers

Digital certificates are the equivalent of tamper-proof photo identification cards. They are based on public-key encryption techniques that verify the identities of the buyer and seller in electronic transactions and prevent documents from being altered after the transaction is completed. Consumers have their own digital certificates stored on their home computers' hard disks. In a transaction, a consumer uses one digital key attached to the certificate that he or she sends to the seller. The seller sends the certificate and his own digital key to a certificate authority, which then can determine the authenticity of the digital signature. Completed transaction documents are stored on a secure hard disk maintained by a trusted third party.

The recipient of an encrypted message uses the certificate authority's public key to decode the digital certificate attached to the message, verifies it as issued by the certificate authority, and then obtains the sender's public key and identification information held within the certificate. With this information, the recipient can send an encrypted reply.

The Downside of Security Measures

Security measures—especially passwords, encryption applications, and firewalls—have a price that relates to more than money: they slow down data communications, and they require user discipline, which is not always easy to maintain. Employees tend to forget their passwords, especially if they must replace them every 30 or 90 days.

Employees are especially annoyed when they have to remember a different password for every system they use; in some companies, there might be four or five different systems, each with its own access control. A simpler solution is an approach called **SSO (single sign-on)**. With SSO, users are required to identify themselves only once before accessing several different systems. However, SSO requires special software that interacts with all the systems in an organization, and the systems must be linked through a network. Not many organizations have installed such software.

CIOs often cite SSO as an effective way to decrease the amount of work their subordinates must do. Such was certainly the case at Philadelphia Gas Works (PGW), a utility company with a staff of 1700 serving over half a million customers. The IT staff received about 20,000 calls per year from employees, about half of which were related to forgotten passwords. The IT staff had to reset these passwords. Since SSO was implemented, the number of calls of this nature decreased to about 10,000 per year.

Encryption slows down communication because the software must encrypt and decrypt every message. Remember that when you use a secure Web site, much of the information you view on your screen is encrypted by the software installed on the site's server, and then decrypted by your browser. All this activity takes time, and the delay only exacerbates the Internet's low download speed during periods of heavy traffic. Firewalls have the same slowing effect; screening every download takes time, which affects anyone trying to access information, including employees, business partners, and consumers.

IT specialists must clearly explain to managers the implications of applying security measures, especially on systems connected to the Internet. The IT specialists and other managers must first determine which resource should be accessed only with passwords and which also require other screening methods, such as firewalls. They must tell employees what impact a new security measure will have on their daily work—and if the measure will adversely affect their work, the specialists must convince the employees that the inconvenience is the price for protecting data. The IT specialists should also continue to work on methods that minimize inconvenience and delays.

Recall the discussion of virtual private networks (VPNs), which enable employees to access ISs using special security software involving passwords and encryption. This approach allows employees to access an intranet only from computers equipped with the proper VPN software and only if they remember passwords. When Wawa Corporation—the convenience store chain discussed in a case in Chapter 13, "Choices in Systems Acquisition,"—implemented a new SAP ERP system, the CIO implemented a one-time keyfob similar to SecurID®. The password changes frequently, and the user does not have to remember it because it appears automatically on the keyfob. There is no need to use VPN software. If someone steals a password, the thief cannot use it for more than a few seconds because it then changes. This enables Wawa employees to access the intranet from any computer in the world.

Information technology can help track down criminals and terrorists, but it also helps criminals and terrorists in their efforts. The technology can help protect privacy and other civil rights, but it can also help violate such rights. The growing danger of terrorism and the continued effort of governments to reduce drug-related and other crimes led to controversial use, or abuse, of IT. In the United States, one particular law with a long name includes controversial provisions that have worried civil libertarians since October 2001. Uniting and Strengthening America by Providing Appropriate Tools Required to Intercept and Obstruct Terrorism Act of 2001, the PATRIOT Act, as it is popularly known, gives law enforcement agencies surveillance and wiretapping rights they did not have before that year. The law permits the FBI to read private files and personal Internet records without informing the suspected citizen and without need for a law enforcement agency to present to the court a probable cause. "Our constitutional freedoms are in jeopardy. Now is the time to restore real checks and balances to the worst sections of the Patriot Act" called a Web posting of the American Civil Liberties Union (ACLU) in 2005, when the law was reconsidered by the U.S. Congress. On the contrary, said many members of Congress, the law should be enhanced to give the FBI even freer hand.

The Electronic Privacy Information Center (EPIC) explains the major concerns with the Act, which made changes to 15 existing laws. The Act gives more power than before to law enforcement agencies in installing pen registers and trap-and-trace devices. A pen register is any device that records outgoing phone numbers. A trap-and-trace device—a caller ID device, for instance—captures and records incoming telephone numbers. Similarly, the Act extends the government's authority to gain access to personal financial information and student information, even if the subject of the investigation is not suspected of wrongdoing. Agents only have to certify that the information likely to be obtained is relevant to an ongoing criminal investigation. In the past, the government had to show to a judge probable cause—a reasonable suspicion that the subject of an investigation is committing or is about to commit a crime. If a government attorney "certifies" that the information collected is likely to be relevant, the judge must grant permission to install the device and collect the information.

The previous federal law referred only to telephones, but the new Act expanded communication tapping to the Internet, because it redefined a pen register as "a device or process which records or decodes dialing, routing, addressing, or signaling information transmitted by an instrument or facility from which a wire or electronic communication is transmitted." This essentially allows law enforcement agencies to record, without probable cause and court supervision, e-mail addresses and URLs. Some jurists opine that this actually allows the agencies to record not only e-mail sender and recipient addresses and Web addresses but also the content of e-mail messages and Web pages.

Even before adoption of the PATRIOT Act, the FBI used "packet sniffing" devices connected to the servers operated by Internet service providers (ISPs). Until 2002, the agency used a custom-built device known as Carnivore, and later started using commercial devices that reportedly perform the same way. The devices are supposed to monitor e-mail traffic of suspects. However, millions of other subscribers use the same servers and therefore are subject to the same surveillance.

When tapping communications, law enforcement agencies need the cooperation of a third party, such as a telephone company or an ISP. In the past, the law limited the definition of such third parties. Now, there is no limitation. Therefore, if a university, public library, municipality, or an airport provides access to the Internet—such as through a hotspot—all users of these services are subject to surveillance. Furthermore, that third party is prohibited from notifying anyone, including unsuspected users, of the surveillance.

Proponents of the Act wanted to leave all its provisions in place and add two more. They would like to allow the FBI to demand records without first obtaining an approval from a prosecutor or a judge. Some would also amend the law to require the U.S. Postal Service to let FBI agents copy information from the outside of envelopes in the mail. The law was not changed, and its term was extended.

Again, we are faced with an old dilemma: How far should we allow our governments to go in their efforts to protect us against crime and terrorism? At what point do we start to pay too much in terms of privacy and civil rights for such protection? And when terrorists strike or threaten to strike, should we give up our liberties for more security?

RECOVERY MEASURES

Security measures might reduce undesirable mishaps, but nobody can control all disasters. According to 2006 statistics of the Federal Emergency Management Agency (FEMA), the cost of weather and other disasters in the period 1986–2005 in the United States was $278 billion. Only $21.6 billion of this damage was caused by terrorism. The other causes were tropical storms, tornadoes, winter storms, earthquakes, and other events. To be prepared for disasters when they do occur, organizations must have recovery measures in place. Organizations that depend heavily on ISs for their daily business often use redundancy; that is, they run all systems and transactions on two computers in parallel to protect against loss of data and business. If one computer is down, the work can continue on the other computer. Redundancy makes the system fault tolerant. However, in distributed systems, doubling every computing resource is extremely expensive, so other measures must be taken.

The Business Recovery Plan

To prepare for mishaps, either natural or malicious, many organizations have well-planned programs in place, called **business recovery plans** (also called *disaster recovery plans, business resumption plans,* or *business continuity plans*). The plans detail what should be done and by whom if critical systems go down. In principle, the systems do not have to be ISs. However, most of the attention and resources in recovery plans are devoted to measures that should be taken when ISs go down or if IS operations become untrustworthy. The U.S. federal government regards business continuity planning as being in the national interest, and the Department of Homeland Security has established a Web site that includes useful information on this topic (*www.ready.gov*).

In business recovery planning, the emphasis should not be on the damage to the organization's assets but to its business. The estimates and measures taken should be to minimize damage to the organization's ability to resume business operations as well as to minimize the damage to operations from the disaster.

Hurricane Katrina, which hit the U.S. Gulf Coast in 2005, was a wake-up call for many executives, reminding them in terrible terms of the need for recovery planning. Concern about disaster recovery has spread beyond banks, insurance companies, and data centers, those traditionally concerned with disaster recovery. Many customer service and retail firms realize that they can easily lose customers if they don't deliver services and products in a timely manner, which is why the terms "business recovery," "business resumption," and "business continuity" have caught on in some circles. In interactive computing environments, when business systems are idle, so are the people who bring in revenue. Employees cannot do their work, customers cannot purchase, and suppliers cannot accept requests for raw materials and services. In addition, companies' reputations can be harmed, and competitive advantage and market share lost.

Experts propose nine steps to develop a business recovery plan:

1. *Obtain management's commitment to the plan.* Development of a recovery plan requires substantial resources. Top management must be convinced of the potential damages that paralysis of information systems might cause. Once management is committed, it should appoint a business recovery coordinator to develop the plan and execute the plan if disaster occurs.
2. *Establish a planning committee.* The coordinator establishes a planning committee comprising representatives from all business units that are dependent on computer-based ISs. The members serve as liaisons between the coordinator and their unit managers. The managers are authorized to establish emergency procedures for their own departments.
3. *Perform risk assessment and impact analysis.* The committee assesses which operations would be hurt by disasters, and how long the organization could continue to operate without the damaged resources. This analysis is carried out through interviews with managers of functional business areas. The committee compiles information regarding maximum allowable downtime, required backup information, and the financial, operational, and legal consequences of extended downtime.

4. *Prioritize recovery needs.* The disaster recovery coordinator ranks each IS application according to its effect on an organization's ability to achieve its mission. **Mission-critical applications**, those without which the business cannot conduct its operations, are given the highest priority. The largest or most widely used system might not be the most critical. Applications might be categorized into several classes, such as:
   - *Critical*: Applications that cannot be replaced with manual systems under any circumstances.
   - *Vital*: Applications that can be replaced with manual systems for a brief period, such as several days.
   - *Sensitive*: Applications that can be replaced with acceptable manual systems for an extended period of time, though at great cost.
   - *Noncritical*: Applications that can be interrupted for an extended period of time at little or no cost to the organization.
5. *Select a recovery plan.* Recovery plan alternatives are evaluated by considering advantages and disadvantages in terms of risk reduction, cost, and the speed at which employees can adjust to the alternative system.
6. *Select vendors.* If it is determined that an external vendor can better respond to a disaster than in-house staff and can provide a better alternative system, then the most cost-effective external vendor should be selected. Factors considered should include the vendor's ability to provide telecommunications alternatives, experience, and capacity to support current applications.
7. *Develop and implement the plan.* The plan includes organizational and vendor responsibilities and the sequence of events that will take place. Each business unit is informed of its responsibilities, who the key contacts are in each department, and the training programs available for personnel.
8. *Test the plan.* Testing includes a walk-through with each business unit, simulations as if a real disaster had occurred, and (if no damage will be caused) a deliberate interruption of the system and implementation of the plan. In mock disasters, the coordinator measures the time it takes to implement the plan and its effectiveness.
9. *Continually test and evaluate.* The staff must be aware of the plan at all times. Therefore, the plan must be tested periodically. It should be evaluated in light of new business practices and the addition of new applications. If necessary, the plan should be modified to accommodate these changes.

The plan should include the key personnel and their responsibilities as well as a procedure to reinstitute interactions with outside business partners and suppliers. Because an organization's priorities and environment change over time, the plan must be examined periodically and updated if necessary. There will be new business processes or changes in the relative importance of existing processes or tasks, new or different application software, changes in hardware, and new or different IS and end users. The plan must be modified to reflect the new environment, and the changes must be thoroughly tested. A copy of the plan should be kept off-site, because if a disaster occurs, an on-site copy might not be available. Many companies keep an electronic copy posted at a server many miles away, so that they can retrieve it from wherever their officer can have Internet access.

Although the threat of terrorism has increased awareness for the need of recovery plans, CIOs often find the tasks of earmarking funds for disaster recovery programs difficult because they cannot show the return on investment (ROI) of such planning. Most companies institute recovery programs only after a disaster or near-disaster occurs. Usually, the larger companies have such programs. Even at companies that do have recovery plans, experts estimate that most plans are never tested. Worse, some experts observed that one out of five recovery plans did not work well when tested.

Recovery Planning and Hot Site Providers

Companies that choose not to fully develop their own recovery plan can outsource it to companies that specialize in either disaster recovery planning or provision of alternative sites. Strohl Systems Group, Inc., EverGreen Data Continuity, Inc., and other companies provide both planning and software for disaster recovery. The software helps create and update records of key people and procedures. Fewer companies provide alternative sites—**hot sites**—chief among them IBM, Hewlett-Packard, and SunGard Availability Services, a division of SunGard. They provide backup and operation facilities to which a client's employees can move and continue operations in case of a disaster.

For example, IBM maintains a business continuity and recovery center in Sterling Forest, New York, 45 miles from midtown Manhattan. The center is equipped with desks, computer systems, and Internet links. Customers can use the duplicate databases and applications maintained for them. The company also provides hotel rooms and air mattresses for people who need to work long hours. As soon as the power went out one summer, the center's diesel-powered generators started up, and it was ready to take in clients' employees. Some clients had secured online systems but no light in the offices. These clients operated the systems from links at the center.

More than 90 percent of U.S. businesses are within 35 miles of a SunGard center. Worldwide, the company maintains redundant facilities totaling 279,000 square meters (3 million square feet), equipped with software and networking facilities to enable a client organization to resume business within hours.

The company collaborates with Cisco, a leading vendor of networking equipment, in what it calls Crisis Management Services. By 2007, the companies had a combined experience of helping organizations in 70 major catastrophes in 49 countries.

Hewlett-Packard's Business Continuity & Availability Services division offers both hot sites and mobile facilities. When a disaster occurs, HP sends a mobile office to a place designated by the client. Each air-conditioned office includes up to 30 desks equipped with computers, telephones, a server, and power generators. Company technicians help load applications and data. The hot site can accommodate up to 1000 client employees. The company says it has 60 recovery facilities worldwide and that by 2007 it had helped clients recover from more than 5000 disasters.

POINT OF INTEREST

The Gap

According to a *Baseline* magazine survey of CIOs, the average dollar proportion of the total IT budget spent on security measures was similar for both large (revenue over $500 million) and small to medium companies (revenue under $500 million): 8 percent for large companies, and 7.4 percent for small and medium ones. However, CIOs of large companies are more confident that their companies have not and will not suffer from security breaches. They also reported greater efforts to secure their systems. For example, to the question "What steps does your company take to protect employee or customer data?", which was followed with a list of possible measures, 77 percent of the small and medium companies said they limited access to personal customer or employee information, and 83 percent of the large companies responded this way. Seventy percent of the large companies said they encrypted personal employee or customer data when transmitted over the Internet or company network, while only 50 percent used encryption in the small and medium companies.

Source: Alter, A.E., "How Secure Are Mid-Market Companies?" *Baseline*, May 17, 2007.

THE ECONOMICS OF INFORMATION SECURITY

Security measures should be dealt with in a manner similar to purchasing insurance. The spending on measures should be proportional to the potential damage. Organizations also need to assess the minimum acceptable rate of system downtime and ensure that they can financially sustain the downtime.

How Much Security Is Enough Security?

From a pure-cost point of view, how much should an organization spend on data security measures? Two types of costs must be considered to answer this question: the cost of the potential damage, and the cost of implementing a preventive measure. The cost of the damage is the aggregate of all the potential damages multiplied by their respective probabilities, as follows:

$$Cost\ of\ potential\ damage = \sum_{i=1}^{n} Cost\ of\ disruption_i \times Probability\ of\ disruption_i$$

where i is a probable event, and n is the number of events.

Experts are usually employed to estimate the cost and probabilities of damages as well as the cost of security measures. Obviously, the more extensive the preventive measures, the smaller the damage potential. So, as the cost of security measures goes up, the cost of potential damage goes down. Ideally, the enterprise places itself at the optimum point, which is the point at which the total of the two costs is minimized, as Figure 14.8 illustrates.

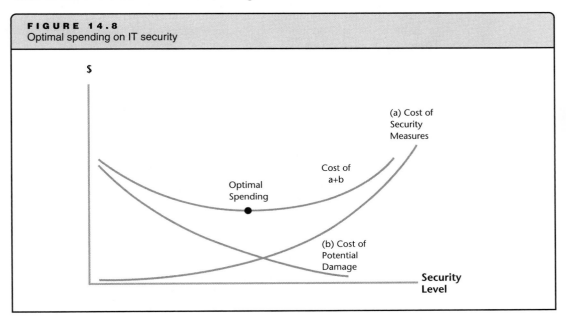

FIGURE 14.8
Optimal spending on IT security

When budgeting for IT security, managers need to define what they want to protect. They should focus on the asset they must protect, which in most cases is information, not applications. Copies of applications are usually kept in a safe place to replace those that get damaged. They should also estimate the loss of revenue from downtime. Then, they should budget sums that do not exceed the value of what the measures protect—information and potential revenues. Even the most ardent IT security advocates agree that there is no point spending $100,000 to protect information that is worth $10,000.

Calculating Downtime

All other factors being equal, businesses should try to install ISs whose downtime is the lowest possible, but if obtaining a system with a higher uptime adds to the cost, they should consider the benefit of greater uptime against the added cost. Mission-critical systems must be connected to an alternative source of power, duplicated with a redundant system, or both. Often, such systems must be up 24 hours per day, 7 days per week.

When the service that the business provides depends on uninterrupted power, the systems are often connected to the grids of two utility companies and an alternative off-grid power source, such as generators. For example, Equinix, a company in Newark and Secaucus, New Jersey, that maintains data centers for large companies uses such an arrangement. Both facilities receive power from two power stations. Even if both utility companies stop supplying electricity, the company's systems are automatically powered by batteries, and shortly after that by diesel generators. Clients can continue to transmit and receive data as if nothing happened.

Recall the discussion of system uptime in Chapter 8, "The Web-Enabled Enterprise." Experts can provide good estimates of the probability that systems will fail, both in terms of power failure in a certain region and for particular applications. Experience in operating certain systems, such as ERP and SCM systems, can teach the IT staff for how many minutes or seconds per year the system is likely to fail. For example, if the uptime of a system is 99 percent ("two nines"), it should be expected to be down 1 percent of the time, and if "time" means 24×7, downtime expectancy is 87.6 hours per year (365 days $\times$ 24 hours $\times$ 0.01). This might be sufficient for a system supporting some human resources operations, but not an airline reservation system or an SCM system of a global company. For these systems, the number of nines must be greater, such as 99.999 percent, in which case there would be only 5.256 minutes of downtime expected per year (365 $\times$ 24 $\times$ 60 $\times$ 0.00001).

More and more ISs are now interfaced with other systems, which makes them a chain or cluster of several interdependent systems. For example, if system A is connected to system B, B depends on A, and the uptime of the systems are 99 percent and 99.5 percent, respectively; the probability of uptime for B is the multiplication of these probabilities, or 98.505 percent. Therefore, you could expect the systems to be down 0.01495 of the time, about 131 hours per year. This is a greater downtime than if system B operated independently. The greater the number of interdependent systems, the greater the expected downtime.

Redundancies, on the other hand, reduce expected downtime. For example, if two airline reservation systems operate in parallel, each can serve all the transactions, and the probabilities of their failures are 2 percent and 3 percent, the probability that the reservation service will be down is 0.06 percent (0.03 $\times$ 0.02), just 0.0006 of the time. This downtime is significantly smaller than the downtime of a service based on either system individually. This is why so many companies rely on redundant power sources and systems, such as duplicate databases, mirrored servers, and duplicate applications, especially when much of their operations are executed online, and even more so when the operations depend on constant online interaction with customers.

There might be no point in spending much money to increase the "nines" of uptime for every system. For example, if the only purpose of an IS is to help access a data warehouse to glean business intelligence (recall the discussions in Chapter 7, "Databases and Data Warehouses," and Chapter 11, "Business Intelligence and Knowledge Management"), spending thousands of dollars to increase its number of nines from 99 to 99.999 is probably not a wise choice. For a data warehouse, if an analysis cannot be performed immediately, it can usually be performed later without serious ramifications.

- The purpose of controls and security measures is to maintain the functionality of ISs, the confidentiality of information, the integrity and availability of data and computing resources, the uninterruptible availability of data resources and online operations, and compliance with security and privacy laws.

- Risks to ISs include risks to hardware, risks to data and applications, and risks to networks.

- Risks to hardware include natural disasters, such as earthquakes, fires, floods, and power failures, as well as vandalism. Protective measures run the gamut from surge protectors to the maintenance of duplicate systems, which make ISs fault tolerant.

- Risks to data and applications include theft of information, identify theft, data alteration, data destruction, defacement of Web sites, computer viruses, worms, and logic bombs, as well as non-malicious mishaps such as unauthorized downloading and installation of software.

- Risks to online operations include denial of service and computer hijacking.

- To minimize disruption, organizations use controls. Controls include program robustness and constraints on data entry, periodic backup of software and data files, access controls, atomic transactions, and audit trails.

- Access controls can be categorized into three groups: what you know, what you have, and who you are. Access controls also include information that must be entered before information resources can be used: passwords, security cards like SecureID®, and biometrics.

- Atomic transactions are an important control that ensures information integrity: either all files involved in a transaction are updated, or no files are updated.

- To protect resources that are linked to the Internet, organizations use firewalls, which are special hardware and software to control access to servers and their contents.

- Encryption schemes scramble messages at the sending end and descramble them at the receiving end. Encryption is also used to authenticate the sender or recipient of a message, verifying that the user is indeed the party he or she claims to be.

- To encrypt and decrypt messages the communicating parties must use a key. The larger the number of bits in the key, the longer it takes to break the encryption. In symmetric encryption, both users use a private, secret key. In asymmetric key encryption, the parties use a public and a private key.

- The public-private key method does not require both parties to have a common secret key before the communication starts. This system is a useful feature that lets consumers and organizations transact business confidentially on the Web.

- SSL, TLS, and HTTPS are encryption standards specially designed for the Web. They are embedded in Web browsers.

- Organizations can purchase public and private keys along with an associated digital certificate from a certificate authority. Digital certificates contain the certificate holder's public key and other information, such as the issue and expiration date of the certificate.

- Many organizations have business recovery plans that are developed and periodically tested by a special committee. The plans identify mission-critical applications and prescribe steps that various employees should take in a disaster.

- A growing number of companies also use the services of organizations that specialize in providing alternative sites, known as hot sites, to continue operations in case of a debilitating event such as a terror attack, natural disaster, or power outage.

- When considering how much to invest in security measures, organizations should evaluate the dollar amounts of the potential damage on one hand, and the cost of security on the other hand. The more that is spent on security, the smaller the potential loss.

- A system that depends on other systems for input has a greater downtime probability than if it is used independently of other systems. Redundant systems significantly reduce downtime probability.

- Governments are obliged to protect citizens against crime and terrorism and therefore must be able to tap electronic communication of suspects. Such practices often collide with individuals' right to privacy.

WORLDWIDE HOST REVISITED

Worldwide Host's Web site has been up and running for several months now. In that time, the TripExpert site has been defaced, experienced a denial-of-service attack, and been hit with an attempt to invade the customer database. Putting its system on the World Wide Web has certainly introduced challenges to Worldwide's IS staff. Let us look at some computer security issues in more depth.

3. The chapter discussed controls on information systems to help secure them. From the description in the opening case, you know that Worldwide Host uses secure servers and physical access controls for its information systems. What other types of controls should it be using to safeguard its systems? Develop a list for Michael Lloyd and his security chief.

What Would You Do?

1. Jason Theodore, Worldwide Host's IS security chief, informs Michael Lloyd that there have been a growing number of Trojan horse attacks that target specific businesses. The perpetrators send e-mails to specific employees who have access to important financial information. The senders disguise themselves as a colleague and ask the employee to go to a Web site or open an attachment that installs a virus that is able to send sensitive financial information back to the perpetrator. Develop a list of recommendations for Michael Lloyd to distribute to employees to help safeguard Worldwide Host.

2. Severe weather is always a concern for IS personnel. If a flood or power outage hit Worldwide Host's offices, it could take down the company's entire operations. Michael Lloyd has asked you to help him develop a disaster recovery plan. What measures would you recommend Worldwide Host take to prepare for and recover from a disaster?

New Perspectives

1. Worldwide Host handles thousands of transactions involving customers' credit cards. The TripExpert. com employees are complaining that the response time for the new system is much slower than their old reservations system. Michael Lloyd knows that this is the result of security measures—use of passwords, encryption and decryption, and screening of transactions. He is meeting with the travel staff to explain the security-response trade-off. Prepare an outline of his speech for him, which discusses the pros and cons of the security measures.

2. Michael Lloyd has been approached by a vendor that provides disaster recovery services at an alternative site. Should Worldwide Host consider use of such a service? If so, for what systems and business functions?

KEY TERMS

access controls, 486	digital certificate, 497	message digest, 495
antivirus software, 482	digital signature, 495	mission-critical application, 501
asymmetric (public key) encryption, 493	DMZ, 490	plaintext, 492
	downtime, 476	proxy server, 490
atomic transaction, 488	encryption, 492	RAID, 485
audit trail, 489	firewall, 490	social engineering, 478
authentication, 492	hijacking, 484	SSO (single sign-on), 498
backup, 485	honeypot, 481	symmetric (secret or private key) encryption, 493
biometric, 487	honeytoken, 481	
blackout, 477	hot site, 502	Transport Layer Security (TLS), 493
brownout, 477	HTTPS, 494	
business recovery plan, 500	identity theft, 478	Trojan horse, 482
certificate authority (CA), 497	information systems auditor, 489	uninterruptible power supply (UPS), 477
ciphertext, 492		
controls, 485	keystroke logging, 477	virus, 481
denial of service (DoS), 484	logic bomb, 482	worm, 482

REVIEW QUESTIONS

1. What are the goals of security measures for ISs? Explain.

2. All the data of your company is concentrated in two databases. All employees use PCs or laptop computers, and all use a corporate network. You are to prioritize protection of the following elements of your company: PCs and laptops, the databases, the corporate network. Which is the element about which you should be most concerned, and why?

3. Data alteration and destruction are dreaded by many IS managers more than any other mishap. Why? Is the threat of Web site defacement as severe as data destruction or alteration? Why or why not?

4. Some companies still make a duplicate copy of disks or tapes and transport them to a remote site as a precaution against loss of data on the original storage media. What is the preferred method of keeping secured copies of data nowadays? Give at least two benefits and one possible drawback of the more recent approach.

5. Comment on the following statement: If your computer is connected to an external communication line, anyone with a similar link can potentially access your systems.

6. What is a honeytoken and how is it used by companies?

7. What is a honeypot and how is it used by businesses?

8. What is the difference between a virus and a worm? Which is potentially more dangerous and why?

9. Why is encryption that uses the public-key method so important in electronic commerce?

10. Assume that you are charged with developing an application to record basketball performance statistics. What limits would you include to ensure that the information entered is reasonable?

11. What is an audit trail? What audit trail information would you have for a shipping record?

12. This chapter gives an example of an atomic transaction. Give another example from any business area.

13. What is the difference between authentication and confidentiality?

14. What are biometric access controls? How are they better than passwords?

15. What is a firewall, and how does it work?

16. What is a DoS? How is it executed, and what is the purpose of zombies in a DoS? What can organizations do to prevent a DoS attack?

17. What is the purpose of business recovery plans?

18. A growing number of companies have implemented business recovery plans, but many still have not developed such plans. What may be the reasons for that?

19. Companies that process credit-card transactions for merchants have their computers vaulted behind concrete walls, iron bars, and heavy steel doors. Employees must enter a code into a keypad to enter the vaults. Yet, every so often information on millions of credit-card accounts is stolen without any physical break-in. How so?

20. In the Blockbuster example of system controls, the cash register displays the message "Do not rent!" when a patron reaches the maximum debt allowed. However, the customer service representative might still rent a videotape to the customer. What would you do to better enforce the chain's policy?

21. A military officer in Colorado orders an item whose part number is 7954. The clerk at the supply center hundreds of miles away receives the order through his computer and ships the item: a ship's anchor, not realizing that Colorado is located hundreds of miles from any ocean. Apparently, the officer wanted to order item number 7945, a fuel tank for a fighter aircraft, but he erred when entering the item's number. What controls would you implement both at the entry system and at the systems employed at the supply center to prevent such mistakes?

22. The average loss in a bank robbery is several thousand dollars, and the culprit has an 85 percent chance of being caught. The average damage in a "usual" white-collar fraud is several tens of thousands of dollars. The average amount stolen in computer fraud against organizations is several hundreds of thousands of dollars, and it is extremely hard to find the culprit. Why is the amount involved in computer fraud so high, and why is it difficult to find the culprits?

23. To prevent unauthorized people from copying data from a database, some companies forbid their employees to come to work with USB flash memory devices and they subject the employees to body searches. Is this an effective measure? Why or why not?

24. The majority of criminals who commit computer fraud are insiders, that is, employees. What measures would you take to minimize insider fraud through ISs?

25. When accessing an information system, would you prefer that your identity be verified with a biometric (such as your palm or fingerprint, or your retinal scan), or with a password? Why?

26. Explain in an intuitive way why the downtime probability of a system that depends on another system is greater than if it were operating independently.

27. Employees often complain about the hurdles they have to pass whenever they need to access data and the slow response of ISs because of firewalls and encryption measures. As a CIO, how would you explain the need for such measures to employees? Would you give them any say in the decision of how to balance convenience and protection of data and applications?

28. Organizations often use firewalls to block employee access to certain Web sites. Do you agree with this practice, or do you think it violates employee privacy?

29. Special software might keep track of Web pages that employees download to their PCs. Do you think this practice violates employee privacy?

30. When financial institutions discover that their ISs (especially databases) have been broken into, they often do not report the event to law enforcement officers. Even if they know who the hacker is, they do what they can to avoid publicity. Why? Should they be forced to report such events?

31. When hackers are caught, they often argue that they actually did a service to the organization whose system they accessed without permission; now, they say, the organization knows its system has a weak point, and it can take the proper steps to improve security. Do you agree with this claim? Why or why not?

32. A CIO tells you, "We regularly review all of the potential vulnerabilities of our information systems and networks. We implement hardware, software, and procedures against any potential event, no matter the cost." What do you say to this executive?

33. Is the potential for identity theft growing? Explain. (*Note*: The question is not about actual identity theft for any period of time; it is about the *potential* of identity theft.)

34. Encryption helps individuals and organizations to maintain privacy and confidentiality, thereby helping protect civil liberties. However, encryption also helps terrorists and criminals hide their intentions. Some governments have laws that forbid nongovernment organizations to use strong encryption software. The idea is to allow people to encrypt their communication, but not strongly enough to prevent the government from decrypting the communication in surveillance of suspected criminals and terrorists. Do you favor such laws, or do you advocate that everybody have access to the strongest encryption software available? Explain.

APPLYING CONCEPTS

35. Search the Web for the full text of HIPAA. Assume you are the CIO of a health insurance company. List and explain five controls that you must implement in your organization's ISs as a result of this law.

36. Research the impact of the Sarbanes-Oxley Act on ISs. Write a two-page report explaining the major controls corporations must incorporate in their ISs to satisfy the Act.

37. Log on to a secure Web site. Figure out which icon you have to click to receive information on the security measures used in the session. Send your professor an e-mail message detailing the site's URL and all the information you obtained: the length of the key that is used for encryption, the type of digital certificate used, the issuer of the digital certificate, the date it was issued and its expiration date, and so forth. Explain each item.

38. Some companies provide free software versions of their firewalls. Research three such firewall applications and compare their characteristics: options to block incoming communication, options to block outgoing communication, ease of learning, ease of use, etc. Make a recommendation for individuals based on your comparison.

HANDS-ON ACTIVITIES

39. Use Excel or another spreadsheet application to show your work when solving the following problem: A company uses three information systems that are linked sequentially: System A feeds System B, and System B feeds System C. Consider the following average uptimes: System A, 98 percent; System B, 97 percent; System C, 95 percent. What is the average expected downtime (as a percentage) of System C?

40. Use Excel or another spreadsheet application to show your work when solving the following problem: To reduce chances of failure, a company has connected all of its vital information systems to electric power from two different utility firms. The probability of failure of electric power from one utility firm is 2 percent. The probability of failure of electric power from the other utility firm is 1.5 percent. What is the probability that these information systems will receive no electric power at all?

41. A CIO states, "Our online transaction system has availability of five nines. However, we have a SaaS (software as a service) contract for using a human resources information system. The HR system has availability of only three nines."

    a. Calculate the minutes of downtime per week for each of these systems.

    b. Explain why the company must have such a high number of nines for one system but can settle for a significantly lower number of nines for the other system.

42. Team up with another student. Research the Web for recovery planning expenditures in your country or worldwide over the past five years. Prepare a table showing the expenditure amounts for each year. Add an analysis that explains the reasons for changes in the expenditures from one year to another.

43. Your team should evaluate the business recovery plan of your school. If there is none, write a plan for the school. If there is one in place, evaluate its weaknesses and make suggestions for improvement. Prepare a 10-minute, software-based presentation of your findings and suggestions.

FROM IDEAS TO APPLICATION: REAL CASES

Good to Be Home Again

To ensure that security measures work well, companies must test the measures periodically. Testing is often outsourced to other companies, organizations that specialize in this field, but some prefer to do the testing in-house. Scottish Re is an example of such do-it-yourself.

Scottish Re is a life reinsurance company, with operations through subsidiaries in the United Kingdom, the United States, and other countries. It was established in 1998 and is headquartered in Bermuda. Reinsurance companies insure insurance companies. Their revenue comes from the premiums that insurance companies pay them to cover part or all of the payment to the insured, or—in the case of life insurance—to an insured's survivors. The company had more than $2.2 billion in revenue in 2006.

When testing ISs for security, an authorized person probes an organization's networks and applications for security vulnerabilities by attempting to exploit them. The testers decide whether a certain vulnerability can be exploited by unauthorized people. Testers can be employees of the organization, but many organizations outsource the testing to a company that specializes in IT security. Scottish Re had outsourced security testing until 2006. Then, it decided to conduct vulnerability tests by itself.

Scottish Re now conducts its own penetration tests using software offered by Core Security Technologies, a company that specializes in such testing applications. The application Scottish Re licensed from the vendor is Core Impact. By using Core Security's application, customers are supposed to gain comprehensive information about their security readiness. The application's purpose is also to help licensees make decisions on how to prioritize vulnerabilities for counteraction (such as patching or redesigning the deployment of networking devices), plan remediation efforts, optimize existing security infrastructure, and make decisions regarding future security products and services. Among Core Technologies' clients are Bloomberg, JPMorgan, Chase, and H&R Block.

Using an outside service to perform security checks typically costs $100,000 per year. Licensing Core Impact costs about $25,000 per year. Considering the labor involved, the tool saves about 30 percent of the total cost. However, moving the activity in-house saves time as well. When using the third party to conduct the tests, Scottish Re employees had to spend 8–10 hours to interpret the tests' results, which they received in a report. It was not easy to determine what a true vulnerability was. Patching what seems to be a vulnerability but actually cannot be exploited is costly. The use of Core Impact eliminated the report review phase. The software clearly informs the users if a vulnerability requires action.

When Scottish Re used the third party to conduct the test, it could not schedule tests. Tests were typically done at the end of each month. Now, company employees can use Core Impact whenever they wish. This enables them to conduct a test whenever they change networked software.

Using the software, Scottish Re IT specialists discovered a vital finding. Although they used Microsoft's patches for firewalls to eliminate the possibility of a hacker taking control of a server, they discovered that the company's own unauthorized employees could compromise unpatched software from the inside despite the firewall patches. The specialists patched the affected servers and established router access control lists to protect several servers that could not be immediately taken down for patching. These are servers that provide services that the company must have available at all times. The control lists make it almost impossible for any unauthorized person to take over the server.

Experts see a downside to in-house penetration testing. They say that the company's security personnel are so engrossed in the company's procedures and security measures that they may lose objectivity. An outside consultant, on the other hand, has no preconceived notions. One expert said it is difficult for an insider to forget everything he or she knows and act like an outsider. He added that even large companies that conduct testing in-house use a safeguard: they occasionally hire the services of outside companies to conduct penetration tests in addition to those conducted by the IT team. Scottish Re no longer hires outside testing. However, the company's internal auditor's staff conduct penetration tests independently of the IT team.

Scottish Re's CIO is content with in-house testing. He indicated that being able to conduct a test whenever his team wished was a great advantage over outsourced tests. This is especially important to him and his team whenever new software is deployed or old software is revised.

Source: Moore, J., "Security Testing: Taking Charge," *Baseline*, March 9, 2007; (www.scottishre.com), August 2007; (www.coresecurity.com), August 2007.

Thinking About the Case

1. What are the advantages of having an organization's own IT team conduct vulnerability tests?

2. What are the advantages of outsourcing vulnerability tests?

3. Why does Scottish Re use the internal audit staff for vulnerability tests? Do you think this is a good substitute for using another company's testing?

4. Why does the Scottish Re IT team use Core Impact after implementing new software or software changes?

Never, Ever Down

When airplanes approach a German airport, their flight is monitored by air-traffic controllers who are government employees. Once an airplane touches the ground, responsibility for its movement is transferred to the airport. Frankfurt International Airport—known as Fraport—is Europe's second largest airport and among the busiest in the world. It serves more than 3.5 million passengers per month and this number is growing. It serves air freight of more than 157,360 metric tons per month, and this number, too, is growing.

Like other airports, Fraport uses a wealth of data for its operations, from managing airplanes between landing and takeoff to displaying flight information for passengers to movement of passenger trams. The data must be available 24/7. If it is not available, the airport can allow only emergency landings. More than 120 software applications are housed on the server to provide access to the data as well as to support ground traffic services and office communications.

Fraport's new data center is buried nine meters (27 feet) underground near a runway. It was built in 2006 to replace a 30-year-old data center. The new data center is a classic example of redundancy. Each server has two power sources and two cooling systems. All servers have mirrored "twins" that are housed in a separate room. Electric power is provided by three different companies from three different and separate power grids. Cooling water comes from redundant chillers that provide 1.4 megawatts of cooling capacity. If anything goes wrong with these cooling sources, the airport can use a refrigerated lake it maintains to provide emergency cooling water.

However, the redundancy of this airport's systems goes beyond servers, power, and cooling. Even the building itself has redundancy. A separate section within the building houses redundant servers. This building-within-a-building will continue to operate and provide current data if the rest of the building is compromised. Fraport AG is the company that manages the airport. Gedas Operational Services is the company that runs the data center under the supervision of Fraport AG.

Such redundancies are not unusual for government facilities but are rare for commercial enterprises. However, Fraport is partly owned by the German federal government and the government of the state of Hesse, where the airport is located. The federal and state governments wished to protect their ownership.

The data center was built fast. One reason for this is the order in which the facility was built. Usually, a company selects servers, then builds a structure to house them and contracts with utility companies to provide power. In this case, the project managers first contracted with the utility companies, then built the facility, and only then selected and purchased the servers. So, when the servers arrived, it took very little time to install them. Experts also observed another important reason for the speed of building the data center: Fraport did not patch its 30-year-old data center. It abandoned it and built a new one from scratch. This eliminated any upgrading of hardware and software. The result was a new, top-of-the-line data center, available within a short time after the design was completed, and which cost significantly less than upgrading the old center would have cost. The center employs 335 people.

To ensure power continuity, the new data center uses American Power Conversion's (APC) InfraStruXure®, an array of facilities integrating power, cooling, server racks, management software, and security services. APC executives hosted executives from Fraport and demonstrated the architecture. One year after the Fraport executives decided to use APC's technology, the data center was completed. An APC executive noted that the speed from decision to a running data center of this magnitude was exceptional. He also noted that what enabled this speedy installation is the ability of InfraStrucXure® to integrate all that is needed in such a facility, from power and cooling to security and services. The racks can simply be rolled into the new building, and the servers can be connected. The system is highly modular. This allows Fraport's CIO to change electric power sources and replace or add servers quite flexibly. Only 35 percent of the physical facility is in use. The rest of the physical plant will accommodate additional hardware, if any is required in the future.

Source: Rash, W., "Keeping the People Moving," *eWeek*, March 28, 2007; (www.apc.com), 2007; "Fraport and T-Systems launch joint venture: New data center controls Frankfurt Airport," *Basman Explore*, December 1, 2006.

Thinking About the Case

1. Fraport's data center uses much redundancy. What is redundant? What is the reason Fraport data center uses so much redundancy?

2. What was the major reason for the short lead time between the decision to build a new data center and the start of operations?

3. Conduct some research about the cost of such a data center. Would a small- or medium-size company be able to afford a data center with so much redundancy?

The Tester

The case studies in this book usually revolve around organizations. This story is about a person, a "soldier" in the war against cybercriminals. Meet Mark Seiden, security tester extraordinaire. "Tell me which information your bank keeps most secret," he challenged a top bank executive four years ago, "and I'll get it anyway." The executive complied. He told Seiden he wanted the identities of clients who were negotiating secret deals, so secret that many people inside the bank referred to them by code names. He also wanted the financial details of some mergers and acquisitions in which his bank was involved. The executive knew that those two types of information were kept under strict electronic locks by the bank.

A week later, Seiden visited the executive again in his office. He gave him a printout of the secret information. He also gave the man photocopies of the floor plans of each bank office and a suitcase filled with backup tapes from which Seiden could reconstruct all the files maintained on the bank's computers. Seiden spent two weekend nights to obtain the information.

Seiden, with 35 years of experience in computer programming, is one of a small group of professional intruders, experts who are paid by corporations to find security loopholes, both in physical facilities and software. Companies hire Seiden to help improve security systems and procedures to protect their ISs and other sensitive corporate assets.

Business is booming for Seiden and his colleagues. As hackers increase their invasion of corporate databases and steal information from a growing number of

organizations, executives are learning fast why it pays to hire the services of such people. Experts say that in the early days of the Internet, breaking into corporate ISs was mainly a matter of showing off one's prowess. Now, it has become a crime of greed. The main targets are personal information and credit-card account numbers. In a survey conducted by the FBI in 2005, 87 percent of the polled corporations said they had routine security audits, an increase from 82 percent in 2004. An analyst for the Gartner Group said that North American corporations spent more than $2 billion on security consultants in 2004, an increase of 14 percent from 2003.

Much of the work such consultants do is not related to hacking. Seiden has a wardrobe of uniforms and other corporate garbs. They include a uniform the FedEx drivers wear, and a windbreaker that Iron Mountain workers wear when they drive their vans to pick up backup files for credit card-processing firms. He also holds a set of tools that help him pick locks, at which he is adept. If you ask, he will tell you that the easiest way to enter a locked room is through the plenum space between the hard ceiling and the tiles underneath, space used for wiring and ventilation. Remove a tile, and you are in a safe room that is no longer so safe.

So how did Seiden get that precious information about the bank? The bank maintains some of the best security software, so Seiden did not even try to crack it. He obtained a badge that the bank often handed out to outside consultants. Wearing the badge, he could enter the room where the bank's computers were housed at headquarters. He noticed that the master keys to the building as well as the building's floor plans were stored in a file cabinet that took him only two minutes to pick. Roaming freely in "safe" rooms, he also found the backup tapes.

He then used social engineering to obtain information. Pretending he was a bank employee, he telephoned the accounting department and asked whom he should contact for assigning a code name to a project. Equipped with the name of the clerk who assigned code names, he went to her office and noticed that she placed sheets with code names in a folder and locked the folder in a file cabinet. Since the office was in a locked area, she had no reason to lock the file cabinet (which Seiden could pick anyway, if he needed to). He later explored the folder and obtained the code names of secret clients and information about confidential mergers and acquisitions that the bank was negotiating.

Seiden agrees that corporations cannot defend themselves against every intrusion, physical or otherwise. He agrees with the analogous construction of a house without windows: it can be built, but nobody would want to live in it. What corporations should do, he says, is ensure that when an intrusion occurs they know about it and take measures to ascertain that this type of intrusion does not happen again.

Sources: Rivlin, G., "The Sniffer vs. the Cybercrooks," *New York Times*, Section 3, p. 1, col. 2, July 31, 2005; (www.msbit.com/mis.html), 2005.

Thinking About the Case

1. The case mentions three different ways of obtaining information illegally. What are they?

2. Why do you think corporations are spending increasing amounts of money on the services of security testers? Can you cite some recent mishaps that would prompt a corporation to do so?

3. Refer to the analogy of "a house without windows." You are the CEO of a large corporation. Give an example of a measure you would never take even if it enhanced security.

access controls—Hardware and software measures, such as user IDs and passwords, used to control access to information systems.

access point (AP)—An arrangement consisting of a device connected to the Internet on one end and to a router on the other end. All wireless devices link to the Internet through the router.

affiliate program—An arrangement by which a Web site promotes sales for another Web site through a link to the seller's site, and for which the affiliate is compensated. There are various schemes of compensation to affiliates.

agile methods—Software development methods that emphasize constant communication with clients (end users) and fast development of code, as well as modifications as soon as they are needed.

algorithm—An sequence of steps one takes to solve a problem. Often, these steps are expressed as mathematical formulas.

antivirus software—Software designed to detect and intercept computer viruses.

applet—A small software application, usually written in Java or another programming language for the Web.

application—A computer program that addresses a general or specific business or scientific need. General applications include electronic spreadsheets and word processors. Specific applications are written especially for a business unit to accommodate special activities.

application program interface (API)—Code in applications that helps link them to other applications. Using operating system APIs enables applications to utilize operating system features.

Application Service Provider (ASP)—A firm that rents the use of software applications through an Internet link. The arrangement is known as Software as a Service (SaaS).

application software—Software developed to meet general or specific business needs.

application-specific software—A collective term for all computer programs that are designed specifically to address certain business problems, such as a program written to deal with a company's market research effort.

Arithmetic Logic Unit (ALU)—The electronic circuitry in the central processing unit of a computer responsible for arithmetic and logic operations.

artificial intelligence (AI)—The study and creation of computer programs that mimic human behavior. This discipline combines the interests of computer science, cognitive science, linguistics, and management information systems. The main subfields of AI are robotics, artificial vision, natural language processors, and expert systems.

assembly languages—Second-generation programming languages that assemble several bytes into groups of characters that are human-readable to expedite programming tasks.

asymmetric (public key) encryption—Encryption technology in which a message is encrypted with one key and decrypted with another.

atomic transaction—A transaction whose entry is not complete until all entries into the appropriate files have been successfully completed. It is an important data entry control. (Atom = Indivisible)

audit trail—Names, dates, and other references in computer files that can help an auditor track down the person who used an IS for a transaction, legal or illegal.

authentication—The process of ensuring that the person who sends a message to or receives a message from another party is indeed that person.

autocategorization—Automatic sorting and indexing of information that is executed by specialized knowledge management software.

automatic taxonomy—A method in knowledge management to organize text and other nonstructured information in classes or categories.

avatar—A pictorial, usually three-dimensional, representation of a person in a software environment. Avatars are used in virtual worlds and other virtual reality environments.

B2B—Business-to-business, a term that refers to transactions between businesses, often through an Internet link.

B2C—Business-to-consumer, a term that refers to transactions between a business and its customers, often through an Internet link.

backbone—The network of copper lines, optical fibers, and radio satellites that supports the Internet.

backup—Periodic duplication of data in order to guard against loss.

backward compatibility—Compatibility of a device with another device that supports only an older standard. For example, USB 2.0 is backward-compatible with computers that support only USB 1.1 devices.

bandwidth—The capacity of the communications channel, practically its speed; the number of signal streams the channel can support, usually measured as number of bits per second. A greater bandwidth also supports a greater bit rate, i.e., transmission speed.

banners—Advertisements that appear on a Web page.

baseband—A communications channel that allows only a very low bit rate in telecommunications, such as unconditioned telephone twisted pair cables.

benchmarking—The measurement of time intervals and other important characteristics of hardware and software, usually when testing them before a decision to purchase or reject.

beta site—An organization that agrees to use a new application for a specific period and report errors and unsatisfactory features to the developer in return for free use and support.

bill of materials (BOM)—A list showing an explosion of the materials that go into the production of an item. Used in planning the purchase of raw materials.

biometric—A unique, measurable characteristic or trait of a human being used for automatically authenticating a person's identity. Biometric technologies include digitized fingerprints, retinal pictures, and voice. Used with special hardware to uniquely identify a person who tries to access a facility or an IS, instead of a password.

bit—Binary digit; either a zero or a one. The smallest unit of information used in computing.

bits per second (bps)—The measurement of the capacity (or transmission rate) of a communications channel.

blackouts and brownouts—Periods of power loss or a significant fall in power. Such events may cause computers to stop working, or even damage them. Computers can be protected against these events by using proper equipment, such as UPS (uninterruptible power supply) systems.

bleeding edge—The situation in which a business fails because it tries to be on the technological leading edge.

blog—A contraction of Web log. A Web site where participants post their opinions on a topic or set of related topics; these postings are listed in chronological order.

Bluetooth—A personal wireless network protocol. It enables wireless communication between input devices and computers and among other devices within 10 meters.

brainstorming—The process of a group collaboratively generating new ideas and creative solutions to problems.

brick-and-mortar—A popular term for companies that use physical structure for doing business directly with other businesses and consumers, such as stores. Often used to contrast with businesses that sell only online.

bridge—A device connecting two communications networks that use similar hardware.

broadband—High-speed digital communication, sometimes defined as at least 200 Kbps. T1, Cable modem, and DSL provide broadband.

Broadband over Power Lines (BPL)—A broadband service provided over electric power lines.

bus—The set of wires or soldered conductors in the computer through which the different components (such as the CPU and RAM) communicate. It also refers to a data communications topology whereby communicating devices are connected to a single, open-ended medium.

business analytics—Software that analyzes business data to help make business decisions, often at the strategic level. An alternative name for business intelligence.

business intelligence (BI)—Information gleaned from large amounts of data, usually a data warehouse or online databases; a BI system discovers not-yet-known patterns, trends, and other useful information that can help improve the organization's performance.

business model—The manner in which businesses generate income.

business planning—The general idea or explicit statement of where an organization wishes to be at some time in the future in terms of its capabilities and market position.

business recovery plan—Organizational plan that prepares for disruption in information systems, detailing what should be done and by whom, if critical information systems fail or become untrustworthy; also called business recovery plan and disaster recovery plan. Also known as business continuity plan.

byte—A standard group of eight bits.

C2C—Consumer-to-consumer business. The term usually refers to Web-based transactions between two consumers via the servers of an organization, such as auctions and sales. eBay is an example of a C2C site.

CASE (Computer-Aided Software Engineering)—Software tools that expedite systems development. The tools provide a 4GL or application generator for fast code writing, facilities for flowcharting or data-flow diagramming, data-dictionary facility, word-processing capability, and other features required to develop and document the new software. The term is much less popular now than it was in the 1980s and early 1990s.

cash management system (CMS)—Information system that helps reduce the interest and fees that organizations have to pay when borrowing money and increases the yield that organizations can receive on unused funds.

central processing unit (CPU)—The circuitry of a computer microprocessor that fetches instructions and data from the primary memory and executes the instructions. The CPU is the most important electronic unit of the computer.

certificate authority (CA)—An organization that issues digital certificates, which authenticate the holder in electronic business transactions.

character—The smallest piece of data in the data hierarchy.

chief information officer (CIO)—The highest-ranking IS officer in the organization, usually a vice president, who oversees the planning, development, and implementation of IS and serves as leader to all IS professionals in the organization.

chief security officer (CSO)—Also called chief information security officer (CISO), the highest-ranking officer in charge of planning and implementing information security measures in the organization, such as access codes and backup procedures.

chief technology officer (CTO)—A high-level corporate officer who is in charge of all information technology needs of the organization. Sometimes the CTO reports to the chief information officer, but in some companies this person practically serves as the CIO.

ciphertext—A coded message designed to authenticate users and maintain secrecy.

circuit switching—A communication process in which a dedicated channel (circuit) is established for the duration of a transmission; the sending node signals the receiving node; the receiver acknowledges the signal and then receives the entire message.

clickstream tracking—The use of software to record the activities of a person at Web sites. Whenever the person clicks a link, the activity is added to the record.

clock rate—The rate of repetitive machine cycles that a computer can perform; also called frequency. Measured in GHz.

closed system—A system that stands alone, with no connection to another system.

coaxial cable—A transmission medium consisting of thick copper wire insulated and shielded by a special sheath of meshed wires to prevent electromagnetic interference. Supports high-speed telecommunication.

co-location—The placement and maintenance of a Web server with servers of other subscribers of the service provider. The servers are co-located in the same facility.

competitive advantage—A position in which one dominates a market; also called strategic advantage.

compiler—A program whose purpose is to translate code written in a high-level programming language into the equivalent code in machine language for execution by the computer.

composite key—In a data file, a combination of two fields that can serve as a unique key to locate specific records.

computer-aided design (CAD)—Special software used by engineers and designers that facilitates engineering and design work.

computer-aided manufacturing (CAM)—Automation of manufacturing activities by use of computers. Often, the information for the activity comes directly from connected computers that were used for engineering the parts or products to be manufactured.

Computerized Numeric Control (CNC)—Control by computers that take data and create instructions that tell robots how to manufacture and assemble parts and products.

conclusion—The *then* component of an *if-then* rule in knowledge representation.

consumer profiling—The collection of information about individual shoppers in order to know and serve consumers better.

control unit—The circuitry in the CPU that fetches instructions and data from the primary memory, decodes the instructions, passes them to the ALU for execution, and stores the results in the primary memory.

controls—Constraints applied to a system to ensure proper use and security standards.

conversion—The process of abandoning an old information system and implementing a new one.

cookie—A small file that a Web site places on a visitor's hard disk so that the Web site can remember something about the visitor later, such as an ID number or username.

cost/benefit analysis—An evaluation of the costs incurred by an information system and the benefits gained by the system.

country-of-destination principle—The legal principle that the party that made an online purchase is entitled to sue the seller in the purchaser's jurisdiction. The European Union has adopted this approach.

country-of-origin principle—The legal principle that the party that sold online is entitled to sue or to have a lawsuit filed against it in its own jurisdiction. No country has formally adopted this legal approach.

CRT (cathode-ray tube)—A display (for a computer or television set) that uses an electronic gun to draw and paint on the screen by bombarding pixels on the internal side of the screen.

custom-designed (tailored) software—Software designed to meet the specific needs of a particular organization or department; also called tailored software.

customer relationship management (CRM)—A set of applications designed to gather and analyze information about customers.

cut-over conversion (flash cut conversion)—A swift switch from an old information system to the new.

dashboard—A graphic presentation of organizational performance. Dashboards display in an easy-to-grasp visual manner metrics, trends, and other helpful information that is the result of processing of business intelligence applications.

data—Facts about people, other subjects, and events. May be manipulated and processed to produce information.

data dictionary—The part of the database that contains information about the different sets of records and fields, such as their source and who may change them.

data flow diagram (DFD)—A graphical method to communicate the data flow in a business unit. Usually serves as a blueprint for a new information system in the development process. The DFD uses four symbols for entity, process, data store, and data flow.

data integrity—Accuracy, timeliness, and relevance of data in a context.

data management module—In a decision support system, a database or data warehouse that allows a decision maker to conduct the intelligence phase of decision making.

data mart—A collection of archival data that is part of a data warehouse, usually focusing on one aspect of the

organization such as sales of a family of products or daily revenues in a geographic region.

data mining—Using a special application that scours large databases for relationships among business events, such as items typically purchased together on a certain day of the week, or machinery failures that occur along with a specific use mode of the machine. Instead of the user querying the databases, the application dynamically looks for such relationships.

data modeling—The process of charting existing or planned data stores and flows of an organization or one of its units. It includes charting of entity relationship diagrams.

data processing—The operation of manipulating data to produce information.

data redundancy—The existence of the same data in more than one place in a computer system. Although some data redundancy is unavoidable, efforts should be made to minimize it.

data warehouse—A huge collection of historical data that can be processed to support management decision making.

data warehousing—Techniques to store very large amounts of historical data in databases, especially for business intelligence.

data word—The number of bits that a CPU retrieves from memory for processing in one machine cycle. When all other conditions are equal, a machine with a larger data word is faster.

database—A collection of shared, interrelated records, usually in more than one file. An approach to data management that facilitates data entry, update, and manipulation.

database administrator (DBA)—The individual in charge of building and maintaining organizational databases.

database approach—An approach to maintaining data that contains a mechanism for tagging, retrieving, and manipulating data.

database management system (DBMS)—A computer program that allows the user to construct a database, populate it with data, and manipulate the data.

debugging—The process of finding and correcting errors in software.

decision support system (DSS)—Information system that aids managers in making decisions based on built-in models. DSSs comprise three modules: data management, model management, and dialog management. DSSs may be an integral part of a larger application, such as an ERP system.

dedicated hosting—An arrangement in which a Web hosting organization devotes an entire server to only the Web site of a single client organization, as opposed to having multiple clients' sites share one server.

denial of service (DoS)—The inability of legitimate visitors to log on to a Web site when too many malicious requests are launched by an attacker. Most DoS attacks are distributed (DDoS).

dialog module—The part of a decision-support system, or any other system, that allows the user to interact with the application. Also called interface.

dial-up connection—A connection to the Internet through a regular telephone and modem. Dial-up connections are slow, as opposed to broadband connections.

digital certificates—Computer files that serve as the equivalent of ID cards.

digital signature—An encrypted digest of the text that is sent along with a message that authenticates the identity of the sender and guarantees that no one has altered the sent document.

digital subscriber line (DSL)—Technology that relieves individual subscribers of the need for the conversion of digital signals into analog signals between the telephone exchange and the subscriber jack. DSL lines are linked to the Internet on a permanent basis and support bit rates significantly greater than a normal telephone line between the subscriber's jack and the telephone exchange.

digital systems—Systems that communicate and process information in a form that follows the binary system of counting and binary methods of representing information, including sound and images.

digital video disc (DVD)—A collective term for several types of high-capacity storage optical discs, used for data storage and motion pictures. Also called digital versatile disc.

dimensional database—A database of tables, each of which contains aggregations and other manipulated information gleaned from the data to speed up the presentation by online processing applications. Also called multidimensional database.

direct access—The manner in which a record is retrieved from a storage device, without the need to seek it

sequentially. The record's address is calculated from the value in its logical key field.

direct attached storage (DAS)—Any data storage device that is directly connected to a computer as opposed to being connected via a communications network. When a disk is contained in the computer box or externally but directly linked to it, it is considered DAS.

disaster recovery plan—*See* business recovery plan.

DMZ—Demilitarized zone, a network of computers and other devices connected to the Internet where visitors are not allowed direct access to other resources connected to the DMZ. DMZs are used to serve visitors while minimizing risk of unauthorized access.

DNS (Domain Name System)—Hardware and software making up a server whose purpose is to resolve domain names (converting them back to IP numbers) and routing messages on the Internet.

domain name—The name assigned to an Internet server or to a part of a server that hosts a Web site.

dot-matrix printer—A printer on which the printhead consists of a matrix of little pins; thus, each printed character is made up of tiny dots.

downloading—The copying of data or applications from a computer to your computer, for example from a source on the Internet to your PC.

downstream—The movement of data bits from another computer to your computer via the Internet. Downstream speed of Internet connection services is usually greater than the upstream speed.

downtime—The unplanned period of time during which a system does not function.

drilling down—The process of finding the most relevant information for executive decision making within a database or data warehouse by moving from more general information to more specific details, such as from performance of a division to performance of a department within the division.

driver—The software that enables an operating system to control a device, such as an optical disc drive or joystick.

dynamic IP address—The IP address assigned to a computer that is connected to the Internet intermittently for the duration of the computer's connection.

dynamic Web page—A Web page whose contents change while the visitor watches it.

e-commerce—Business activity that is electronically executed between parties, such as between two businesses or between a business and a consumer.

economic order quantity (EOQ)—The optimal (cost-minimizing) quantity of a specific raw material that allows a business to minimize overstocking and save cost without risking understocking and missing production deadlines.

effectiveness—The measure of how well a job is performed.

efficiency—The ratio of output to input; the greater the ratio, the greater the efficiency.

electronic funds transfer (EFT)—The electronic transfer of cash from an account in one bank to an account in another bank.

electronic product code (EPC)—A product code embedded in a radio frequency identification (RFID) tag. Similar to the older UPC.

EMI (electromagnetic interference)—Unwanted disturbance in a radio receiver or electrical circuits caused by electromagnetic radiation from an external source. Fiber-optic cable is not susceptible to EMI.

employee knowledge network—Software that facilitates search of relevant knowledge within an organization. The software points an employee with need for certain information or expertise to coworkers who might have such information or expertise.

encryption—The conversion of plaintext to an unreadable stream of characters, especially to prevent a party that intercepts telecommunicated messages from reading them. Special encryption software is used by the sending party to encrypt messages, and by the receiving party to decipher them.

enterprise applications—Applications that fulfill a number of functions together, such as inventory planning, purchasing, payment, and billing.

enterprise resource planning (ERP) system—An information system that supports different activities for different departments, assisting executives with planning and running different interdependent functions.

entity—Any object about which an organization chooses to collect data.

entity relationship diagram (ERD)—One of several conventions for graphical rendition of the data elements involved in business processes and the logical relationships among the elements.

EPC (electronic product code)—The electronic equivalent of a universal product code (UPC), commonly embedded in an RFID (radio frequency identification) tag.

ergonomics—The science of designing and modifying machines to better suit people's health and comfort.

Ethernet—The design, introduced and named by Xerox, for the contention-based data communications protocol.

European Article Number (EAN)—A European standard of product code, similar to UPC but containing more information.

expert system (ES)—A computer program that mimics the decision process of a human expert in providing a solution to a problem. Current expert systems deal with problems and diagnostics in narrow domains. An ES consists of a knowledge base, an inference engine, and a dialog management module.

external data—Data that are collected from a wide array of sources outside the organization, including mass communications media, specialized newsletters, government agencies, and the Web.

extranet—A network, part of which is the Internet, whose purpose is to facilitate communication and trade between an organization and its business partners.

fault tolerance—The ability of a system to continue to function despite a catastrophe or other usually disruptive events. Fault tolerance systems are usually redundant.

feasibility studies—A series of studies conducted to determine if a proposed information system can be built, and whether or not it will benefit the business; the series includes technical, economic, and operational feasibility studies.

Fiber to the Home (FTTH)—The connection of a home to the Internet through optical fiber technology. Often, other services, such as television and landline phone, are also provided through the same medium.

field—A data element in a record, describing one aspect of an entity or event. Referred to as attribute in relational databases.

File Transfer Protocol (FTP)—Software that allows the transfer of files over communications lines.

firewall—Hardware and software designed to control access by Internet surfers to an information system, and access to Internet sites by organizational users.

first mover—A business that is first in its industry to adopt a technology or method.

fixed wireless—A network of fixed transceivers to facilitate connection to the Internet. Requires line of sight between transceivers.

flash drive—A storage device containing flash memory. Flash drives are used in numerous electronic devices and often are designed to connect to a computer through a USB port.

flash memory—A memory chip that can be rewritten and can hold its content without electric power. Thumb drives, as well as ROM, are made of flash memory.

foreign key—In a relational database: a field in a table that is a primary key in another table. Foreign keys allow association of data between the two files.

frame relay—A high-speed packet switching protocol used on the Internet.

fulfillment—Picking, packing, and shipping after a customer places an order online.

general-purpose application software—Programs that serve varied purposes, such as developing decision-making tools or creating documents; examples include spreadsheets and word processors.

geographic information system (GIS)—Information system that exhibits information visually on a computer monitor with local, regional, national, or international maps, so that the information can easily be related to locations or routes on the map. GISs are used, for example, in the planning of transportation and product distribution, or the examination of government resources distributed over an area.

Gigabit Ethernet—A network protocol often used in local area networks (LANs) supporting up to 1 Gbps.

global information system—Any information system that crosses national borders.

Global Trade Item Number (GTIN)—A number that uniquely identifies products and services. The GTIN is a global standard succeeding the EAN and UPC.

glocalization—The planning and designing of global Web sites so that they also cater to local needs and preferences.

group decision support system (GDSS)—Decision support system for a group of people rather than an individual. Often, a GDSS serves remote workers through the Internet, and provides mechanisms for bringing up ideas, discussing them, voting, and concluding a decision.

groupware—Any of several types of software that enable users of computers in remote locations to work together on the same project. The users can create and change documents and graphic designs on the same monitor.

hard disk—A stack of several rigid aluminum platters coated with easily magnetized substance to record data. Usually installed in the same box that holds the CPU and other computer components, but may be portable.

hardware—All physical components of a computer or computer system.

hijacking—In the context of networks, computers that are remotely taken advantage of by people who were not authorized to do so by the lawful owner. The computer is "hijacked" after a controlling application was surreptitiously installed on the computer's hard disk. Hijacked computers are exploited to participate in spamming or DDoS attacks.

honeypot—A duplicate database on a server connected to the Internet to trace an intruder. The server is dedicated specifically for detection of intrusions and is not productive. The honeypot is there to be attacked in lieu of a productive server. The traces can be used to improve security measures and possibly catch the intruder.

honeytoken—A bogus record in a database on a honeypot or productive server that is likely to draw an intruder's attention. If the intruder changes the record, the security officers know that the server has been attacked and can fix vulnerabilities.

host—A computer that contains files and other resources that can be accessed by "clients," computers link to it via a network.

hot site—A location where a client organization hit by a disaster can continue its vital operations. The structure—often underground—is equipped with hardware and software to support the client's employees.

hotspot—An area, usually of 300-feet radius, in which a wireless device can connect to the Internet. The hotspot is created by installing an access point consisting of a device connected to the Internet on one end and to a router on the other end. All wireless devices link to the Internet through the router.

HTTPS—The secure version of HTTP.

hub—In networking, a device connecting several computers or other electronic devices.

Hypertext Markup Language (HTML)—A programming language for Web pages and Web browsers.

Hypertext Transfer Protocol (HTTP)—Software that allows browsers to log on to Web sites.

Hypertext Transfer Protocol Secure—*See* HTTPS.

identity theft—The criminal practice of obtaining enough personal information to pretend to be the victim, usually resulting in running up that person's credit cards or issuing new credit cards under that person's name.

IEEE 802.11—A standard for wireless communication. Several other IEEE 802.x standards have been approved by the Institute of Electrical and Electronics Engineers.

imaging—The transformation of text and graphical documents into digitized files. The document can be electronically retrieved and printed to reconstruct a copy of the original. Imaging has saved much space and expense in paper-intensive business areas.

impression—In Web advertising, the event of an ad displayed on a surfer's monitor.

inference engine—The part of an expert system that links facts and relationships in the knowledge base to reach a solution to a problem.

information—The product of processing data so that they can be used in a context by human beings.

information system (IS)—A computer-based set of hardware, software, and telecommunications components, supported by people and procedures, to process data and turn it into useful information.

information technology (IT)—Refers to all technologies that collectively facilitate construction and maintenance of information systems.

ink-jet printer—Inexpensive type of printer that sprays ink to create the printed text or pictures of a computer-generated document.

input—Raw data entered into a computer for processing.

input device—A tool, such as a keyboard or voice recognition system, used to enter data into an information system.

instant messaging (IM)—The capability for several online computer users to share messages in real time; also called chatting online.

intelligent agent—A sophisticated program that can be instructed to perform services for human beings, especially on the Internet.

internal memory—The memory circuitry inside the computer, communicating directly with the CPU. Consists of RAM and ROM.

Internet Protocol (IP) address—A unique number assigned to a server or another device that is connected to the Internet for identification purposes. Consists of

32 bits. The newer IPv6 protocol contains 128 bits, allowing many more unique IP addresses.

Internet service provider (ISP)—An individual or organization that provides Internet connection, and sometimes other related services, to subscribers.

interpreter—A programming language translator that translates the source code, one statement at a time, and executes it. If the instruction is erroneous, the interpreter produces an appropriate error message.

intranet—A network using Web browsing software that serves employees within an organization.

join table—In relational database manipulation, a table created by linking—that is, joining—data from multiple tables.

just-in-time (JIT)—The manufacturing strategy in which suppliers ship parts directly to assembly lines, saving the cost of warehousing raw materials, parts, and subassemblies.

key—A field in a database table whose values identify records either for display or for processing. Typical keys are part number (in an inventory file) and Social Security number (in a human resources file). In computer security, a specific series of bits is used to decipher encrypted information.

keystroke logging—Automatically recording the keystrokes of a computer user. The logging is done by special software, usually surreptitiously with the intention of later using secret access codes.

knowledge base—The collection of facts and the relationships among them that mimic the decision-making process in an expert's mind and constitute a major component of an expert system.

knowledge management—The combination of activities involved in gathering, sharing, analyzing, and disseminating knowledge to improve an organization's performance.

LAN (local area network)—A computer network confined to a building or a group of adjacent buildings, as opposed to a wide area network.

language translator—Software that translates high-language (source) code into machine (object) code. Compilers and interpreters are translators.

late mover—An organization that adopts a technology or method after competitors have adopted it.

liquid crystal display (LCD)—A flat-panel computer monitor in which a conductive-film-covered screen is filled with a liquid crystal whose molecules can align in different planes when charged with certain electrical voltage, which either blocks light or allows it to pass through the liquid. The combination of light and dark produces images of characters and pictures.

load balancing—The transfer of visitor inquiries from a busy server to a less busy server.

logic bomb—A destructive computer program that is inactive until it is triggered by an event taking place in the computer, such as the deletion of a certain record from a file. When the event is at a particular time, the logic bomb is referred to as a time bomb.

machine cycle—The steps that the CPU follows repeatedly: fetch an instruction, decode the instruction, execute the instruction, and store the result.

machine language—Binary programming language that is specific to a computer. A computer can execute a program only after the program's source code is translated to object code expressed in the computer's machine language.

magnetic disk—A disk or set of disks sharing a spindle, coated with an easily magnetized substance to record data in the form of tiny magnetic fields.

magnetic tape—Coated polyester tape used to store computer data; similar to tape recorder or VCR tape.

magnetic-ink character recognition (MICR)—A technology that allows a special electronic device to read data printed with magnetic ink. The data are later processed by a computer. MICR is widely used in banking. The bank code, account number, and the amount of a check are printed in magnetic ink on the bottom of checks.

mainframe computer—A computer larger than a midrange computer but smaller than a supercomputer.

management information system (MIS)—A computer-based information system used for planning, control, decision making, or problem solving.

manufacturing execution system—An information system that helps pinpoint bottlenecks in production lines.

manufacturing resource planning (MRP II)—The combination of MRP with other manufacturing-related activities to plan the entire manufacturing process, not just inventory.

many-to-many relationship—In databases, a relationship between two tables whereby every record in a table can be associated with several records in the other table.

master production schedule (MPS)—The component of an MRP II system that specifies production capacity to meet customer demands and maintain inventories.

material requirements planning (MRP)—Inventory control that includes a calculation of future need.

m-commerce—Mobile commerce, enabled by advances in technology for mobile communications devices.

metadata—Information about the data in a database, often called data dictionary.

microcomputer—The smallest type of computer; includes desktop, laptop, and handheld computers.

microprocessor—An electronic chip that contains the circuitry of either a CPU or a processor with a dedicated and limited purpose, for example, a communications processor.

microwaves—Short (high frequency) radio waves. Used in telecommunications to carry digital signals.

midrange computer—A computer larger than a microcomputer but smaller than a mainframe.

migration—The move from old hardware or software to new hardware or software. Migrating a legacy system is the process of adapting the old system to work more efficiently or more effectively, especially when interfacing it with other systems.

MIPS—Millions of instructions per second; an inaccurate measure of computer speed.

mirror server—An Internet server that holds the same software and data as another server, which may be located thousands of miles away.

mission-critical applications—Applications without which a business cannot conduct its operations.

Mobile Broadband Wireless Access (MBWA)—IEEE 801.20 standard to support continuous wireless connection while moving in vehicles.

model—A representation of reality.

model management module—A collection of models that a decision-support system draws on to assist in decision making.

modem (modulator/demodulator)—A communications device that transforms digital signals to analog telephone signals, and vice versa, for data communications over voice telephone lines. The term is widely used for all devices that connect a computer to a wide area network, such as the Internet, even if the device does not modulate or demodulate.

modulation—The modification of a digital signal (from a computer) into an analog signal (for a phone line to transmit).

multicore processor—A processor that contains more than one central processing unit. Each core is equivalent to a CPU.

multidimensional database—*See* dimensional database.

multimedia software—Software that processes and displays various forms of information: text, sound, pictures, and video.

multiprocessing—The mode in which a computer uses more than one processing unit simultaneously to process data.

multitasking—The ability of a computer to run more than one program seemingly at the same time; it enables the notion of windows in which different programs are represented.

multithreading—Computer technology that allows more than one stream (thread) of processing at the same time.

network—A combination of a communications device and a computer or several computers, or two or more computers, so that the various devices can send and receive text or audiovisual information to each other.

network administrator—The individual who is responsible for the acquisition, implementation, management, maintenance, and troubleshooting of computer networks throughout the organization.

network interface card (NIC)—Circuitry embedded or installed in a computer to support proper linking of the computer to a network.

network-attached storage (NAS)—An arrangement of storage devices linked to computers through a network.

neural network—An artificial intelligence computer program that emulates the way in which the human brain operates, especially its ability to learn.

node—A device connected to at least one other device on a network.

nonimpact printer—A printer that creates an image on a page without pressing any mechanism against the paper; includes laser, ink-jet, electrostatic, and electrothermal printers.

notebook computer—A computer as small as a book, yet with computing power similar to that of a desktop microcomputer.

object code—Program code in machine language, immediately processable by the computer.

object-oriented database—A database, in which data are part of an object, that is processed using object-oriented programs.

object-oriented programming (OOP) language—A programming language that combines data and the procedures that process the data into a single unit called an "object," which can be invoked from different programs.

OC (optical carrier)—A family of several very high-speed technologies using optical fibers. Usually, the standard is marked as OC-3, OC-12, OC-48, etc.

offshoring—Outsourcing work to employees in other countries.

one-to-many relationship—In a database, a relationship between two tables such that each record in the one table can be associated with several records in the other table but each record in the other table can be associated with only one record in the first table.

online analytical processing (OLAP)—A type of application that operates on data stored in databases and data warehouses to produce summary tables with multiple combinations of dimensions. An OLAP server is connected to the database or data warehouse server at one end and to the user's computer at the other.

open source software—Software whose source code can be accessed by the general public.

open system—A system that interfaces and interacts with other systems.

operating system (OS)—System software that supports the running of applications developed to utilize its features and controls peripheral equipment.

optical disc—A disc on which data are recorded by treating the disc surface so it reflects light in different ways; includes CD and DVD.

optical tape—A storage device that uses the same principles as a compact disc.

organizational culture—An umbrella term referring to the general tone of a corporate environment.

output—The result of processing data by the computer; usually, information.

output device—A device, usually a monitor or printer, that delivers information from a computer to a person.

outsourcing—Buying the services of an information service firm that undertakes some or all of the organization's IS operations.

packaged software—General-purpose applications that come ready to install from a magnetic disk, CD, or file downloaded from a vendor's Web site.

packet—Several bytes that make up a part of a telecommunicated message.

packet switching—A telecommunications method whereby messages are broken into groups of fixed amounts of bytes, and each group (packet) is transmitted through the shortest route available. The packets are assembled at the destination into the original message.

parallel conversion—Using an old information system along with a new system for a predetermined period of time before relying only on the new one.

parallel processing—The capacity for several CPUs in one computer to process different data at the same time.

parameters—The categories that are considered when following a sequence of steps in problem solving.

peer-to-peer file sharing—Software applications that enable two Internet users to send and receive to each other. The technology is highly objectionable to organizations that sell copyrighted materials because the software promotes violation of copyrights.

peer-to-peer LAN—A local area network (LAN) in which no central device controls communications.

personal area network (PAN)—A network of devices typically within a small radius that enables a user to use two or more devices wirelessly, such as wireless keyboard and mouse.

personal digital assistant (PDA)—A small handheld computer. Many PDAs require the use of a special stylus to click displayed items and to enter handwritten information that is recognized by the computer. An increasing number of PDAs also serve as mobile phones, music players, and GPS devices.

phased conversion—Implementing a new information system one module at a time.

phishing—The criminal practice of luring Internet users to provide their personal information via e-mail or the Web. Phishing almost always results in fraud or identity theft.

piloting—A trial conversion in which a new information system is introduced in one business unit before introducing it in others.

pixel—The smallest picture element addressable on a monitor, short for "picture element." In an LCD monitor, it is a triad of three transistors controlling the colors of red, green, and blue that can be switched on and off and kept on with varying amounts of electricity to produce various colors and hues. In a CRT monitor, the triad is made of phosphorous dots that are excited by an electron gun.

plaintext—An original message, before encryption.

plasma display—A flat panel display with gas (plasma) between two layers of glass. When excited by electric power, the gas gives off light in various colors.

plug-and-play—The ability of an operating system to recognize a new attachment and its function without a user's intervention.

podcasting—The practice of posting sound files at a Web site for automating downloading and playing by subscribers.

port—A socket on a computer to which external devices, such as printers, keyboards, and scanners, can be connected. Also, software that enables direct communication of certain applications with the Internet.

primary key—In a file, a field that holds values that are unique to each record. Only a primary key can be used to uniquely identify and retrieve a record.

process—Any manipulation of data, usually with the goal of producing information.

productivity—Efficiency, when the input is labor. The fewer labor hours needed to perform a job, the greater the productivity.

programming—The process of writing software.

programming languages—Sets of syntax for abbreviated forms of instructions that special programs can translate into machine language so a computer can understand the instructions.

project management—The set of activities that is performed to ensure the timely and successful completion of a project within the budget. Project management includes planning activities, hiring and managing personnel, budgeting, conducting meetings, and tracking technical and financial performance. Project management software applications facilitate these activities.

proprietary software—Software owned by an individual or organizations. The owner can control licensing and usage terms of the software. Nonproprietary software is not owned by anyone and is free for use.

protocol—A standard set of rules that governs telecommunication between two communications devices or in a network.

prototyping—An approach to the development of information systems in which several analysis steps are skipped, to accelerate the development process. A "quick and dirty" model is developed and continually improved until the prospective users are satisfied. Prototyping has evolved into agile development methods.

proxy server—A computer that serves as an intermediary between two servers on the Internet, often for the purpose of security or filtering out certain information.

public-key encryption—Encryption technology in which the recipient's public key is used to encrypt and the recipient's private key is used to decrypt.

pure-play—A business operating with clients only via the Web, as opposed to operating via stores or other physical facilities.

query—A request for information, usually addressed to a database.

radio frequency identification (RFID)—Technology that enables identification of an object (such as product, vehicle, or living creature) by receiving a radio signal from a tag attached to the object.

RAID (Redundant Array of Independent Disks)—A set of magnetic disk packs maintained for backup purposes. Sometimes RAIDs are used for storing large databases.

RAM (random access memory)—The major part of a computer's internal memory. RAM is volatile; that is, software is held in it temporarily and disappears when the machine is unplugged or turned off, or it may disappear when operations are interrupted or new software is installed or activated. RAM is made of microchips containing transistors. Many computers have free sockets that allow the expansion of RAM.

rapid prototyping—Using software and special output devices to create prototypes to test design in three dimensions.

reach percentage—The percentage of Web users who have visited a site in the past month, or the ratio of visitors to the total Web population.

record—A set of standard field types. All the fields of a record contain data about a certain entity or event.

reengineering—The process by which an organization takes a fresh look at a business process and reorganizes it to attain efficiency. Almost always, reengineering includes the integration of a new or improved information system.

relational database model—A general structure of a database in which records are organized in tables (relations) and the relationships among tables are maintained through foreign keys.

relational operation—An operation that creates a temporary table that is a subset of the original table or tables in a relational database.

repeater—A device that strengthens signals and then sends them on their next leg toward their next destination.

request for information (RFI)—A request to vendors for general, somewhat informal, information about their products.

request for proposal (RFP)—A document specifying all the system requirements and soliciting a proposal from vendors who might want to bid on a project or service.

resolution—The degree to which the image on a computer monitor is sharp. Higher resolution means a sharper image. Resolution depends on the number of pixels on the screen and the dot pitch.

return on investment (ROI)—A financial calculation of the difference between the stream of benefits and the stream of costs over the life of an information system; often used as a general term to indicate that an investment in an information system is recouped or smaller than the cost the system saves or the increase in revenue it brings about.

reverse auction (name-your-own-price auction)—An online auction in which participants post the price they want to pay for a good or service, and retailers compete to make the sale; also called a name-your-price auction.

RFI (radio frequency interference)—The unwanted reception of radio signals that occurs when using metal communication lines. Optical fibers are not susceptible to RFI.

ROM (read-only memory)—The minor part of a computer's internal memory. ROM is loaded by the manufacturer with software that cannot be changed. Usually, ROM holds very basic system software, but sometimes also applications. Like RAM, ROM consists of microchips containing transistors.

router—A network hub, wired or wireless, that ensures proper routing of messages within a network such as a LAN and between each device on that network and another network, such as the Internet.

RSS—Really Simple Syndication, a type of application using XML for aggregating updates to blogs and news posted at Web sites.

SaaS (Software as a Service)—An arrangement in which an application software provider (ASP) enables a client's employees to use software through communication lines, often through the Internet. Payment is determined by the software made available, number of users, and the contract length of time.

Safe Harbor—A list of U.S. corporations that have agreed to conform to European Union data protection laws with regard to EU citizens. The arrangement enables the corporations to continue to do business with European companies.

scalability—The ability to adapt applications as business needs grow.

schema—The structure of a database, detailing the names and types of fields in each set of records, and the relationships among sets of records.

search advertising—Placing ads at search engine Web sites.

semistructured problem—An unstructured problem with which the decision maker may have had some experience. Requires expertise to resolve.

sensitivity analysis—Using a model to determine the extent to which a change in a factor affects an outcome. The analysis is done by repeating *if-then* calculations.

sequential storage—A file organization for sequential record entry and retrieval. The records are organized as a list that follows a logical order, such as ascending order of ID numbers, or descending order of part numbers. To retrieve a record, the application must start the search at the first record and retrieve every record, sequentially, until the desired record is encountered.

server—A computer connected to several less powerful computers that can utilize its databases and applications.

service-level agreement—A document that lists all the types of services expected of an outsourcing vendor as well as the metrics that will be used to measure the degree to which the vendor has met the level of promised services. Usually, the client makes the list.

shared hosting—An arrangement by which the Web sites of several clients are maintained by the hosting vendor on the same server.

social engineering—Deceptive methods that hackers use to entice people to release confidential information such as access codes and passwords. Often, the crooks misrepresent themselves as technicians who need one's password for fixing a problem in a network.

software—Sets of instructions that control the operations of a computer.

solid state disk (SSD)—Flash memory that serves as external storage medium as if it were a hard disk.

source code—An application's code written in the original high-level programming language.

source data input device—A device that enables data entry directly from a document without need for human keying. Such devices include bar-code readers and optical character readers.

speech recognition—The process of translating human speech into computer-readable data and instructions.

spyware—A small application stored surreptitiously by a Web site on the hard disk of a visitor's computer. The application tracks activities of the user, including visits to Web sites, and transmits the information to the operator's server.

SSO (single sign-on)—Enabling employees to access several information systems by using a single password.

static IP address—An IP address permanently associated with a device.

storage—(1) The operation of storing data and information in an information system; (2) Any non-RAM memory, including internal and external hard disks, flash memory, and optical discs.

storage area network (SAN)—A device that enables multiple networked computers to save data on a group of disks located in a special area.

storage service provider (SSP)—A firm that rents storage space for software through an Internet link.

strategic advantage—A position in which one dominates a market; also called competitive advantage.

strategic information system—Any information system that gives its owner a competitive advantage.

structured problem—A problem for whose solution there is a known set of steps to follow. Also called a programmable problem.

Structured Query Language (SQL)—The data definition and manipulation language of choice for many developers of relational database management systems.

stylus—A penlike marking device used to enter commands and data on a computer screen.

subsystem—A component of a larger system.

GLOSSARY

suite—A group of general software applications that are often used in the same environment. The strengths of the different applications can be used to build a single powerful document. Current suites are usually a combination of a spreadsheet, a word processor, and a database management system.

supercomputer—The most powerful class of computers, used by large organizations, research institutions, and universities for complex scientific computations and the manipulation of very large databases.

supply chain—The activities performed from the purchase of raw material to the shipping of manufactured goods and collecting for their sale.

supply chain management (SCM)—The coordination of purchasing, manufacturing, shipping, and billing operations, often supported by an enterprise resource planning system.

support—The maintenance and provision for user help on an information system.

switching costs—Expenses that are incurred when a customer stops buying a product or service from one business and starts buying it from another.

symmetric (secret or private key) encryption—Encryption technology in which both the sender and recipient of a message use the same key for encryption and decryption.

synergy—From Greek "to work together." The attainment of output, when two factors work together, that is greater or better than the sum of their products when they work separately.

system—An array of components that work together to achieve a common goal or multiple goals.

system clock—Special circuitry within the computer control unit that synchronizes all tasks.

system administrator—A computer professional who manages and maintains an organization's operating systems. Often referred to as "sys admin."

system requirements—The functions that an information system is expected to fulfill and the features through which it will perform its tasks.

system software—Software that executes routine tasks. System software includes operating systems, language translators, and communications software. Also called support software.

systems analysis—The early steps in the systems development process, to define the requirements of the proposed system and determine its feasibility.

systems analyst—An IT professional who analyzes business problems and recommends technological solutions.

systems design—The evaluation of alternative solutions to a business problem and the specification of hardware, software, and communications technology for the selection solution.

systems development life cycle (SDLC)—The oldest method of developing an information system, consisting of several phases of analysis and design, which must be followed sequentially.

systems integration—Interfacing several information systems.

systems thinking—The approach of thinking of an organization in terms of its suborganizations or systems; a framework for problem solving and decision making.

table—A set of related records in a relational database.

tablet computer—A full-power personal computer in the form of a thick writing tablet.

targeted marketing—Promoting products and services to the people who are most likely to purchase them.

TCP/IP (Transmission Control Protocol/Internet Protocol)—A packet-switching protocol that is actually a set of related protocols that can guarantee packets are delivered in the correct order and can handle differences in transmission and reception rates.

technology convergence—The combining of several technologies into a single device, such as mobile phone, digital camera, and Web browser.

telecommunications—Communications over a long distance, as opposed to communication within a computer, or between adjacent hardware pieces.

throughput—A general measure of the rate of computer output.

time to market—The time between generating an idea for a product and completing a prototype that can be mass-manufactured; also called engineering lead time.

touch screen—A computer monitor that serves both as input and output device. The user touches the areas of a certain menu item to select options, and the screen senses the selection at the point of the touch.

trackball—A device similar to a mouse, used for clicking, locking, and dragging displayed information; in this case, the ball moves within the device rather than over a surface.

track pad—A device used for clicking, logging, and dragging displayed information; the cursor is controlled by moving one's finger along a touch-sensitive pad.

transaction—A business event. In an IS context, the record of a business event.

transaction processing system (TPS)—Any system that records transactions.

transmission rate—The speed at which data are communicated over a communications channel.

Transport Layer Security (TLS)—The successor of Secure Sockets Layer (SSL), the software in the Web browser responsible for secure communication.

Trojan horse—A malicious piece of software hidden with a benign and legitimate software that one downloads or agrees to otherwise accept and install on one's computer. The Trojan horse then causes damage.

twisted pair cable—Traditional telephone wires, twisted in pairs to reduce electromagnetic interference.

Unicode—An international standard to enable the storage and display of characters of a large variety of languages—such as Asian, Arabic, and Hebrew—on computers.

Unified Modeling Language (UML)—An extensive standard for graphically representing elements of programming, specifically accommodating programming in object-oriented languages and Web technologies.

Uniform Resource Locator (URL)—The address of a Web site. Always starts with *http://* but does not have to contain *www*.

uninterruptible power supply (UPS)—A device that provides an alternative power supply as soon as a power network fails.

Universal Product Code (UPC)—A code usually expressed as a number and series of variable width bars that uniquely identifies the product by scanning.

universal serial bus (USB)—A ubiquitous socket that enables the connection of numerous devices to computers.

unstructured problem—A problem for whose solution there is no pretested set of steps, and with which the solver is not familiar—or is only slightly familiar—from previous experience.

upstream—The movement of data from your computer to another computer via a network, usually the Internet. Upstream speed through the services of Internet providers is typically lower than the downstream speed.

uptime—The percentage of time (so much time per year) that an information system is in full operation.

USB drive—Any storage device that connects to a computer through a USB socket, but especially flash drives.

user application development—Development of corporate applications by employees rather than IT professionals.

utilities—Programs that provide help in routine user operations.

value-added network (VAN)—A telecommunications network owned and managed by a vendor that charges clients periodic fees for network management services.

videoconferencing—A telecommunications system that allows people who are in different locations to meet via transmitted images and speech.

virtual memory—Storage space on a disk that is treated by the operating system as if it were part of the computer's RAM.

virtual private network (VPN)—Hardware and software installed to ensure that a network path that includes the Internet enables employees of the same organization or employees of business partners to communicate confidentially. The hardware and software create an impression that the entire communication path is private.

virtual private server—Part of a server that serves as an Internet server for a client of a Web hosting company, while other clients share the same physical server.

virtual reality (VR)—A set of hardware and software that creates images, sounds, and possibly the sensation of touch that give the user the feeling of a real environment and experience. In advanced VR systems, the user wears special goggles and gloves.

virtual world—A mimicked world created on the Web with software. Subscribers use avatars to move around and communicate in an imaginary world. With proper software, designers can "develop" islands and other plots of land for various uses.

virus—Destructive software that propagates and is activated by unwary users; a virus usually damages applications and data files or disrupts communications.

visual programming language—A programming language that provides icons, colors, and other visual elements from which the programmer can choose to speed up software development.

VoIP (Voice over Internet Protocol)—Technologies that enable voice communication by utilizing the Internet instead of the telephone network.

GLOSSARY

Web hosting—The business of organizations that host, maintain, and often help design Web sites for clients.

Web page authoring tools—Software tools that make Web page composition easier and faster than writing code by providing icons and menus.

Webmaster—The person who is in charge of constructing and maintaining the organization's Web site.

what if analysis—An analysis that is conducted to test the degree to which one variable affects another; also called sensitivity analysis.

wide area network (WAN)—A network of computers and other communications devices that extends over a large area, possibly comprising national territories. Example: the Internet.

Wi-Fi—A name given to the IEEE 802.11 standards of wireless communication. Wi-Fi technologies are used in hotspots and in home and office networks. Wi-Fi is usually effective within a radius of 300 feet.

WiMAX—The IEEE 802.16 standard for wireless networking with a range of up to 50 km (31 miles). (WiMAX stands for the organization that promotes that standard, Worldwide Interoperability for Microwave Access.)

wireless LAN (WLAN)—A local area network that uses electromagnetic waves (radio or infrared light) as the medium of communication. In recent years almost all WLANs have been established using Wi-Fi.

work order—A numbered (or otherwise uniquely coded) authorization to spend labor and other resources on the manufacturing of a product or rendering of a service. Usually, work orders are opened within a project number. The systems of project number and work orders helps track costs and activities related to an assignment in an organization, typically one in the manufacturing sector.

workstation—A powerful microcomputer providing high-speed processing and high-resolution graphics. Used primarily for scientific and engineering assignments.

worm—A rogue program that spreads in a computer network. Unlike other computer viruses, worms do not need human intervention to spread.

XHTML—A standard that combines HTML standards and XML standards.

XML (Extensible Markup Language)—A programming language that tags data elements in order to indicate what the data mean, especially in Web pages.

yield management software—Software that helps maximize the capacity of airline seats and hotel rooms by analyzing which variables affect purchasing of such services and in what way.

SUBJECT INDEX